Waste Treatment in the Service and Utility Industries

Advances in Industrial and Hazardous Wastes Treatment Series

Advances in Hazardous Industrial Waste Treatment (2009)
Edited by Lawrence K. Wang, Nazih K. Shammas, and Yung-Tse Hung

Waste Treatment in the Metal Manufacturing, Forming, Coating, and Finishing Industries (2009)
Edited by Lawrence K. Wang, Nazih K. Shammas, and Yung-Tse Hung

Heavy Metals in the Environment (2009)
Edited by Lawrence K. Wang, Jiaping Paul Chen, Yung-Tse Hung, and Nazih K. Shammas

Handbook of Advanced Industrial and Hazardous Wastes Treatment (2010)
Edited by Lawrence K. Wang, Yung-Tse Hung, and Nazih K. Shammas

Decontamination of Heavy Metals: Processes, Mechanisms, and Applications (2012)
Jiaping Paul Chen

Remediation of Heavy Metals in the Environment (2017)
Edited by Jiaping Paul Chen, Lawrence K. Wang, Mu-Hao Sung Wang, Yung-Tse Hung, and Nazih K. Shammas

Waste Treatment in the Service and Utility Industries (2017)
Edited by Yung-Tse Hung, Lawrence K. Wang, Mu-Hao Sung Wang, Nazih K. Shammas, and Jiaping Paul Chen

RELATED TITLES

Handbook of Industrial and Hazardous Wastes Treatment (2004)
Edited by Lawrence K. Wang, Yung-Tse Hung, Howard H. Lo, and Constantine Yapijakis

Waste Treatment in the Food Processing Industry (2006)
Edited by Lawrence K. Wang, Yung-Tse Hung, Howard H. Lo, and Constantine Yapijakis

Waste Treatment in the Process Industries (2006)
Edited by Lawrence K. Wang, Yung-Tse Hung, Howard H. Lo, and Constantine Yapijakis

Hazardous Industrial Waste Treatment (2007)
Edited by Lawrence K. Wang, Yung-Tse Hung, Howard H. Lo, and Constantine Yapijakis

Tratamiento de los Residuos de la Industria del Procesado de Alimentos (2008)
Edited by Lawrence K. Wang, Yung-Tse Hung, Howard H. Lo, and Constantine Yapijakis
Translated by Alberto Ibarz Ribas

Waste Treatment in the Service and Utility Industries

Edited by
Yung-Tse Hung
Lawrence K. Wang
Mu-Hao Sung Wang
Nazih K. Shammas
Jiaping Paul Chen

CRC Press is an imprint of the
Taylor & Francis Group, an **informa** business

CRC Press
Taylor & Francis Group
6000 Broken Sound Parkway NW, Suite 300
Boca Raton, FL 33487-2742

© 2017 by Taylor & Francis Group, LLC
CRC Press is an imprint of Taylor & Francis Group, an Informa business

No claim to original U.S. Government works

Printed on acid-free paper

International Standard Book Number-13: 978-1-4200-7237-2 (Hardback)

This book contains information obtained from authentic and highly regarded sources. Reasonable efforts have been made to publish reliable data and information, but the author and publisher cannot assume responsibility for the validity of all materials or the consequences of their use. The authors and publishers have attempted to trace the copyright holders of all material reproduced in this publication and apologize to copyright holders if permission to publish in this form has not been obtained. If any copyright material has not been acknowledged, please write and let us know, so we may rectify in any future reprint.

Except as permitted under U.S. Copyright Law, no part of this book may be reprinted, reproduced, transmitted, or utilized in any form by any electronic, mechanical, or other means, now known or hereafter invented, including photocopying, microfilming, and recording, or in any information storage or retrieval system, without written permission from the publishers.

For permission to photocopy or use material electronically from this work, please access www.copyright.com (http://www.copyright.com/) or contact the Copyright Clearance Center, Inc. (CCC), 222 Rosewood Drive, Danvers, MA 01923, 978-750-8400. CCC is a not-for-profit organization that provides licenses and registration for a variety of users. For organizations that have been granted a photocopy license by the CCC, a separate system of payment has been arranged.

Trademark Notice: Product or corporate names may be trademarks or registered trademarks, and are used only for identification and explanation without intent to infringe.

Visit the Taylor & Francis Web site at
http://www.taylorandfrancis.com

and the CRC Press Web site at
http://www.crcpress.com

Contents

Preface ix
Editors xi
Contributors xiii

Chapter 1 Safety and Control for Toxic Chemicals and Hazardous Wastes 1

Nazih K. Shammas and Lawrence K. Wang

Chapter 2 Biological Treatment of Poultry Processing Wastewater 27

Nazih K. Shammas and Lawrence K. Wang

Chapter 3 Treatment of Wastewater, Storm Runoff, and Combined Sewer Overflow by Dissolved Air Flotation and Filtration 65

Lawrence K. Wang, Mu-Hao Sung Wang, and Nazih K. Shammas

Chapter 4 Waste Treatment and Management in Chlor-Alkali Industries 99

Hamidi Abdul Aziz, Miskiah Fadzilah Ghazali, Mohd. Suffian Yusoff, and Yung-Tse Hung

Chapter 5 Dissolved Air Flotation (DAF) for Wastewater Treatment 145

Puganeshwary Palaniandy, Hj. Mohd Nordin Adlan, Hamidi Abdul Aziz, Mohamad Fared Murshed, and Yung-Tse Hung

Chapter 6 Restaurant Waste Treatment and Management 183

Jerry R. Taricska, Jaclyn M. Taricska, Yung-Tse Hung, and Lawrence K. Wang

Chapter 7 Treatment of Textile Industry Waste 207

Siew-Teng Ong, Sie-Tiong Ha, Siew-Ling Lee, Pei-Sin Keng, Yung-Tse Hung, and Lawrence K. Wang

Chapter 8 BOD Determination, Cleaning Solution Preparation, and Waste Disposal in Laboratories 285

Mu-Hao Sung Wang, Lawrence K. Wang, and Eugene De Michele

Chapter 9 Principles, Procedures, and Heavy Metal Management of Dichromate Reflux Method for COD Determination in Laboratories 297

Mu-Hao Sung Wang, Lawrence K. Wang, and Eugene De Michele

Chapter 10 Treatment and Management of Metal Finishing Industry Wastes 307

Nazih K. Shammas and Lawrence K. Wang

Chapter 11 Environment-Friendly Activated Carbon Processes for Water and Wastewater Treatment 353

Wei-chi Ying, Wei Zhang, Juan Hu, Liuya Huang, Wenxin Jiang, and Bingjing Li

Chapter 12 Treatment of Wastes from the Organic Chemicals Manufacturing Industry 373

Debolina Basu, Sudhir Kumar Gupta, and Yung-Tse Hung

Chapter 13 Management, Recycling, and Disposal of Electrical and Electronic Wastes (E-Wastes) 389

Lawrence K. Wang and Mu-Hao Sung Wang

Index 417

Preface

Waste Treatment in the Service and Utility Industries is one of the books in the *Advances in Industrial and Hazardous Wastes Treatment Series*. Municipal wastewaters are quite similar with respect to characteristics of organic and inorganic pollutants. Consequently, their treatment processes are quite similar in various countries. In contrast, the characteristics of wastewaters generated from service and utility industries vary from one industry to another, and would require different treatment processes, depending on the particular industry producing the wastewater. The standards for discharging various types of effluents from service and utility industries into the public sewer systems or to receiving waters are determined not only by the concentration of biochemical oxygen demand (BOD), chemical oxygen demand (COD), and total suspended solids (TSS) but also by the concentration of organic and inorganic elements, which may vary in different countries. It is very important for environmental professionals, government officials, educators, and the public to have current knowledge and experiences in wastewaters generated from service and utility industries.

We are happy to work with CRC Press/Taylor & Francis Group to develop the *Advances in Industrial and Hazardous Wastes Treatment Series*, one of which is this book contributed by a group of environmental scientists, engineers, and educators from several countries in the world, who are experts in the relevant subjects of study. Since the area of waste treatment in the service and utility industries is rather broad, collective contributions are selected to better represent the most updated and complete knowledge.

Waste Treatment in the Service and Utility Industries covers most recently updated information in the subject area. Chapters 1 through 4 address treatment of toxic chemicals and hazardous wastes, poultry processing wastewater, strom runoff and combined sewer overflow, and chlor-alkali wastewaters. Applications of dissolved air flotation for storm water treatment and wastewater treatment are described in Chapter 3 and Chapter 5, respectively. Treatment of restaurant waste, textile waste, metal finishing waste, organic wastes, and electrical and electronic wastes is presented in Chapters 6, 7, 10, 12, and 13, respectively. Chapters 8 and 9 describe BOD and COD determinations and waste management methods in laboratories. Application of activated carbon for water and wastewater treatment is presented in Chapter 11.

This book can be used as a reference book for environmental professionals for learning and practice. Readers in environmental, civil, chemical, and public health engineering and science as well as governmental agencies and nongovernmental organizations will find valuable information in this book to trace, follow, duplicate, or improve on a new or existing industrial hazardous waste treatment practice.

The editorial team and the authors of this book thank many people who have encouraged and supported them during the preparation of this book. Our family members, colleagues, and students have done a great job in supporting us during the preparation of the text. Mr. Joseph Clements, senior editor at CRC Press, has provided strong support to the team for many years. Without all these people, the completion of this book would have been impossible.

Yung-Tse Hung
Lawrence K. Wang
Mu-Hao Sung Wang
Nazih K. Shammas
Jiaping Paul Chen

Editors

Yung-Tse Hung has been a professor of civil engineering at Cleveland State University since 1981. He is a fellow of the American Society of Civil Engineers, and has taught at 16 universities in 8 countries. His research interests and publications have been involved with biological processes and industrial waste treatment. Dr. Hung is credited with over 470 publications and presentations on water and wastewater treatment. He received his BSCE and MSCE from National Cheng-Kung University, Taiwan, and his PhD from the University of Texas at Austin. He is the editor of *International Journal of Environment and Waste Management, International Journal of Environmental Engineering,* and *International Journal of Environmental Engineering Science.*

Lawrence K. Wang has over 28 years of experience in facility design, environmental sustainability, natural resources, resources recovery, global pollution control, construction, plant operation, and management. He has expertise in water supply, air pollution control, solid waste disposal, water resources, waste treatment, and hazardous waste management. He is a retired dean/director/vice president of the Lenox Institute of Water Technology, Krofta Engineering Corporation, and Zorex Corporation, respectively, in the United States. Dr. Wang is the author of over 700 papers and 45 books and is credited with 24 U.S. patents and 5 foreign patents. He received his BSCE from National Cheng-Kung University, Taiwan, his two MS degrees from the Missouri University of Science and Technology and the University of Rhode Island, and his PhD degree from Rutgers University. Currently, he is the chief series editor of the *Advances in Industrial and Hazardous Wastes Treatment* series (CRC Press/Taylor & Francis Group) and the *Handbook of Environmental Engineering* series (Springer).

Mu-Hao Sung Wang has been an engineer, an editor, and a professor serving private firms, governments, and universities in the United States and Taiwan for over 25 years. She is a licensed professional engineer and a diplomate of the American Academy of Environmental Engineers. Her publications have been in the areas of water quality, modeling, environmental sustainability, waste management, NPDES, flotation, and analytical methods. Dr. Wang is the author of over 50 publications and holds 14 U.S. and foreign patents. She received her BSCE from National Cheng Kung University, Taiwan, her MSCE from the University of Rhode Island, and her PhD from Rutgers University. She is the co-series editor of the *Handbook of Environmental Engineering* series (Springer) and a member of AWWA, WEF, NEWWA, NEWEA, and OCEESA.

Nazih K. Shammas is an environmental consultant and a professor for over 45 years. He is an ex-dean/director of the Lenox Institute of Water Technology and an advisor to Krofta Engineering Corporation. Dr. Shammas is the author of over 250 publications and 15 books in the field of environmental engineering. He has experience in environmental planning, curriculum development, teaching, and scholarly research and expertise in water quality control, wastewater reclamation and reuse, physicochemical and biological processes, and water and wastewater systems. He received his BE degree from the American University of Beirut, Lebanon, his MS from the University of North Carolina at Chapel Hill, and his PhD from the University of Michigan.

Jiaping Paul Chen has been a professor in the environmental engineering program at the National University of Singapore (NUS) since 1998. His research interests are physicochemical treatment of water and wastewater and modeling. He has published 3 books, more than 100 journal papers and book chapters with citation of above 4600, and H-index of 39 (ISI). He holds seven patents in the areas of adsorption and membrane technologies and ballast water management systems. He has received various honors and awards, including the Sustainable Technology Award from the IChemE, guest professor of the Huazhong University of Science and Technology and Shandong University of China, and Distinguished Overseas Chinese Young Scholar of the National Natural Science Foundation of China. He has been recognized as an author of highly cited papers (Chemistry and Engineering) of ISI Web of Knowledge. Professor Chen received his ME degree from Tsinghua University and his PhD degree from Georgia Tech.

Contributors

Hj. Mohd. Nordin Adlan
School of Civil Engineering
Universiti Sains Malaysia
Nibong Tebal, Penang, Malaysia

Hamidi Abdul Aziz
School of Civil Engineering
Universiti Sains Malaysia
Nibong Tebal, Penang, Malaysia

Debolina Basu
Department of Civil Engineering
Motilal Nehru National Institute of Technology Allahabad
Allahabad, Uttar Pradesh, India

Eugene De Michele
Technical Services and Committee Liaison
Water Environment Federation
Alexandria, Virginia

Miskiah Fadzilah Ghazali
School of Civil Engineering
Universiti Sains Malaysia
Nibong Tebal, Penang, Malaysia

Sudhir Kumar Gupta
Centre for Environmental Science and Engineering
Indian Institute of Technology Bombay
Mumbai, Maharashtra, India

Sie-Tiong Ha
Faculty of Science
Universiti Tunku Abdul Rahman
Kampar, Perak, Malaysia

Juan Hu
State Environmental Protection Key Laboratory of Environmental Risk Assessment and Control on Chemical Process
East China University of Science and Technology
Shanghai, People's Republic of China

Liuya Huang
State Environmental Protection Key Laboratory of Environmental Risk Assessment and Control on Chemical Process
East China University of Science and Technology
Shanghai, People's Republic of China

Yung-Tse Hung
Department of Civil and Environmental Engineering
Cleveland State University
Cleveland, Ohio

Wenxin Jiang
State Environmental Protection Key Laboratory of Environmental Risk Assessment and Control on Chemical Process
East China University of Science and Technology
Shanghai, People's Republic of China

Pei-Sin Keng
Department of Pharmaceutical Chemistry
International Medical University
Kuala Lumpur, Malaysia

Siew-Ling Lee
Ibnu Sina Institute for Scientific and Industrial Research
Universiti Teknologi Malaysia
Skudai, Johor, Malaysia

Bingjing Li
Shanghai LIRI Technologies Co. Ltd.
Shanghai, People's Republic of China

Mohamad Fared Murshed
School of Civil Engineering
Universiti Sains Malaysia
Nibong Tebal, Penang, Malaysia

Siew-Teng Ong
Faculty of Science
Universiti Tunku Abdul Rahman
Kampar, Perak, Malaysia

Puganeshwary Palaniandy
School of Civil Engineering
Universiti Sains Malaysia
Nibong Tebal, Penang, Malaysia

Nazih K. Shammas
Lenox Institute of Water Technology
Lenox, Massachusetts
and
Krofta Engineering Corporation
Lenox, Massachusetts

Jaclyn M. Taricska
JEA
Jacksonville, Florida

Jerry R. Taricska
Hole Montes, Inc.
Naples, Florida

Lawrence K. Wang
Lenox Institute of Water Technology
Lenox, Massachusetts
and
Krofta Engineering Corporation
Lenox, Massachusetts
and
Zorex Corporation
Newtonville, New York

Mu-Hao Sung Wang
Lenox Institute of Water Technology
Lenox, Massachusetts

Wei-chi Ying
State Environmental Protection Key
　Laboratory of Environmental Risk
　Assessment and Control on Chemical
　Process
East China University of Science and
　Technology
Shanghai, People's Republic of China

Mohd. Suffian Yusoff
School of Civil Engineering
Universiti Sains Malaysia
Nibong Tebal, Penang, Malaysia

Wei Zhang
State Environmental Protection Key
　Laboratory of Environmental Risk
　Assessment and Control on Chemical
　Process
East China University of Science and
　Technology
Shanghai, People's Republic of China

1 Safety and Control for Toxic Chemicals and Hazardous Wastes

Nazih K. Shammas
Lenox Institute of Water Technology and Krofta Engineering Corporation

Lawrence K. Wang
Lenox Institute of Water Technology, Krofta Engineering Corporation, and Zorex Corporation

CONTENTS

1.1 Introduction 2
1.2 The Problem 2
1.3 Risk Assessment and Risk Management 7
1.4 Legislation and Regulation 8
    1.4.1 Toxic Substance Control Act 8
    1.4.2 Resources Conservation and Recovery Act 9
        1.4.2.1 Essential Requirements for Waste Generators 9
        1.4.2.2 Essential Requirements for Transporters 10
        1.4.2.3 Essential Requirements for TSDFs 10
        1.4.2.4 Inadequacies in the RCRA Regulations 10
    1.4.3 Comprehensive Environmental Response, Compensation, and Liability Act 11
    1.4.4 Clean Air Act and Amendments 13
1.5 Management and Implementation Strategies 14
1.6 Waste Treatment and Disposal Technologies 16
1.7 Accident Prevention and Emergency Response 16
1.8 Education and Training 20
1.9 Conclusion and Recommendations 20
Acknowledgments 21
Appendix   New York State Multimedia Pollution Prevention Policy, Regulation, and Programs 21
References 23

Abstract

When toxic chemicals and hazardous wastes contaminate the environment, all organisms are exposed to potential risks. Subtle harm such as human disease or ecological disruption caused by toxic chemicals and hazardous wastes, may not become apparent until it is too late to. Furthermore, significant amounts of toxic pollutants are cycled and recycled in the environment, not only from sources but also from waste treatment and disposal activities. Environmental issues, particularly concerns about toxic chemicals and hazardous wastes, are receiving more attention in policy, programming and media in the United States and in other developed countries. Every year, thousands of leaks and spills of toxic chemicals and hazardous wastes go unreported, mainly because they are not detected. Pollution from toxic chemicals and hazardous wastes is more extensive and difficult to manage than what was *previously* believed. Balancing environmental protection, economic growth, and budgetary constraints encourages a manage-for-results approach. This chapter includes the following topics: (a) legislation and regulation for toxic chemicals and hazardous wastes in the United States, (b) management and implementation of control technologies, including treatment and handling, and (c) current strategies for prevention of chemical accidents, emergency response, and risk identification and evaluation. The chapter briefly deals with the problems of toxic and hazardous wastes and the need for education and training of personnel. Education is the key to achieving the vital goal of a healthy environment. Lack of well-trained professionals in the environmental discipline impedes the management of toxic chemicals and hazardous wastes as much as funding limitations do.

1.1 INTRODUCTION

The issue of toxic chemicals and hazardous wastes began to gain public attention in the United States in 1978, when the Love Canal in Niagara Falls, NY, was evacuated after it became known that Occidental Petroleum had dumped 22,000 tons of toxic chemicals there. Recently, in Flint, MI (2015/2016), the presence of high levels of lead in the water supply adversely affected the health of children [1].

To be able to develop effective action and realistic proposals for protection from toxic waste, one needs to understand the critical issues and real problems regarding toxic chemicals and hazardous wastes. The objectives of managing toxic wastes include the following:

1. Examine the critical environmental issues due to toxic chemicals and hazardous wastes.
2. Prepare an inventory of high-priority actions to develop legislation and regulation, program planning, and implementation of chemical safety measures.
3. Formulate realistic proposals to deal with critical issues.
4. Propose new projects that will help restore, protect, and improve environmental quality and reduce health risks.
5. Raise the level of understanding and commitment to action.

The chapter includes the following topics:

1. Legislation and regulation for toxic chemicals and hazardous wastes in the United States.
2. Management and implementation of control technologies, including treatment and handling.
3. Current strategies for prevention of chemical accidents, emergency response, and risk identification and evaluation.

The chapter briefly deals with the problems of toxic and hazardous wastes and the need for education and training of personnel. Education is the key to achieving the vital goal of a healthy environment. Lack of well-trained professionals in the environmental discipline impedes management of toxic chemicals and hazardous wastes as much as funding limitations do.

1.2 THE PROBLEM

Environmental issues, particularly concerns about toxic chemicals and hazardous wastes, are receiving more attention in policy, programming, and media in the United States and in other developed countries. Every year, thousands of leaks and spills of toxic chemicals and hazardous wastes go unreported, mainly because they are not detected. Pollution from toxic chemicals and hazardous wastes is more extensive and difficult to manage than what was previously believed. Balancing environmental protection, economic growth, and budgetary constraints encourages a manage-for-results approach.

Modern conveniences would not be possible without the production of thousands of different chemicals. Most of these chemicals are organic compounds derived from petroleum, while the remaining are inorganic in nature, such as ammonia, chlorine, and metals (see Table 1.1) [2]. These

TABLE 1.1
Highly Hazardous Chemicals, Toxics, and Reactives (OSHA)

Chemical Name

Acetaldehyde
Acrolein (2-propenal)
Acrylyl chloride
Allyl chloride
Allylamine
Alkylaluminums
Ammonia, anhydrous
Ammonia solutions (>44% ammonia by weight)
Ammonium perchlorate
Ammonium permanganate
Arsine (also called arsenic hydride)
Bis(chloromethyl) ether
Boron trichloride
Boron trifluoride
Bromine
Bromine chloride
Bromine pentafluoride
Bromine trifluoride
3-Bromopropyne (also called propargyl bromide)
Butyl hydroperoxide (tertiary)
Butyl perbenzoate (tertiary)
Carbonyl chloride (see phosgene)
Carbonyl fluoride
Cellulose nitrate (concentration >12.6% nitrogen)
Chlorine
Chlorine dioxide
Chlorine pentrafluoride
Chlorine trifluoride
Chlorodiethylaluminum (also called diethylaluminum chloride)
1-Chloro-2,4-dinitrobenzene
Chloromethyl methyl ether
Chloropicrin
Chloropicrin and methyl bromide mixture
Chloropicrin and methyl chloride mixture
Commune hydroperoxide
Cyanogen
Cyanogen chloride
Cyanuric fluoride
Diastole peroxide (concentration >70%)
Diazomethane
Dibenzoyl peroxide
Diborane
Dibutyl peroxide (tertiary)
Dichloro acetylene
Dichlorosilane
Diethylzinc
Diisopropyl peroxydicarbonate

(Continued)

TABLE 1.1 *(Continued)*
Highly Hazardous Chemicals, Toxics, and Reactives (OSHA)

Chemical Name

Dilauroyl peroxide
Dimethyldichlorosilane
1,1-Dimethylhydrazine
Dimethylamine, anhydrous
2,4-Dinitroaniline
Ethyl methyl ketone peroxide (also methyl ethyl ketone peroxide; concentration >60%)
Ethyl nitrite
Ethylamine
Ethylene fluorohydrin
Ethylene oxide
Ethyleneimine
Fluorine
Formaldehyde (formalin)
Furan
Hexafluoroacetone
Hydrochloric acid, anhydrous
Hydrofluoric acid, anhydrous
Hydrogen bromide
Hydrogen chloride
Hydrogen cyanide, anhydrous
Hydrogen fluoride
Hydrogen peroxide (52% by weight or more)
Hydrogen selenide
Hydrogen sulfide
Hydroxylamine
Iron, pentacarbonyl
Isopropylamine
Ketene
Methacrylaldehyde
Methacryloyl chloride
Methacryloyloxyethyl isocyanate
Methyl acrylonitrile
Methylamine, anhydrous
Methyl bromide
Methyl chloride
Methyl chloroformate
Methyl ethyl ketone peroxide (concentration >60%)
Methyl fluoroacetate
Methyl fluorosulfate
Methyl hydrazine
Methyl iodide
Methyl isocyanate
Methyl mercaptan
Methyl vinyl ketone
Methyltrichlorosilane
Nickel carbonyl (nickel tetracarbonyl)
Nitric acid (94.5% by weight or more)

(Continued)

TABLE 1.1 *(Continued)*
Highly Hazardous Chemicals, Toxics, and Reactives (OSHA)

Chemical Name

Nitric oxide
Nitroaniline (para nitroaniline)
Nitromethane
Nitrogen dioxide
Nitrogen oxides (NO; NO_2; N_2O_4; N_2O_3)
Nitrogen tetroxide (also called nitrogen peroxide)
Nitrogen trifluoride
Nitrogen trioxide
Oleum (65%–80% by weight; also called fuming sulfuric acid)
Osmium tetroxide
Oxygen difluoride (fluorine monoxide)
Ozone
Pentaborane
Peracetic acid (concentration >60% acetic acid; also called peroxyacetic acid)
Perchloric acid (concentration >60% by weight)
Perchloromethyl mercaptan
Perchloryl fluoride
Peroxyacetic acid (concentration >60% acetic acid; also called peracetic acid)
Phosgene (also called carbonyl chloride)
Phosphine (hydrogen phosphide)
Phosphorus oxychloride (also called phosphoryl chloride)
Phosphorus trichloride
Phosphoryl chloride (also called phosphorus oxychloride)
Propargyl bromide
Propyl nitrate
Sarin
Selenium hexafluoride
Stibine (antimony hydride)
Sulfur dioxide (liquid)
Sulfur pentafluoride
Sulfur tetrafluoride
Sulfur trioxide (also called sulfuric anhydride)
Sulfuric anhydride (also called sulfur trioxide)
Tellurium hexafluoride
Tetrafluoroethylene
Tetrafluorohydrazine
Tetramethyl lead
Thionyl chloride
Trichloro (chloromethyl) silane
Trichloro (dichlorophenyl) silane
Trichlorosilane
Trifluorochloroethylene
Trimethyoxysilane

Source: Occupational Safety and Health Administration (OSHA), Listing of toxic and reactive highly hazardous chemicals, OSHA, 58 FR 35115, June 30, 1993. www.osha.gov/pls/oshaweb/owadisp.show_document?p_table=STANDARDS&p_id=10647, January, 2016. Accessed April 2, 2017.

TABLE 1.2
Health Effects of Contaminants Frequently Found at RCRA Corrective Action Sites

Substance	Potential Health Effects
Arsenic	Carcinogenic to humans (skin, lung, bladder, liver) • stomach and intestinal irritation, nausea, vomiting • decreased production of red and white blood cells • damage to blood vessels • skin changes • abnormal heart rhythm
Benzene	Carcinogenic to humans (leukemia) • harmful to bone marrow, decreased red blood cells, anemia • vomiting, stomach irritation • drowsiness, dizziness, rapid heart rate, headaches, tremors, convulsions, unconsciousness
Cadmium	Likely to be carcinogenic to humans • kidney, bone, and lung damage • stomach irritation, vomiting, diarrhea • birth defects in some animal studies
Chloroform	Likely to be carcinogenic to humans • liver and kidney damage • skin sores • dizziness, fatigue, headaches • reproductive and birth defects in rats and mice
Lead	Likely to be carcinogenic to humans • damage to the brain and nervous system (adults, children, unborn children) • miscarriage, premature births, neonatal mortality due to decreased birth weight, decreased male fertility • diminished learning abilities in children • increased blood pressure • kidney damage
Mercury	Brain, kidney, and lung damage • serious harm to neural development of fetuses and young children • chest pains, nausea, vomiting, diarrhea • skin rashes and eye irritation • increased blood pressure and heart rate • irritability, sleep disturbances, tremors, coordination problems, changes in vision and hearing, memory problems
Perchlorate	Inhibition of iodine uptake • hypothyroidism, which may adversely affect the skin, heart, lungs, kidneys, gastrointestinal tract, liver, blood, neuromuscular system, nervous system, skeleton, reproductive system, and numerous endocrine organs
Polychlorinated Biphenyls (PCBs)	Likely to be carcinogenic to humans • liver damage • skin rashes and acne • decreased birth weight • short-term behavioral and immune system impacts in children exposed via breast milk
Polycyclic Aromatic Hydrocarbons (PAHs)	Likely to be carcinogenic to humans • irritation of skin, lungs, and stomach • reproductive and birth defects in animal studies
Tetrachloroethylene (PCE)	Likely to be carcinogenic to humans • dizziness, headaches, sleepiness, confusion, nausea, difficulty speaking and walking, unconsciousness
Trichloroethylene (TCE)	Carcinogenic to humans • liver, kidney, and nervous system damage • impaired immune system and heart function • impaired fetal development • skin rashes, lung irritation, headaches, dizziness, nausea, unconsciousness

Source: Federal Register, Comprehensive Environmental Response, Compensation, and Liability Act (CERCLA or Superfund) 42 U.S.C. s/s 9601 et seq. (1980), United States Government, Public Laws. www.epa.gov/superfund/learn-about-superfund, January, 2014. Accessed April, 2017.

chemicals are important in many facets of daily life, from preserving food to expediting transportation and communication. Some chemicals are largely unknown to the general public, but they are extremely important in research and manufacturing processes. Most chemicals are not harmful if used properly, while some can be extremely harmful if people are exposed to them, even in minute amounts (Table 1.2) [3].

When toxic chemicals and hazardous wastes contaminate the environment, all organisms are exposed to potential risks. Subtle harm, such as human disease or ecological disruption caused by toxic chemicals and hazardous wastes, may not become apparent until it is too late. Furthermore, significant amounts of toxic pollutants are cycled and recycled in the environment, not only from sources but also from waste treatment and disposal activities.

Many segments of our society generate hazardous waste. Waste generators include industries, commercial establishments, agricultural and mining activities, hospitals, research laboratories, and households. But industries are by far the largest source of hazardous wastes. In the United States, chemical and allied companies produce the highest amount (60%) of all industrial wastes. Plastics are a big concern as they are petroleum-based, halogenated compounds. Pesticides have organic, phosphate, and chromium compounds. Pharmaceuticals often contain organic solvents, heavy metals, and salts. Other waste generators include petroleum producers (with phenols), metal industries, leather processors, and textile companies. Waste comes in four forms: solid, liquid, sludge, and gas. Waste may change from one form to another, depending on factors such as temperature, pressure, and storage conditions.

Advancement in industrial technology has resulted in environmental abuse that has led to high social costs. These social costs can now be measured because of technological advancement in computers, which has made it possible for us to access databases from various national and international organizations for cost assessments.

Pollution problems result mostly from mismanagement caused by a variety of reasons: carelessness, indifference, or ignorance, as well as lack of measurement and monitoring of instruments and methods to provide baseline or background data to serve as a technical basis for regulatory action and engineering control, and as a verification of adverse effects on health and environment.

In addition,

> Collusion between industry and science may be harming our health; the rift centers around the best way to measure the health effects of chemical exposures. The debate may sound arcane, but the outcome could directly affect our health. It will shape how government agencies regulate chemicals for decades to come: how toxic waste sites are cleaned up, how pesticides are regulated, how workers are protected from toxic exposure and what chemicals are permitted in household items. Those decisions will profoundly affect public health: the rates at which we suffer cancer, diabetes, obesity, infertility, and neurological problems like attention disorders and lowered IQ. [4].

A substantial use of chemicals is essential to meet the social and economic goals of the world community. The current best practice demonstrates that chemicals can be used widely in a cost-effective manner and with a high degree of safety. However, a great deal remains to be done to ensure sound management of toxic chemicals in the environment, within the principles of sustainable development and improved quality of life for humankind. The two major problems in implementing this practice, particularly in developing countries, are (a) lack of sufficient scientific information to assess risks involved in the use of a wide range of chemicals and (b) lack of resources to assess chemicals for which data are available [5].

1.3 RISK ASSESSMENT AND RISK MANAGEMENT

In recent years, toxic chemicals [Dichlorodiphenyltrichloroethane (DDT), Polychlorinated biphenyls (PCBs), dioxins, and Chlorofluorocarbons (CFCs)] and hazardous wastes have been major issues affecting the environment. Although the issues differ, the major concerns regarding public health and safety remain the same. Society is constantly faced with the following fundamental questions:

1. What are the risks associated with certain chemicals and wastes?
2. How serious are they?
3. What are acceptable risks?
4. How well can they be assessed and managed?

We assess chemical risk by estimating the levels and duration of exposure to chemical toxicity or hazard. Exposures that are lethal to plants and animals can usually be determined easily. However, estimating the risk associated with low concentrations of chemical compounds suspected to cause cancer or mutations is difficult and controversial. Chemical risk assessment attempts to identify the contaminants of concern and their pathways, to determine their exposure concentrations, and to evaluate their toxic effects. It is used to determine the adverse effect on health that might result from exposure to a given chemical or chemical compound.

There are four types of releases in all industrial facilities, including chemical industries that use, manufacture, or store toxic chemicals: (a) releases from limited process upsets, (b) emissions from process vents, (c) fugitive emissions, and (d) accidental, sudden, large releases.

The primary methods for expressing toxicity are:

1. IDLH: immediately dangerous to life and health.
2. LLC: low lethal concentration.
3. LC_{50}: 50% lethal concentration.
4. PEL: permissible exposure limit.
5. STEL: short-term exposure limit.

Risk management involves assessing risk and reducing it to acceptable levels. It is used to determine the acceptability of various levels of risk by taking into account the cost, among other factors. Risk management requires value judgments that integrate social, economic, and political concerns with scientific risk assessment. Essentially, both risk assessment and risk management assist individuals in deciding whether to take a certain drug, smoke cigarettes, or drink diet soft drinks. They also help in deciding whether ethylene dibromide (EDB) should be used as a fumigant or whether saccharin should be used as a food additive.

The U.S. Environmental Protection Agency (U.S. EPA) has selected 17 priority chemicals from its Toxic Release Inventory (TRI), based on health and environmental effects, potential for exposure, production volume, and potential for reducing releases. The 17 priority chemicals are benzene, cadmium and its compounds, carbon tetrachloride, chloroform, chromium and its compounds, cyanides, dichloromethane, lead and its compounds, mercury and its compounds, methyl ethyl ketone, methyl isobutyl ketone, nickel and its compounds, tetrachloroethylene, toluene, trichloroethane, trichloroethylene, and xylene(s).

1.4 LEGISLATION AND REGULATION

In the United States, with the enactment of the Toxic Substance Control Act (TSCA) [6] and the Resource Conservation and Recovery Act (RCRA) in 1976, the U.S. EPA began to regulate toxic chemicals and hazardous wastes [7]. Until then, only pesticides were regulated under the 1972 Federal Insecticide, Fungicide, and Rodenticide Act (FIFRA) [8].

The growing concern that exposure to toxic chemicals possibly causes cancer in humans has provided the impetus for regulatory control over the chemical industries that produce toxic chemicals and hazardous wastes. The ultimate goal is to ensure that all toxic chemicals and hazardous wastes are handled, treated, and disposed of properly.

The legal structures to control toxic chemicals and hazardous wastes usually represent a compromise between the public health and welfare on one hand, and technical, economic, and political factors on the other hand. The manner in which this compromise is achieved largely depends on the situation existing in each jurisdiction across the country.

1.4.1 TOXIC SUBSTANCE CONTROL ACT

The TSCA [6] regulates existing and new chemical substances. It applies primarily to manufacturers, distributors, processors, and importers of chemicals. The TSCA can be divided into five parts as follows:

1. *Testing*: Under TSCA, Section 4, the U.S. EPA can require product testing of any substance that may present an unreasonable risk of injury to health or to the environment. Some testing standards have been proposed.
2. *Inventory and Pre-manufacture Notification* (PMN): The U.S. EPA has published an inventory of existing chemicals. An unlisted substance is considered "new" and requires PMN to the U.S. EPA at least 90 days before the chemical can be manufactured, shipped, or sold (TSCA, Section 5). The U.S. EPA may reject PMN because of insufficient data, negotiate for suitable data, and prohibit manufacture or distribution until risk data become available. The U.S. EPA may completely ban the product from the market or may review the product data for an additional 90 days.
3. *Reporting and Record-keeping*: Section 8(a) deals with general reporting. Section 8(c) calls for records of significant adverse effects of toxic substance on human health and on the environment. It requires that records of alleged adverse reaction be kept for a minimum of five years. Section 8(d) authorizes the U.S. EPA to compel manufacturers, processors, and distributors of certain listed chemicals to submit to the U.S. EPA lists of health and safety

studies conducted by, known to, or ascertainable by them. Studies include individual files, medical records, and daily monitoring reports. Section 8(e) requires action upon discovery of certain data.

4. *Regulation under Section 6*: The U.S. EPA can impose Section 6 rule if there is reason to believe that the manufacture, processing, distribution, use, or disposal of a chemical substance or mixture causes, or may cause, an unreasonable risk of injury to health or to the environment. Regulatory action ranges from labeling requirements to complete prohibition of the product. Section 6 rule requires an informal rulemaking, a hearing, and a cost–benefit analysis.
5. *Imminent Hazard*: A chemical substance or mixture can be declared as imminently hazardous if it can cause an imminent and unreasonable risk of serious or widespread injury to health or to the environment. When such a condition prevails, the U.S. EPA is authorized by TSCA, Section 7, to commence a judicial action against any responsible person or organization in the U.S. District Court. Remedies include seizure of the chemical or other relief, including notice of risk to the affected population or recall, replacement, or repurchase of the substance.

1.4.2 Resources Conservation and Recovery Act

The RCRA [7] was signed on October 21, 1976, and subsequently amended in 1980, 1984, 2002, and 2008 [3]. The major statutory restrictions or prohibitions include the following:

1. Placement of any bulk hazardous waste in salt domes, salt bed formations, underground mines, or caves is prohibited until the facility receives a permit.
2. Landfilling of bulk or noncontained liquid hazardous waste or free liquids contained in hazardous waste is prohibited.
3. Placement of any nonhazardous liquid in a landfill operating under interim status or a permit is prohibited unless the operation will not endanger groundwater drinking sources.
4. Land disposal of solvents and dioxins is prohibited unless human health and the environment will not be endangered.
5. Land disposal of wastes listed in Section 3004(d)(2) is prohibited unless human health and the environment will not be endangered.
6. New units, lateral expansions, and replacement of existing units at interim status landfills and impoundments need to have double liners and leachate collection systems; waste piles require a single liner.

The U.S. EPA compiled a list of more than 200 hazardous commercial chemical products and chemical intermediates by generic names. Substances became hazardous wastes when discarded. If a commercial substance is on the list, its off-specification species and spill residues are also considered hazardous. The overall listing further includes acutely toxic commercial chemical wastes.

1.4.2.1 Essential Requirements for Waste Generators

The essential requirements for waste generators under the RCRA (40CFR Part 262) are listed here [7]:

1. *Identification*: Hazardous wastes must be identified by lists, test methods for hazardous characteristics, or experience, and must be assigned waste identification numbers.
2. *Notification*: No later than 90 days after a hazardous waste is identified or listed, a notification is to be filed with the U.S. EPA or with an authorized state. The U.S. EPA identification number must be assigned.
3. *Manifest System*: The manifest system must be implemented, and the prescribed procedures for tracking and reporting shipments must be followed.

4. *Packing*: Packaging, labeling, marking, and placarding requirements prescribed by the Department of Transportation (DOT) regulations must be implemented.
5. *Annual Report*: Waste generators are required to submit an annual report by March 1.
6. *Exception Reports*: When a generator does not receive a signed copy of manifest from the designated treatment, storage, and disposal facility (TSDF) within 45 days, the generator sends an Exception Report to the U.S. EPA, which includes a copy of manifest and a letter describing efforts made to locate waste and findings.
7. *Accumulation*: When waste is accumulated for less than 90 days, generators shall comply with special requirements, including contingency plan, prevention plan, and staff training.
8. *Permit for Storage More than 90 Days*: If hazardous wastes are retained on-site for more than 90 days, a generator is subject to all requirements of TSDFs including the need for RCRA permits.

1.4.2.2 Essential Requirements for Transporters
The essential requirements for transporters under the RCRA are listed here [7]:

1. *Notification*: Same as for generators.
2. *Manifest System*: The transporter must fully implement the manifest system. The transporter signs and dates the manifest, returns one copy to the generator, ensures that the manifest accompanies the waste, obtains the date and signature of the TSDF or the next receiver, and retains one copy of the manifest.
3. *Delivery to TSDF*: The waste is delivered only to the designated TSDF or an alternative.
4. *Record Retention*: The transporter retains copies of the manifest signed by the generator, himself, and the accepting TSDF or receiver, and he keeps these records for a minimum of 3 years.
5. *Discharges*: If discharges occur, notice is given to the National Response Center. Appropriate immediate action is taken to protect health and the environment, and a written report is sent to the DOT.

1.4.2.3 Essential Requirements for TSDFs
The essential requirements for TSDFs under the RCRA are listed here:

1. *Notification*: Same as for generators.
2. *Interim Status*: These facilities include TSDFs, on-site hazardous waste disposal; on-site storage for more than 90 days; in-transit storage for more than 10 days; and the storage of hazardous sludges, listed wastes, or mixtures containing listed wastes intended for reuse.
3. *Interim Status Facility Standards*: The prescribed standards and requirements are met and include the following: general information, waste analysis plan, security, inspection plan, personnel training, handling requirements, preparedness and prevention, contingency planning and emergency procedures, records and reports, manifest system, operating logs, annual and other reports, groundwater monitoring, closure and post-closure plans, financial requirements, containers, tanks, surface impoundments, piles, all treatment processes, landfills, and underground inspection.
4. *Permit*: Facilities with interim status must obtain an RCRA Part B permit.

1.4.2.4 Inadequacies in the RCRA Regulations
Although RCRA is quite detailed in coverage and is considered one of the best legislations in the United States, there are inadequacies and loopholes in the regulations:

1. Regulatory structure is unnecessarily complex, and there are no incentives for compliance or better waste management.
2. Regulations have too many cross-references.

3. Regulations complicate the permit process. Voluminous details about the means of carrying out the regulations result in a slow and costly process.
4. Regulations insufficiently take into consideration important features. There is little flexibility for the important characteristics of a particular facility, surrounding population, underlying groundwater aquifers, changing atmospheric conditions, varied waste properties, and the facility's compliance history.
5. Priority setting has not been realistic. Consequently, there is a widespread perception of inadequate progress in improving environmental quality.
6. Differences exist between U.S. EPA and the state agencies, including priorities and stringencies of enforcement.
7. Public education such as hearings and communicating risks posed from exposure have been ineffective.
8. Vital data and analytical techniques are lacking. Major technical and scientific uncertainties still exist. Data management systems need improvement.

1.4.3 Comprehensive Environmental Response, Compensation, and Liability Act

The Comprehensive Environmental Response, Compensation, and Liability Act, otherwise known as CERCLA or Superfund, provides a federal "Superfund" program to clean up uncontrolled or abandoned hazardous waste sites as well as accidents, spills, and other emergency releases of pollutants and contaminants into the environment [9]. Through CERCLA, the U.S. EPA was given authority to seek out those parties responsible for any release and ensure their cooperation in the cleanup.

The U.S. EPA cleans up sites when potentially responsible parties cannot be identified or located, or when they fail to act. Through various enforcement tools, the U.S. EPA obtains private party cleanup through orders, consent decrees, and other small party settlements. It also recovers costs from financially viable individuals and companies once a response action has been completed.

The U.S. EPA is authorized to implement the Act in all 50 states and U.S. territories. Superfund site identification, monitoring, and response activities in states are coordinated through the state environmental protection or waste management agencies.

The Superfund Amendments and Reauthorization Act (SARA) of 1986 reauthorized CERCLA to continue cleanup activities around the country. Several site-specific amendments, definitions clarifications, and technical requirements were added to the legislation, including additional enforcement authorities.

Over the past 30 years, the U.S. EPA and its state partners have built an efficient and successful RCRA Corrective Action Program (see Table 1.3), one that oversees the cleanup of a wide variety of contaminated sites, including many with high risks. The location of RCRA Corrective Action 2020 Universe sites is shown in Figure 1.1.

By the end of fiscal year 2012, of the 3747 sites in the 2020 Universe, the program had [10]:

- Met the human exposures EI (Environmental Indicator) at 3041 sites (more than 81%), covering 13.6 million acres.
- Met the groundwater EI at 2691 sites (more than 71%), covering 7.2 million acres;
- Reached final remedy construction at 1762 sites (more than 47%), covering 2.1 million acres.

Since 1980, Congress has mandated that several programs be established to clean up contaminated land. Four main programs are listed below, with the year of establishment, features that distinguish one program from another, and the current number of acres addressed by each program.

- *Superfund (1980)*: Sites that are abandoned, bankrupt, or have multiple responsible parties and uncontrolled releases of hazardous substances (3.9 million acres).
- *RCRA Corrective Action (1984)*: Sites with viable owners or operators that have treated, stored, or disposed of hazardous waste since 1980 and released hazardous constituents to the environment (18 million acres).

TABLE 1.3
RCRA Timeline for Corrective Action Program

Milestones/Timeline

1984	RCRA amended by Congress, creating the RCRA Corrective Action Program
1985–1992	Universe assessed; sites prioritized; states authorized; investigations begun; regulations (Subpart S) proposed
1994	Environmental Indicators established
1996	Subpart S regulations not finalized but issued as guidance; RCRA/CERCLA parity policy; public participation manual
1998	Post-closure regulations; risk-based clean closure guidance; remediation waste guidance
1999–2001	RCRA Cleanup Reforms implemented; GPRA goals established; Land Revitalization Initiative begun; Groundwater Handbook released
2002	RCRA 2020 Vision announced; first National Corrective Action Conference
2003	Corrective Action Completion guidance
2005	1st set of GPRA goals exceeded for 1714 highest-priority Corrective Action facilities (2005 Baseline)
2008	2nd set of GPRA goals exceeded for 1968 highest-priority Corrective Action facilities (2008 Baseline)
2009	"Strategies for Meeting the 2020 Vision" training delivered in all 10 regions
2011	GAO report praises program success but warns about resource constraints and challenges; NESCA issued
By 2020	EPA will strive to achieve environmental indicators and construct remedies at 95% of the 3747 facilities on the 2020 Universe

Source: Federal Register, Comprehensive Environmental Response, Compensation, and Liability Act (CERCLA or Superfund) 42 U.S.C. s/s 9601 et seq. (1980), United States Government, Public Laws. www.epa.gov/superfund/learn-about-superfund, January, 2014. Accessed April, 2017.

Safety and Control for Toxic Chemicals and Hazardous Wastes

FIGURE 1.1 Location of RCRA Corrective Action 2020 Universe Sites. Although not shown, the 2020 Universe also includes 6 sites in Alaska, 13 in Hawaii, 51 in Puerto Rico, 1 in the U.S. Virgin Islands, 5 in Guam, and 2 in the Pacific Trust Territories. (From US EPA, RCRA Corrective Action: Case Studies Report, EPA 530-R-13-002, April 2013.)

- *Underground Storage Tanks (1984)*: Leaking underground storage tanks (507,000 acres).
- *Brownfields (1998)*: Contaminated sites (Superfund, RCRA, or other) restored to usable property with assistance from a Brownfields grant (164,000 acres).

1.4.4 CLEAN AIR ACT AND AMENDMENTS

The Clean Air Act (CAA) [11] Amendments of 1990 have made great changes in the way U.S. industry operates. These changes include manufacturing processes, design and material, and products, in relationship with regulations and the economics of doing business.

The Title III of the Air Toxics of the CAA Amendments lists 189 chemicals to be regulated and includes a procedure for the U.S. EPA to add and delete chemicals from the list. The U.S. EPA is establishing a list of major sources and area source categories that need to be regulated. A major source is generally defined as a stationary source located within a contiguous area that emits or has the potential to emit 10 or more tons per year of any hazardous air pollutants, or 25 or more tons per year of any combination of hazardous air pollutants.

The CAA Amendments authorize the U.S. EPA to promulgate accident prevention regulations. In 1992, the U.S. EPA listed 100 extremely hazardous air pollutants along with threshold values. Some of the listed pollutants include chlorine, anhydrous ammonia, methylchloride, ethylene oxide, vinylchloride, methyl isocyanate, hydrogen cyanide, ammonia, hydrogen sulfide, toluene diisocyanate, phosgene, bromine, anhydrous hydrogen chloride, hydrogen fluoride, anhydrous sulfur dioxide, and sulfur trioxide.

The U.S. EPA has established regulations for detection and control of accidental releases. Owners or operators of industrial facilities that handle these extremely hazardous substances are required to prepare risk management plans that identify and prevent potential accidental releases. The information must be made open to the public. The Act established a Chemical Safety Board to investigate accidents and establish reporting and other regulations. The Act also calls for enhanced monitoring and compliance certifications when requested by the U.S. EPA.

Under the CAA Amendments, every industry is required to submit a written emergency response plan. The information should include person(s) to be notified, evacuation plans, standards of procedure for various types of chemical emergencies, and types of personnel training. Monitoring equipment will be maintained to detect pollutants in general and in emergencies, and to determine the bounds and extent of contamination. Cleanup materials include compounds capable of neutralizing acids and bases. Those actively involved in cleanup activities are required to have a self-contained breathing apparatus and possibly chemical-resistant suits.

Ideally, laws and regulations should be simple, reasonable, implementable, and affordable. However, the U.S. government tries to be specific and explicit about what is required. The resulting thick regulatory package is basically a complex legal document. Such laws and regulations can defeat their intended purpose. The U.S. legal and administrative overhead in hazardous waste work is extremely high. For example, legal and administrative efforts, instead of technical research or reclamation and engineering work, have depleted a large percentage of Superfund [9] monies for remediation of hazardous waste disposal sites.

1.5 MANAGEMENT AND IMPLEMENTATION STRATEGIES

About 50 years ago, most industries did not employ an "environmental manager." In fact, most industries did not know they added to environmental problems. Currently, however, industries realize they must understand and meet their company's responsibility to ensure good environmental, health, and safety practices. If they do not, they might face enormous negative economic impact. Consequently, industries recognize the need for well-trained and informed environmental managers.

In the past, regulatory leaders and business managers were isolated from environmental decisions that required an in-depth understanding of scientific and technological issues. Improper handling of toxic chemicals and mismanagement of hazardous wastes have posed serious environmental pollution problems. All too often, the emphasis of waste management is on pollution removal, that is, collecting and treating pollutants, and removing pollutants from the waste stream before discharging them into the environment. Removal technology may not be the most environmentally efficient method of waste management. Only in recent years has waste management shifted from pollution removal to pollution prevention through waste minimization in homes, businesses, industries, and private and public institutions. Industries should be encouraged to substitute pollution prevention practices for installation of pollution control devices.

A vital mutual dependence exists between hazardous waste regulation and the development of a detoxification and treatment facility. Without vigorous regulatory control of hazardous waste from its generation to its final disposal, short-term economic pressure will continue to favor cheap, environmentally unsound handling, transportation, and disposal. At the same time, regulation cannot succeed without proper treatment technologies and disposal facilities. The final step in hazardous waste management includes destruction, detoxification, or isolation of wastes from the environment. Regulatory control can track wastes from its generation to final disposal, but cannot destroy or detoxify these wastes if treatment facilities do not exist.

Concurrent development of regulation and treatment capacity will provide other countries with a comprehensive system for hazardous waste management. It will also prevent creation of further uncontrolled sites that leak hazardous materials into the environment and threaten public health.

The multimedia approach for environmental management and pollution prevention is highly recommended. It manages the quality of air, water, and land as a whole. The benefits of a multimedia approach are it (a) avoids the possibility of shifting waste or transferring pollutants from one medium to another, resulting in secondary pollution, (b) better coordinates regulatory efforts, and (c) improves risk assessment.

Pollution prevention will sustain economic development by reducing waste and conserving resources. It uses materials, processes, equipment, or products that avoid, reduce, or eliminate wastes or toxic releases through reduction of toxins and closed-loop recycling. All industries should emphasize on pollution prevention over treatment and disposal of waste. Pollution prevention includes:

1. Reducing the quantity and/or toxicity of pollutants generated by production processes by using source reduction or waste minimization, and process modification.

2. Eliminating pollutants by substituting nonpollutant chemicals or products (e.g., material substitution, changes in product specification).
3. Recycling waste materials (e.g., tracking by-products, reuse, and reclamation).

For example, the industry is phasing out methyl chloride, a high-volume commodity chemical, as it poses health risks and other environmental concerns, and using a substitute. Its replacement is likely to be *N*-methyl-2-pyrrolidone, which is noncarcinogenic, nonmutagenic, and does not bioaccumulate. In instances when the industry cannot substitute, it should use on-demand generation to form toxic chemicals. The on-demand generation and immediate consumption occur on the customer's site, eliminating storage and transport of hazardous substances such as vinyl chloride, arsine, phorgene, and formaldehyde [12].

Chemical industries must reduce the costs of production, preserve the environment, and remain competitive in the marketplace. They must meet government regulations for pollution control and remain responsive to public concerns over environmental hazards. Meanwhile, chemical industries must not only treat its generated waste and develop cleaner manufacturing processes, but also develop products that produce less environmental impact [13].

Chemical industries must begin to view pollution as a waste, a symptom of inefficiency in production. When they begin to recognize pollution as a cost of production, they will invest to improve processes, to increase efficiency, and to reduce pollution and waste. When regulatory agencies provide economic incentives for such improvements, industries will affect changes quickly.

Chemical industries in the United States have begun to use sophisticated laboratory instruments in the industrial process stream and to connect new instrumental tools to the state-of-the-art microcomputer technology and data control of process streams. Process efficiency will benefit the environment. This approach will enhance the competition in the chemical and material industry while furthering environmental protection.

The National Environmental Policy Act (NEPA) [14] detailed guidelines and procedures to implement the policies of these eight environmental systems: environmental impact assessment, sustainable economic development, liquid waste discharging fee, responsibility to achieve NEPA's goals, monitoring and assessment, permit/licensing, prevention and control, and an implementation schedule.

Early overall planning and adoption of a risk base will help manage a facility's toxic chemicals and hazardous wastes most efficiently. Assessing the source and fate of chemicals in the process of chemical plants will help prevent pollution (e.g., by studying pollution release of a chemical plant's integrated system, as shown in Figure 1.2).

A successful multimedia pollution prevention program requires the regulator to address three major issues: (a) database information, (b) science and technology, and (c) manpower and

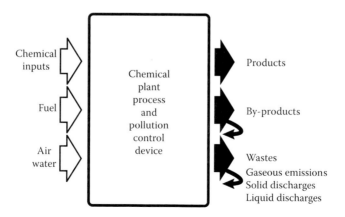

FIGURE 1.2 Pollution release study of a chemical plant integrated system.

financial resources. Appendix at the end of this chapter briefly describes New York State's multimedia pollution prevention policy, regulation, and programs.

Environmental management and technological progress should focus on utilization of high efficiency, low-consumption energy, and production technologies using multimedia pollution prevention and control program. Furthermore, innovative training programs at all levels need to be developed and all chemical plants need to be encouraged to replace outdated processes, equipment, and techniques.

1.6 WASTE TREATMENT AND DISPOSAL TECHNOLOGIES

Pollution prevention through waste minimization is the first priority of waste management practice. Since it is possible to only minimize a certain percentage of all wastes generated from various sources, waste treatment and disposal technologies are still needed. Treatment technology can be grouped into the following categories: physical, chemical, biological, and thermal. Both the public and the private sectors utilize these processes to some extent. However, treatment processes have had only limited applications in hazardous waste management because of economic constraints, and, in some cases, because of technological constraints [15–17].

Physical treatment processes concentrate brine wastes and remove soluble organics and ammonia from aqueous wastes [18]. Industries extensively use processes such as flocculation [19,20], sedimentation [21], and filtration [22], which primarily separate precipitated solids from the liquid phase. Ammonia stripping can remove ammonia from certain hazardous waste streams. Carbon sorption [23] is able to remove many soluble organics from aqueous waste streams. Evaporation can concentrate brine wastes to minimize the cost of final disposal.

Chemical treatment processes are also a vital part of proper hazardous waste management. Neutralization is carried out in part by reacting acid wastes with alkaline wastes [24]. Sulfide precipitation is required to remove toxic metals like arsenic, cadmium, mercury, and antimony [25]. Oxidation–reduction processes treat cyanide and chromium(VI)-bearing wastes [26,27].

Biological treatment processes can also biodegrade organic wastes [28–30]. They operate effectively only within narrow ranges of flow, composition, and concentration variations. Generally, biological treatment processes should be used only when the organic waste stream is diluted and fairly constant in its composition. Systems that provide the full range of biodegradation facilities usually require large land areas. Toxic substances present a constant threat to biological organisms.

Thermal treatment methods are able to destroy or degrade solid or liquid combustible hazardous wastes [31]. Incineration is the standard process used throughout the industry to destroy organic liquid and solid wastes. Concerns over toxic emissions from incinerators and ash disposal problems have made the incineration process unpopular. Pyrolysis, another thermal process, converts hazardous wastes into more useful products, such as fuel gas and coke.

Current disposal methods vary, depending on the form of the waste stream (solid or liquid), transportation costs, and local ordinances. Dumps and landfills have been utilized for all types of hazardous wastes; ocean disposal [32] and deep-well injection [33] have been used primarily for liquid hazardous wastes. These methods are no longer permitted to operate in the United States unless the regulatory agencies approve them.

Land disposal will continue to be required in the foreseeable future, although detoxification may be necessary [34–36]; hence, capacity needs will always be a concern. Finally, all forms of waste management must allow for industrial growth, and thus the need for additional land disposal capacity must be recognized. By reducing per unit product generation rates, waste minimization can provide at least a partial answer to these problems.

1.7 ACCIDENT PREVENTION AND EMERGENCY RESPONSE

The Bhopal calamity in India has profoundly affected the chemical industry. Since December 1984, many chemical plants, concerned with safety, have now increased their efforts to prevent, prepare for, and respond to release of accidental chemicals [2]. Table 1.4 summarizes potential categorical causes of chemical accidents. Two essential approaches will minimize chemical accidents: (a) chemical safety audit, which may be divided into three major phases as described in Figure 1.3. Required background information is listed in Table 1.5, and (b) chemical risk assessment, which has already been discussed in the earlier section.

Although the U.S. EPA does not play a central role in chemical accident prevention, its programs probably have an ancillary effect on reducing the incidence of sudden, hazardous releases. For example, under the TSCA, the U.S. EPA evaluates the hazards of chemical products and intermediates, and then it restricts or imposes controls on market entry, manufacture, and use of chemicals that present unreasonable risks. To strengthen this chemical regulatory program, the U.S. EPA proposed a rule that would require manufacturers and importers of substances such as those included in the TSCA Chemical Substance Inventory, to report current data regarding the production volume, plant site, and site-limited status of each substance. Besides this, compliance with emission standards under the CAA reinforces the integrity of systems designed to limit routine chemical releases to the atmosphere, thus reducing the likelihood of an accidental release [37].

TABLE 1.4
Major Causes of Chemical Accidents

1. Lack of or insufficient knowledge on plant and chemical safety, including but not limited to:
   - Safety data
   - Failure rates by category
   - Potential faults
   - Most realistic events
   - Damage functions
   - Control strategies
   - Emergency response organization, resources, reporting, and response strategies/plans
2. Lack of or insufficient knowledge on plant's hazards and operability (HAZOP)
3. Lack of or insufficient knowledge on chemical process quantitative risk assessment (CPQRA)
4. Poor management practices resulting from judgement calls which are often based on insufficient knowledge about the plant safety problems
5. Lack of knowledge on integrated operations and management of plant processes and facilities
6. Lack of or incomplete systematic repair, maintenance, and parts replacement approach, including:
   - System inspection and audits
   - Testing and quality assurance/quality control
   - Orderly implementation activities
   - System tracking and orderly scheduling
   - Early identification and implementation and corrective action(s)
   - Definition of authority, functions, and responsibilities of plant personnel
7. Lack of or incomplete inventory of emergency response resources
8. Lack of personnel training, including materials on:
   - Chemical safety
   - Work place safety and health
   - Use and inventory of personal protective equipment
   - Knowledge on inventory of emergency response resources
   - Internal and external lessons learned (case studies)
9. Lack of an optimal management/tracking system for spare parts

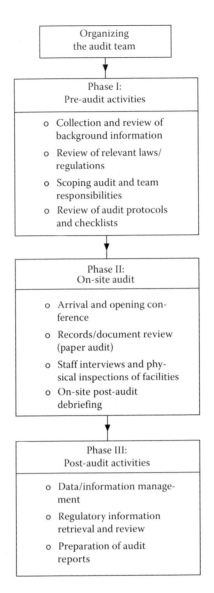

FIGURE 1.3 Chemical safety audit phases.

NEPA developed policies, programs, and procedures to prevent leaks and spills of toxic chemicals and to reduce generation of hazardous wastes. The policies specify the role of industries and provincial and local governments and emphasize their role in accident prevention and emergency response to leaks and spills of toxic chemicals and hazardous wastes. In the case of emergency services, either the local government or industry is the first line of response to cope with regulations and procedures.

To perform accident prevention and emergency response activities, a working group should be formed including high officials or their representatives, who will have the primary responsibility to manage overall activities. The functions of this working group include the following:

1. Determining the availability of technical expertise and financial resources necessary to address the program activities concerning toxic chemicals and hazardous wastes.
2. Providing resources including personnel, supplies, and equipment as necessary to restore and maintain essential services and fulfill the planned responsibilities.

TABLE 1.5
Required Background Information

Facility Background

- o   Maps of facility and building/shop location and environmental and geographic features (production units, storage tanks, discharge pipes, waste disposal sites, etc.)
- o   Geology/hydrogeology of the area
- o   Environmental setting (climate conditions, land use, floodplain, wetlands, historical sites, and sensitive receptors at and near the facility)
- o   Names and phone numbers of facility officials
- o   Records of changes in facility conditions since previous audits

Audit Reports and Other Relevant Records

- o   Federal and state compliance reports
- o   Correspondence between the facility and local, state and federal agencies on safety, hygiene, health, and environmental matters
- o   Citizens' complaints, follow-up studies, and findings
- o   Audit records, reports, and correspondence on past incidents or violations
- o   EPA, State, and consultant studies and reports
- o   RCRA reports, CERCLA submittals, NPDES permit records
- o   OSHA reports
- o   Records of hazardous substance spills

Plant Processes, Layouts, Configurations, and Chemical Inventory

- o   Production processes and their interrelationship
- o   Plant layouts and configurations (including reactors, processors, piping and storage tanks)
- o   Plant design drawings
- o   Inventory of chemicals used, manufactured, stored, and transported
- o   Process-monitoring data and reports

Pollutant and Waste Generation, Control, Storage, Transportation, Treatment, and Disposal

- o   Description and design data for pollution control systems and process operations
- o   Sources and characterization of wastewater discharges, hazardous wastes, emissions, types of treatment and disposal
- o   Waste storage, treatment and disposal areas
- o   Waste/spill contingency plans
- o   Bypasses, diversions, and spill containment facilities
- o   Self-monitoring data and reports

Chemical Emergency Response Systems and Plans

- o   Emergency response organization
- o   Emergency response plan
- o   Emergency response policies, resources (eg., fire protection, fire fighting, etc.) and practices

Requirements, Regulations, and Limitations

- o   Permit applications, draft or existing permits, registrations, approvals, and applicable federal, state, and local regulations and requirements
- o   Draft permits or information on draft permit terms which are different from current conditions
- o   Conditioned or unconditioned exemptions and waivers
- o   Receiving stream water quality standards, ambient air standards, State Implementation Plans, protected uses
- o   RCRA notification and Part A and Part B permit applications

3. Appointing appropriate managers to compile and maintain records, documents, and other items necessary to substantiate provincial or local claims for national assistance.
4. Designating responsibilities to program and operate units in accordance with their expertise to effectively carry out accident prevention and emergency response obligations.

1.8 EDUCATION AND TRAINING

Successful management and implementation of environmental protection programs require well-trained environmental professionals who are fully informed in the principles and practices of such programs. The programs need to develop in professionals a deep appreciation of the necessity for multimedia pollution prevention of toxic chemicals and hazardous wastes. New instructional materials and tools should incorporate emerging concepts in the existing curricula of elementary and secondary education, colleges and universities, and training institutions. Computerized automation offers much hope [38]. The demand for competent environmental professionals is large, and it will grow considerably over the following decades. Government agencies need to conduct a variety of activities to achieve three main educational objectives:

1. Ensure an adequate number of high-quality environmental professionals.
2. Encourage individuals and groups to undertake careers in environmental fields and to stimulate all institutions to participate more fully in training environmental professionals.
3. Generate a database that can improve the environmental literacy of the general public, especially through the media.

Industries and universities must teach the new environmental ethic—new clean technologies and new products are essential to protect the environment. For example, chemical plants have usually been developed to maximize reliability, productivity, product quality, and profitability at the cost of chronic emissions and effluents, inadequate waste treatment, and disposal. Engineers and managers in these chemical plants must learn to design and operate the process and manage waste treatment and disposal facilities to produce less waste and less toxicity in the waste and improve the production of goods and services. The traditional market prices do not reflect the subsequent costs of waste disposal or the uncompensated environmental and health damages. Thus, the responsibility of waste generators must go beyond those from production and marketing to social costs of pollution as well.

NEPA should develop an innovative education program of chemical safety and hazardous waste management at all levels of need in accordance with the other elements of the NEPA plan and programs. The essential qualities of environmental education should also include integrity, respect, caring, and responsibility. We need to broaden environmental education so that these essential qualities will enable people to act responsibly when faced with ethical problems.

1.9 CONCLUSION AND RECOMMENDATIONS

Environmental protection will be a never-ending battle against pollution unless we prevent pollution sources from developing now. Success of implementing multimedia pollution prevention programs will require an adequate number of well-trained environmental professionals and a shift in attitude and perception. We must restrict the use of toxic chemicals in places where they might enter into our air, water, or land. We must ban the disposal of untreated hazardous wastes on the land or in the ocean, where it threatens our health or the environment. By taking precautions today, we will be making an important investment in a safer and cleaner environment tomorrow. The following are our recommendations to attain the above objective:

1. Expanding and accelerating international assessment of chemical risks.
2. Harmonization of classification and labeling of chemicals.
3. Information exchange on toxic chemicals and chemical risks.
4. Establishment of risk-reduction programs.
5. Strengthening of national capabilities and capacities for management of chemicals.
6. Prevention of illegal international trafficking of toxic and dangerous products.

7. Improvement in measuring and monitoring capabilities as well as in establishing guidelines for chemical risk assessment using source–pathway–effect relationships.
8. Making health, safety, and environmental considerations a priority in design, operation, and maintenance of chemical plants.
9. Development and production of environmentally acceptable products that can be transported, used, and disposed of safely, and counseling customers on the safe use, transportation, and disposal of chemical products.
10. Construction of regional hazardous waste detoxification and disposal facilities to meet the urgent needs of such facilities in these industrialized regions.
11. Developing an innovative education program with training materials that are in harmony with the other elements of NEPA's chemical safety and hazardous waste management programs.
12. Conducting or supporting research on the health, safety, and environmental effects of chemical products, processes, and waste materials.
13. Encouragement of Chinese, Indian, and Brazilian professionals to participate in chemical safety and hazardous waste management programs.
14. Establishment of an advisory panel to improve communication, coordination, and cooperation among agencies of environmental concern to convince decision makers to lend their support and provide these agencies with adequate resources to protect public health and environment.

ACKNOWLEDGMENTS AND DEDICATION

The authors dedicate this chapter to the loving memory of Dr. Thomas T. Shen. Dr. Shen was president of the East-American Chapter of the Phi Tau Phi Scholastic Honor Society of America. He was also president of the Overseas Chinese Environmental Engineers and Scientists Association (OCEESA), Cleveland, OH. He received his PhD degree in Environmental Engineering from Rensselaer Polytechnic Institute in Troy, NY. Dr. Shen had nearly four decades of experience in environmental pollution prevention and control. He served as a senior research scientist with the New York State Department of Health and Department of Environmental Conservation. He taught at Columbia University Graduate School and was a member of the U.S. EPA Science Advisory Board. He also served as a consultant to the United Nations (WHO and UNDP) and was one of the technical reviewers for President Bush's 1991 and 1992 Annual Environment and Conservation Challenge Awards. In 1993, he received Taiwan's National Industrial Waste Minimization Excellent Performance Award and the Air and Waste Management Association's Frank Chamber Scientific Award. Dr. Shen is listed in the Who's Who in America and was a Diplomat of the American Academy of Environmental Engineers (AAEE). Dr. Shen authored and coauthored several books, book chapters, and more than a hundred articles, reports, documents, and training manuals for short courses. Dr. Shen devoted his entire life to the safety and control of toxic chemicals and hazardous wastes and to the environmental technology exchange among the United States, mainland China, and Taiwan. Joyce Shen Shavers, one of his daughters, edited this chapter. Dr. Shen is missed by all of his family members, former colleagues, and professional friends.

APPENDIX   NEW YORK STATE MULTIMEDIA POLLUTION PREVENTION POLICY, REGULATION, AND PROGRAMS

In 1987, New York state legislators passed the preferred statewide hazardous waste practices hierarchy law, which is used to guide all hazardous waste policies and decisions. It consists of four parts:

1. Source reduction.
2. Reuse, recycling, and recovery.

3. Treatment, detoxification, and other destruction methods.
4. Disposal of the treated waste in a safe manner.

The policy on toxic chemicals and hazardous waste in New York State is to reduce (a) the use and release of toxic chemicals and (b) the generation of hazardous wastes at the source across all environmental media. A waste reduction impact statement (WRIS) was established to analyze the potential for reducing generation and/or toxicity of waste across all media. WRIS is required as a permit condition.

In 1990, another law was passed that requires facilities to reduce emission of hazardous wastes and toxic substances into the environment to the maximum possible extent, through changes or modifications of a technically feasible and economically practicable waste minimization process or operation. The law also requires waste generators to prepare, implement, and submit a hazardous waste minimization plan (HWMP). The plan must be updated biannually, with annual status reports. The HWMP will be used in place of WRIS.

Multimedia Pollution Prevention Regulation

The New York State Department of Environmental Conservation (NYSDEC) has a policy to reduce the generation of hazardous waste and the release of hazardous substances at the source across all environmental media. A lateral bill authorized NYSDEC to develop regulations requiring reporting of toxic substances and levying fees on certain industries to discourage the generation of hazardous waste and the release and discharge of toxic substances into air, water, and soil.

The regulation requires facilities with 10 or more employees that produce 25,000 lb of toxic chemicals or use over 10,000 lb of such chemicals annually to report to the NYSDEC of each toxic chemical meeting these threshold levels. The regulation also imposes a fee of $75 per 1000 lb or fraction thereof for each toxic chemical released into the environment, provided that such chemicals are required to be reported to NYSDEC. To implement the multimedia approach, NYSDEC will issue integrated permits for facilities meeting certain criteria.

Technical Assistance

In the spring of 1989, NYSDEC published a Waste Reduction Guidance Manual. The purpose of the manual is to promote New York State's four-part waste management hierarchy, to provide some measure for waste reduction efforts, and to ensure that New York State has adequate hazardous waste disposal capacity for the next 10 years. The manual was used statewide for training at workshops.

Since 1988, NYSDEC has held three conferences on Hazardous Waste Reduction annually. The conferences provided a forum for members of the industry, government, and educational institutions to share their knowledge and experiences and to promote waste minimization methods.

NYSDEC also prepares fact sheets on successful waste reduction activities from industries in New York State. The success stories identify process and chemical changes implemented by the industries and the corresponding reduction and benefits.

NYSDEC maintains a Waste Reduction Information Clearinghouse, which provides four main databases: referral, library, regulations, and hotline. The referral database contains information on consultants, transporters, laboratories, waste facilities, mobile treatment facilities, equipment and service vendors, and recyclers. The library database contains entries on case studies and abstracts dealing directly with pollution prevention issues. The regulations database contains all New York State solid and hazardous waste regulations. Air and water regulations were included later. The hotline database was created to track incoming calls. It has become a valuable source for tracking down previous problems of a similar nature to facilitate faster handling of all inquiries.

Safety and Control for Toxic Chemicals and Hazardous Wastes

WASTE EXCHANGE PROGRAM

The program is designed to encourage and assist industry to reduce, reuse, recycle, and exchange hazardous wastes. To implement the program, NYSDEC is responsible for the following activities:

1. Review and compiling R&D information on methods and technologies.
2. Develop technical information on methods and economic means to reduce and recycle industrial materials.
3. Maintain relevant data systems.
4. Research available markets for recycled materials.
5. Maintain a waste exchange publication for industrial materials.

RESEARCH AND DEVELOPMENT

The R&D program focuses on new and improved methods and technologies that help to perform the following activities:

1. Contribute to reduction in the quantity or toxicity of wastes generated at the source.
2. Recover, recycle, and reuse hazardous and toxic substances from waste streams.
3. Treat waste streams to render them less or nonhazardous.
4. Provide safe and permanent remedies for cleanup of inactive waste sites or emergency spills.

TRI REPORTING

The TRI reporting requirements began in 1988. As of 1990, a total of 894 large facilities in New York State reported releasing 85.5 million pounds of hazardous wastes into the environment. The total amount reported was almost 25% below the 1988 figures for the same reported chemicals and categories.

REFERENCES

1. Krajicek, D. J., Flint is part of a pattern: 7 Toxic assaults on communities of color, Alternet and Salon. Com. www.salon.com/2016/01/26/the_hideous_racial_politics_of_pollution_partner, January 26, 2016. Accessed January, 2017.
2. Occupational Safety and Health Administration (OSHA), Listing of toxic and reactive highly hazardous chemicals, OSHA, 58 FR 35115, June 30, 1993. www.osha.gov/pls/oshaweb/owadisp.show_document?p_table=STANDARDS&p_id=10647, January, 2016. Accessed April 2, 2017.
3. U.S. Environmental Protection Agency, Amendments to RCRA, U.S. EPA. http://search.epa.gov/epasearch/epasearch?querytext=Amendments+to+RCRA&site=archive&typeofsearch=epa&result_template=archive.ftl, 2016.
4. Editor Feature, Why the United States leaves deadly chemicals on the market, In These Times. http://inthesetimes.com/article/18504/epa_government_scientists_and_chemical_industry_links_influence_regulations, November 2, 2015. Accessed January, 2017.
5. United Nations Environment Programme (UNEP), Environmentally sound management of toxic chemicals including prevention of illegal international traffic in toxic and dangerous products. www.unep.org/Documents.Multilingual/Default.asp?DocumentID=52&ArticleID=67, February, 2016. Accessed January, 2017.
6. Toxic Substance Control Act (TSCA), 15 U.S.C. §2601 et seq. (1976) U.S. Government, Public Laws. www.epa.gov/laws-regulations/summary-toxic-substances-control-act, February, 2016. Accessed January, 2017.
7. U.S. EPA, Resource Conservation and Recovery Act (RCRA)—Orientation Manual, U.S. Environmental Protection Agency, Report # EPA530-R-02-016, Washington, DC, January, 2003.
8. Federal Insecticide, Fungicide, and Rodenticide Act (FIFRA), 7 U.S.C. §136 et seq. (1996), U.S. Government, Public Laws. www.epa.gov/laws-regulations/summary-federal-insecticide-fungicide-and-rodenticide-act, February, 2016. Accessed February, 2017.

9. Federal Register, Comprehensive Environmental Response, Compensation, and Liability Act (CERCLA or Superfund) 42 U.S.C. s/s 9601 et seq. (1980), United States Government, Public Laws. www.epa.gov/superfund/learn-about-superfund, January, 2014. Accessed April, 2017.
10. U.S. Environmental Protection Agency, RCRA Corrective Action: Case Studies Report, U.S EPA 530-R-13-002, 32 pp, April, 2013. Accessed February, 2017
11. Federal Register, Clean Air Act (CAA), 33 U.S.C. ss/1251 et seq. (1977), U.S. Government, Public Laws, Full text is at www3.epa.gov/npdes/pubs/cwatxt.txt, May, 2014. Accessed April 2, 2017.
12. Ember, L. R., Strategies for reducing pollution at the source are gaining ground. *Chemical & Engineering News*, 69(27), 7–16, 1991.
13. Ling, J. T., *The Industry's Environmental Challenge for the 1990's and beyond*, Keynote Address, the Environmental Technology Exposition and Conference, Las Vegas, Nevada, March 13–16, 1991.
14. The National Environmental Policy Act (NEPA), 42 U.S.C. §4321 et seq. (1969), U.S. Government, Public Laws. www.epa.gov/laws-regulations/summary-national-environmental-policy-act, February, 2016. Accessed February, 2017.
15. Shen, T. T., *Hazardous Waste Incineration*, Air Pollution Control Association, Pittsburgh, PA, 1982.
16. Cheremisinoff, N. P., Treating waste water, *Pollution Engineer*, 22(9), 60–65, 1990.
17. Pojasek, R. B., New and promising ultimate disposal options. In: Pojasek, R. B. (Ed), *Toxic and Hazardous Waste Disposal*, Vol. 4, Ann Arbor Science Publishers Inc., Ann Arbor, MI, 1980.
18. Wang, L. K., Hung, Y. T., and Shammas, N. K. (Eds), *Physicochemical Treatment Processes*, Humana Press, Totowa, NJ, 723 p. 2005.
19. Shammas, N. K., Physicochemically-enhanced pollutants separation in wastewater treatment, *Proceedings of International Conference: Rehabilitation and Development of Civil Engineering Infrastructure Systems—Upgrading of Water and Wastewater Treatment Facilities*, Organized by The American University of Beirut and University of Michigan, Beirut, Lebanon, June 9–11, 1997.
20. Shammas, N. K., Coagulation and flocculation. In: Wang, L. K., Hung, Y. T., and Shammas, N. K. (Eds), *Physicochemical Treatment Processes*, Humana Press, Totowa, NJ, 2006.
21. Shammas, N. K., Kumar, I., Chang, S., and Hung, Y. T., Sedimentation. In: Wang, L. K., Hung, Y. T., and Shammas, N. K. (Eds), *Physicochemical Treatment Processes*, Humana Press, Totowa, NJ, 2005.
22. Krofta, M., Miskovic, D., Shammas, N. K., Burgess, D., and Lampman, L. K., An innovative multiple stage flotation-filtration low cost municipal wastewater treatment system, *IAWQ 17th Biennial International Conference*, Budapest, Hungary, July 24–30, 1994.
23. Hung, Y-T, Lo, H. H., Wang, L. K., Taricska, J. R., and Li, K. H., Granular activated carbon. In: Wang, L.K., Hung, Y. T., and Shammas, N.K. (Eds), *Physicochemical Treatment Processes*, Humana Press, Totowa, NJ, 2005.
24. Goel, R. K., Flora, J. R. V., and Chen, J. P., Flow equalization and neutralization. In: Wang, L. K., Hung, Y. T., and Shammas, N. K. (Eds), *Physicochemical Treatment Processes*, Humana Press, Totowa, NJ, 2005.
25. Shammas, N. K. and Wang, L. K., Treatment of metal finishing wastes by sulfide precipitation. In: Chen, J. P., Wang, L. K., Shammas, N. K., Wang, M-H, and Hung, Y-T (Eds), *Remediation of Heavy Metals in the Environment*, CRC Press, Boca Raton, FL, 2016.
26. Shammas, N. K., Yuan, P., Yang, J., and Hung, Y. T. Chemical oxidation. In: Wang, L. K., Hung, Y. T., and Shammas, N. K. (Eds), *Physicochemical Treatment Processes*, Humana Press, Totowa, NJ, 2005.
27. Wang, L. K., Vaccari, D., Li, Y., and Shammas, N. K. Chemical precipitation. In: Wang, L. K., Hung, Y. T., and Shammas, N. K. (Eds), *Physicochemical Treatment Processes*, Humana Press, Totowa, NJ, 2005.
28. Shammas, N. K., *Biological Technologies for Wastewater Treatment and Reuse—A Primer for Selection*, Keynote Address, The 2nd PKC Water Reuse Forum on Biological Treatment for Water Reuse, Riyadh, May 16–17, 2011.
29. Wang, L. K., Pereira, N., Hung, Y. T., and Shammas, N. K., *Biological Treatment Processes*, Humana Press, Totowa, NJ, 2009.
30. Wang, L. K., Shammas, N. K., and Hung, Y. T., *Advanced Biological Treatment Processes*, Humana Press, Totowa, NJ, 2009.
31. Shammas, N. K. and Wang, L. K., Incineration and combustion of hazardous wastes. In: Wang, L. K., Hung, Y. T., and Shammas, N. K. (Eds), *Handbook of Advanced Industrial and Hazardous Wastes Treatment*, CRC Press, Boca Raton, FL, 2010.
32. Tay, K-L., Osborne, J., and Wang, L. K., Ocean disposal technology assessment. In: Wang, L. K., Shammas, N. K., and Hung, Y-T. (Eds), *Biosolids Engineering and Management*, Humana Press and Springer, Totowa, NJ, 2008.

33. Shammas, N. K. and Wang, L. K., Hazardous waste deep-well injection. In: Wang, L. K., Hung, Y. T., and Shammas, N. K. (Eds), *Handbook of Advanced Industrial and Hazardous Wastes Treatment*, CRC Press, Boca Raton, FL, 2010.
34. Shammas, N. K. and El-Rehaili, A., *Wastewater Engineering*, Textbook on Wastewater Treatment Works and Maintenance of Sewers and Pumping Stations, General Directorate of Technical Education and Professional Training, Institute of Technical Superintendents, Riyadh, Kingdom of Saudi Arabia, 1988.
35. Wang, L. K., Hung, Y. T., and Shammas, N. K. (Eds), *Handbook of Advanced Industrial and Hazardous Wastes Treatment*, CRC Press, Boca Raton, FL, 2010.
36. Shammas, N. K. and Wang, L. K., Hazardous waste landfill. In: Wang, L. K., Hung, Y. T., and Shammas, N. K. (Eds), *Handbook of Advanced Industrial and Hazardous Wastes Treatment*, CRC Press, Boca Raton, FL, 2010.
37. U.S. Environmental Protection Agency, *Strategies for Health Effects Research*, Science Advisory Board, U.S. EPA, Report No. SAB-EC-88-040D, Washington, DC, September 1988.
38. Shen, T. T., *Education Aspects of Multimedia Pollution Prevention*, Presented at the International Pollution Prevention Conference, Washington, DC, June 10–13, 1990.

2 Biological Treatment of Poultry Processing Wastewater

Nazih K. Shammas
Lenox Institute of Water Technology and
Krofta Engineering Corporation

Lawrence K. Wang
Lenox Institute of Water Technology, Krofta Engineering
Corporation, and Zorex Corporation

CONTENTS

2.1	Introduction	29
2.2	Description of Poultry Processing Industry	30
	2.2.1 Description of Poultry Processing Operations	31
	2.2.2 Poultry First Processing Operations	31
	2.2.2.1 Receiving Areas	31
	2.2.2.2 Killing and Bleeding	32
	2.2.2.3 Scalding and Defeathering	33
	2.2.2.4 Evisceration	33
	2.2.2.5 Chilling	35
	2.2.2.6 Packaging and Freezing	35
	2.2.3 Poultry Further Processing Operations	35
2.3	Planning for Wastewater Treatment	36
	2.3.1 Wastewater Surveys and Waste Minimization	36
	2.3.2 Initial Planning for a Wastewater Treatment System	37
	2.3.3 Selection of a Poultry Wastewater Treatment Process	38
	2.3.3.1 Activated Sludge Processes	38
	2.3.3.2 Trickling Filters	40
	2.3.3.3 Lagoons	42
	2.3.4 Upgrading Existing Lagoons	42
2.4	Operating a Wastewater Treatment System	43
	2.4.1 Daily Checklist	46
	2.4.2 Weekly Checklist	46
	2.4.3 Monthly Checklist	47
	2.4.4 Yearly Checklist	47
2.5	Case History: The Original Gold Kist Wastewater Facilities	47
	2.5.1 Site Selection	47
	2.5.2 Wastewater Survey and Criteria	48
	2.5.3 Selection of the Treatment Process	49
	2.5.4 The Flow Diagram	50
	2.5.5 Design Criteria	50
	2.5.6 Future Expansion Provisions	53
	2.5.7 Waste Treatment System Costs	54
	2.5.8 Operating Arrangements	54

2.6   Case History: Current Expansion at Gold Kist Project History 54
    2.6.1   Current Wastewater Loads 55
    2.6.2   Current Operating Difficulties 55
        2.6.2.1   Condenser Cooling Water 55
        2.6.2.2   Hydraulic Overflows 57
        2.6.2.3   Solids Control 57
        2.6.2.4   Odor Problems 58
    2.6.3   Proposed Wastewater Treatment System Loads 58
        2.6.3.1   Hydraulic Loads 58
        2.6.3.2   Biological Loads 59
    2.6.4   Review of Component Adequacy 59
        2.6.4.1   Lift Station 59
        2.6.4.2   By-Products Collector 59
        2.6.4.3   Aeration Tanks 60
        2.6.4.4   Final Clarifier 61
        2.6.4.5   Aerobic Digester 61
        2.6.4.6   Sludge Drying Beds 61
        2.6.4.7   Stabilization Pond 61
        2.6.4.8   Outfall Sewer and Cl_2 Facilities 62
    2.6.5   Proposed Modifications 62
        2.6.5.1   Air Supply 62
        2.6.5.2   Plant Hydraulics 62
        2.6.5.3   By-Products Reclamation System 62
        2.6.5.4   Sludge Drying Facilities 62
        2.6.5.5   Condenser Cooling Water 63
        2.6.5.6   Grease Removal 63
Acronyms 63
References 63

Abstract

The quantity of wastewater discharged from processing operations in a poultry plant may range from 5 to 10 gal/bird, with 7 gal being a typical value. Poultry processing wastewater is typically organic in character, high in biochemical oxygen demand (BOD) than domestic wastewater, and contains high amount of suspended solids and floating material such as scum and grease. Since wastewater from poultry processing plants is typically organic, it responds well to treatment by biological methods. In biological waste treatment systems, microorganisms use the polluting constituents of the wastewater as food for survival and growth. Wastewater treatment systems in the poultry processing industry usually provide primary and secondary treatment and may or may not include tertiary treatment. The secondary treatment required for discharging wastewater to a stream may be in the form of an activated sludge or trickling filter system, a system of lagoons, or an irrigation system. Each of these methods of biological treatment has been tried with varying degrees of success. Activated sludge systems that may be applied to poultry plant wastes include conventional activated sludge, activated sludge using step aeration, high-rate activated sludge, extended aeration activated sludge, and the contact stabilization process. Anaerobic lagoons, aerobic lagoons, and a combination of an anaerobic lagoon and an aerobic lagoon may be used for secondary treatment of poultry wastes. All poultry wastewater effluents should be chlorinated before discharge into the receiving stream. All of the aforementioned processes are described in this chapter and represent the commonly used waste treatment processes. The use of other systems, such as microfiltration and certain chemical processes, while possible, is generally not prevalent due to high initial and operating costs.

2.1 INTRODUCTION

Wastewater treatment or processing for removal of harmful constituents in poultry wastewater is required to protect the quality of existing water resources. Today, almost every lake or stream in the United States is regulated in terms of the quality and quantity of effluents that may be discharged into it. Federal, state, or local agencies have the authority to establish water quality standards for effluents and to enforce these standards. Consequently, one of the best sources of guidance in terms of wastewater treatment in a poultry plant is the authority having jurisdiction over discharge into the receiving stream. The authority having jurisdiction shall determine the quality of the effluent discharged and can offer appropriate guidance based on his or her experience on this subject.

The quantity of wastewater discharged from processing operations in a poultry plant may range from 5 to 10 gal/bird, with 7 gal being a typical value [1]. Poultry processing wastewater is typically organic in character and high in biochemical oxygen demand (BOD) than domestic wastewater and contains a high amount of suspended solids and floating material such as scum and grease. Table 2.1 shows some characteristics of wastewater from a poultry plant.

Since wastewater from poultry processing plants is typically organic, it responds well to treatment by biological methods [2,3]. In biological waste treatment systems, microorganisms use the polluting constituents of the wastewater as food for survival and growth. The primary microorganisms encountered in wastewater treatment are bacteria, fungi, algae, protozoa, rotifers, and crustaceans. Bacteria can only assimilate soluble food and may or may not require oxygen depending on whether the bacteria are aerobic or anaerobic. Fungi can only absorb soluble food but are strictly aerobes; that is, they need oxygen to survive. Algae utilize primarily inorganic compounds and sunlight for energy, and in this process, oxygen is given off as a by-product. Protozoa are single-celled animals and use bacteria and algae as their primary source of energy. Rotifers are multicellular animals that use bacteria and algae as their primary source of food. Crustaceans are complex multicelled animals with hard shells. The microscopic forms of crustaceans use higher forms of microorganisms as their source of food. Protozoa, rotifers, and crustaceans grow only in an aerobic environment [2].

Microorganisms find their place in the carbon cycle at the elemental levels of conversion of residual organic carbon to carbon dioxide. Briefly, the carbon cycle consists of green plants utilizing inorganic carbon in the form of carbon dioxide and converting it to organic carbon by using energy from sunlight through photosynthesis. Animals consume the resulting plant tissue and convert a part of it to animal tissue and carbon dioxide. Plant and animal tissue and other residual organic carbon compounds are then oxidized back to inorganic carbon dioxide by microorganisms.

Aerobic degradation of wastewater constituents is a biochemical reaction in which living cells assimilate food for energy and growth in the presence of dissolved oxygen. About one-third of the organics are oxidized, providing energy to synthesize the remainder into additional living cells. The end products of these biochemical reactions are carbon dioxide and water. When oxygen is absent from the reaction, the degradation is anaerobic and the end products are organic acids, aldehydes, ketones, and alcohols. Special bacteria, called methane formers, metabolize about 80% of the organic matter to form methane and carbon dioxide, with the remaining 20% forming additional living cells. These biochemical principles are used in wastewater treatment systems to render wastewater streams resulting from poultry processing facilities suitable for discharge into a stream [2].

Wastewater treatment systems in the poultry processing industry usually provide primary and secondary treatment and may or may not include tertiary treatment. Primary treatment consists of screening, comminutors, and primary sedimentation, or flotation for removal of solid and particulate matter.

The secondary treatment required for discharging wastewater to a stream may be in the form of an activated sludge or trickling filter system, a system of lagoons, or an irrigation system. Each of these methods of biological treatment has been tried with varying degrees of success.

TABLE 2.1
Characteristics of Poultry Plant Wastewater

Analysis	Unit	Range	Average
pH		6.3–7.4	6.9
DO	mg/L	0–2.0	.5
BOD$_5$	mg/L	370–620	473.0
Suspended solids	mg/L	120–296	196.0
Total solids	mg/L	NA	650.0
Volatile solids	mg/L	NA	486.0
Fixed solids	mg/L	NA	164.0
Settleable solids	mg/L	15–20	17.5
Grease	mg/L	170–230	201.0

Activated sludge systems that may be applied to poultry plant wastes include conventional activated sludge, activated sludge using step aeration, high-rate activated sludge, extended aeration activated sludge, and the contact stabilization process. Anaerobic lagoons, aerobic lagoons, and a combination of an anaerobic lagoon and an aerobic lagoon may be used for secondary treatment of poultry wastes. All poultry wastewater effluents should be chlorinated before discharge into the receiving stream.

All of the aforementioned processes are described in this chapter and represent the commonly used waste treatment processes. The use of other systems, such as microfiltration and certain chemical processes, while possible, is generally not prevalent due to high initial and operating costs.

2.2 DESCRIPTION OF POULTRY PROCESSING INDUSTRY

Poultry processing includes the slaughter of poultry and small game animals (e.g., quails, pheasants, and rabbits), slaughter of exotic poultry (e.g., ostriches), and the processing and preparing of these products and their by-products. Slaughtering is the first step in processing poultry into consumer products. Poultry slaughtering (first processing) operations typically encompass the following steps:

1. Receiving and holding of live animals
2. Stunning prior to slaughter
3. Slaughter
4. Initial processing

Poultry first processing facilities are designed to accommodate this multistep process. In most facilities, the major steps are carried out in separate rooms.

In addition, many first processing facilities further process carcasses, producing products that might be breaded, marinated, or partially or fully cooked. Many first processing facilities also include rendering operations that produce edible products such as fat, and inedible products, primarily ingredients for animal feeds, including pet foods.

The U.S. Census of Manufacturers reported 260 companies engaged in poultry slaughtering. These companies own or operate 470 facilities, employ 224,000 employees, and produce 50 billion 2014 USD in value of shipments. The poultry slaughtering sector has relatively few facilities with fewer than 20 employees; a few very large facilities dominate the sector. Almost 50% of the sector employment and over 40% of the value of shipments were accounted for by 75 facilities that employ more than 1000 workers each. Eighty percent of employment and 74% of total shipments are produced by facilities that employ more than 500 workers. Yet these facilities compose only 36% of the poultry processing industry [4].

Poultry processing is largely concentrated in the southeastern states. Arkansas and Georgia have the largest number of facilities and the highest employment and value of shipments. Alabama and North Carolina rank third and fourth in all these measures. California is the only state in the top 10 poultry-producing states that is not in the southeast. California ranks 10th in terms of employment and value of shipments, and 8th in terms of the number of facilities.

2.2.1 Description of Poultry Processing Operations

Poultry processing begins with the assembly and slaughter of live birds and can end with the shipment of dressed carcasses or continue with a variety of additional activities. Poultry processing operations are classified as first or further processing operations or as an integrated combination. First processing operations include those operations that receive live poultry and produce a dressed carcass, either whole or in parts. In this classifications system, first processing operations simply produce dressed whole or split carcasses or smaller segments for sale to wholesale distributors or directly to retailers. First processing operations supply products for further processing activities such as breading, marinating, and partial or complete cooking, which can occur on- or off-site.

U.S. Environmental Protection Agency (USEPA) considers the reduction of whole poultry carcasses into halves, quarters, or smaller pieces, which might be with or without bone and might be ground as part of first processing when performed at first processing facilities. Consequently, USEPA considers cutting, boning, and grinding operations to be further processing operations when performed at facilities not engaged in first processing activities [4].

2.2.2 Poultry First Processing Operations

Common to all, poultry first processing operations is a series of operations necessary to transform live birds into dressed carcasses. Figure 2.1 illustrates these operations, and the following sections describe them.

2.2.2.1 Receiving Areas

Birds are transported to processing plants in scheduled delivery time so that all the birds are processed on the same day. Live-bird holding areas are usually covered and have cooling fans to reduce bird weight loss and mortality during hot weather conditions [5].

Broiler chickens are typically transported to processing plants in cage modules stacked on flatbed trailers. Each cage module can hold about 20 average-sized broiler chickens. The cage modules are removed from the transport trailer and tilted using a forklift truck to empty the cage. Alternatively, tilting platforms can be used to empty the cage modules after they have been removed from the transport trailer. When the cage module tilts, the lower side of the cage opens and the birds slide onto a conveyor belt, which moves them into the hanging area inside the plant. In the hanging area, the live birds are hung by their feet on shackles attached to an overhead conveyer system, commonly referred to as the killing line that moves the birds into the killing area. The killing line moves at a constant speed, and up to 8000 birds/h (133 birds/min) can be shackled in a modern plant, although, in practice, this number is much lower because workers cannot unload broilers fast enough to fill every shackle [6]. Cage modules are also used to transport ducks, geese, and fowls.

Turkeys are usually transported in cages permanently attached to flatbed trailers. The cages are emptied manually in a live-bird receiving area outside the confines of the processing plant. Turkeys are unloaded manually to minimize bruising. They are more susceptible than broilers to bruising during automatic unloading because of their heavier weight and irregular body shape. Turkeys are then immediately hung on shackles attached to an overhead conveyer system that passes through the unloading area into the processing plant [5].

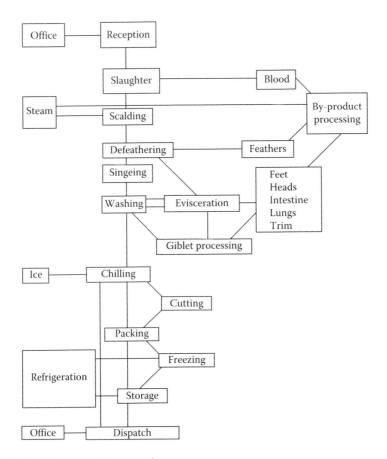

FIGURE 2.1 Poultry first processing operations.

Following the unloading process, cages and transport trucks might be washed and sanitized to prevent disease transmission among grower operations. The washing and sanitizing of cages and trucks is common in the turkey industry, but not in the broiler chicken industry.

2.2.2.2 Killing and Bleeding

Almost all birds are rendered unconscious through stunning just prior to killing. Some exemptions are made for religious meat processing (e.g., kosher, halal). Stunning immobilizes the birds to increase killing efficiency, cause greater blood loss, and increase defeathering efficiency. Stunning is performed by applying a current of 10–20 mA per broiler and 20–40 mA per turkey for approximately 10–12 s [5]. Poultry are killed by severing the jugular vein and carotid artery or, less typically, by debraining. Usually a rotating circular blade is used to kill broilers, while manual killing is often required for turkeys because of their varying size and body shape. Decapitation is not performed because it decreases blood loss following death [7].

Immediately after being killed, broilers start bleeding as they pass through a "blood tunnel" designed to collect blood to reduce wastewater BOD and total nitrogen concentration. The blood tunnel is a walled area designed to confine and capture blood splattered by muscle contractions, following the severing of the jugular vein and carotid artery. The blood collected is processed with recovered feathers in the production of feather meal, a by-product feedstuff used in livestock and poultry feeds as a source of protein. On average, broilers are held in the tunnel from 45 to 125 s for bleeding, with an average time of 80 s; turkeys are held in the tunnel from 90 to 210 s, with an average time of 131 s. Blood loss approaches 70% in some plants, but, generally speaking, only 30%–50% of a broiler's blood is lost in the killing area. Depending on plant operating conditions, blood is collected in troughs and transported to a rendering facility by vacuum, gravity, or pump systems, or it is allowed to congeal on the plant floor and collected manually. Virtually all plants collect blood for rendering on- or off-site and thereby limit the amount of blood present in the wastewater [4].

2.2.2.3 Scalding and Defeathering

After killing and bleeding, birds are scalded by immersing them in a scalding tank or by spraying them with scalding water. Scalding is performed to relax feather follicles prior to defeathering. Virtually all plants use scald tanks because of the high water usage and inconsistent feather removal associated with spray scalding. Scalding tanks are relatively long troughs of hot water into which the bleeding birds are immersed to loosen their feathers. Depending on the intended market of the broilers, either soft (semi-scald) or hard scalding is used. Soft scalding is used for the fresh, chilled market, whereas hard scalding is preferred for the frozen sector [8]. The difference between these two types of scalding techniques lies in the scalding temperature used. Soft scalding is performed at about 53°C (127F) for 120s; it loosens feathers without subsequent skin damage. Hard scalding is performed at 62°C–64°C (144–147F) for 45s; it loosens both feathers and the first layer of skin. Sometimes, chemicals are added to scald tanks to aid in defeathering by reducing surface tension and increasing feather wetting. The U.S. Department of Agriculture (USDA) requires that all scald tanks have a minimum overflow of 1L (0.26 gal)/bird to reduce the potential for microbial contamination [5].

Because scalding and mechanical defeathering do not completely remove duck and goose feathers, immersion in a mixture of hot wax and rosin follows. After this mixture partially solidifies, the remaining feathers are removed [7].

The next stage is automated defeathering, which is done by machines with multiple rows of flexible, ribbed, rubber fingers on cylinders that rotate rapidly across the birds. The abrasion caused by this contact removes the feathers and occasionally the heads of the birds. At the same time, a continuous spray of warm water is used to lubricate the bird and flush away feathers as they are removed. Feathers are flumed to a screening area using scalding overflow for dewatering prior to processing for feather meal production. Different defeathering machines might be used for different types of birds [4].

Following defeathering, pinfeathers might be removed manually because they are still encased within the feather shaft and thus are resistant to mechanical abrasion. After pinfeather removal, birds pass through a gas flame that singes the remaining feathers and fine hairs. Next, feet and heads are removed. Feet are removed by passing them through a cutting blade, and heads are removed by clamps that pull necks upward. Removing the head from a bird is advantageous because the esophagus and trachea are removed with it. Removing the head also loosens the crop and lungs for easier automatic removal during evisceration [8]. At this point, the blood, feathers, feet, and heads of broilers are collected and sent to a rendering facility, where they are transformed into by-product meal. Chicken feet might also be collected for sale, primarily in export markets.

After removal of the feet, the carcasses are rehung on shackles attached to an overhead conveyer, known as an evisceration line, and washed in enclosures using high-pressure cold water sprays prior to evisceration. The purpose of this washing step is to sanitize the outside of the bird before evisceration to reduce microbial contamination of the body cavity. This transfer point is often referred to as the point separating the "dirty" and "clean" sections of the processing plant [6]. The killing line conveyor then circles back, and the shackles are cleaned before they return to the unloading bay.

2.2.2.4 Evisceration

Evisceration is a multistep process that begins with removing the neck and opening the body cavity. Then, the viscera are extracted but remain attached to the birds until they are inspected for symptoms of disease. Next, the viscera are separated from the bird, and the edible components (hearts, livers, and gizzards) are harvested. The inedible viscera, known as offal, are collected and combined with heads and feet for subsequent rendering. Entrails are sometimes left attached for religious meat processing (e.g., Buddhist, Confucianist). Depending on the plant design, a wet or dry collection system is used. Wet systems use water to transport the offal by fluming it to a screening area for dewatering before rendering. Dry systems, which are not common, use a series of conveyor belts or vacuum or compressed air stations for offal transport.

Automation of the evisceration process varies depending on plant size and operation. A fully automated line can eviscerate approximately 6000 broilers per hour [8]. The type of equipment available for plant use varies by location and manufacturer. Many parts of the process can be performed manually, especially for turkeys. Though a fully automated evisceration line can be used for broilers, the variation in size among turkeys makes automation more difficult. Female turkeys (hens) are significantly smaller than male turkeys (toms).

When broilers first enter the evisceration area, they are rehung on shackles by their hocks to a conveyor line that runs directly above a wet or dry offal collection system [6]. The birds' necks are disconnected by breaking the spine with a blade that applies force just above the shoulders. As the blade retracts, the neck falls downward and hangs by the remaining skin while another blade removes the preen gland from the tail. The preen gland produces oil that is used by birds for grooming and has an unpleasant taste for humans [5]. Next, a venting machine cuts a hole with a circular blade around the anus for extraction of the viscera. Great care must be taken not to penetrate the intestinal lining of a broiler because the resulting fecal contamination results in condemnation during inspection [4].

Following venting, the opening of the abdominal wall is enlarged to aid in viscera removal. At this point, all viscera are drawn out of the broiler by hand, with the aid of scooping spoons, or, more commonly, by an evisceration machine. The evisceration machine immobilizes the broiler and passes a clamp through the abdominal opening to grip the visceral package. Once removed, this package is allowed to hang freely to aid in the inspection process. Every bird must be inspected by the USDA inspector or the USDA-supervised plant worker for symptoms of disease or contamination before being packaged and sold. The inspector checks the carcass, viscera, and body cavity to determine wholesomeness with three possible outcomes: pass, conditional, and fail. If the bird is deemed conditional, it is hung on a different line for further inspection or to be trimmed of unwholesome portions. Failed birds are removed from the line and disposed of, usually by rendering [7]. The viscera are removed from the birds that have passed inspection and are pumped to a harvesting area where edible viscera are separated from inedible viscera. A giblet harvester is used to collect the edible viscera, including heart, liver, neck, and gizzard, and to prepare each appropriately. The heart and liver are stripped of connective tissue and washed. The gizzard is split, its contents are washed away, its hard lining is peeled off, and it is given a final wash. The minimum giblet washer flow rate required by USDA is 1 gallon of water for every 20 birds processed. Meanwhile, the inedible viscera, including intestines, proventriculus, lower esophagus, spleen, and reproductive organs, are extracted and sent to a rendering facility. Finally, the crop and lungs are mechanically removed from each bird. The crop is pushed up through the neck by a probe, and the lungs are removed by vacuum. A final inspection is required to ensure the carcass is not heavily bruised or contaminated, and then the carcass is cleaned [4]. Bruised birds are diverted to salvage lines for recovery of parts.

The second carcass washing of the broilers is very thorough. Nozzles are used to spray water both inside and outside the carcass. These high-pressure nozzles are designed to eliminate the majority of remaining contaminants on both the carcass and the conveyor line, and the water is often mixed with chlorine or other antimicrobiological chemicals. From this area, the conveyor system travels to the chilling area [4].

Kosher and halal poultry producers pack the birds (inside and out) in salt for 1 h to absorb any residual blood or juices. The birds are then rinsed and shipped to kosher/halal meat distributers. On an average day, a typical kosher poultry facility (generating approximately 2 Mgal of wastewater/day) would use approximately 80,000 lb of salt in its operations [4]. Industry has stated that most kosher operations are in urban areas with sewer connections.

2.2.2.5 Chilling

After birds have been eviscerated and washed, they are chilled rapidly to slow the growth of any microorganisms present to extend shelf life and to protect quality [5]. USDA regulations require that broilers be chilled to 4°C (40 F) within 4 h of death and turkeys within 8 h of death. Most poultry processing plants use large chilling tanks containing ice water; very few use air chilling. Several types of chilling tanks are used, including (1) a large enclosed drum that rotates about a central axis, (2) a perforated cylinder mounted within a chilling vat, and (3) a large open chilling tank containing a mechanical rocker to provide agitation. In all cases, birds are cascaded forward with the flow of water at a minimum overflow rate per bird specified by SDA guidelines for specified cuts of poultry [9].

Most poultry plants use two chilling tanks in series, a pre-chiller and a main chiller. The direction of water flow is from the main chiller to the pre-chiller, which is opposite to the direction of carcass movement. Because water and ice are added to only the main chiller, the water in the pre-chiller is somewhat warmer than that in the main chiller. Most plants chlorinate bird chiller makeup water to reduce potential carcass microbial contamination. The USDA requires 0.5 gal (2 L) of overflow/bird in the chillers [4]; the typical flow is about 0.75 gal (3 L)/bird [5]. The effluent from the first chiller is usually used for fluming offal to the offal screening area.

USDA requires a pre-chiller water temperature of less than 18°C (65 F), and temperature values typically range between 7°C and 12°C (45 F and 54 F) [7]. Agitation makes the water a very effective washer, and the pre-chiller often cleans off any remaining contaminants. Most broiler carcasses enter the pre-chiller at about 38°C (100 F) and leave at a temperature between 30°C and 35°C (86 F and 95 F). The cycle lasts 10–15 min, and water rapidly penetrates the carcass skin during this period [5]. Water weight gained in the pre-chiller is strictly regulated and monitored according to poultry classification and final destination of the product as proposed by USDA. Cut-up and ice-packed products are allowed to retain more water than their whole carcass pack or whole frozen counterparts [9].

The main chill tank's water temperature is approximately 4°C (39 F) at the entrance and 1°C (34 F) at the exit because of the countercurrent flow system. Broiler carcasses stay in this chiller for 45–60 min and leave the chill tank at about 2°C–4°C (36–39 F). Air bubbles are added to the main chill tanks to enhance heat exchange. The bubbles agitate the water and prevent a thermal layer from forming around the carcass. If not agitated, water around the carcass would reach thermal equilibrium with the carcass and retard heat transfer [5].

If air chilling is used, it normally involves passing the conveyor of carcasses through rooms of air circulating at between −7°C and 2°C for 1–3 h. In some cases, water is sprayed on the carcasses, increasing heat transfer by evaporative cooling [5]. Giblets, consisting of hearts, livers, gizzards, and necks, are chilled similarly to carcasses, though the chilling systems for giblets are separate and smaller.

2.2.2.6 Packaging and Freezing

After the birds are chilled, they are packed as whole birds or processed further. Whole birds are sold in both fresh and frozen forms. Chickens are primarily sold as fresh birds, and turkeys are primarily sold as frozen birds. Fresh birds not sold in case-ready packaging are packed in ice for shipment to maintain a temperature of 0°C (32 F). Poultry sold frozen is cooled to approximately −18°C (0 F) [6].

2.2.3 POULTRY FURTHER PROCESSING OPERATIONS

Further processing can be as simple as splitting the carcass into two halves or as complex as producing a breaded or marinated, partially or fully cooked product. Therefore, further processing might involve receiving, storage, thawing, cutting, deboning, dicing, grinding, chopping, canning, and final product preparation. Final product preparation includes freezing, packaging, and shipping. Further processing might be performed after first processing in an integrated operation, or it might be performed at a separate facility. Further processing is a highly automated process designed to transform eviscerated broiler carcasses into a wide variety of consumer products. Depending on the type of product being produced, plant production lines might overlap, especially for producing cooked, finished products [4].

2.3 PLANNING FOR WASTEWATER TREATMENT

2.3.1 Wastewater Surveys and Waste Minimization

Planning for a wastewater treatment facility for a poultry plant begins with a survey of the wastewater sources within the plant. An industrial wastewater survey in an existing poultry processing facility would consist of determining the volume and characteristics of the composite wastewater discharge. The survey may be as simple as measuring flow and taking a composite sample at a single point or may be as complex as measuring flows and sampling each source of wastewater discharged. The latter has the advantage that each point within the plant may be studied to determine the possibilities available for reducing the volume of wastewater and pollution at the source.

Recommended techniques to minimize generation of wastewater include the following [10–12]:

1. Adoption of the attitude that waste load reduction is one of the best business decisions a manager can make.
2. Training of employees in the concepts of pollution prevention, and showing them how to perform their jobs in a way that will cut waste loads in the plant.
3. Removal of solid organic waste from transport equipment before rinsing and washing. Organic materials should be collected separately for recycling.
4. Use of grids and screens for recovering solids.
5. Collection of solids from the floor and equipment by sweeping and shoveling the material into containers before actual cleanup begins; ensuring that water hoses are not used as brooms.
6. Ensuring that leakage from animal by-product storage containers is avoided (e.g., preventive maintenance and corrosion inspection).
7. Improvement of blood collection by ensuring that all birds are properly stunned and by installing a blood collection system; coagulated blood from the floor and walls should be removed before they are washed down.
8. Considering the use of steam scalding of birds to avoid excessive wastewater generation from scalding tanks.
9. Where scalding tanks are used, ensuring the entry of birds to the scalding tank does not cause overflow of the tank liquid; drippings from birds leaving the scalding tank and from overflows should be collected and reused in the scalding tank.
10. Regular adjustment of evisceration machinery to reduce accidental release of fecal matter due to the rupture of birds' intestinal tract (resulting in the need for frequent rinsing).
11. Where feasible, transportation of organic material using vacuum pumps instead of water transport.
12. Application of appropriate tank and equipment cleaning procedures; cleaning-in-place (CIP) procedures are useful to reduce chemical, water, and energy consumption in cleaning operations.
13. Choosing cleaning agents and application rates that do not have adverse impacts on the environment, or on wastewater treatment processes and sludge quality for agricultural application.

For a new poultry processing facility where wastewater flow streams do not exist, the wastewater for the proposed plant must be synthesized based on the experiences of similar existing processing plants. Using this method for determining wastewater quantity and character requires great care to ensure that all waste constituents are included in the synthetic wastewater sample and that the

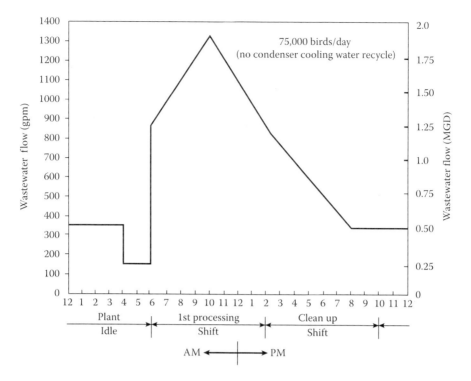

FIGURE 2.2 Actual daily hydraulic load from a poultry processing plant. (From USEPA, Upgrading existing poultry processing facilities to reduce pollution, US Environmental Protection Agency, Technology Transfer EPA-62513-73-001, Washington, DC, 1973.)

constituents are included in proportions that will be truly representative of the waste from the proposed facility. It is suggested that an experienced engineer be retained to prepare a study that will determine the properties of the design influent [13,14].

The volume of wastewater, in general, will vary with bird production, increasing with increased bird production and decreasing with low production. In processing and production, more than 90% of clean water is converted into wastewater. The basic functions of water at plants include cleaning the animal channels during the process and afterwards, cooling the channels after taking out the internals (guts), and cleaning the installation, in general. Wastewater generated at the plant has a high concentration of colloidal suspended solids and waste, BOD_5 (biochemical oxygen demand for 5 days), grease and oils, and coloring [15]. Poultry processing plants, as many other food processing units, typically are high water users. For broilers, 5–10 gal of water is used to process one 5 lb, average-sized chicken. When processing turkey, the volume of water is even higher, with the average weight of slaughtered turkeys exceeding four times that of a chicken [16]. Poultry plants may discharge as much as 65 lb BOD_5/1000 broilers processed.

Figure 2.2 shows how the volume of waste discharged varies with time at one poultry plant. Some of the characteristics of wastewater that should be determined include suspended solids, biochemical oxygen demand (BOD_5), toxic substances, grease and fats, dissolved solids, solid matter, temperature, pH, color, and septicity [4].

2.3.2 Initial Planning for a Wastewater Treatment System

Criteria for the design of a treatment plant are available from the authority having jurisdiction for control of discharge to the receiving stream and must be evaluated by an engineer experienced in the design and engineering of wastewater treatment plants for poultry processing facilities.

Wastewater from many poultry processing plants is discharged to publicly owned treatment works (POTWs). These treatment plants must remove most of the pollutants (waste load) before the water is discharged to a public waterway. Treating wastewater costs money, and most treatment works charge according to the volume of water treated. In addition, they commonly charge extra (apply a surcharge) if the waste load exceeds certain preset levels [1,12]. The type of treatment and the type, number, and size of components may be selected when the required treatment efficiency, in terms of removal of contaminants, has been established.

Costs, both capital costs and operating costs, must be determined in the preliminary planning phase of the project. This phase should result in a report describing the location of the waste treatment plant, the nature of the wastes, all the components of the proposed wastewater treatment system, the provisions proposed for future expansion, the anticipated removal efficiency, the character of effluent, and the estimated capital and operating costs. The preliminary report performs three primary functions:

1. It may be used in discussions with the authorities early in the project.
2. It provides the cost data essential to establish the economic feasibility of the project.
3. It serves as a basis for the preparation of working drawings and construction contract documents.

2.3.3 Selection of a Poultry Wastewater Treatment Process

The secondary treatment processes commonly used in the biological treatment of poultry wastes are various forms of the activated sludge process, standard and high rate trickling filters, and aerobic and anaerobic lagoons [17–19]. In the past, aerobic and anaerobic lagoons have been employed in the majority of private installations, with activated sludge plants as a second choice. Trickling filters have been used mainly in plants treating both municipal and poultry wastes. With the exception of anaerobic lagoons, all of the aforementioned processes provide complete treatment and achieve about 70%–90% reduction in the influent BOD_5 and an estimated 80%–95% removal of suspended solids [1]. Each of the systems to be discussed has its advantages and disadvantages, and, in general, the treatment requirements will dictate, to some degree, the particular system selected. The main differences between the systems are construction and land costs. The major costs for activated sludge and trickling filter plants are construction and operating costs, whereas the major expense for lagoons is land utilization costs.

2.3.3.1 Activated Sludge Processes

There are four general types of activated sludge processes that are used in poultry waste treatment [1]:

1. Conventional
2. High rate
3. Extended aeration
4. Contact stabilization

All of the above use the activated sludge theory, whereby aerobic bacteria assimilate the organic matter present in the waste stream for cellular growth and in that way provide for the waste stream purification. The common elements of all activated sludge processes are an activated sludge floc, a mixing and aeration chamber, and a clarification or separation tank [1,20].

2.3.3.1.1 Conventional Activated Sludge
In the conventional activated sludge process, the waste stream, following primary treatment, is mixed with a proportional amount of the returned settled sludge from the final clarifier and enters

Biological Treatment of Poultry Processing Wastewater

the head of the aeration basin. In general, the aeration basin is designed to provide a detention time of 6–8 h. Mixing and aeration are uniform along the tank and are provided for by mechanical mixers and/or pressurized air diffusers. Following aeration, the mixed liquor is settled in a clarifier, the clear supernatant being discharged to the receiving water and the concentrated sludge being proportionately returned and wasted (see Figure 2.3). One modification of the conventional process is step aeration, where the waste stream and/or return sludge enters through a number of inlets along the aeration basin rather than at a common inlet. A second modification is tapered aeration, where aeration along the tank is varied. The advantages of the conventional activated sludge process are as follows:

1. They require lower capital costs than for equivalent trickling filter plants.
2. There are low hydraulic head losses.
3. High-quality effluent is obtained.

The disadvantages are listed here:

1. There are higher mechanical operating costs than for equivalent trickling filter plants.
2. It requires skilled operators.
3. It does not respond well to shock loads.
4. It generates a large volume of sludge that needs to be disposed of.
5. Problems in sludge settling are sometimes encountered.

2.3.3.1.2 High-Rate Activated Sludge

The main differences between the high-rate process and the conventional process are the smaller detention period in the aeration basin and a smaller return sludge rate (see Figure 2.4). The advantages of this system are as follows:

1. The capital costs are lower than for the conventional process.
2. The sludge generated is much denser, resulting in a less voluminous sludge to be dispensed with.
3. The operating costs are less because of the shorter detention time.

The main disadvantage is the lower quality of the effluent.

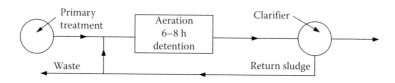

FIGURE 2.3 Conventional activated sludge process.

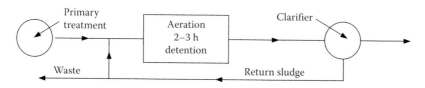

FIGURE 2.4 High-rate activated sludge process.

2.3.3.1.3 Extended Aeration

In the extended aeration process, the aeration basin provides for 24–30 h of detention time (see Figure 2.5). The advantages of this process are as follows:

1. A very high-quality effluent is obtained.
2. Less manpower time is required to operate the process.
3. Smaller volumes of sludge are generated.

The main disadvantages of this process are the high capital investment required and the possibility of sludge settling problems.

2.3.3.1.4 Contact Stabilization

In the contact stabilization process, the waste stream does not undergo primary clarification but is mixed with the return sludge and enters the aeration basin directly (see Figure 2.6). Since the mode of treatment is by adsorption and absorption, a detention period of only 30 min is provided. After settling, the concentrated sludge is stabilized by separate aeration before being proportionately returned to the waste stream. The main advantages of this process are

1. Low capital investment.
2. Low operating costs.
3. Ability to handle shock loads and variations in flow.

BOD reductions of ±90% and suspended solids removals of ±90% have been reported.

2.3.3.2 Trickling Filters

The biological mechanism involved in treating wastewater by percolation through trickling filters assimilates organic matter into cellular growth by aerobic bacteria. Unlike the activated sludge process, where the biological process takes place in a "fluid bed," the biological activity in a trickling filter is conducted on the filter medium by surface fauna. Portions of the bacterial fauna are continually sloughing off into the wastewater stream and are removed in the final clarifier. Trickling filters can achieve upward of ±90% in both reduction of BOD_5 and removal of suspended solids. There are two types of trickling filters—standard rate and high rate. The number of filters in series determines the "stage" of the filter system. The primary elements of a trickling filter are the

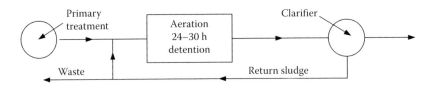

FIGURE 2.5 Extended aeration process.

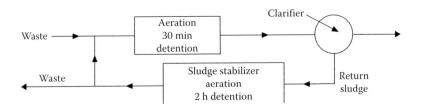

FIGURE 2.6 Contact stabilization process.

covered or uncovered containing structure, the filter medium, the waste flow distribution system, and the subdrain collection system. The typical mediums used are rocks, slag, and honeycombed cellular modules of synthetic construction. Distribution systems are spray nozzles attached to fixed or rotating manifolds. The subdrainage system may consist of tiles, concrete, or synthetic drainage tiles [1,21].

2.3.3.2.1 Standard Rate Trickling Filters

The main distinction of the standard rate trickling filter is the low BOD loading rate (see Figure 2.7). Its main advantages are

1. Production of a high-quality effluent
2. Low operating costs
3. Highly skilled operating personnel not needed
4. Resistant to shock loadings and variations in flow

The main disadvantages are high capital costs and requirement of considerable land space. Flies and insects are sometimes a problem but are usually controllable.

2.3.3.2.2 High Rate Trickling Filters

The main distinction of the high rate trickling filter is the high BOD_5 loadings (see Figure 2.8). These loadings may be as much as twice as great as those of a standard rate filter. Although, for a single pass, BOD_5 reductions are only about 60%–70%, recirculation increases the total reduction. The advantages are as follows:

1. Its versatility of treating high-strength wastes.
2. Its resistance to shock loads and variations in flow.
3. Considerable reduction in the area required.
4. Problems with flies and insects usually eliminated.

The main disadvantage is the high-power requirements caused by recirculation.

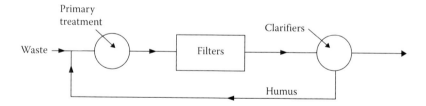

FIGURE 2.7 Standard rate trickling filter.

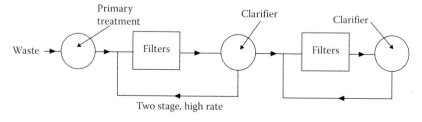

FIGURE 2.8 High rate trickling filter.

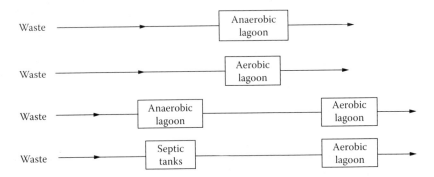

FIGURE 2.9 Schematics of various lagoon systems. (From USEPA, Upgrading existing poultry processing facilities to reduce pollution, US Environmental Protection Agency, Technology Transfer EPA-62513-73-001, Washington, DC, 1973.)

2.3.3.3 Lagoons

Lagoons are presently the most common method of treating poultry waste at private waste treatment installations. This prevalence is primarily due to the availability of low-cost land when most of the poultry processing facilities were constructed, since both aerobic and anaerobic lagoons require very large allocations of land. Successful operation of these lagoons depends to a high degree on favorable climatic conditions—warm, clear, and sunny [1,22]. Depending on the geological site conditions, the lagoon may require lining of the bottom. In general, the waste stream is pretreated by mechanical screens to remove offal and feathers. The biological mechanisms responsible for the purification process have been discussed earlier. Mechanical aerators and diffused air systems are used in aerobic lagoons. A properly designed aerobic lagoon can be expected to achieve about a 90% reduction in BOD_5 and suspended solids removal. When anaerobic lagoons are used, expected removals are only about 70%–80%. Odor problems are frequently associated with anaerobic lagoons, although chemical additives can usually control the problem. In the past, flies, insects, and excessive growth of bordering vegetation have been a problem [6]. Figure 2.9 contains some common flow schematics of lagoon systems.

As an example of the variety of schemes that may be incorporated into the design of poultry waste treatment systems, one such plant has combined, as a single treatment system, the extended aeration and aerobic lagoon concepts to affect considerable cost savings plus a reduction in land requirements. This system, which could appropriately be called an "aerated lagoon," has been in use for many years and reportedly achieves upward of a 90% reduction in influent BOD_5 and suspended solids removal of better than 80% [1]. Since the waste treatment system has been such a success and water shortage is also a problem at this plant, the USEPA has supported a pilot plant investigation to determine the feasibility of recycling the treatment plant effluent for use as process water.

2.3.4 Upgrading Existing Lagoons

The problems associated with the operation of lagoon systems depend heavily on the local climatic conditions. Odors are a common problem, always present in anaerobic systems and occurring whenever anaerobic conditions develop in aerobic lagoons. Since there is no control of the biomass, the need for frequent removal of suspended solids and the maintenance of minimum dissolved oxygen levels are troublesome. Major constituents of the suspended solids present in lagoon effluents are algae. In the absence of sunlight, algae require molecular oxygen for endogenous respiration. During periods that are characterized by low-intensity sunlight radiation, the algae may deplete the dissolved oxygen below the minimum requirements, resulting in algae degradation and biochemical oxygen demand. Stricter effluent criteria, based on the total pounds or maximum concentration rather than a percentage removal of influent loading, ultimately forced the abandonment of conventional lagoon systems to other alternatives or the upgrading of the existing lagoons [1].

To effect any substantial improvement in the effluent quality, any steps to upgrade a lagoon system must be oriented toward control of the biomass. Because algae do not form dense flocs, clarification—either alone or combined with chemical flocculation—has not been successful. Chemical flocculation [23] followed by air flotation [24] has had success, but the application of this method to the entire effluent stream is economically expensive. Conversion of an existing lagoon system to one of the forms of activated sludge systems previously discussed is both advantageous and economically feasible. In the activated sludge process, nutrients present in the waste stream are used in the synthesis of the biomass. Since the biomass is ultimately separated and removed from the waste stream, the degree of algae synthesis in the effluent stream is limited.

To provide sufficient control of the biomass while converting to an activated sludge system, air supply and solids removal equipment must be provided. Generally, these will require the installation of blowers, air distribution systems, floating aerators, clarifiers, concentrators, and strainers. As in all biological systems, some solids will have to be disposed of and considerable attention should be given to reclaiming the solids as feed meal or fertilizer. Maximum effort should be exerted to use existing lagoons as much as possible. In some cases, portions of the lagoons can be converted to serve as multipurpose units such as clarifiers or aerobic digesters, as well as aeration basins. Lagoons can also be used as "polishing" ponds capable of supporting fish life and providing the capability for recycling the treated wastewater.

Generally, once poultry processing operations have commenced, more reliable information can be made available on flows, loadings, temperatures, and waste characteristics, which can be used to establish good design criteria for the upgrading of lagoon systems. Similarly, it is often found that changes in existing process procedures can result in reduced flows and loadings, thereby reducing the need to size equipment.

Construction for the upgrading of lagoon systems is best accomplished by a "staged" sequence. New construction should be scheduled when plant production and the wastewater flow are at a minimum. In this way, the staged concept may allow plant personnel to perform much of the required construction. Also, present effluent criteria can be satisfied while a future water use and waste treatment program is established. Staged construction allows for distributing the capital costs for treatment facilities over a period of time, resulting in minimization of upgrading costs. Staged construction could consist of initial installation of improved aeration systems, then later installation of a clarification device, followed by final installation of a system for recirculation of process water from a polishing pond.

The costs for upgrading existing lagoon systems are often significantly less than those which would be required for equivalent new wastewater treatment systems. In short, to minimize both present and future treatment costs, the design for upgrading an existing lagoon system should incorporate, wherever possible, the existing facilities, process water conservation, and the flexibility to provide for present or future solids reclamation and treated wastewater recycling. Included are some typical flow schematics (Figures 2.10 through 2.12) showing various methods of upgrading existing lagoons.

2.4 OPERATING A WASTEWATER TREATMENT SYSTEM

When the waste treatment plant is designed, constructed, and finally placed on stream, it is the responsibility of the waste treatment plant operator to ensure that the equipment performs as intended. The intent of the waste treatment plant is to reduce the pollution potential of the wastewater discharge to below specified limits. The operator must control the flow streams within the plant to achieve optimum efficiency; this may require, as in the case of activated sludge systems, that the rate at which sludge is wasted and air and chlorine are applied be adjusted to assimilate the contaminants emanating from the processing plant. The pollution load or contaminants emanating from the plant depend upon the particular processes used within the plant, the bird production, and

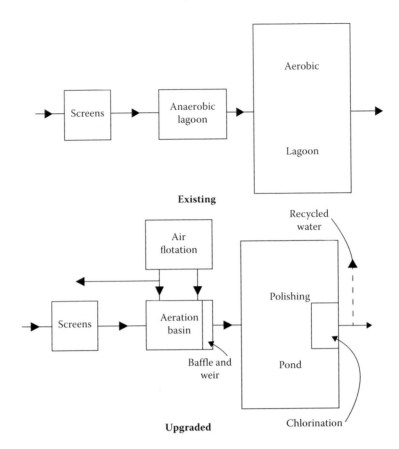

FIGURE 2.10 Example of upgrading a lagoon system. (From USEPA, Upgrading existing poultry processing facilities to reduce pollution, US Environmental Protection Agency, Technology Transfer EPA-62513-73-001, Washington, DC, 1973.)

the plant personnel. The waste treatment plant operator must learn to anticipate and recognize when the influent has changed in character and make the necessary adjustments in the waste treatment plant flow streams before the quality of the effluent has declined sufficiently to constitute a violation or affect the quality of the receiving stream. Good indicators from which to judge the operation of the waste treatment plant are the appearance, color, and smell of sludge and wastewater in the various components, and the general odor at the plant. A successfully operating waste treatment plant has little or no unpleasant odor [25–30].

Maintaining good records of the results of tests and plant operation cannot be overemphasized. Results of record tests, as well as operating tests, indicate how well the plant is functioning. The authority having jurisdiction over the discharge to the receiving stream is responsible to the public for the proper functioning of the waste treatment plant. Where good records are maintained, the authorities, engineers, and others may have ready access to the history of plant operation. This history is valuable in providing assistance in evaluating plant performance and analyzing problems that may occur.

The successful waste treatment plant is also a well-maintained waste treatment plant. Routine inspection and maintenance procedures must be developed for each waste treatment plant; however, a few basic guidelines may be set forth. A manufacturer's instruction manual and shop drawings should be furnished for each piece of equipment in the plant. The operator should be intimately familiar with these manuals and their contents. These manuals give complete information on lubrication, adjustments, and other equipment maintenance. The waste treatment plant should be visited at least once each day. Daily, weekly, monthly, and yearly checklists for inspection and

Biological Treatment of Poultry Processing Wastewater

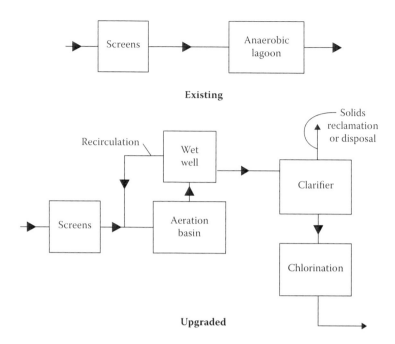

FIGURE 2.11 Example of upgrading an anaerobic lagoon. (From USEPA, Upgrading existing poultry processing facilities to reduce pollution, US Environmental Protection Agency, Technology Transfer EPA-62513-73-001, Washington, DC, 1973.)

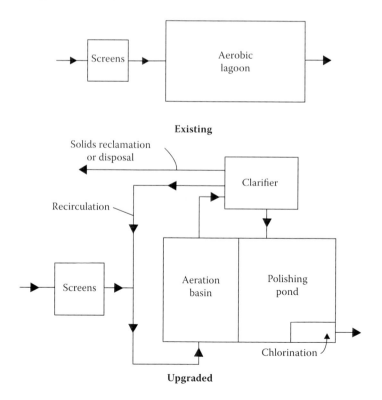

FIGURE 2.12 Example of upgrading an aerobic lagoon. (From Carawan, R. E., Merka, B., Poultry processors: You can reduce waste load and cut sewer surcharges, North Carolina Cooperative Extension Service, CD-22, March 1966, www.bae.ncsu.edu/programs/extension/publicat/wqwm/cd22.html, 2016.)

maintenance should be developed and followed to ensure proper waste treatment plant operation and maintenance. The following checklists were suggested for an activated sludge waste treatment plant treating wastewater from a poultry processing plant [1]. These checklists are not intended to be complete, but to serve as aids for an operator developing his or her own checklists.

2.4.1 Daily Checklist

1. Check the air compressors.
   a. Check lubrication.
   b. Check motor, bearings, and compressors for overheating.
   c. Check air filter for fouling.
   d. Change the compressors "in service." (Set up a rotation schedule to use all of the compressors, including the standby, on a daily switchover schedule so that all compressors are operated the same amount of time.)
   e. Check the "cold" compressor to be put into service for lubrication, free rotation, clean inlet filter, and so forth.
   f. Start the cold compressor and put it on the line before shutting down the one(s) in service (start-up procedure, lubrication, operation, etc.), so that you can satisfy yourself that it starts and operates properly and is performing satisfactorily.
   g. Allow the cold compressor to operate while you are at the plant and recheck it before leaving.
2. Check the chlorinator.
   a. For operation, setting
   b. Chlorine supply
3. Observe the air pattern in all tanks. Adjust it if necessary.
4. Check the final clarifier tank for operation and flow.
5. Check return sludge airlift pump for operation.
6. Check skimmer and scum remover at final clarifier and by-products collector for operation.
7. Check by-products collector for operation and flow.
8. Check telescoping valve for proper sludge and grease removal.
9. Observe sludge appearance in all tanks. Investigate and correct any deficiencies.
10. Observe final clarification and by-products collector tanks. Correct any deficiencies.
11. Observe the condition of raw wastewater entering the plant.
12. Observe the operation of the froth spray system (if used). Clean it if necessary.
13. Check the operation of the comminutor, motor, and bearing for overheating.
14. Rake screenings from bypass bar screen; remove them in covered receptacles for burial on site.
15. Skim floating solids from final clarifier, by-products collector, and digester supernatant decant chamber; place in covered receptacles.
16. Hose down tank and compartment walls, weirs, clarifier center well, raw wastewater inlet boxes, and channels to maintain a clean plant.
17. Correct any plant deficiencies noted.
18. Perform tests—dissolved oxygen (DO), sludge, and chlorine residual—required.
19. Recheck the operation of the cold compressor.

2.4.2 Weekly Checklist

1. Use a coarse brush on a handle to brush down accumulation of algae and other foreign matter that hosing down will not clean up; follow by hosing.
2. Check the air compressors and sludge collector mechanism for oil level and lubrication. (Follow the manufacturer's instructions, using the lubrication oils and greases

recommended; and observe and schedule oil changes, bearing lubrication, and other maintenance, as recommended by the manufacturer.)
3. Make a test of the sludge solids in the extended aeration tanks and in the aerobic digester. Make corrections, if necessary, by adjusting the sludge return to the extended aeration tanks and the aerobic digester.
4. Make a DO test of the final effluent. Correct the amount of air, if necessary.
5. Check the chlorine residual in the plant effluent. Correct the chlorine feed rate if necessary.
6. Open and close all plant valves momentarily to be sure they are operating freely.
7. Maintain the premises (rake, mow, etc.).

2.4.3 MONTHLY CHECKLIST

1. Check and observe the raw wastewater flow rate through the plant.
2. Make a settleable solids (Imhoff cone) test of the raw wastewater entering the plant; record.
3. Make an Imhoff cone test of the plant effluent from the chlorine contact chamber; record.
4. Collect a sample of the plant influent and a sample of the plant effluent for a BOD_5 analysis at an independent laboratory.
5. Remove, inspect, and clean the inlet air filter screens for the air compressor units.
6. Remove the belt guard from the sludge collector drive; check belt tension; grease and service shear pin coupling; and check vent plugs.
7. Remove, clean, and check the surge relief valve(s) upon the discharge of each air compressor.

2.4.4 YEARLY CHECKLIST

1. Clean, touch up, and paint all items requiring attention.
2. Inspect, flush, and clean bearings; lubricate and overhaul all items of equipment.
3. Proceed with any plant modernization or improvement that needs attention.

Keep ahead of the waste treatment plant. Attend it daily. Good housekeeping and maintenance are signs of a good plant operation. A dirty, unkempt plant is a poorly operated one. Ninety percent of plant odors are caused from poor housekeeping.

2.5 CASE HISTORY: THE ORIGINAL GOLD KIST WASTEWATER FACILITIES

2.5.1 SITE SELECTION

Giffels Associates was retained by the Gold Kist Poultry Processing people, then known as the Cotton Producers Association, to undertake a study of waste treatment problems in connection with a proposed new plant to be built in Suwannee County, FL [1]. The location was predicated by the fact that the cooperative farm members were experiencing declining returns on tobacco farming and the Cotton Producers Association elected to bring new agricultural business into the area for the benefit of its farmer members.

Poultry was considered the most logical choice since the firm wished to expand its poultry processing facilities in any event and the area was suited to the rearing, hatching, and dispatching of a large quantity of broilers.

The site finally selected was adjacent to the Suwannee River, which offered some degree of diluting water for the treated industrial waste and also offered a sandy site, on which facilities could be readily built, located over a very adequate groundwater supply to furnish the water for poultry processing.

A site of 100 acres was selected for the project, giving sufficient room for construction of the poultry processing plant, the hatchery, and the required waste treatment facilities, while providing a buffer zone between the plant and adjoining property owners. In addition, the proximity to the interstate highway system offered a ready means for shipment of dressed poultry, and nearby rail access at the town of Live Oak ensured adequate facilities for the receipt of grain and other feed materials.

2.5.2 Wastewater Survey and Criteria

The plant was designed originally to process 50,000 birds/day on one shift. It is currently being expanded to a two-shift operation, with a capacity of up to 130,000 birds/day, which will be described later.

The original wastewater treatment system required facilities to treat poultry processing wastes from 50,000 birds, processed over a period of 8 h, plus wastewaters originating from a second-shift cleanup operation. In addition, all facilities had to be designed to accommodate future expansion up to, or exceeding, twice the then anticipated production rate. In addition to the normal poultry processing wastes, the sanitary wastes from a plant population of about 225 persons and wastewater from the condensers on the feed meal cookers had to be included in the total wastewaters to be treated.

An investigation was made at the existing Canton, GA, poultry processing operation operated at that time by Gold Kist, and the wastes were examined with respect to what loads could be expected in the new Florida facility. Flow quantities were measured at Canton and found to be approximately 12–13 gal/bird; however, operations at that time were considered lax with respect to water conservation, and it was decided that the flow from the new plant would probably be about 10 gal/bird. Subsequent operations proved this figure to be correct; however, diligence is always required to maintain water use at that level, with water consumption practices in the plant being constantly monitored.

An analysis of the wastewater being discharged at Canton indicated the characteristics shown in Table 2.2.

Since this total load was contained in approximately 20% more water than it was assumed would be used at Live Oak, the figures were increased by one-third for design purposes because of the expected higher concentration of contaminants in the wastewaters.

In addition to the poultry processing loads, about 200 gpm of water was to be run through the jet condensers on the by-product cookers. This water collected the grease and vapors from the feed cooking operations and also had to be treated in the industrial waste treatment facilities.

When summarized, the daily load on the plant was expected to be, and later proved to be, about 700,000 gal/day, with a BOD_5 loading of about 2620 lb/day. It should be recognized that this waste treatment plant, therefore, was equivalent to a public wastewater treatment plant for a town of approximately 7000 persons with respect to its hydraulic flow, and to a town of approximately 15,000 to 16,000 persons with respect to its biological load.

TABLE 2.2
Characteristics of the Wastewater Discharged at Canton

Parameter	Range (mg/L)	Average (mg/L)
pH, units	6.3–7.4 units	6.9 units
DO	0–2	0.5
BOD_5	370–620	473
Suspended solids	120–296	196
Total solids	—	650
Volatile solids	—	486
Fixed solids	—	154
Settleable solids	15–20	18
Grease content	170–230	201

Since the wastewater flow from the plant personnel was very small compared with the process waste flow, it was completely ignored in the original design, with the only consideration being given to proper disinfection of the effluent prior to discharge to the Suwannee River.

Flow variations throughout the day were also measured at Canton, with considerable variations being noted; however, a no-flow condition never existed, even during periods of no operations, such as weekends. In other words, even with the plant idle, some flow was experienced due to water consumption that could not be shut off.

It was determined that the minimum flow rate for the new wastewater stream would be approximately 150 gpm, with a maximum flow rate of about 730 gpm and an average flow rate of about 490 gpm.

An important consideration was the fact that a 20-mesh vibrating screen in the processing plant ahead of the process sewer retained the bird entrails, offal, feathers, heads, flushings, and other material, so that those by-products were sent directly to the feed meal recovery cookers and the wastewater sent to the process sewer did not contain significantly large solids and was relatively free of feathers.

2.5.3 Selection of the Treatment Process

Once the wastewater treatment criteria had been established, the decision had to be made as to what type of waste treatment process was suitable. The Florida State regulatory agencies said that lagoons, if built in the area, would have to be fully lined because of earlier problems with groundwater contamination in the area. They would not accept anything except expensively lined lagoons. Furthermore, the criteria that they set with respect to the discharge to the Suwannee River were such that a complete lagoon treatment system would be taxed to achieve the desired results unless the lagoons were made almost ridiculously large. Consideration was therefore given to using a lagoon merely as a tertiary polishing device. The front end of the system and the primary and secondary waste treatment facilities would have to be biological treatment facilities capable of achieving pollutant reduction to a point where the subsequent tertiary pond would be minimum in size and would also ensure meeting the strict effluent requirements of the State of Florida. Those requirements were not over 10 mg/L BOD_5 and zero settleable or suspended solids with rather strict requirements with respect to turbidity and color.

It cannot be overemphasized, as can be seen from these very strict requirements, that in considering any industrial waste treatment process, working closely with state and regulatory agencies is essential before any design work actually begins.

Consideration was given to trickling filters and to various modifications of the activated sludge process [20,21]. In the end, considering costs, reliability, and other matters, the extended aeration modification of the activated sludge process was considered the only possible solution. While this process is often frowned upon in large municipal work because of its high costs, it must be remembered that in an industrial waste treatment facility, the operating expenses are written off for tax purposes, whereas capitalization costs cannot be.

The extended aeration process, as it evolved, offered the following advantages: The initial capitalization cost, although high, was not appreciably higher than the cost of other processes considered when sized to meet the strict effluent criteria. Again, although operating costs were higher, due to the increased amount of compressed air required for the process, highly skilled operating techniques are not always required to keep the system operating near peak efficiency. In other words, exact lab control is not required.

In addition, the usual 24-h aeration period required for extended aeration processes was not conducive to smoothing out the variations in flow and biological loading that were going to be experienced in this waste treatment facility. Furthermore, occasional overloads, either hydraulic or biological, would not cause alarming reduction in waste treatment efficiencies. The process itself, when properly sized, is almost foolproof compared with what happens to overloaded trickling filters and a conventional activated sludge system with short detention periods. Certain waste treatment processes, when overloaded by 10%, result in waste treatment efficiency dropping off substantially, maybe as much as 50%. A 10% overload on an extended aeration system would cause little reduction in waste treatment efficiency.

As a further consideration, the sludge resulting from this process, being thoroughly degraded, is easy to dispose of on drying beds, is seldom inclined to cause odor problems, and is small in volume compared with what might be expected from a trickling filter or activated sludge plant designed and operated by conventional methods.

2.5.4 The Flow Diagram

Accordingly, the system was set up somewhat conservatively consisting of an initial clarifier and skimming device, which was used as a by-products collector mechanism, followed by aeration tanks, a final clarifier, and the tertiary pond. Sludge collected from the final clarifier was pumped by an airlift and recirculated to the head of the extended aeration tank or to the parallel aerobic digester. This digester is used for long-term aeration of grease and excess solids returned from the final clarifier. This aerobic digester should offer the key to successful biological degradation of any grease remaining from the poultry processing since it aerates its contents for a period of approximately 10 days prior to discharge to the clarifier and thence to the final pond.

Plant sanitary wastes were added downstream from the by-products collector and were introduced directly into the head end of the aeration tanks, thereby eliminating any possibility of human wastes contaminating feed meal that is fed to the birds later.

During subsequent expansion considerations, this factor was felt not to be of significant importance and secondary sludge was also considered for reclamation.

2.5.5 Design Criteria

The primary settling tank was originally sized by somewhat conventional means, with a surface settling rate of 1000 gal/ft^2/day being used as criteria for this work. For the final design, the rate was reduced to about 800 gal/ft^2/day by increasing the by-products collector size to 40 ft to permit slightly better grease and solids removal.

The criteria for the extended aeration tanks were based on two factors. First, it was believed necessary to maintain the conventionally accepted basis of 1000 ft^3/20 lb of applied BOD$_5$/day; it was also believed necessary to maintain at least the 24-h minimum detention period normally used in extended aeration systems. As it turned out, the 24-h criterion governed and, in effect, turned out to be somewhat longer than 24 h because the hour flow was considered to be the process waste flow plus the recirculated sludge, which was returned at 50% of the total flow rate to the plant. Normally, the 24-h period is computed using only the total wastewater flow; it does not include the recirculated sludge. It was necessary to begin, therefore, with a total of 36 h of detention based on process flow only. This period may be considered almost excessive by some authorities; however, since the plant was dealing with the problem of slow biological grease degradation and a need to achieve ±98% of BOD$_5$ removal, and since there was also a plan to expand the plant sometime in the future, it was felt that this was not an unreasonable criterion to use for the original plant design. Total aeration tank capacity, as determined by this method, was 140,000 ft^3. It was elected to split this required capacity into two tanks so that one could be operated with a responsible degree of treatment while the other was being maintained or cleaned.

Compressed air was furnished to the tank on the basis of 15,000 ft^3 air/day/lb of applied BOD$_5$. The job was not unique in that the equipment selected for air diffusion consisted of conventional swing-out diffusers used with air headers in Y walls between the tanks. Since the tank was designed based on hydraulic loading rather than BOD$_5$ loading, it actually turned out that plant had only 13 lb of BOD$_5$ applied/1000 ft^3 of tank capacity. Again, this was quite conservative but, as it turned out, quite successful. As a general philosophy in the design of industrial waste treatment facilities, it is necessary to be conservative because there is no assured means of determining that process loads will not increase once successful production lines are put into operation. This proved to be the case at Gold Kist.

Subsequent to the extended aeration tanks, a final clarifier was provided, which was also designed somewhat conservatively, although not to the extent used on the aeration tanks because of the lack of a polishing pond to follow this unit. The final clarifier was designed on the basis of a surface settling rate of 730 gal/ft^2/day and a weir overflow rate of 360 gal/ft/day. Sludge collected by the final clarifier, which was equipped with a skimming arm as well as a scraping mechanism, was returned by an airlift to the head of the system.

This minimized maintenance and operation problems because at this point, the only motors needed for the system were those on the blowers and the drive mechanisms for the by-products collector and final clarifiers. Additional sludge pumps were not required.

Again, when it came down to the final design, this tank diameter was increased slightly so these criteria were even less. There was a constant effort to create a system that could be somewhat overloaded without causing a significant reduction in waste treatment efficiency.

It was strongly suspected that it might be possible to process up to 10,000 birds/h on occasion, which indicated that a one-shift operation could run as high as 70,000 to 80,000 birds/shift if the processing plant were run at a maximum rate. This actually did happen with production rates approaching 10,000 birds/h and the single-shift operation extending to 9 or 10 h/day.

The aerobic digester, which was provided adjacent to the aeration tank, had a volume of 46,500 ft^3. Air was supplied in exactly the same fashion as to the aeration tank on the basis of 20 ft^3/min of air/1000 ft^3 of digester capacity. A supernatant decant well was provided at the end of the digester, with clarified liquor rising slowly and very quiescently since the overflow rate can be controlled by the operator based on the amount of sludge returned to the aerobic digester, with the remaining amount of return sludge being sent to the aeration tanks. This supernatant decant well was actually nothing but a timber baffle with a hopper bottom built into the end of the aerobic digester.

Sludge from the aerobic digester was directed to the sludge drying beds by the simple means of a fire hose, using the head on the aerobic digester to force the sludge to the drying beds. As a general rule of thumb, the amount of solids generated by the extended aeration system is about 0.5 lb/day/lb of BOD$_5$ treated in the tanks.

The sludge drying beds were perhaps the weak point in the system. Although north Florida seems to be entirely made up of deep sand beds, somehow, unwittingly, the only spot of clay in Suwannee County was picked on which to build the drying beds. Therefore, although subdrainage systems were provided to dry the sludge, it did not dry as readily as hoped for. Further complications were involved in the fact that, at least initially, the outlet from the underdrains in the sludge drying beds was completely filled over by some incidental grading operations done by a contractor. The old rule, if anything can go wrong on a project it will, seemed to apply here.

Even though difficulties were experienced with the sludge drying beds, they were not serious because the nature of the sludge produced by this waste treatment process is such that it is inoffensive, dries readily, and in a pinch can just be spread out on the ground. It is not troublesome with respect to odors, and can always be scraped up and hauled away from almost any place and stored temporarily, as long as it is allowed to dry. The warm Florida weather helped in this matter.

Sealing the final tertiary pond subsequent to the clarifier was accomplished rather inexpensively. Asphalt liners, rubber linings, and other membranes were considered, and the cost always proved to be somewhat shocking, so the designers ended up with a rather simple system. A thin layer of Visqueen (a brand name of polyethylene sheeting material) was placed over the entire bottom of the lagoon in small overlapping sections and trucks full of sand dumped directly on the Visqueen. The sand was spread by hand, making a 1-ft layer of earth and sand on top of the Visqueen to protect it. The system may not be 100% watertight, but the designers were pleased to note that before any water was added by the waste treatment process to the lagoon, it did collect and store rainwater and was ready to go at the time the poultry processing facility went onstream.

The inlet to the square 4-acre pond was located in its geometric center. This was done so that any inadvertent solids carryover from the final clarifier would settle out in the central area of the pond, and if there was an odor problem resulting from this, it would be at least 200 ft from any point on the shore. The best way to control an odor is to keep it as far away from people's noses as possible.

Effluent from the pond was subjected to 20 min of detention in a chlorination pond. A chlorinator was provided, which was capable of handling from 20 to 100 lb of gaseous chlorine per day. The expected required rate was approximately 42 lb/day which left a 0.5 mg/L residual in the effluent.

The effluent was chlorinated by electrolytic cells generating sodium hypochlorite, and in retrospect, it was wise to install such a system in a facility such as this to eliminate the hazards involved in using gaseous chlorine.

The air compressor facility for the initial waste treatment plant consisted of three, 3000 ft^3/min turbo compressors, each with a 75 hp motor delivering air against 6 lb of pressure. Actually, two compressors were required to run full time to furnish air to the aeration tanks and airlift. The third compressor was merely a standby.

In summary, the wastewater treatment plant components consisted of the following devices, which are detailed in Figure 2.13 [1]:

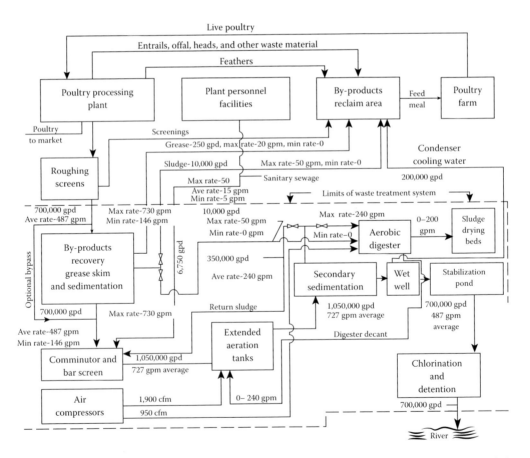

FIGURE 2.13 Waste treatment diagram of the facility at Live Oak, FL. (From USEPA, Upgrading existing poultry processing facilities to reduce pollution, US Environmental Protection Agency, Technology Transfer EPA-62513-73-001, Washington, DC, 1973.)

1. The system used a by-product collector tank a conventional circular settling tank with motor-driven sludge collection and skimming mechanisms, 40 ft in diameter with a 7-ft side water depth, and a detention time of approximately one and a half hour.
2. Extended aeration tanks consisted of two concrete tanks each 170 ft long and 26 ft wide with a water depth of 15 ft and a freeboard of 2 ft. Each tank contained 95 air diffusers and piping to supply 150 ft^3/min of air to the mixed liquor. A valve port between the two tanks permitted equalized loading on each tank if required. The detention time was 24 h based on process flow plus a 50% allowance for sludge recirculation.
3. The aerobic digester was constructed next to the aeration tanks; therefore, it had an equal length of 170 ft but it was only 21 ft wide, again with a water depth of 15 ft. The supernatant decant well at the end of the tank, which was used to contain the solids in the tank, had a surface settling rate of less than 800 gal/ft^2/day and an overflow rate of 240 gal/min to the clarifier.
4. The final settling tank was again a conventional collection mechanism with a motor-driven scraper and skimming mechanism. It was made 44 ft in diameter with an 8-ft side water depth and a 9-ft center depth. Detention time in this tank was about 2 h, which is the maximum that could probably be used without running into septicity problems in the stored sludge.
5. The final stabilization pond or tertiary device was 4 acres in size and had a detention time of approximately 10 days. The depth was 5 ft, which is about the maximum that can be used in an aerobic pond, simply because depths in excess of 5 ft tend to become septic at the bottom due to lack of sunlight and poor oxygen transfer.
6. The chlorination facility was simply a small building with a wall-mounted chlorinator, scales, and ancillary devices. It was placed at the far end of the ponds, convenient to the chlorine detention pond.
7. The air facility used three turbo compressors rather than positive displacement rotary compressors for the simple reason that they were believed less expensive and more reliable, and certainly less noisy. Judgment has proved this fact, and turbo compressors, as opposed to positive displacement air compressor devices, have been made standard waste treatment equipment wherever possible. Their one drawback is that the pressure they can develop is somewhat limited and they cannot be discharged into really deep aeration tanks. About the best that can be done is to operate with about 13 ft of water pressure and the aeration diffuser devices.

2.5.6 Future Expansion Provisions

The entire facility was arranged to permit future expansion if necessary, and piping was valved and placed so that another by-product collection tank could be added if needed. Another final settling tank also could be added, if required, as well as more air compressors.

In addition, there was space on-site to double the stabilization pond and add additional chlorination detention ponds if necessary. It cannot be overemphasized that one must have a site that is big enough for the waste treatment facilities and 100% expansion if necessary. It was possible to accomplish 100% expansion without doubling up on these facilities, as will be discussed later. At the time, however, doubling of the facilities seemed to be the practical way to handle the problem.

No expansion provisions were provided in the original aeration tank installation since, in effect, it had been overdesigned and it was felt that in a pinch, a conventional activated sludge operation could be adopted, which would require less tank volume but more sophisticated control. In such a case, only 6 h or 8 h of aeration time would be necessary.

The facility was originally designed to achieve BOD$_5$ reduction in the following fashion. Of the 2600 lb of BOD$_5$ applied/day, 780 lb would be removed in the by-products collector—30% of the applied load. This is a conventional criterion for devices of this type. It was expected that the subsequent extended aeration tanks would be able to remove 90% of the remaining 1820 lb of applied BOD_5. Again, this seems to have been a reasonable assumption since 90% removal using long-term aeration is quite feasible.

The stabilization pond also was capable of removing another 200 lb of BOD_5/day. The rate of 50 lb/acre/day is allowed in the State of Florida. As one goes farther north, colder temperatures inhibit biological actions in lagoons; therefore, in the northern states, 20–30 lb of BOD_5 loading/acre/day is all that is allowed.

Totaling up the removal of BOD_5 in all components, it was possible to indicate, in theory at least, that 100% of BOD_5 would be removed. Although it was known that this was impossible, 97%–98% BOD_5 removal was expected, which is very good efficiency. After the plant went into operation, these results were achieved, and perhaps even a little better, which justified the selection of the process and the conservative design approach.

2.5.7 Waste Treatment System Costs

The total cost of the wastewater treatment facility in terms of 2014 USD was estimated to be 1.8 million USD. It may have cost somewhat more, however, since the construction costs for the waste treatment facilities were lumped in with the construction costs for the on-site hatchery and poultry processing plant. The exact cost could not be determined, but in any event, the cost was less than 2 million USD.

Amortizing the costs over a 10-year life expectancy of the facilities, it was found that the original investment of 1,585,000 USD plus an allowance for interest on that investment made the total cost of the facility 2,056,000 USD. Based on processing 250,000 birds per week at a dressed-out weight of 2.5 lb/bird, it turned out that the capitalization cost for waste treatment for this plant was approximately 0.63 cent/lb of finished dressed poultry.

Operating costs were estimated to be approximately 7 man-h/week of labor, 7 days of power for the two blowers running continuously, and 7 days' worth of chlorine, which was estimated to cost about 38 USD/day. Total weekly operating costs came out to be about 2330 USD, which indicated an operating cost of about 0.38 cent/lb of finished product. Therefore, the total waste treatment costs of dressed poultry appeared to be in the neighborhood of 1 cent/lb of finished dressed poultry.

2.5.8 Operating Arrangements

In addition to an adequate design for wastewater treatment facilities, another important matter must be mentioned. The true success of any waste treatment operation is based as much on operating skill as on the design and capacity of the system. As engineers responsible for the design of this system, the consultants were most fortunate to have an owner who elected to go out and hire a skilled, competent man to run these facilities. He was also assigned the job of running the by-product reclamation system and cookers. Thus, he was assigned a job in which, if he reclaimed his by-products effectively, his wastewater loads were less and he was, therefore, in complete control of his own destiny. If he messed up on one job, he would not be able to straighten it out on the other.

To help him in his job, an operating manual was prepared, which described the plant, its flow stream and theory, its initial start-up procedures, and its normal operating procedures. It also contained checklists for maintenance and for equipment operation. The importance of an operating manual prepared by the design engineer cannot be minimized.

2.6 CASE HISTORY: CURRENT EXPANSION AT GOLD KIST PROJECT HISTORY

The original facility was designed to treat wastewaters resulting from the processing of 50,000 birds per day plus a nearly insignificant sanitary wastewater flow from the plant personnel facilities. The rationale by which hydraulic, biological, and solids loadings were determined has been described previously. Also described were the basic criteria used for determining facility component sizes and the provisions for future expansion as foreseen at that time. During final design, however, the sedimentation facilities were increased in size.

This somewhat arbitrary increase in the size of the by-products collector tank (primary settling) to 40 ft in diameter and the final clarifier (secondary settling) to 44 ft in diameter is the principal reason, along with a conservative original design, why the system has continued to function well under recent substantial overloads. For these same reasons, the modifications now required are minimal [1].

2.6.1 CURRENT WASTEWATER LOADS

Once poultry processing start-up difficulties had been overcome, it was economically practical to increase hourly production rates and the one-shift processing time to 9 h or 10 h. This resulted in daily processing rates of 72,000 to 80,000 birds and measured hydraulic loads of about 1 million gallons per day (MGD). This roughly 50% increase in production caused flow rates ranging from 1.7 to 0.4 MGD on processing days. During weekends (supposedly no-flow conditions), 0.2 MGD flows were recorded (see Figures 2.14 and 2.15).

Original flows were expected to be 10 gal/bird (processing and cleanup) +4 gal/bird for by-products cooker–condenser cooling waters, for a total flow of 700,000 gal/day. Actual measured wastewater flows have averaged 13 gal/bird and totaled over 1 MGD. In addition, laboratory tests showed the properties of the wastewaters to be above average during processing hours (see Table 2.3).

2.6.2 CURRENT OPERATING DIFFICULTIES

As the loading on the wastewater treatment passed the design criteria, problems arose in the operation of the facility [1]. These problems were met as they arose by the certified wastewater plant operator hired by the owner. By skillful operation of the facility, consistent results were obtained. The proposed revisions included measures to correct system deficiencies that became apparent and included a new package laboratory permitting the operator to exercise better control.

2.6.2.1 Condenser Cooling Water

While an allowance of 0.2 MGD was made in the original design for use of process water to condense feed meal cooker vapors, even more water was required. Problems were encountered from the very beginning, with feathers in the process waste stream clogging the condenser cooling water pumps. Although it was originally intended that the recirculation well be located downstream of the secondary sedimentation clarifier, it was installed between the roughing screens and the by-products recovery tank (primary settling). This decision was based on terrain and plant configuration and in hindsight was unwise.

Although ample process wastewater was available during processing hours to meet the anticipated 140-gpm demand of the feed meal cooker condensers, the demand proved even higher and the ensuing difficulties precluded the use of any process wastewater. In addition, the low process waste discharge flows during cleanup and no-processing hours caused the wastewater stream to heat up to an unacceptable temperature (120 F +) even when condenser–cooker water pumps were operable. At these temperatures, successful condensing of the odorous cooker vapors was incomplete.

As a result, it was necessary to connect an additional unmetered process waterline into the cooker vapor condensers with a nearly 24-h constant water flow and resulting discharge to the process sewer. This flow is believed to be almost entirely responsible for the increase in flow above the metered process water consumptions of 10 gal/bird. In other words, about a 0.30 MGD increase (3.5 gal/bird) in process wastewater flow above that expected (10 gal/bird) has occurred due to this processing complication. When added to the average daily metered process water consumptions of 740,000 gal/day (processing and cleanup for approximately 75,000 birds/day), the approximate 1 MGD flow that was measured was confirmed.

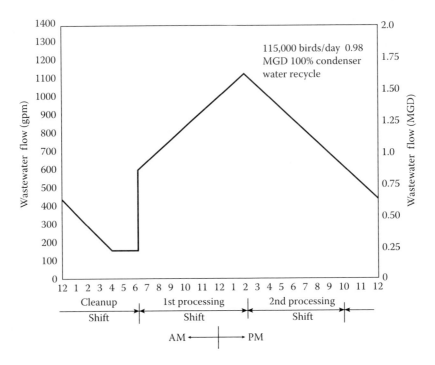

FIGURE 2.14 Future average daily hydraulic load on Gold Kist Facilities. (From USEPA, Upgrading existing poultry processing facilities to reduce pollution, US Environmental Protection Agency, Technology Transfer EPA-62513-73-001, Washington, DC, 1973.)

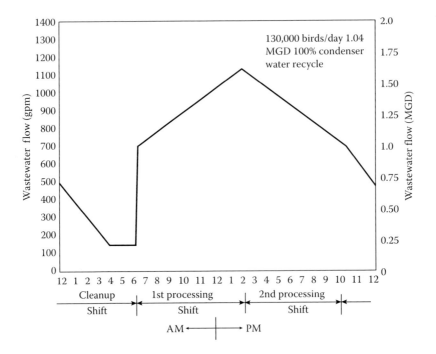

FIGURE 2.15 Future maximum daily hydraulic load on Gold Kist Facilities. (From USEPA, Upgrading existing poultry processing facilities to reduce pollution, US Environmental Protection Agency, Technology Transfer EPA-62513-73-001, Washington, DC, 1973.)

TABLE 2.3
Wastewater Properties

Item	Original Assumptions (mg/L)	Measured Properties (mg/L)[a]
Total solids	650	910
Suspended solids	200	300
BOD$_5$	470	420

[a] A large part of the increase in solids is due to the fact that the by-products collector sludge has often contained too much water for consistently economical reclaim. When pumped back over the by-products reclaim screens in the processing plant, many fine solids pass through back into the process waste sewer.

Later, condenser water pumps were successfully relocated from their original location to the process waste sump in the waste treatment facility. Several days of clog-free operation were then possible. During these periods, metered process water demands dropped by 0.1 MGD to 640,000 gal/day, or only an average of 8.5 gal/bird.

Such examples prove that a wastewater estimate of 10-gal/bird processed is ample provided all by-product cooker–condenser cooling water can be taken from the process wastewater stream. The successful operation of the wastewater treatment facilities under the proposed two-shift, 115,000-bird load is predicated on this assumption and will be discussed further herein.

2.6.2.2 Hydraulic Overflows

At times of peak processing, slight overflows have spilled onto the ground from the aeration tank effluent troughs, and aeration tank weirs have flooded out. The original anticipated design flow rate between the aeration tanks and final clarifier was 0.7 MGD (490 gpm), +50% (peak flow allowance), +0.5 MGD (350 gpm) sludge recirculation, or approximately 1.55 MGD (980 gpm). A system head curve for the gravity flow line ($C=100$) between the aeration tanks and the final clarifier revealed only 1.2 ft of the elevation differential required. Effluent trough overflow was considered, at that time (original design), to be unlikely.

However, recent peak flow rates of nearly 1.7 MGD plus sludge recirculation have caused a flow rate of over 2 MGD in the aeration tank discharge line. At this flow, about 2.5 ft of elevation differential is required. Since the design provided for 3.5 ft of elevation difference between the aeration tank water surface and the final clarifier water surface, recent overflows seemed unexplainable.

Recent checking of "as built" elevations, however, reveal only about a 2.5-ft difference in elevation; apparently, a construction error was made. Later discussion contained herein will describe the corrective work required.

2.6.2.3 Solids Control

The plant operator has been plagued by problems relating to solids control. Not only has he had difficulty with maintaining optimum MLSS (mixed liquor suspended solids) ratios, but also with removing sludge from the aerobic digester to the sludge drying beds. These problems have generally been caused by the following complications [1]:

1. Uncanny intuition on the part of the designers placed the sludge drying beds in the only pocket of clay in an otherwise all-sand site. This resulted in poor performance.

2. Earth spoil from construction done subsequent to start-up was placed in the only available low spot on the site. Unfortunately, this also buried the outlet end of the sludge bed subdrainage system.
3. The sludge drawoff pipe from the aerobic digester was placed on the wrong side of the decant baffle. Instead of being able to draw the concentrated liquor from the quiescent side of the baffle in a steady stream, it was necessary to shut off the digester air supply and draw only whatever sludge settled in the vicinity of the outlet pipe. Time for this was limited due to a rapid decrease in the digester content dissolved oxygen residuals.

Proposals for correction of these deficiencies follow herein.

2.6.2.4 Odor Problems

During weekend periods of low or near-zero flows, oxygen levels in the by-products collector (primary settling tank) and lift station sumps often disappear, and septicity, with its resulting odors, occurs. Proposed revisions (described later) have corrected these conditions.

In addition, frequent power outages (three to four times a year) have presented problems with the air compressors, which require a manual restart sequence. Aeration tank contents become offensive by Monday morning when such outages occur on Friday night or Saturday. System improvements have provided for automatic air compressor restart.

2.6.3 Proposed Wastewater Treatment System Loads

2.6.3.1 Hydraulic Loads

Actual process water flows recently measured during processing periods have ranged from 51,000 to 40,000 gal/h. Therefore, during these times, there was an average flow rate (metered process water) of about 45,500 gal/h or a 1.1 MGD rate. The average bird processing rate during these times has been 9400 birds/h; so processing seems to require an average of 5 gal/bird.

Assuming the actual use to be 6 gal/bird for the sake of conservative design, and based on occasionally reached processing rates of 10,000 birds/h, the maximum process wastewater flow rates expected may reach 10,000 birds × 6 gal/bird/60 = 1,000 gpm, or about a 1.4 MGD rate.

Recently measured cleanup water consumptions have ranged from 239,000 to 294,000 gal/day, or an average of 266,000 gal/day. The average daily number of birds processed during this time has been 76,000 birds/day, with a range of from 72,000 to 80,000 birds/day. This indicates cleanup water demands of about 3.5 gal/bird. This flow generally occurs in the 6 h following the end of the poultry processing work, the initial flow roughly equaling the process flow, and gradually decreasing thereafter to supposedly near-zero flow about 6 h later. Hence, the average flow during the 6-h cleanup period is about 500 gpm, or about a 0.7 MGD rate.

Data gathered indicate minimum weekend flows of about a 0.2 MGD (140 gpm) rate. This is presumed to be a completely shutdown flow rate and is due to sanitary flows, infiltrations, and other miscellaneous water consumption that cannot be reduced further. These data plus the unmetered cooker condenser flow estimate have been used to produce the hourly flow rates shown in Figure 2.2. Figures 2.14 and 2.15 indicate expected average and maximum daily flows from the increased poultry processing. These future hourly flow rates and total daily flows are based on the following premises [1]:

1. The processing of 115,000 birds/day in 16 h will result in an average production rate of 7200 birds/h. This rate is somewhat lower than at present, when an average of 75,000 birds are processed in 8 h (9400 birds/h).
2. Peak processing rates of 10,000 birds/h will again be possible and will occur on days when production reaches the proposed maximum of 130,000 birds/day; peak flow rates of 1000 gpm of process wastewater will be generated at such times.

3. Process water demands will not exceed 6 gal/bird and cleanup demands will not exceed 4 gal/bird. These rates seem readily obtained based on present operating experience.
4. Wastewater flows will be minimized by discontinuing the use of unmetered process water for by-products cooker vapor condensing. All condenser water demands will be met by using untreated process waste supplemented when required by recirculation of by-products collector tank (primary settling) contents and aeration tank contents. At these times, those waste treatment system components will be, in effect, heat exchanger devices. Anticipated maximum water temperatures in those components will not exceed 120 F and should not, therefore, be detrimental to biological processes.

It was demonstrated by Figures 2.14 and 2.15 that although total wastewater flows will increase by 5%–12%, peak flow rates will actually be reduced by 16%.

2.6.3.2 Biological Loads

The original design anticipated a BOD_5 load of 2600 lb/day while processing 50,000 birds/day, or 0.052 lb of BOD_5/bird. If cooker–condenser water recirculation had been continuously practiced instead of using fresh unmetered water, this BOD_5 load should have been contained in 50,000 gal (4,170,000 lb) of water/day, which would then have had an average BOD_5 concentration of 624 mg/L. With additional water for cooker vapor condensation, the flow was estimated to be 700,000 gal/day. At this time, the average BOD_5 concentration in the raw waste would have been 446 mg/L. Laboratory tests made on wastewater flows found an average influent BOD_5 of 350 and 397 mg/L. These tests indicate that the originally anticipated BOD_5 assumptions were quite adequate.

Influent grab samples during processing hours had a peak influent BOD_5 of 550 mg/L. This increased to 650 mg/L during those hours when wastage of by-products collector sludge through by-products area screens was practiced instead of being sent to the cookers for reclamation. These figures are similar to those BOD_5 concentrations assumed during the original design work.

The monthly operating reports indicate gradually increasing loads on the waste treatment facilities as production increased. Near the end of the second year of operation, it became necessary to run the third standby blower continuously to satisfy oxygen demands.

2.6.4 Review of Component Adequacy

2.6.4.1 Lift Station

The existing lift station contains three separate sumps for process water, sanitary wastewater, and by-products. Although total flows are expected to increase slightly, maximum flow rates should decrease so that no changes are required in this facility. However, to ensure an ample cooker vapor condenser-cooling water supply, a new 1000-gpm nonclog pump will be installed where space was reserved for the future third-process waste pump. The discharge for this pump will run by means of a new 6-in. force main back to the by-products area.

2.6.4.2 By-Products Collector

Under present conditions, this 40-ft-diameter tank has an average surface settling rate of 765 gal/ft^2/day and an average weir loading rate of 7650 gal/ft/day. Short-term peak flow rates have probably reached twice these values.

With the increased production on a two-shift basis, the average surface settling rate is expected to increase to 805 gal/ft^2/day, and the average weir loading rate to 8050 gal/ft/day. Since peak flow rates will be reduced by about 200 gpm, and since this facility has averaged a 39% reduction in BOD_5, it is believed that the increased loads can be handled without modifications. The BOD_5 removal rate is undoubtedly enhanced by the substantial grease removal in this tank.

2.6.4.3 Aeration Tanks

The extended aeration process utilized in these tanks has been very successful in meeting the present system overloads. With an assumed average BOD_5 influent concentration of 450 mg/L, a 35% reduction of BOD_5 in the by-products collector, and a 0.93 MGD flow, the present loading on the aeration tank is approximately 2260 lb of BOD_5/day. The present detention time in the aeration tank is 27 h (2×525,000 gal/930,000 gal/day = 1.13 day). With the standby blower presently on at all times, the air supply to the tanks has been increased by about 50% over the original design, and it is now about 1425 cfm to each tank, or a total of 2850 cfm.

Based on these figures, current aeration tank operating data are illustrated in Table 2.4.

Proposed modifications to the aeration tanks will consist of approximately doubling the number of Sparjers on each header and increasing the air supply to 1900 cfm of air to each aeration tank.

With 100% condenser cooling using process waste and at peak processing rates (130,000 birds/day), and once again assuming 0.052 lb of BOD_5/bird, tank data are developed as shown in Table 2.5. (This is again predicated on 35% of BOD_5 removal in the by-products collector.)

If this were a conventional activated sludge application, certain part of the preceding data would tend to indicate that the aeration system had been pushed to a practical limit; however, the long detention time and ample air supply should ensure continued operation as an extended aeration system. Under these conditions, however, the settling properties of the sludge in the final clarifier should be improved (less light, endogenous cellular material). But waste sludge volumes may increase by several times over what has been encountered up to this time. Improved sludge wastage facilities are described later herein.

TABLE 2.4
Current Aeration Tank Operation Data

Applied BOD_5 per day	2260 lb
Daily flow	0.93 MGD
Detention time	27 h
Tank volume per pound of applied BOD_5 per day	62 ft^3 or 16.2 lb of BOD_5/1000 ft^3
Air supply per pound of applied BOD_5 per day	1830 ft^3
Air supply per 1000 ft^3 of tank capacity	20 cfm

TABLE 2.5
Peak Aeration Tank Operation Data

Applied BOD_5 per day	4400 lb
Daily flow	1.05 MGD
Detention time	24 h
Tank volume per pound of applied BOD_5 per day	32 ft^3 or 31.5 of BOD_5/1000 ft^3
Air supply per pound of applied BOD_5 per day	1250 ft^3
Air supply per 1000 ft^3 of tank capacity	27 cfm

2.6.4.4 Final Clarifier

The performance of this unit will probably actually improve under the additional loading. Settling properties of the sludge contained in the mixed liquor received from the aeration tanks should be improved and the 24-h aeration period almost precludes sludge bulking under normal operation. This 44-ft-diameter unit is currently operating with an average surface settling rate of 612 gal/ft^2/day, and a weir loading rate of 680 gal/ft/day. These will increase to 645 gal/ft^2/day and 710 gal/ft/day, respectively, but again, as in the case of the by-products collector, peak flow rates will be less.

The return sludge airlift that forms a part of this component was originally designed to return over 350 gpm of sludge to the head end of the aeration tanks and/or aerobic digester. This amounted to 70% of the anticipated average hydraulic load (700,000 gal/day = 490 gpm). Under present conditions (0.93 MGD), it is returning sludge at a rate of ±50% of the total flow through the facility. Since the new average hydraulic load is only slightly greater, the airlift should still be adequate.

2.6.4.5 Aerobic Digester

The aerobic digester, as originally designed, was sized on the basis of 4.5 ft^3/capita based on an average of the hydraulic and biological population equivalents. The resulting capacity of 46,500 ft^3 was supplied with 20 cfm of air/1000 ft^3 of capacity to ensure adequate mixing and dissolved oxygen residuals. The decant well surface area (256 ft^2) was considered capable of retaining all solids in the system if sludge were wasted to the digester at a rate not in excess of 10,000 gal/day (375 gal/ft^2/day). Actually, required sludge wastage flow rates were considered as probably being considerably less.

With the future applied load of 4400 pounds of BOD$_5$/day on the aeration facilities, about 2200 lb/day of solids may be generated. If these were contained in a sludge having a water content of 99.8%, about 1.1 million lb or 132,000 gal/day would have to be wasted to the aerobic digester. Under these flow conditions (92 gpm), the surface settling rate on the decant well of the aerobic digester would be only 515 gal/ft^2/day; the weir loading rate would be 8200 gal/ft/day; and the digester detention time would be approximately 3 days. With a regular daily sludge drawoff, the aerobic digester is still believed to be quite adequate.

2.6.4.6 Sludge Drying Beds

The original sludge drying beds provided had a total area of approximately 20,000 ft^2. On the basis of the original biological population equivalent, this area amounted to 1.33 ft^2/capita. Although the plant has been operating under a ±50% overload condition, the area has still proved ample; however, the method by which digester sludge is conveyed to the beds has been less than satisfactory.

The fact that the sludge is the end product of an extended aeration process and has been further subjected to several days of additional aerobic digestion has resulted in small quantities of a readily dried product with no offensive odors.

One additional 10,000-ft^2 drying bed with underdrains will be constructed, and the existing beds and underdrainage system improved to provide more capacity and eliminate previously discussed operating problems.

2.6.4.7 Stabilization Pond

The stabilization or polishing pond has an area of 193,000 ft^2, or 4.4 acres. High dissolved oxygen content in the pond has enabled bass and bream to flourish, and the pond has been the site of company fishing contests. When power failures have occurred and septic conditions have arisen in the aeration tanks, the pond has prevented such conditions from reaching the Suwannee River until the treatment processes were again in order.

Although, at this time, no change is contemplated in the pond, aeration may have to be added to the pond if the increased wastewater load causes anaerobic conditions to develop. This could be accomplished by means of submerged perforated polyethylene air headers or floating surface aerators placed near the central inlet.

2.6.4.8 Outfall Sewer and Cl₂ Facilities

Since total flows are only expected to increase by 9%–12% and peak flows will actually be reduced, the existing facilities should still be adequate.

2.6.5 PROPOSED MODIFICATIONS

2.6.5.1 Air Supply

The number of aeration Sparjers in the aeration tanks will be approximately doubled. To furnish the additional air required, two more 75-hp turbo compressors will be added. During operations, four will be running and one will be kept as a standby. The motor control center near the lift station was originally designed to accommodate two more future blowers, so only power connections from the motor control center to the blower motors are needed. The motor control center will also have a "motor minder" added to it, which will provide automatic sequential blower restarts in the event of a power failure. This device will also protect the 75-hp motors against low voltage and single phasing.

2.6.5.2 Plant Hydraulics

Although peak flow rates will be reduced, the possibility of aeration tank effluent trough overflows may still exist. A new head box and an increase in the aeration tank discharge line size to 18 in. from 14 in. should prevent any further problem.

2.6.5.3 By-Products Reclamation System

The original system consisted of a collection sump in the facility lift station and a pump which returned the sludge and grease collected by the primary settling (by-products) tank to the cookers in the plant. Considerable sand and grit in the sludge caused clogging in the return line, which was sized for future flows but in which present velocities failed to keep the grit and sand in suspension. In addition, the by-products reclaimed often contained too much water for economical recovery. Some system improvements include the following:

1. Replace the present by-products pump with a 200-gpm, 35-ft head-float-actuated nonclog centrifugal pump. This will return by-products to the cookers with sufficient velocity to prevent clogging of the return line.
2. Install a secondary grease tank in the by-products area of the plant. This tank will receive grease from the by-products tank skimmer box. This tank will permit reutilization of the pump to draw grease from the bottom of the tank and discharge to a separate grease cooker for separate rendering.
3. As an alternative to this scheme, possible use of a centrifuge to thicken the sludge (20% solids) before discharge to the cookers is being considered. If it proves practical to reclaim secondary as well as primary sludge, this alternative will appear to be more attractive.

2.6.5.4 Sludge Drying Facilities

A modification to the sludge withdrawal system in the decant well of the aerobic digester is required. An adjustable sludge drawoff pipe should permit the operator to withdraw sludge from that level in the decant well that contains the thickest sludge.

A new 100-gpm sludge drawoff pump on brackets secured to the side of the aerobic digester will discharge to either one of the existing drying beds or to the new third sludge drying bed. As an alternative, the discharge can be sent to the by-products sump. Should the secondary sludge be amenable for reuse as poultry food, there will be no need for sludge drying beds.

2.6.5.5 Condenser Cooling Water

The success of the waste treatment facility operation under the increased loading anticipated is dependent on 100% reuse of process wastewater for by-product cooker vapor condenser cooling. To ensure an ample water supply, an 800- to 1000-gpm, 50-psi, 75-hp pump will be installed in the facility lift station. Taking its suction from the process sump, it will discharge back through a new force main to the by-products cooker vapor condensers.

During periods of low process waste flow, the drain valves from the aeration tanks and aerobic digesters will be opened sufficiently to maintain an adequate water supply to the condenser cooling water pump. This will, in effect, convert the primary settling tank (by-products collector) and aeration tanks to heat exchangers. Since some process waste (200-gpm minimum) always flows, it is felt that the over 1 MG of aerated mixed liquor in the aeration tanks will not become heated to a point detrimental to biological processes. Calculations indicate cooling water temperatures will remain below 120 F for successful cooker vapor condensation, and such temperatures are not expected to injure biological process.

2.6.5.6 Grease Removal

By-products handling is, on occasion, hampered by excessive grease in the reclaimed feed meal (conveyor fouling, clogging, etc.). At the option of the poultry processor, a grease receiving tank will be installed next to the by-products collector. Scum and grease from this tank will be collected separately and pumped to a separate cooker.

ACRONYMS

BOD	Biochemical oxygen demand
BOD_5	5-Day biochemical oxygen demand
CIP	Cleaning-in-place
DO	Dissolved oxygen
gpm	Gallons per minute
MGD	Million gallons per day
POTWs	Publicly owned treatment works
USDA	US Department of Agriculture
USEPA	US Environmental Protection Agency

REFERENCES

1. USEPA, Upgrading existing poultry processing facilities to reduce pollution, US Environmental Protection Agency, Technology Transfer EPA-62513-73-001,Washington, DC (1973).
2. Wang, L. K., Pereira, N., Hung, Y. T. (eds), and Shammas, N. K. (Consulting Editor), *Biological Treatment Processes*, Humana Press, Totowa, NJ (2009).
3. Wang, L. K., Shammas, N. K., and Hung, Y. T. (eds.), *Advanced Biological Treatment Processes*, Humana Press, Totowa, NJ (2009).
4. USEPA, Technical development document for the final effluent limitations guidelines and standards for the meat and poultry products, US Environmental Protection Agency, EPA-821-R-04-011, Washington, DC (2008).
5. Sams, A. R., *Poultry Meat Processing*, CRC Press, Boca Raton, FL (2001).
6. Wilson, A., *Wilson's Practical Meat Inspection*, 6th edition, Blackwell Science, Ltd, Malden, MA (1998).
7. Stadelman, W. J., *Egg and Poultry-Meat Processing*, Ellis Horwood Ltd, New York (1988).
8. Mead, G. C., *Processing of Poultry*, Elsevier Science Publishing Co., Inc., New York (1989).
9. U.S. Department of Agriculture, Food Safety Inspection Service, Guidelines for specified cuts of poultry, USDA, FSIS, February 26, 1986, www.fsis.usda.gov/oppde/rdad/fsisdirectives/7110%2D1.pdf (2016).

10. Ecologix Environmental Systems, Dairy, meat & poultry industry wastewater treatment, www.ecologixsystems.com/industry-meat-poultry.php (2016).
11. International and Finance Corporation, Environmental, health, and safety guidelines for poultry processing, IFC, World Bank Group, April 30, 2007, http://documents.worldbank.org/curated/en/937061486569846211/text/112695-WP-ENGLISH-Poultry-Production-PUBLIC.txt (2016).
12. Carawan, R. E. and Merka, B., Poultry processors: You can reduce waste load and cut sewer surcharges, North Carolina Cooperative Extension Service, CD-22, March 1966, https://foodprocessing.ncsu.edu/documents/poultry_reduce_waste_load.pdf
13. Kiepper, B. H., Characterization of poultry processing operations, wastewater generation, and wastewater treatment using mail survey and nutrient discharge monitoring methods, MS Thesis, The University of Georgia (2003).
14. Plumber, H. S. and Kiepper, B. H., Impact of poultry processing by-products on wastewater generation, treatment, and discharges. In: *Proceedings of the 2011 Georgia Water Resources Conference*, The University of Georgia, Athens, GA, April 11–13, 2011.
15. Process Cooling, Wastewater treatment for chicken processing plants, September 1, 2003, www.processcooling.com/articles/83926-wastewater-treatment-for-chicken-processing-plants?v=preview (2016).
16. McMahon, J., Streamlining wastewater treatment in poultry processing, *Food Quality Magazine*, December/January, www.foodqualityandsafety.com/article/streamlining-wastewater-treatment-in-poultry processing/ (2007).
17. Rowenvironmental, Poultry processing plant wastewater treatment system, www.rowenvironmental.com/Rowenvironmental_Service.html. Rowenvironmental, Pittsburg, TX (2016).
18. The Poultry Site, Assessment of biological nitrogen removal in poultry processing facilities, August 26, 2011, www.thepoultrysite.com/articles/2146/assessment-of-biological-nitrogen-removal-in-poultry-processing-facilities (2016).
19. Smithson Mills, Inc., Development options for small-scale poultry processing facilities in Georgia, SMI, Asheville, NC (2012).
20. Wang, L. K., Wu, Z., and Shammas, N. K., Activated sludge processes. In: *Biological Treatment Processes*, Wang, L. K., Shammas, N. K., and Hung, Y. T. (eds), Humana Press, Totowa, NJ, 207–279 (2009).
21. Wang, L. K., Wu, Z., and Shammas, N. K., Trickling filters. In: *Biological Treatment Processes*, Wang, L. K., Shammas, N. K., and Hung, Y. T. (eds), Humana Press, Totowa, NJ, 371–428 (2009).
22. Shammas, N. K., Wang, L. K., and Wu, Z. Waste stabilization ponds and lagoons. In: *Biological Treatment Processes*, Wang, L. K., Shammas, N. K., and Hung, Y. T. (eds), Humana Press, Totowa, NJ, 315–370 (2009).
23. Shammas, N. K., Coagulation and flocculation. In: *Physicochemical Treatment Processes*, Humana Press, Totowa, NJ, 103–139 (2005).
24. Aulenbach, D. B., Shammas, N. K., Wang, L. K., and Marvin, R. C., Algae removal by flotation. In: *Flotation Technology*, Wang, L. K., Shammas, N. K., Selke, W. A., and Aulenbach, D. B. (eds), Humana Press, Totowa, NJ, 363–397 (2010).
25. Aries Chemical, Magnesium hydroxide for biological treatment of wastewater: Activated sludge and nitrification processes at a chicken processing plant, www.arieschem.com/water-treatment-success/using-magnesium-hydroxide.php (2016).
26. Hydro Flow, Poultry processing wastewater treatment system process description, http://www.hydroflotech.com/Typical%20Applications/Poultry%20Wastewater%20Tretment%20System/Poultry%20Wastewater%20Treatment%20System.htm (2016).
27. Yordanov, D., Preliminary study of the efficiency of ultrafiltration treatment of poultry slaughterhouse wastewater, *Bulgarian Journal of Agricultural Science*, 16 (6), 700–704 (2010).
28. McMahon, J., Streamlining wastewater treatment in poultry processing: Michigan Turkey & Lyco manufacturing team up to reduce BOD and TSS levels, Michigan Turkey Producers, MI, www.miturkey.com (2014).
29. Minnesota Technical Assistance Program, Wastewater reduction options for poultry processing plants. Fact sheet, University of Minnesota (2014).
30. Williams, C. L., Poultry waste management in developing countries, Poultry Development Review, Food and Agriculture Organization of the United Nations (UN FAO), www.fao.org/docrep/013/al715e/al715e00.pdf (2014).

3 Treatment of Wastewater, Storm Runoff, and Combined Sewer Overflow by Dissolved Air Flotation and Filtration

Lawrence K. Wang
Lenox Institute of Water Technology, Krofta Engineering Corporation, and Zorex Corporation

Mu-Hao Sung Wang
Lenox Institute of Water Technology

Nazih K. Shammas
Lenox Institute of Water Technology and Krofta Engineering Corporation

CONTENTS

3.1	Introduction	66
3.2	Combined Sewer Overflow Treatment	67
	3.2.1 CSO Treatment by Physicochemical Technologies	67
	3.2.2 CSO Treatment by Biological Technologies	68
	3.2.3 CSO Treatment by Disinfection Technologies	68
3.3	Wastewater Treatment	68
	3.3.1 Wastewater Treatment by Physicochemical Technologies	68
	3.3.2 Wastewater Treatment by Biological Technologies	69
3.4	Dissolved Air Flotation and Filtration Processes	70
	3.4.1 Dissolved Air Flotation	70
	3.4.2 DAF with Filtration	71
	3.4.3 Six-Month Investigation	72
	3.4.4 Summary and Conclusions	75
3.5	Case History: Treatment of an Oil Company Stormwater Runoff	76
	3.5.1 Pilot Flotation and Filtration System	76
	3.5.2 Evaluation of Treatment by Floatation and Filtration	76
	3.5.3 Posttreatment by GAC Adsorption	77
	3.5.4 Design and Operation of Full-Scale Wastewater Treatment System	79
	3.5.4.1 Oil–Water Separators	79
	3.5.4.2 DAF Clarifier	82
	3.5.4.3 Optional GAC Carbon Beds	82
	3.5.4.4 Summary of CSO Liquid Stream Treatment	83
	3.5.5 Management of Waste Sludge	84
	3.5.5.1 Waste Sludge and Residues	84
	3.5.5.2 Analysis of DAF Floated Sludge	85

	3.5.5.3	Thickening and Dewatering of Floated Sludge 86
	3.5.5.4	Extraction Procedure Toxicity of DAF Floated Sludges 88
3.6	Cost-Effectiveness of Stormwater Quality Controls... 88
	3.6.1	Offline Storage-Release Systems.. 88
		3.6.1.1 Surface Storage 89
		3.6.1.2 Deep Tunnels 89
	3.6.2	Swirl Concentrators 89
	3.6.3	Screens.. 89
	3.6.4	Sedimentation Basins 90
	3.6.5	Disinfection... 90
	3.6.6	Best Management Practices 90
		3.6.6.1 Detention Basins 90
		3.6.6.2 Retention Basins 91
		3.6.6.3 Infiltration Trenches.. 91
		3.6.6.4 Infiltration Basins... 91
		3.6.6.5 Sand Filters 92
		3.6.6.6 Water Quality Inlet 92
		3.6.6.7 Grassed Swales 92
		3.6.6.8 Assessment of BMP Control Performance 93
	3.6.7	O&M Costs for Controls.. 93
Nomenclature.. 95
References... 95

Abstract

Various conventional and innovative physiochemical, biological, and disinfection technologies for treating combined sewer overflow (CSO) and wastewater are reviewed and discussed. Special emphasis is placed on introduction and investigation of innovative dissolved air flotation (DAF) technology and DAF-filtration (DAFF) technology. The cost-effectiveness of stormwater quality control equipment (offline storage-release systems, swirl concentrators, screen, sedimentation basins, disinfection equipment, etc.), best management practices (BMP), and control costs are presented. A case history involving the removal of soluble arsenic (+5) from an oil company's storm run-off water by DAF, DAFF, granular activated carbon adsorption (GAC), and ion exchange (IX) processes was introduced in detail for the purpose of illustration. The best pretreatment unit was DAFF clarifier consisting of both flotation and filtration. DAFF consistently removed over 90% of arsenic, turbidity, and color, and over 50% of chemical oxygen demand (COD) and oil and grease. Using either DAFF or DAF for pretreatment, and then using either GAC or IX for second-stage treatment, the soluble arsenic in the storm water can be totally removed. The waste sludge generated from the DAF or DAFF cell did not possess any characteristic of extraction procedure toxicity (EPT) and thus was not hazardous. The dewatered sludge met the New Jersey Department of Environmental Protection's limits on cadmium, chromium, copper, nickel, lead, and zinc for ultimate land application.

3.1 INTRODUCTION

Combined sewer overflow (CSO) is a discharge of untreated wastewater from a combined sewer system at a point prior to the headwork of publicly owned treatment works. CSOs generally occur during wet weather (rainfall or snowmelt). During periods of wet weather, these systems become overloaded, bypass treatment works, and discharge directly to receiving waters. Combined sewer systems serve roughly 772 communities with about 40 million people in the United States. Most communities with CSOs are located in the Northeast and Great Lakes Regions [1].

The CSO Control Policy of the U.S. Environmental Protection Agency (U.S. EPA), published on April 19, 1994 [2], is the national framework for control of CSOs. The Policy provides guidance on how communities with combined sewer systems can meet Clean Water Act [3] goals in as flexible and cost-effective a manner as possible. The Policy contains four fundamental principles to ensure that CSO controls are cost-effective and meet local environmental objectives:

1. Clear levels of control to meet health and environmental objectives.
2. Flexibility to consider the site-specific nature of CSOs and find a cost-effective way to control them.
3. Phased implementation of CSO controls to accommodate a community's financial capability.
4. Review and revision of water quality standards during the development of CSO control plans to reflect the site-specific wet weather impacts of CSOs.

The first milestone under the CSO Policy was the January 1, 1997, deadline for implementing minimum technology-based controls (the "nine minimum controls") [4]. The nine minimum controls are measures that can reduce the prevalence and impacts of CSOs and that are not expected to require significant engineering studies or major construction. Communities with combined sewer systems are also expected to develop long-term CSO control plans that will ultimately provide for full compliance within the Clean Water Act, including attainment of water quality standards [5].

The high cost of traditional wastewater and CSO treatment has led to the need for developing alternative treatment technologies. However, sound design and operating criteria for innovative technologies are less available than those for well-established conventional technologies because there are not enough published engineering data to provide a basis for practical design.

There are two innovative technologies that are discussed in this chapter: dissolved air flotation (DAF) and combined dissolved air flotation and filtration (DAFF). The objectives of this chapter are (1) to review the current technologies for treatment of wastewater and CSO prior to their discharge to receiving waters, such as rivers, lakes, and seas; (2) to evaluate the feasibility of treating a typical domestic wastewater by DAF under both dry-to-wet and warm-to-cold weather conditions; (3) to evaluate the feasibility of treating the same typical domestic wastewater by both DAF and DAFF under dry-to-wet and warm-to-cold weather conditions; (4) to discuss the possible application of tertiary treatment by granular-activated carbon (GAC), when and if desired; (5) to present the management options for the handling of the generated sludges; and (6) to discuss the cost-effectiveness of stormwater quality controls.

3.2 COMBINED SEWER OVERFLOW TREATMENT

3.2.1 CSO Treatment by Physicochemical Technologies

Because of the adverse and intense flow conditions and unpredictable shock-loading effects, it has been difficult to adapt existing treatment methods to storm-generated overflows, especially the microorganism-dependent biological processes. Physical/chemical treatment techniques have shown more promise than biological processes in overcoming storm shock-loading effects. To reduce capital investments, projects have been directed toward high-rate operations approaching maximum loading. Storm-flow treatment methods demonstrated by a program sponsored by the U.S. Environmental Protection Agency (U.S. EPA) include physical, physical–chemical, wetlands, biological, and disinfection. These processes, or combinations of these processes, can be adjuncts to the existing wastewater treatment plants or serve as remote satellite facilities at the outfall [6–10].

Physicochemical treatment processes with or without chemicals, such as micro- and fine screens, swirl degritters, high-rate filters (HRFs), sedimentation, and DAF, have been successfully demonstrated by many environmental engineers. Physicochemical processes are important for storm-flow treatment because they are adaptable to automated operation, rapid startup and shutdown, high-rate operation, and resistance to shock loads. U.S. EPA states that the high-rate processes such as DAF, micro- and fine mesh screening, and HRF are ready for municipal application.

The microstrainer conventionally designed for polishing secondary wastewater plant effluent has successfully been applied to CSOs, and high-rate applications have given TSS (total suspended solids) removals higher than 90% rate. Full-scale microscreening units were demonstrated in two locations. In Syracuse, NY, TSS removals of about 50% were achieved. A Cleveland, Ohio, pilot study using 6 in. columns showed high potential for treating CSOs by a fine-screening HRF system. A large-scale (30 in. diameter) fine-screening HRF pilot system was evaluated in New York City for the dual treatment of dry- and wet-weather flows. Removals of TSS and BOD were 70% and 40%, respectively.

Results from a 5 MGD screening and DAF demonstration pilot plant in Milwaukee, WI, indicate that more than 70% removals of BOD and TSS are possible. By adding chemical coagulants, 85%–97% phosphate reduction can be achieved as an additional benefit. Screening-DAF prototype systems (20 and 40 MGD) have been demonstrated. Removals of TSS and BOD were 70% and 50%, respectively. Treatment processes such as microscreens and DAF are now being used by municipalities.

The swirl degritter has also been developed for grit removal. The small size, high efficiency, and absence of moving parts offer economical and operational advantages over conventional degritting facilities. A full-scale demonstration of a (16 ft diameter and 11 MGD design flow rate) swirl degritter has been completed in Tamworth, Australia. Removal efficiencies confirmed laboratory results. Compared with a conventional grit chamber, construction costs are halved, and operation and maintenance costs are considerably lower.

3.2.2 CSO Treatment by Biological Technologies

The biological processes such as trickling filtration, contact stabilization, rotating biological contactors (RBC), and lagoons have been demonstrated. They have had positive evaluation, and with the exception of long-term storage lagoons, they all must operate conjunctively with dry-weather flow plants to supply biomass, and require some form of flow equalization.

3.2.3 CSO Treatment by Disinfection Technologies

Because disinfectant and contact demands are great for storm flows, research has centered on high-rate applications by static and mechanical mixing, higher disinfectant concentrations, and more rapid oxidants, that is, chlorine dioxide, ozone, and ultraviolet (UV) light.

Demonstrations in Rochester and Syracuse, NY; East Chicago, Indiana; and Philadelphia, PA, indicate that adequate reduction of fecal coliforms can be obtained with contact times of 2 min or less by induced mixing. Dosing with a contact time of less than 10 s was conducted in New York City.

A hypochlorite batching facility is still in use in New Orleans, LA, to protect swimming beaches in Lake Ponchartrain. U.S. EPA supported the development of a brine hypochlorite generator now being used in industry.

3.3 WASTEWATER TREATMENT

3.3.1 Wastewater Treatment by Physicochemical Technologies

It is apparent that any technologies feasible for wastewater treatment will also be feasible for CSO treatment because CSO is mainly a combination of domestic wastewater and storm water.

There are large varieties of unit operations and processes that are suitable for domestic wastewater treatment. However, only certain unit operations and processes are repeatedly applied in wastewater treatment, as they have proven to be successful in accomplishing the required regulatory standard limits for wastewater discharge.

A physicochemical wastewater treatment system generally consists of the following process components [11–13]:

1. Preliminary wastewater handling and treatment facilities (pumps, screens, grinders, grit chambers, samplers, etc.)
2. Chemical feeders and rapid mixing tanks
3. Flocculation basins (with 45–60 min detention time)
4. Sedimentation tanks (with 2–4 h detention time), or DAF clarifiers (with 5–15 min detention time)
5. Filtration beds
6. GAC beds
7. Chlorination chambers (with 30–60 min retention time)
8. Effluent structure

Physicochemical treatment is mainly applied in industrial wastewater treatment applications due to the toxic substances present in the wastewater (heavy metals, odor-causing substances, oil, grease, foaming substances, etc.), which could not be handled by biological means. Also, the presence of a high percentage of insoluble BOD can justify a physicochemical treatment. In these cases, a non-biological process is technically and economically feasible even when sedimentation is adopted for clarification. Niagara Falls Wastewater Treatment Plant in New York state adopts this physicochemical treatment technology for treating its combined domestic and industrial wastewater 72 MGD flow.

Dissolved air flotation has been developed as a replacement of sedimentation clarification, which rendered the capital and operations and maintenance (O&M) costs of a physicochemical treatment system to become affordable. This innovative physicochemical treatment system involving the use of DAF alone and the use of DAF + DAFF will be discussed in detail in this chapter.

3.3.2 WASTEWATER TREATMENT BY BIOLOGICAL TECHNOLOGIES

A conventional biological wastewater treatment system generally consists of the following components incorporating physical [11–13] and biological processes [14,15]:

1. Preliminary wastewater handling and treatment facilities (pumps, screens, grinders, grit chambers, samplers, etc.)
2. Primary sedimentation clarifiers (2–4 h detention time) or DAF clarifiers (5–15 min detention time)
3. Biological reactors:
   - Activated sludge (typically about 6 h detention time)
   - Extended aeration (typically about 24 h detention time)
   - Trickling filters (large area and long detention time)
   - Rotating biological contactors (long detention time)
4. Secondary sedimentation clarifiers (2–4 h detention time) or DAF clarifiers (5–15 min detention time)
5. Optional tertiary filtration
6. Chlorination chambers
7. Effluent structure

Biological processes are capable of successfully treating wastewater containing mainly biodegradable pollutants, especially in dissolved form, by either removing or reducing them to an acceptable level before its discharge to receiving waters. The application of a biological process to municipal domestic wastewater treatment is common, although the cost is not low.

Again DAF has been developed as a replacement of sedimentation clarification, which brought down the capital and O&M costs of a biological treatment system. Such innovative biological systems involving the use of primary DAF clarification and/or the use of secondary DAF or DAFF clarification are also discussed in detail in this chapter.

3.4 DISSOLVED AIR FLOTATION AND FILTRATION PROCESSES

3.4.1 Dissolved Air Flotation

The primary DAF unit used in the investigation (see Figure 3.1) was 4 ft (1.20 m) in diameter, less than 2 ft (0.56 m) in depth, and constructed with stainless steel [16]. The DAF unit was installed at Lee Wastewater Treatment Facility in Lee, MA, for a 6-month period.

A pressure pump draws from the DAF clarifier effluent at a regulated flow to be used as recycle water. This water flow is pressurized and fed into the air dissolving tube (ADT), saturated with compressed air, and released (decompressed) into the influent stream as it enters the primary flotation unit. The recycle flow was 20%–25% of the influent flow. After pressure release, the pressurized aerated water from ADT (or other air dissolving means) is mixed with the chemically treated raw water just before the raw water inlet. The combined ADT effluent and the chemically treated raw water enter the primary flotation unit where the generated fine bubbles attach to the suspended particles in the wastewater and float them to the wastewater surface as floated sludge. The spiral scoop removes the floated sludge and discharges it to the sludge drain pipe. The subnatant of the primary flotation unit becomes the DAF effluent [17–19].

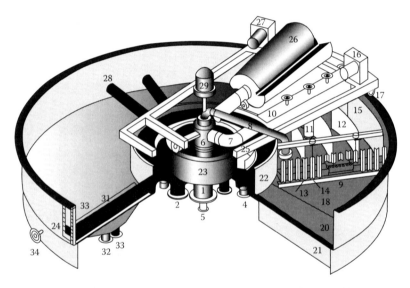

1. Raw water inlet
2. Clarified water outlet
3. Floated sludge outlet
4. Clarified water recycle outlet
5. Pressurized water inlet
6. Rotary joint
7. Rubber pipe connection
8. Pressurized water piping
9. Pressurized water distribution header
10. Raw water distribution header
11. Distribution header outlet pipes
12. Flow control channels
13. Turbulence reduction baffles
14. Adjustable height baffle attachment
15. Flow control channel outer wall
16. Rotating carriage gearmotor drive
17. Carriage drive wheel
18. Wheel support RIM
20. Tank wall
21. Tank floor support structure
22. Rotating clarified water containment wall
23. Sludge well
24. Level control overflow weir
25. Rotating carriage structure
26. Revolving spiral scoop
27. Spiral scoop gearmotor drive
28. Clarified water extraction pipes
29. Electrical slip ring
30. Tank window
31. Sediment removal sump
32. Final drain
33. Sediment purge outlet
34. Level control adjustment handwheel

FIGURE 3.1 Details of the flotation cell (DAF). (From Krofta, M., Wang, L.K., *Flotation Engineering*, Technical manual no. Lenox/1-06-2000/368, Lenox Institute of Water Technology, Lenox, MA, 2000.)

3.4.2 DAF with Filtration

The secondary DAFF unit used in this investigation (see Figure 3.2) was also installed at Lee Wastewater Treatment Facility in Lee, MA, for a 6-month period.

The secondary DAFF unit provides both flotation and filtration clarification [16]. The main floatation tank is 5 ft (1.52 m) in diameter with 3 ft (0.90 m) effective water depth (flotation zone). Beneath the flotation tank are 23 filter segments utilizing dual media 7 in. of anthracite (1.0 mm effective diameter) on top of 7 in. of sand (0.35 mm effective diameter). The unit is provided with a carriage-mounted backwash system with suspended hood, which allows for backwashing of each of the filter segments individually while the balance of the segments are providing filtration. The backwash waste stream can be directed to return to the base of the internal flocculation tank for reprocessing by flotation, or diverted to waste. Floated solids are collected by the spiral scoop and directed to the central sludge cone for discharge from the DAFF clarifier. A portion of the clarified

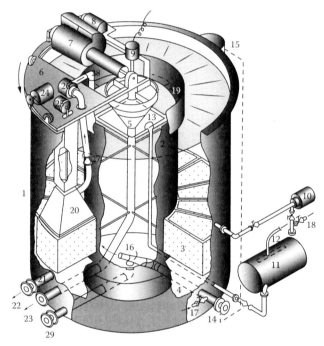

1. Outside tank
2. Inside flocculation tank
3. Sandbed assembly with screen
4. Tank bottom
5. Sludge collection funnel
6. Moveable carriage assembly
7. Sprial scoop
8. Scoop variable speed drive
9. Electrical rotary contact
10. Pressure pump
11. Air dissolving tube
12. Compressed air addition point
13. Aerated water distribution pipes
14. Raw water inlet regulating valve
15. Tank level control sensor
16. Raw water inlet jet nozzle
17. Coagulant addition point
18. Polyelectrolyte addition point
19. Deflector ring into flotation tank
20. Backwash hood assembly
21. Clarified water pipeline
22. Clarified water flow regulating valve
23. Floated sludge discharge pipe
24. Main carriage drive
25. Motor to lift backwash hood assembly
26. Backwash suction pump
27. Check valve (baskflow preventor)
28. Dirty backwash water recycle pipe
29. Drain line

FIGURE 3.2 Details of flotation/filtration cell (DAFF). (From Krofta, M., Wang, L.K., *Flotation Engineering*, Technical manual no. Lenox/1-06-2000/368, Lenox Institute of Water Technology, Lenox, MA, 2000.)

water is directed to a pressure pump for feeding to the ADT, saturated with compressed air, and returned as recycle flow to be released (decompressed) near the top of the flocculation tank as the source of the pressurized and aerated water stream for flotation [20–22].

3.4.3 Six-Month Investigation

The 6-month (27 weeks) pilot test was conducted from June 8 to December 7. A flow diagram of the existing Lee Wastewater Treatment Facility plus the location of the DAF and DAFF process equipment is shown in Figure 3.3.

The modular clarification system containing primary DAF and secondary DAFF in sequence was located on-site in a trailer next to the Lee Wastewater Treatment Facility's grit chamber. The domestic wastewater was degritted (grit chamber) before it was pumped to the primary DAF and secondary DAFF clarification units in the trailer. The effluents from both DAF and DAFF were sampled for analyses before being returned together with the floated sludge to the Lee Wastewater Treatment Facility for disposal.

The DAF and DAFF flow rate was 40 gpm and the recycle rates for the primary DAF and secondary DAFF clarifiers were 25% and 20%, respectively. The hydraulic loading rates were 5.2 gpm/ft^2 for the DAF unit and 3.5 gpm/ft^2 for the DAFF unit. Table 3.1 shows an overview of the operational parameters applied.

A flocculant was needed in the flotation unit (DAF) to remove the larger suspended material. For the secondary flotation–filtration unit (DAFF), a coagulant/flocculant dosing combination was needed to achieve the best removals. Table 3.2 shows the optimal chemicals and dosage ranges.

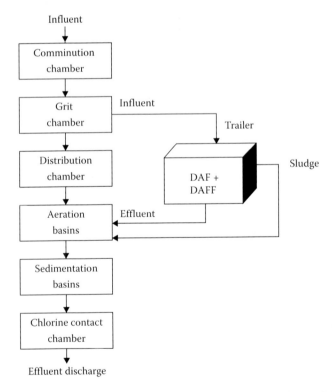

FIGURE 3.3 Flow diagram of Lee WWTF with pilot plant location. (From Wang, L.K., et al., Treatment of domestic sewage and combined sewer overflow by dissolved air flotation and filtration, *International Conference of Water, Asia 2000*, India, September 2000; Krofta, M., et al., Treatment of domestic sewage and combined sewer overflow by dissolved air flotation and filtration, Technical report no. Lenox/Juiy-23-2000/388, July 23, 2000.)

TABLE 3.1
Operational Parameters of Treatment Units

Parameter	DAF	DAFF
Tank diameter, m (ft)	1.20 (4)	1.52 (5)
Tank depth, m (ft)	0.56 (2)	0.90 (3)
Influent flow, L/min (gal/min)	151 (40)	151 (40)
Recycle flow, L/min (gal/min)	38 (10)	30 (8)
ADT pressure in/out, bar (psi)	5.9/5.2 (86/76)	6.5/5.3 (96/78)
ADT air feed, L/min (SCFH)	2.36 (5)	2.36 (5)
Carriage, min/rev	1:50	—
Scoop speed, rpm	2	—
Backwash flow, L/min (gal/min)	—	39.7 (10.5)
Backwash time, s	—	50
Backwash frequency	—	1:12
Backwash recycle, %	—	100
Head loss, cm (in.)	—	25.4–45.7 (10–18)
Hydraulic loading rate, L/min/m^2 (gal/min/ft^2)	215.71 (5.3)	142.5 (3.5)

Source: Wang, L.K., et al., Treatment of domestic sewage and combined sewer overflow by dissolved air flotation and filtration, *International Conference of Water, Asia 2000*, India, September 2000.

TABLE 3.2
Chemical Dosage after Optimization

Unit	Chemical	Dosage Range (mg/L)
DAF	Magnifloc 496C	1–2
DAFF	Ferric sulfate	50–60
DAFF	Magnifloc 496C	1–2

Source: Wang, L.K., et al., Treatment of domestic sewage and combined sewer overflow by dissolved air flotation and filtration, *International Conference of Water, Asia 2000*, India, September 2000.

Laboratory analysis was conducted daily on composite samples of raw influent, primary DAF unit effluent, and secondary DAFF unit effluent. Testing consisted of pH, TSS, chemical oxygen demand (COD), and biological oxygen demand (BOD). Testing for phosphorous, TKN, NO_3–N, NO_2–N, NH_3–N, and Sludge TS (on DAF and DAFF samples) was performed on a weekly basis. The composition of the raw municipal wastewater (collected after the grit chamber) is presented in Table 3.3.

In the course of the pilot study, adjustment to the chemical dosage had to be made due to variations in flow and influent composition. Various chemicals were tried during the study, which explains the occurrence of some inconsistent values in the results. Overall, removal of efficiencies do not show any relation to climate. Since the 6-month testing period took place during the summer and winter months, variations in the influent wastewater quality are expected.

The large variations in the early stage of this study were due to the extreme fluctuations in the incoming raw wastewater feed. These fluctuations were due to the flow variations in the sewer collection mains. Cleaning of these sewer collection mains had also been performed by the Town of Lee during this study period. Especially the second through sixth week showed significant absolute and relative peak values.

It can be seen that primary flotation DAF reduced these parameters very efficiently to a constant low level, regardless of how high the influent value was. The overall performance of the modular

TABLE 3.3
Raw Influent (Grit Chamber Effluent) Wastewater Parameters

Parameter	Low	Average	High
pH	6.97	7.29	8.11
TSS, mg/L	98	315	1191
COD, mg/L	251	585	1196
BOD, mg/L	116	247	497
Total P, mg/L	1.6	5.2	12.8
TKN, mg/L	25	40	77
NO_3–N, mg/L	0.7	2.4	5.3
NO_2–N, mg/L	0.011	0.020	0.049

Source: Wang, L.K., et al., Treatment of domestic sewage and combined sewer overflow by dissolved air flotation and filtration, *International Conference of Water, Asia 2000*, India, September 2000.

Note: Temperature (°C) = 15–20.

dissolved air flotation–filtration system (DAF + DAFF) was very good, effectively reducing TSS (99.6%), COD (74%), and BOD (76%) levels in a relatively short treatment time.

Removal efficiencies of total P reached an average value of 83%, most of it being reduced by secondary flotation–filtration (DAFF) through the precipitation of phosphorus by the application of a coagulant.

The aim of primary DAF and secondary DAFF was for clarification, and they were not designed for nitrogen removal. Any reductions in TKN, NH_3–N, NO_2–N, and NO_3–N were coincidental accomplishments.

The efficient removal of nitrate–nitrogen was mainly achieved by secondary DAFF clarification and reached an overall average reduction of 65%. NH_3–N was effectively reduced by 56%. Average removals for TKN and NH_3–N were 36% and 18%, respectively.

A summary of the results from the pilot study is presented in Table 3.4, showing the average results from the entire 6-month testing period.

TABLE 3.4
Evaluation of DAF/DAFF for Wastewater Treatment: Summary of Test Results

Parameter	Raw Low	Raw Average	Raw High	DAF Low	DAF Average	DAF High	DAFF Low	DAFF Average	DAFF High	Total Reduction Average
pH	6.9	7.32	8.1	6.8	7.40	7.9	6.2	7.30	7.8	—
TSS, mg/L	98	315	1191	15	73	240	ND	4	23	99.6%
COD, mg/L	251	586	1196	142	290	475	74	145	342I	74%
BOD_5, mg/L	116	247	497	68	122	222	30	59	127	76%
Total P, mg/L	1.60	5.20	12.80	1.10	4.10	5.80	0.04	0.98	3.30	83%
TKN, mg/L	25.0	40.0	77.0	23.0	32.0	61.0	8.4	27.0	50.0	36%
NH_3–N, mg/L	20	23	31	17	21	26	10	19	23	18%
NO_2–N, mg/L	0.011	0.020	0.049	0.010	0.050	0.370	ND	0.020I	0.060	56%
NO_3–N, mg/L	0.70	2.40	5.30	0.10	1.50	10.60	ND	0.74	8.60	65%
Sludge TS, mg/L	—	—	—	1,800	12,400	30,000	200	4,400	33,300	—

Source: Wang, L.K., et al., Treatment of domestic sewage and combined sewer overflow by dissolved air flotation and filtration, *International Conference of Water, Asia 2000*, India, September 2000.

3.4.4 Summary and Conclusions

Since the CSO is a mixture of wastewater and storm water, the wet weather flow in this investigation is a good simulation of CSO. Hence, a successful treatment would indicate the feasibility of treating domestic wastewater as well as CSO by DAF and/or DAFF. Based on the information and review presented earlier, the following conclusions are presented [23,24]:

1. A compilation of the U.S. EPA program's best research efforts in CSO treatment and control over its 18 years' duration has concluded that physical/chemical treatment techniques have shown more promise than biological processes in overcoming storm shock-loading effects. To reduce capital investments, projects have been directed toward high-rate operations approaching maximum loading.
2. Physicochemical treatment processes with or without chemicals, such as micro- and fine screens, swirl degritters, HRF, sedimentation, and DAF, have been successfully demonstrated in CSO management. Physicochemical processes are adaptable to automated operation, rapid startup and shutdown, high-rate operation, and resistance to shock loads.
3. The U.S. EPA states that the high-rate processes such as DAF, micro- and fine mesh screening, and HRF are ready for municipal application.
4. The results from a 5 MGD screening and DAF demonstration pilot plant in Milwaukee indicate that greater than 70% removal of BOD and TSS is possible. By adding chemical coagulants, 85%–97% phosphate reduction can be achieved as an additional benefit. Screening-DAF prototype systems (20 and 40 MGD) have been demonstrated. Removals of TSS and BOD were 70% and 50% respectively.
5. The swirl degritter has been developed for grit removal in CSO treatment. The small size, high efficiency, and absence of moving parts offer economical and operational advantages over conventional degritting facilities. A full-scale demonstration of a swirl degritter (16 ft diameter and 11 MGD flow rate) has been completed in Tamworth, Australia. Removal efficiencies confirmed laboratory results. Compared with a conventional grit chamber, construction costs are halved, and operation and maintenance costs are considerably lower.
6. High disinfectant concentrations and more rapid oxidants, that is, chlorine dioxide, ozone, and UV, light are considered effective in CSO management. Both ozone and UV can be readily used in conjunction with DAF and DAFF processes.
7. The 72-MGD physicochemical wastewater treatment system as practiced at Niagara Falls Wastewater Treatment Facility, Niagara Falls, NY, can be further improved by using DAF (5–15 min detention time) to replace sedimentation (2–4 h detention time) for clarification.
8. A conventional wastewater treatment system as practiced at the 1-MGD Lee Wastewater Treatment Facility, Lee, MA, and many other municipalities around the world, can be easily improved by using DAF primary clarification (instead of sedimentation) and DAF or DAFF secondary clarification (instead of sedimentation secondary clarification).
9. The 6-month on-site investigation at the Lee Wastewater Treatment Facility demonstrated the feasibility of using primary DAF clarification alone for primary treatment of municipal wastewater; 76.6% of TSS, 48% of COD, 49% of BOD_5, 25% of total P, and 20% of TKN were removed from raw municipal wastewater by DAF when using a polymer (1–2 mg/L dosage).
10. The feasibility of using secondary flotation–filtration clarification (DAFF) for secondary municipal wastewater treatment has also been established at the Lee Wastewater Treatment Facility by its excellent reductions of TSS (98.2%), COD (48%), BOD_5 (51%), total P (75%), and TKN (22%).
11. The combination of primary and secondary DAFF clarification was able to reduce 99% of TSS, 74% of COD, and 76% of BOD_5 from Lee municipal wastewater. Therefore, this could be an excellent technology for emergency municipal wastewater treatment, wastewater

treatment facility upgrading (replacing existing sedimentation clarifiers), storm runoff treatment, or commercial wastewater pretreatment.
12. A master research work completed by Ms. Marika S. Holtorff [25] at the Lenox Institute of Water Technology concluded that a package plant consisting of primary DAF, 4-h activated sludge (4000 mg/L MLSS), and DAFF removed 99.9% TSS, 90% BOD, 98% ammonia nitrogen, and 94% phosphate.
13. GAC would be an excellent tertiary treatment for removal of any toxic substances from wastewater or CSO. The addition of GAC filtration as a tertiary treatment stage can produce a colorless and odorless effluent, free of bacteria and viruses, with a BOD of less than 1 mg/L and a COD of less than 20 mg/L [26]. The application of GAC is successfully incorporated in the physicochemical treatment design of the Niagara Falls Wastewater Treatment Facility.
14. The feasibility of removing soluble arsenic (+5) and other conventional pollutants from combined storm runoff and process wastewater by oil–water separation, DAF, filtration, and GAC was fully demonstrated for the Imperial Oil Company in Morganville, NJ [27].

3.5 CASE HISTORY: TREATMENT OF AN OIL COMPANY STORMWATER RUNOFF

The feasibility of removing pollutants from combined stormwater runoff and process wastewater by oil–water separation, DAF, filtration, and GAC adsorption was pilot-tested for an oil blending company in a Northeastern State. This first step was followed by the design, construction, and operation of the company's stormwater pollution control system [27,28].

3.5.1 Pilot Flotation and Filtration System

The pilot plant, as shown in Figure 3.4, consists of a rapid mixing chamber, a rectangular static hydraulic flocculator ($L \times W \times H$ = 102 in. × 16 in. × 10 in.) (259.08 cm × 40.64 cm × 25.40 cm), a DAF unit (diameter 0.91 m [3 ft], depth 55.88 cm [22 in]), and three sand filters (28 cm of sand as filter bed). Table 3.5 shows the typical pilot plant operational conditions. The pilot plant influent is a combined storm runoff water and process water, pretreated by the existing API oil–water separators, and spiked with soluble arsenic when necessary.

Both flotation cell and filter effluents were sampled for various analyses including arsenic, COD, oil and grease (O&G), and pH. The sludge produced from the DAF unit was collected in a storage tank for suspended solids analysis and proper sludge thickening/dewatering.

Three pilot plant demonstrations were conducted as follows:

1. *Experiment one*: Continuous treatment of oil–water separator effluent (As 0.015 mg/L) by DAF and sand filtration at 9 gpm for 2.5 h.
2. *Experiment two*: Continuous treatment of contaminated oil–water separator effluent (spiked with 0.5 mg/L of soluble arsenic; As = 0.518 mg/L) by DAF and sand filtration at 9 gpm, for 2.5 h.
3. *Experiment three*: Continuous treatment of contaminated oil–water separator effluent (spiked with 1.0 mg/L of soluble arsenic; As = 1.010 mg/L) by DAF and sand filtration at 9 gpm, for 2.5 h.

3.5.2 Evaluation of Treatment by Floatation and Filtration

The results of the three pilot plant demonstration experiments for the pollutant removal efficiencies of DAF and sand filtration are shown in Table 3.6.

The results clearly demonstrate that for removal of arsenic from the wastewater, the arsenic removal efficiency by DAF ranged from 33.3% to 90.1%, depending on the influent arsenic

Treatment of Wastewater, Storm Runoff, and Combined Sewer Overflow

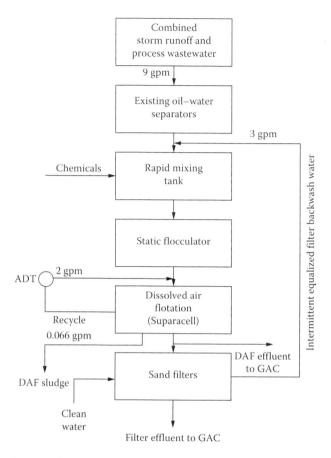

FIGURE 3.4 Flow diagram of pilot plant. (From Wang, L.K., et al., *Water Treat.*, 9, 223–233, 1994.)

concentration. The arsenic removal efficiency of flotation and filtration combination, however, was always over 90%.

There was a significant difference in water clarity between the DAF and DAFF effluents. The former was yellowish and turbid, while the latter was crystal clear. Quantitatively, DAF removed 30.0%–45.7% of turbidity and 40.8%–50.5% of color, while DAFF removed over 93% of both turbidity and color (see Table 3.6). The treatment efficiency of DAF and sand filtration for COD and O&G removals was also consistently higher than that of DAF alone.

In summation, DAF alone proved to be a sufficient treatment if the aim is to remove arsenic, assuming influent arsenic concentration is below 1000 μg/L and the desired effluent arsenic concentration is equal to or less than 200 μg/L. If the influent arsenic concentration is occasionally higher than 1000 μg/L and/or the effluent arsenic concentration requirement is more stringent (equal to or less than 100 μg/L As), the use of DAF and sand filtration system is required. As demonstrated below, GAC beds can be used as a posttreatment if total arsenic removal is required.

3.5.3 Posttreatment by GAC Adsorption

Six various combinations of samples from effluents of Experiments One, Two, and Three designated A, B, C, D, E, and F, as shown in Table 3.7, were prepared for arsenic removal by a continuous carbon adsorption process. The efficiency of carbon adsorption for removal of color, O&G, and soluble chemical oxygen demand (SCOD) was also explored.

TABLE 3.5
Operational Conditions of the Pilot Plant

Characteristic	Value, Metric Unit (English Unit)
Chemicals	15 mg/L, sodium aluminate (as Al_2O_3)
	15 mg/L, ferric chloride or ferric sulfate (as Fe)
Influent flow	39.62 m^3/h (9 gpm)
DAF effluent flow	39.33 m^3/h (8.934 gpm)
DAF sludge flow	0.29 m^3/h (0.066 gpm)
DAF recycle flow	8.8, m^3/h (2 gpm)
Filter backwash	13.21 m^3/h (3 gpm)
Mixing detention time	0.425 min between filter backwashes
	0.318 min during filter backwash
Flocculation detention time	7.84 min between filter backwashes
	5.88 min during filter backwash
Flotation (DAF) detention time	7.12 min between filter backwashes
	5.59 min during filter backwash
DAF overflow	4.28 m^3/h/m^2 (1.75 gpm/ft^2) between backwashes
	5.42 m^3/h/m^2 (2.22 gpm/ft^2) during backwash
Pressure of air dissolving tube	90 psig
Air flow rate	0.02832 m^3/h (0.386 gal/h)
Sand filtration rate	4.88, m^3/h/m^2 (2.0 gpm/ft^2) between backwashes
	6.54 m^3/h/m^2 (2.68 gpm/ft^2) during backwash
Sand specifications	Effective size = 0.85 mm
	Uniformity coefficient = 1.55
Sand bed depth	28 cm (11 in. of sand)
Sand filter backwash rate	28.12 m^3/h/m^2 (11.5 gpm/ft^2)

Source: Wang, L.K., et al., Treatment of domestic sewage and combined sewer overflow by dissolved air flotation and filtration, *International Conference of Water, Asia 2000*, India, September 2000.

A 3 cm ID × 100 cm Plexiglas column was filled with 20 × 40 GAC. The column was fitted at the base with a flow distributor, which also served to retain the granular carbon in the column. The carbon was thoroughly wetted before introduction into the column to ensure against the entrapment of air in the carbon bed.

Five and the half liters each of the six feed solutions were separately processed by passing them through the carbon column at a controlled flow rate of 70.65 mL/min (2 gpm/ft^2). Each test was continued for a period of 45 min. Ten min after initiation of flow and at least five times (every 10 min) during the testing period, effluent samples were collected for determination of arsenic and other parameters shown in Tables 3.7 and 3.8.

Table 3.7 indicates that although DAF and DAFF are excellent pretreatment processes for arsenic removal, GAC adsorption is an efficient second-stage treatment unit for complete arsenic removal. The effluent arsenic concentrations were consistently zero in all six tested samples. Table 3.8 indicates that GAC adsorption is an excellent second-stage treatment unit for color and SCOD removal. The oil and grease removal, however, is not significant.

A synthetic storm water containing 0.5 mg/L of soluble arsenic (As) was prepared using soluble arsenic stock solution and tap water for granular carbon adsorption breakthrough study. Table 3.9 shows the characteristics of the tap water that was used in the preparation of synthetic storm water.

TABLE 3.6
Treatment Efficiency of Dissolved Air Flotation and Sand Filtration

Treatment Efficiency	Experiments One	Two	Three
Influent arsenic, mg/L	0.015	0.518	1.010
Removal by flotation, %	33.3	79.5	90.1
Removal by flotation/filtration, %	100.0	90.3	90.6
Influent turbidity, NTU	3.4–3.6	3.5	3.0
Removal by flotation, %	35.7	45.7	30.0
Removal by flotation/filtration, %	93.0	93.7	93.3
Influent color, unit	48–49	49	50
Removal by flotation, %	50.5	40.8	43.0
Removal by flotation/filtration, %	98.9	97.9	98.0
Influent O&G, mg/L	29.2	29.3	28.5
Removal by flotation, %	52.7	51.5	43.2
Removal by flotation/filtration, %	69.2	61.1	74.7
Influent COD, mg/L	85	86	83
Removal by flotation, %	37.6	46.5	32.5
Removal by flotation/filtration, %	58.8	53.5	51.8
Influent pH, unit	7.0–7.1	7.1–7.1	7.2–7.2
After flotation, unit	6.9–7.0	7.1–7.2	6.9–7.1
After flotation/filtration, unit	6.9–7.0	7.0–7.1	7.0–7.1

Source: Wang, L.K., et al., *Water Treat.*, 9, 223–233, 1994.
Notes: Chemical flotation treatment: Sodium aluminate, 15 mg/L as Al_2O_3, and either ferric chloride or ferric sulfate, 15 mg/L as Fe; Sand specifications: ES = 0.85 mm, UC = 1.55, depth = 11 in.

Tap water was used instead of distilled water to simulate storm water containing soluble arsenic as well as common mineral substances.

The synthetic storm water was continuously passed through the 100 cm activated carbon column at a flow rate of 70.65 mL/min (10.47 mL/min/cm^2). The column effluent was periodically collected at 10 min intervals for residual arsenic (As) determination. The experimental procedure is similar to that of the previous carbon adsorption posttreatment study, except that the combined storm runoff and process water containing 0.5 mg/L of soluble arsenic was used and the period for the GAC breakthrough study was 900 min instead of 45 min.

The granular carbon adsorption breakthrough curve is shown in Figure 3.5. The effluent arsenic concentration remained at zero value (arsenic-free effluent) up to 600 min (10 h) of operation, after which the effluent arsenic concentration started to rise. The technical feasibility of using GAC for removal of soluble arsenic from water is thus positively demonstrated.

3.5.4 Design and Operation of Full-Scale Wastewater Treatment System

3.5.4.1 Oil–Water Separators

The oil blending company in a northeastern state adopted the process system that was developed in the pilot plant study. The full-scale system for treatment of combined storm runoff and process wastewater is illustrated in Figure 3.6.

The influent CSO is pumped to three American Petroleum Institute (API) oil–water separators connected in series (see Table 3.10). The API has standard design procedures for gravity differential types of oil–water separators. The design is based on detaining wastewater for a period of time sufficient to allow free oil to rise above the tank outlet and be removed at the surface.

TABLE 3.7
Granular Activated Carbon Adsorption Test Results

	GAC Test No./ Sample No. and Description	Arsenic, mg/L As Before GAC	After GAC
A.	Combined storm runoff and process wastewater, pretreated by oil–water separator, (Initial As = 0.015 mg/L), and treated by dissolved air-flotation	0.01	0
B.	Combined storm runoff and process wastewater, pretreated by oil–water separator, (Initial As = 0.015 mg/L), then treated by dissolved air flotation and sand filtration	0	0
C.	Combined storm runoff and process wastewater, pretreated by oil–water separator, spiked with 0.5 mg/L soluble As (Initial As = 0.518 mg/L), then treated by dissolved air flotation	0.106	0
D.	Combined storm runoff and process wastewater, pretreated by oil–water separator, spiked with 0.5 mg/L soluble As (Initial As = 0.518 mg/L), then treated by dissolved air flotation and sand filtration	0.05	0
E.	Combined storm runoff and process wastewater, pretreated by oil–water separator, spiked with 1.0 mg/L soluble As (Initial As = 1.010 mg/L), then treated by dissolved air flotation	0.10	0
F.	Combined storm runoff and process wastewater, pretreated by oil–water separator, spiked with 1.0 mg/L soluble As (Initial As = 1.010 mg/L), then treated by dissolved air flotation and sand filtration	0.095	0

Source: Wang, L.K., et al., *Water Treat.*, 9, 223–233, 1994.
Note: Residual As concentration of GAC effluent was zero in the entire testing period.

TABLE 3.8
Removal of Color, O&G, and SCOD by Granular Activated Carbon Adsorption

Test or Sample No.	Before GAC Adsorption				After GAC Adsorption			
	pH Unit	Color Unit	O&G mg/L	SCOD mg/L	pH Unit	Color Unit	O&G mg/L	SCOD mg/L
A	6.9	25	13.8	53	6.3	0	9.6	12
B	6.9	1	9	35	6.3	0	0	28
C	7.1	30	14.2	46	6.4	0	10.0	15
D	7.0	1	11.4	40	6.3	0	9.5	23
E	6.9	32	16.2	56	6.5	0	10.2	14
F	7.0	2	7.2	40	6.6	0	2.5	25

Source: Wang, L.K., et al., *Water Treat.*, 9, 223–233, 1994.

TABLE 3.9
Characteristics of Tap Water

Parameters	Water Quality
Temperature, F	57
pH	7.2
Turbidity, NTU	0.3
Color, unit	0
Aluminum, mg/L	0.07
Total Alkalinity, mg/L CaCO$_3$	60.0
Humic Substances, mg/L	0.8
Total Hardness, mg/L CaCO$_3$	65.0
Total Suspended Solids, mg/L	4.0
Calcium, mg/L	18.0
Copper, mg/L	0.0
Sodium, mg/L	2.2
Iron, mg/L	0.08
Manganese, mg/L	0.00
Silica, mg/L SiO$_2$	2.5
Sulfate, mg/L SO$_4$	10.0
Chloride	5.5
Specific Cond. micromhos/cm	90.0
Nitrogen-Ammonia, mg/L	0.0
Nitrogen-Nitrate, mg/L	0.0
Nitrogen-Nitrite, mg/L	0.0

Source: Wang, L.K., et al., *Water Treat.*, 9, 223–233, 1994.

The separators' capabilities were evaluated by computing the retention time in each chamber. Retention time is based upon the free oil rising above the chamber's baffle opening to be removed. A corresponding flow rate was then computed.

Among the oils handled by the oil blending company, the maximum specific gravity is 0.9071. Using the API equation for the rate of rise of a 0.015 cm diameter oil globule, this specific gravity and an operating water temperature of 40 F yield an oil rise rate of 0.12 ft/min (3.65 cm/min).

To be removed, oil must rise above the baffle opening at a chamber's discharge end. Oil–water separator No. 1 has a center wall with an eight in. (20.32 cm) deep trough passing beneath it. The wall's bottom is at the same elevation as the tank bottom; theoretically, there should be no free oil passing from this chamber. Any oil passing this point would have to rise 18 in. (45.72 cm) to be removed by the next baffle system. At the oil rise rate to be removed (0.12 ft/min), the time to rise 18 in. is 12.5 min (1.5 ft/0.12 ft/min = 12.5 min).

Assuming the center wall divides the volume of oil–water separator No 1 equally, its second chamber has a volume of 938 ft^3 (26,556 L) at a 4 ft operating depth. With a detention time of 12.5 min, oil entering the separator would rise sufficiently to be removed. The flow rate corresponding to this detention time is 75 ft^3/min.

oil–water separators No. 2 and No. 3 both have spaced baffles and chamber volumes that yield allowable flow rates of 75 ft^3/min (2.12 m^3/min).

The influent wastewater flows, in series, through nine baffles groups while passing through the three oil–water separators. Baffles prevent short circuiting due to turbulent high flows. Operating the three oil–water separators at a maximum flow rate of 1.27 m^3/min (45 ft^3/min) makes it possible to adequately treat the CSO. A safety factor of 1.67 has been considered.

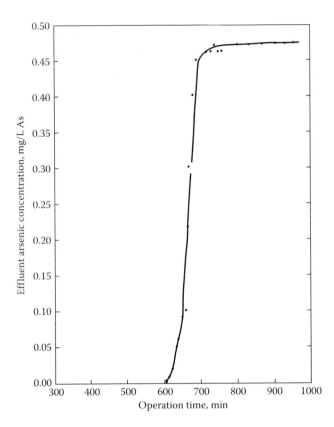

FIGURE 3.5 Granular carbon adsorption breakthrough curve for arsenic removal. (From Wang, L.K., et al., *Water Treat.*, 9, 223–233, 1994.)

3.5.4.2 DAF Clarifier

A DAF clarifier (diameter 8 ft; 2.44 m) was purchased by the oil company for the treatment of their oil–water separator effluent at 50–75 gpm (average 60 gpm). The DAF clarifier is equipped with chemical feeders for sodium aluminate (5 mg/L as $Na_2Al_2O_4$) and ferric chloride (12.7 mg/L as $FeCl_3$), and an inline mixer flocculator.

The DAF effluent has met the federal/state NPDES effluent discharge criteria since the unit was installed. The effluent has been discharged to the receiving water without the need of a continuous posttreatment by GAC.

3.5.4.3 Optional GAC Carbon Beds

Two commercial GAC adsorbers were purchased and installed by the oil company for possible posttreatment of DAF effluent, when necessary. Each GAC adsorber is 116.8 cm in diameter and 152.4 cm in height and holds 878 L of GAC adsorbent.

Each GAC adsorber employs an innovative liquid collection system that promotes even flow distribution for effective wastewater treatment, maximum GAC utilization efficiency, and low hydraulic resistance. Adsorber construction is a double epoxy/phenolic or double phenolic lined steel as standard. The vessel is cross-linked high-density polyethylene with PVC and fiberglass in the interior for corrosion resistance. Standard adsorber units have full open heads for replacing GAC.

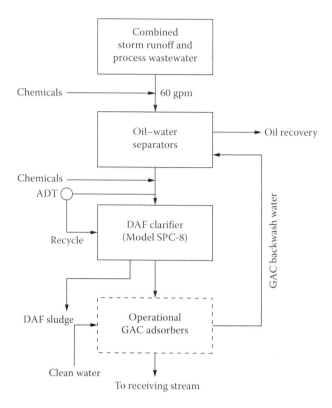

FIGURE 3.6 Flow diagram of full-scale CSO treatment system. (From Wang, L.K., et al., *Water Treat.*, 9, 223–233, 1994.)

TABLE 3.10
Oil–Water Separators

Oil–Water Separators	Dimensions	Volume
Separator No. 1	50'0"L × 10'0"W × 8'0"H	3583 ft^3
Separator No. 2	40'3"L × 10'3"W × 8'9"H	2507 ft^3
Separator No. 3	40'9"L × 08'4"W × 9'9"H	2562 ft^3

Source: Wang, L.K., et al., *Water Treat.*, 9, 223–233, 1994.

3.5.4.4 Summary of CSO Liquid Stream Treatment

Based on an extensive pilot plant study and 3-year full-scale operation, the following conclusions are drawn:

1. For treatment of normal combined storm runoff and process wastewater generated from an oil blending company, a wastewater treatment system consisting of oil–water separators and DAF clarifier will be sufficient to meet the federal/state's NPDES effluent discharge criteria.
2. If the storm runoff and process wastewater are contaminated by arsenic, then ferric chloride or ferric sulfate shall be used as a coagulant (in conjunction with an alkalinity supplement, such as sodium aluminate, or equivalent) for arsenic removal in the DAF unit.

3. If the contaminant level of combined wastewater is much higher than normal, posttreatment of DAF effluent by GAC or by filtration-GAC may be necessary.

3.5.5 Management of Waste Sludge

This section presents and discusses the sludge management technologies. Initially, the O&G in the CSO was removed by three API oil–water separators in series. The separated oil was in virgin form, and was therefore skimmed off, dried, and reused.

The major waste sludge generated from the wastewater treatment system was the DAF floated sludge, which amounted to 0.7% of influent wastewater flow. No coliform bacteria could survive in the DAF floated sludge, which contained arsenic. Therefore, microbiological stabilization of DAF sludges was not required. The DAF sludge did not posses any characteristic of Extraction Procedure Toxicity and was not hazardous. Thus, the DAF liquid sludge, which had 0.85%–0.96% consistency, needs only to be concentrated. The filtrate would be recycled to DAF for reprocessing, and the dewatered sludge could be handled by land application [29].

3.5.5.1 Waste Sludge and Residues

Figure 3.7 illustrates the oil company's wastewater treatment system, waste sludge and residue production, and the waste sludge management plan.

There are only three sources of waste sludge and residues:

1. The floated oil from oil–water separators
2. The floated sludge from DAF clarifier
3. The spent GAC from the stand-by GAC adsorbers.

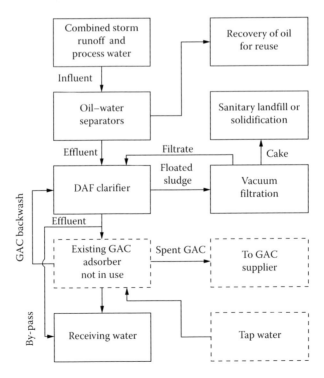

FIGURE 3.7 CSO waste sludge management system. (From Wang, L.K., et al., Management of waste sludge for Imperial Oil Company, *Proceedings of the 44th Industrial Waste Conference*, Lafayette, IN, 667–673, 1990.)

The virgin oil recovered by the oil–water separators is reused. Disposal of spent GAC will not be a problem when GAC adsorbers are in use. The GAC supplier agreed to take the spent GAC back for regeneration and reuse, if necessary. The analysis performed by the oil company during GAC operation indicated that the spent GAC did not need to be manifested out as a hazardous waste and, therefore, could be disposed of in an ordinary garbage dump with no special precautions or restrictions.

The only problem is the floated wet sludge from the DAF clarifier.

3.5.5.2 Analysis of DAF Floated Sludge

The three composite DAF floated sludge samples, Sludge A, Sludge B, and Sludge C, were collected from stormwater treatment Experiment One, Experiment Two, and Experiment Three, respectively.

Each composite sludge was analyzed for its total solids, total fixed solids, total volatile solids, TSS, water content, total coliform, cadmium, chromium, copper, nickel, lead, zinc, mercury, iron, aluminum, and arsenic, in accordance with Standard Methods for the Examination of Water and Wastewater [30]. Total coliform contents were determined by the Millipore Filtration technique. The concentrations of cadmium, chromium, copper, nickel, lead, zinc, mercury, iron, aluminum, and arsenic in the three sludge samples were all measured with a Perkin Elmer Model Atomic Absorption Spectrophotometer.

Analytical results of the three DAF floated sludge samples are shown in Table 3.11. The floated sludges can be thickened and dewatered by adequate sludge treatment facilities, such as flotation thickening, vacuum filtration, centrifugation, and filter-belt press. Table 3.12 indicates the heavy metal limitations of dewatered dry sludge for disposal by land application that have been established by the state's Department of Environmental Protection. It can be seen that the dewatered dry sludges (Table 3.11) meet the state limits on cadmium, chromium, copper, nickel, lead, and zinc for land application. The iron and aluminum contents in the dewatered dry sludge were not considered to be toxic. The state limit on arsenic in the dewatered dry sludge was not established.

TABLE 3.11
Analysis of Composite Floated Sludges Generated from DAF Treatment of CSO

| | Sludge Characteristics | | |
Parameters	Sludge A	Sludge B	Sludge C
Total Solids, mg/L	8546	9319	9568
Total Fixed Solids, mg/L	6082	6184	6174
Total Volatile Solids, mg/L	2464	3135	3394
Total Suspended Solids, mg/L	7956	8766	9004
Total Dissolved Solids, mg/L	590	553	564
Water Content, percent	99.15	99.07	99.04
Total Coliform, #/100 mL	0	0	0
Cadmium, mg/kg dry sludge	1.44	1.51	2.20
Chromium, mg/kg dry sludge	138.36	135.65	129.70
Copper, mg/kg dry sludge	322.16	480.25	519.03
Nickel, mg/kg dry sludge	16.79	30.79	31.32
Lead, mg/kg dry sludge	18.90	37.11	30.16
Zinc, mg/kg dry sludge	1255.10	2285.85	2352.67
Mercury, mg/kg dry sludge	173.90	271.11	236.43
Iron, mg/kg dry sludge	249382.64	464578.96	341647.33
Aluminum, mg/kg dry sludge	30470.14	76284.42	67633.41
Arsenic, mg/kg dry sludge	52.66	14367.31	14407.30

Source: Wang, L.K., et al., Management of waste sludge for Imperial Oil Company, *Proceeding of the 44th Industrial Waste Conference*, Lafayette, IN, 667–673, 1990.

TABLE 3.12
Heavy Metal Limitations of Dewatered Dry Sludge

Heavy Metals	State Limits for Land Application (mg/kg dry sludge)
Cadmium	25
Chromium	1000
Copper	1000
Nickel	200
Lead	1000
Zinc	2500

Sources: Department of Environmental Protection; Wang, L.K., et al., Management of waste sludge for Imperial Oil Company, *Proceedings of the 44th Industrial Waste Conference*, Lafayette, IN, 667–673, 1990.
Note: 1 mg/kg dry sludge = 1 dry weight basis.

3.5.5.3 Thickening and Dewatering of Floated Sludge

The feasibility of using an Auto-Vac filter for thickening and dewatering of the floated sludges generated from the DAF clarifier was explored. The filter was designed to remove virtually all suspended solids from liquid slurries and is ideally suited for difficult to handle sludges. The filter continuously removed the solids, thereby eliminating binding problems.

The vacuum filter was ready-to-go; the operator just directed the liquid to be filtered into the unit, brought in power to the control panel, and started it up. The vacuum filter consisted of a stainless steel drum covered with a polypropylene cloth mounted on a hollow shaft. The filter drum was immersed in a "pan," and the hollow shaft was connected to a vacuum pump (see Figure 3.8).

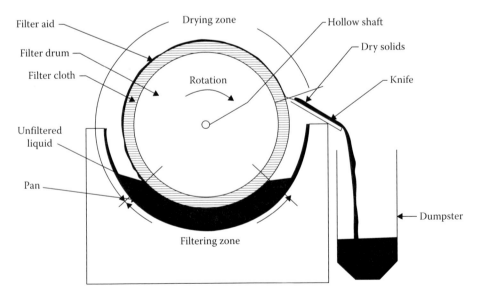

FIGURE 3.8 Auto-Vac vacuum filter. (From Wang, L.K., et al., Management of waste sludge for Imperial Oil Company, *Proceedings of the 44th Industrial Waste Conference*, Lafayette, IN, 667–673, 1990.)

To precoat the filter, filter aid and water were fed into the pan. The vacuum drew the filter aid on the cloth, building up a thick layer. The water passed through the filter aid and was out the hollow shaft. This precoat operation took about 15 min and was performed daily.

Next, the liquid to be filtered was fed into the pan. The vacuum drew the liquid through the filter aid and out of the shaft. The solids were left on the surface of the filter aid where they were air-dried for two-thirds of the revolution of the filter drum. This process removed virtually all of the free moisture. The solids were continuously removed from the surface of the filter aid by a Stellite knife blade advancing with lathelike precision, leaving a clean surface for rapid filtration. The dry solids were deposited into a dumpster for ultimate disposal.

The Alar Auto-Vac filter drum has been designed so that the entire pressure-drop occurs across the filter aid. Consequently, the Auto-Vac filters perform at high rates (reducing capital costs and space requirements) and produce drier cakes (reducing ultimate disposal costs).

The data for the filtration and analysis of the composite DAF Sludges A, B, and C are included in Tables 3.13 and 3.14, respectively. The filtrate TSS of the sludges were 5, 2, and 0 mg/L, respectively. The filtrate from the Auto Vac precoat vacuum filtration unit is recycled to the DAF clarifier for processing, as shown in Figure 3.7. The sludge cakes produced from the Auto-Vac filter have total solids in the range of 21%–22%, and their water contents in the range of 78%–79%. This indicates that sludge thickening and dewatering by vacuum filtration is technically feasible.

TABLE 3.13
Data for the Filtration of the Composite DAF Sludges A, B, and C

	Sludge A	Sludge B	Sludge C
Measured filtration rate, gal/h/ft^2	41	27	25
Filtrate appearance	Clear	Clear	Clear
Filtrate pH	6.5	6.5	6.5
Solids appearance	Dry	Dry	Dry
Cake penetration	None	None	None
Cake release	OK	OK	OK
Pre-coat liquid	Water	Water	Water

Source: Wang, L.K., et al., Management of waste sludge for Imperial Oil Company, *Proceedings of the 44th Industrial Waste Conference*, Lafayette, IN, 667–673, 1990.

TABLE 3.14
Analysis of Sludge Cake and Filtrate from Auto-Vac Filter

Sludge Samples	Sludge Cake Total Solids (%)	Sludge Cake Water Content (%)	Filtrate TSS (mg/L)
Sludge A (from Exp. one)	21.93	78.07	5
Sludge B (from Exp. two)	21.09	78.91	2
Sludge C (from Exp. three)	21.87	78.13	0

Source: Wang, L.K., et al., Management of waste sludge for Imperial Oil Company, *Proceedings of the 44th Industrial Waste Conference*, Lafayette, IN, 667–673, 1990.

TABLE 3.15
Extraction Procedure Toxicity Study for DAF Floated Sludges

Heavy Metals and Trace Elements	Sludge A E.P. Toxicity Leachate (mg/L)	Sludge B E.P. Toxicity Leachate (mg/L)	Sludge C E.P. Toxicity Leachate (mg/L)	EPA Maximum Leachate Concentration (mg/L)
Arsenic	<0.01	<0.01	<0.01	5.0
Barium	<0.02	<0.02	<0.02	100.0
Cadmium	<0.001	<0.001	<0.001	1.0
Chromium	0.02	0.03	0.03	5.0
Lead	<0.02	<0.02	<0.02	5.0
Mercury	<0.002	<0.002	<0.004	0.2
Selenium	<0.01	<0.01	<0.01	1.0
Silver	<0.002	<0.002	<0.002	5.0

Source: Wang, L.K., et al., Management of waste sludge for Imperial Oil Company, *Proceedings of the 44th Industrial Waste Conference,* Lafayette, IN, 667–673, 1990.

3.5.5.4 Extraction Procedure Toxicity of DAF Floated Sludges

The extraction procedure (EP) is designed to simulate the leaching a waste will undergo if disposed of in a sanitary landfill. It is a laboratory test in which a representative sample of a waste is extracted with distilled water maintained at a pH of 5 using acetic acid. The EP extract is then analyzed to determine if any of the thresholds established for the eight elements (arsenic, barium, cadmium, chromium, lead, mercury, selenium, silver), four pesticides (Endrin, Lindane, Methoxychlor, Toxaphene), and two herbicides (2,4,5-trichloro phenoxy propionic acid and 2,4-dichloro phenoxy acetic acid) have been exceeded. If the EP extract contains any one of the above substances in an amount equal to or exceeding the levels specified in 40 CFR 261.24, the waste possesses the characteristic of Extraction Procedure Toxicity and is a hazardous waste.

The Extraction Procedure Toxicity data of Sludges A, B, and C are documented in Table 3.15. All three floated sludges (A, B, and C) did not exhibit the characteristic of EP toxicity due to the fact that the concentrations of arsenic, barium, cadmium, chromium, lead, mercury, selenium, and silver in the extracts of Sludges A, B, and C were all lower than the respective maximum leachate concentration allowable by the U.S. EPA.

3.6 COST-EFFECTIVENESS OF STORMWATER QUALITY CONTROLS

Stormwater quality control is used to reduce pollutant loadings from urban runoff events. In most cases, the volume and peak flow of the event has a direct bearing on the discharge quality. Some facilities where the local regulatory focus was on peak flow reduction are now being reevaluated for quality control as well. Predominant stormwater quality controls are outlined in the following sections, and available cost information on them is provided [31].

3.6.1 OFFLINE STORAGE-RELEASE SYSTEMS

Storage-release systems are designed to intercept effluent and retain it for a predetermined period prior to its discharge into receiving waters. Before the effluent is released from the storage unit, it has undergone some physical settling, and, perhaps some biological treatment. The two main types of storage systems evaluated here are surface storage and deep tunnels.

3.6.1.1 Surface Storage

Surface storage units are offline storage, at or near the surface, and are typically made of concrete. Typically, large diameter culverts are used. The best source of empirical cost data on surface storage can be found in a U.S. EPA Manual [32], which relates cost as a function of size, or volume of the facility. This relationship has been updated to 2015 USD [33,58,59] and is found in Equation 3.1.

$$C = 6.28V^{0.826}, \quad (3.1)$$

where
   C = construction cost in millions, 2015 USD
   V = volume of storage system, MG (where $0.15 \leq V \leq 30$ MG).

3.6.1.2 Deep Tunnels

Deep tunnels, bored into bedrock, have been used increasingly in urban areas because space is unavailable for surface storage units. Although they function similarly to surface storage units, it is difficult to add biological treatment enhancements or baffling to tunnels. U.S. EPA [32] is currently the best source of data on the cost of deep tunnels.

This source relates cost as a function of size, or storage volume. This relationship has been updated to 2015 USD and is expressed in Equation 3.2.

$$C = 8.6V^{0.795}, \quad (3.2)$$

where
   C = construction cost in million, 2015 USD
   V = volume of storage system, MG (where $1.8 \leq V \leq 2000$ MG)

3.6.2 SWIRL CONCENTRATORS

Swirl concentrators use centrifugal force and gravitational settling to remove the heavier sediment particles and floatables from urban runoff. They are typically used in CSO situations, but may also be used in general urban runoff events [32]. These devices alone do not provide any means to reduce peak discharge, but they are commonly used in conjunction with some form of storage, and their performance varies [34].

The best source of data on swirl concentrators is currently U.S. EPA [32], which relates cost as a function of size, or, in this case, design flow. This relationship has been updated to 2015 USD and is expressed in Equation 3.3.

$$C = 0.3Q^{0.611}, \quad (3.3)$$

where
   C = construction cost in millions, 2015 USD
   Q = design flow rate, MGD (where $3 \leq Q \leq 300$ MGD)

3.6.3 SCREENS

Coarse screens are used to remove large solids and some floatables from CSO discharges. U.S. EPA [32] is the best current source of available cost data. Cost is expressed as a function of size, or design flow. This relationship has been updated to 2015 USD and is shown in Equation 3.4.

$$C = 0.118Q^{0.843}, \quad (3.4)$$

where
    C = construction cost in millions, 2015 USD
    Q = design flow rate, MGD (where $0.8 \leq Q \leq 200$ MGD)

3.6.4 Sedimentation Basins

Sedimentation basins detain storm water to allow physical settling prior to its discharge. These basins are usually baffled to eliminate short circuiting of the flow. U.S. EPA [32] is the best current source of cost data on sedimentation basins. This source relates cost as a function of size, or design flow. The relationship has been updated to 2015 USD and is expressed in Equation 3.5.

$$C = 0.387 Q^{0.688}, \tag{3.5}$$

where
    C = construction cost in millions, 2015 USD
    Q = design flow rate, MGD (where $1 \leq Q \leq 500$ MGD)

3.6.5 Disinfection

Disinfection is used to kill pathogenic bacteria prior to a CSO discharge. The best current source of data on disinfection (chlorination without dechlorination) is U.S. EPA [32]. This source relates cost as a function of size, or design flow. This relationship has been updated to 2015 USD and is expressed in Equation 3.6.

$$C = 0.223 Q^{0.464}, \tag{3.6}$$

where
    C = construction cost in millions, 2015 USD
    Q = design flow rate, MGD (where $1 \leq Q \leq 200$ MGD)

3.6.6 Best Management Practices

The term "Best Management Practices" (BMPs) is used for any practice meant to control and manage the quality or quantity of urban runoff [34]. This definition delineates stormwater BMPs as structural and nonstructural. Structural BMPs include such devices as detention basins, retention basins, infiltration trenches, or basins. They are typically constructed as part of the urban development process to mitigate the deleterious effects of urban runoff. A key BMP, minimizing the directly connected impervious area, is not included in this analysis, as less data are available on its cost [34]. The more typical, nonstructural BMPs include such activities as street sweeping and public education on the disposal of pollutants, for example, oils. These methods are more difficult to assess.

3.6.6.1 Detention Basins

Detention basins are storage basins designed to empty after each storm. These basins are most common in rapidly developing urban areas. They use an undersized outlet, which causes water to back up and fill the basin [35]. The rate of discharge depends upon the outlet size and is usually set by local standard. Detention basins attenuate the peak runoff from the developed area. These basins perform well in controlling local water quantity impacts of urban runoff. If the outlet is designed appropriately, water quality can also be controlled to some extent. The best current source of cost information is Young et al. [36], who give cost as a function of storage volume as shown in Equation 3.7.

$$C = 76,035 V^{0.69}, \tag{3.7}$$

where
  C = construction cost, 2015 USD
  V = volume of basin, MG

The construction costs have been updated to 2015 USD. Land costs were excluded. The basis for this relationship is a study done for the Metropolitan Washington Council of Governments [37].

3.6.6.2 Retention Basins

Retention basins are similar to detention basins, except that the permanent pool is increased. By increasing the permanent pool (i.e., the point at which discharge occurs) in the storage volume (and typically increasing the storage size as well), increased physical and biological treatments occur due to the longer residence time in the basin.

These types of basins are called retention basins, or wet ponds. The amount of physical storage available is determined by the difference between the height set as the permanent pool volume and the height above the top of the weir or outlet structure available, or freeboard. Because cost depends upon volume, retention basins are more costly in controlling the same amount of peak discharge as a dry detention basin from a quantity standpoint.

The best available cost data on retention basins is found in Young et al. [36], who give cost as a function of the total volume of the pond (not the available storage). This relationship is shown by Equation 3.8.

$$C = 84,287 V^{0.75}, \tag{3.8}$$

where
  C = construction cost, 2015 USD
  V = volume of pond, MG

The construction costs have been updated to 2015 USD. Land costs were excluded. The basis for this relationship is a study done for the Metropolitan Washington Council of Governments [37].

3.6.6.3 Infiltration Trenches

Infiltration is the process of runoff water soaking into the ground. Since infiltrated water is removed from surface waters, it represents a complete control for that fraction of stormwater that can be infiltrated [35]. An infiltration trench is used in areas where space is a problem. It usually consists of excavating a void volume, lining the volume with filter fabric to keep out fine material, installing conveyance piping, and filling the void with gravel or crushed stone. The trench's performance depends greatly upon the soil characteristics of the area, and operating and maintenance practices [34].

The best available cost data on infiltration trenches are found in Young et al. [36], who give cost as a function of the total volume of the trench. This relationship is shown by Equation 3.9.

$$C = 217 V^{0.63}, \tag{3.9}$$

where
  C = construction cost, 2015 USD
  V = volume of trench, ft^3

The construction costs have been updated to 2015 USD. Land costs were excluded.

3.6.6.4 Infiltration Basins

Infiltration basins are similar to retention ponds; however, they are typically used in flatter terrain, and discharge only in low frequency events. Permeable soils underlying the basin and high rates of evapotranspiration are the major prerequisites for using these basins. The water typically can only leave via percolation into the groundwater, or evapotranspiration. Performance in buffering runoff water quality is high; however, from a quantity standpoint, a large land area must be used to control significant runoff events. A major disadvantage is the high maintenance involved due to clogging of the basin.

The best available cost data on infiltration basins are found in Young et al. [36], who give cost as a function of the total volume of the basin. This relationship is:

$$C = 21.1V^{0.69}, \qquad (3.10)$$

where
C = construction cost, 2015 USD
V = volume of infiltration basin, ft^3

The construction costs have been updated to 2015 USD. Land costs were excluded. The basis for this relationship is a study done by Schueler [38].

3.6.6.5 Sand Filters

Sand filters remove sediment and pollutants from runoff. Usually the filters have a presettling chamber to induce settling of the larger solids that would typically clog the sand filter itself. The filtered outflow is collected, rather than infiltrated, and either discharged or treated further. Performance of these systems is typically good in space-limited areas and in arid climates [36].

The best available cost data on sand filters are found in Young et al. [36], who give cost as a function of the total impervious surface area draining to the filter. This relationship is found in Equation 3.11.

$$C = KA, \qquad (3.11)$$

where
C = construction cost, 2015 USD
A = impervious surface, acres
K = constant, ranging from 15,300 to 30,600

The construction costs have been updated to 2015 USD [33]. Land costs were excluded. The basis for this relationship is a study done for the Metropolitan Washington Council of Governments [39].

3.6.6.6 Water Quality Inlet

Water quality inlets are inlets modified for the control of some solids, oil, and grease. These are sometimes referred to as oil and grit separators. According to Urbonas [34], the performance of these devices has not been very good.

The best available cost data on water quality inlets are found in Young et al. [36]. Updated to 2015 USD [33,58,59], the costs range from $9900 USD to $29,700 USD. The basis for this relationship is a study done by Schueler [38]. The data behind this relationship were not reported.

3.6.6.7 Grassed Swales

Grassed swales are vegetated channels used in lieu of the traditional concrete curb and gutter typical of urban areas. Pollutants are removed through filtration by vegetation, settling, and infiltration into the soil [36]. The performance of these systems is highly variable. The use of swales is not recommended in dense urban areas where space is at a premium, or in commercial/industrial areas where contamination of groundwater can occur due to oils and grease in the effluent [34].

The best available cost data on grassed swales are found in Young et al. [36], where cost is found to vary as follows:

$$C = KL, \qquad (3.12)$$

TABLE 3.16
BMP Pollutant Removal Ranges

			Removal Range (%)				
BMP	TSS	Total P	TKN	Zinc	Lead	BOD_5	Bacteria
Grass buffer strip	10–20	0–10	0–10	0–10	N/A	N/A	N/A
Grass lined swale	20–40	0–15	0–15	0–20	N/A	N/A	N/A
Infiltration basin	0–98	0–75	0–70	0–99	0–99	0–90	75–98
Percolation trench	98	65–75	60–70	95–98	N/A	90	98
Retention pond	91	0–79	0–80	0–71	9–95	0–69	N/A
Extended detention	50–70	10–20	10–20	30–60	75–90	N/A	50–90
Wetland basin	40–94	(−)4–90	21	(−)29–82	27–94	18	N/A
Sand filters (fraction flowing through filter)	14–96	5–92	(−)129–84	10–98	60–80	60–80	N/A

Source: US EPA, Costs of urban stormwater control, EPA-600/R-02/021, US Environmental Protection Agency, Cincinnati, OH, January 2002.

where
  C = construction cost, 2015 USD
  L = length of swale, ft
  K = constant, 7–19

The construction costs have been updated to 2015 USD [33,58,59]. No land costs were included in this analysis. These costs can be significant because an increased right-of-way is needed to include the swale. The basis for this relationship is a study done by Schueler [40].

3.6.6.8 Assessment of BMP Control Performance

An overall assessment of BMP control performance can be found in Table 3.16 [34]. The table lists expected removal ranges for TSS, total phosphorus, total nitrogen, zinc, lead, BOD, and bacteria, compiled from several different sources. Urbonas [34], however, cautions the use of the table alone and argues that the definition of "effectiveness is fundamentally flawed, as it is typically a snapshot in time, and ignores the performance of the control over time, and the variability of maintenance to the control." For example, porous pavement is excellent at removal of solids, but is certainly not designed to do so and will clog very quickly if a high solids loading is applied to it.

Urbonas [34] advocates a more design-oriented approach in assessing control performance. An example of this approach is found in Table 3.17. While subjective, this approach does provide the designer with enough information to evaluate the control under a wider range of conditions than the regulatory approach found in Table 3.16. However, much more work needs to be done in this area to properly assess the expected benefits of the BMP control in question.

3.6.7 O&M COSTS FOR CONTROLS

O&M cost data for controls are only available for a limited number of CSO-type controls, that is, sedimentation, disinfection, and screens. CSO-type controls are expected to be significantly more expensive in terms of operating and maintenance costs than those controls that handle only storm water; however, no data were available for non-CSO-type controls. These relationships can be found in Figure 3.9 from U.S. EPA [31]. For a complete cost/benefit analysis of

TABLE 3.17
An Assessment of Design Robustness Technology for Several BMPs

		Removal of Constituents in Stormwater		
Structural BMP	Hydraulic Design[a]	TSS	Dissolved	Overall Design Robustness
Swale	High	Low–moderate	None–low	Low
Buffer (filter) strip[b]	Low–moderate	Low–moderate	None–low	Low
Infiltration basin[c]	Low–high	High	Moderate–high	Low–moderate
Percolation trench	Low–moderate	High	Moderate–high	Low–moderate
Extended detention (dry)	High	Moderate–high	None–low	Moderate–high
Retention pond (wet)	High	High	Low–moderate	Moderate–high
Wetland	Moderate–high	Moderate–high	Low–moderate	Moderate
Media filter	Low–moderate	Moderate–high	None–low	Low–moderate
Oil separator	Low–moderate	Low	None–low	Low
Catch basin inserts	Uncertain	N/A	N/A	N/A
Monolithic porous pavement[b]	Low–moderate	Moderate–high	Low–high[c]	Low
Modular porous pavement[b]	Moderate–high	Moderate–high	Low–high[c]	Low–moderate

Source: US EPA, Costs of urban stormwater control, EPA-600/R-02/021, US Environmental Protection Agency, Cincinnati, OH, January 2002.

[a] Weakest design aspect, hydraulic or constituent removal, governs overall design robustness.
[b] Robustness is site-specific and very much maintenance-dependent.
[c] Low-to-moderate whenever designed with an underdrain and not intended for infiltration.

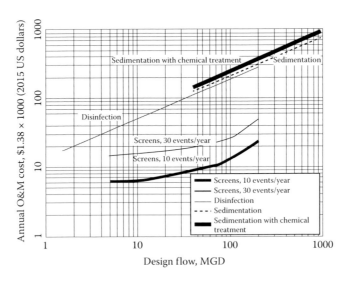

FIGURE 3.9 Operation and maintenance costs for CSO controls. (From US EPA, Costs of urban stormwater control, EPA-600/R-02/021, US Environmental Protection Agency, Cincinnati, OH, January 2002.)

each control, one needs operating and maintenance costs to complete a life-cycle cost analysis (LCA). LCA is done by bringing all controls to the same design life (by including replacements as necessary), amortizing the control over the same period, and including in this annual cost the annual operating and maintenance cost for each control. LCA is then compared with the benefits of the control [41–46].

Additional technical information regarding various treatment technologies (chemical treatment, oil–water separation, DAF, DAFF, GAC, UV, etc.), costs and governmental rules, regulations, and codes for design, and construction of storm water and CSOs systems can be found from the literature [47–61].

NOMENCLATURE

A Impervious surface, acres
K Constant
L Length of swale, ft
C Construction cost, 2015 USD
Q Design flow rate, MGD
V Volume, MG or ft^3

REFERENCES

1. US EPA. What are CSOs, and why are they important? http://cfpub.epa.gov/npdes/search.cfm (2017).
2. US EPA. Combined sewer overflow (CSO) control policy. *Federal Register*, Vol. 59, No. 75, pp. 1–12, April 19 (1994).
3. US EPA. Clean Water Act (CWA), 33 U.S.C. ss/1251 et seq. (1977), Federal Register. www.access.gpo.gov/uscode/title33/chapter26_.html, May (2002).
4. US EPA. Combined sewer overflows guidance for nine minimum controls. EPA 832-B-95-003, US Environmental Protection Agency, Washington, DC, May (1995).
5. US EPA. Guidance: Coordinating CSO long-term planning with water quality standards reviews. EPA-833-R-01-002, US Environmental Protection Agency, Washington, DC, July 31 (2001).
6. US EPA. A planning and design guidebook for combined overflow control and treatment. US Environmental Protection Agency, Washington, DC (1982).
7. US EPA. Combined sewer overflow management fact sheet pollution prevention. EPA 832-F-99-038, US Environmental Protection Agency, Washington, DC, September (1999).
8. US EPA. Combined sewer overflow management fact sheet: Sewer separation. EPA 832-F-99-041, US Environmental Protection Agency, Washington, DC, September (1999).
9. Ohio EPA. Storm water program fact sheet. State of Ohio Environmental Protection Agency, Columbus, OH, April (2003).
10. US EPA. After the storm: A citizen's guide to understanding stormwater. EPA 833-B-03-002, US Environmental Protection Agency, Washington, DC, January (2003).
11. L. K. Wang, Y. T. Hung, and N. K. Shammas (Eds). *Physicochemical Treatment Processes*, Humana Press, Totowa, NJ, pp. 526–671 (2005).
12. L. K. Wang, Y. T. Hung, and N. K. Shammas (Eds). *Advanced Physicochemical Treatment Processes*, Humana Press, Totowa, NJ, pp. 203–260 (2006).
13. L. K. Wang, Y. T. Hung, and N. K. Shammas (Eds). *Advanced Physicochemical Treatment Technologies*, Humana Press, Totowa, NJ, pp. 295–390 (2007).
14. L. K. Wang, N. Pereira, and Y. T. Hung, (Eds), and N. K. Shammas (Consulting Editor). *Biological Treatment Processes*, The Humana Press, Totowa, NJ (2008).
15. L. K. Wang, N. K. Shammas, and Y. T. Hung (Eds). *Advanced Biological Treatment Processes*, The Humana Press, Totowa, NJ (2008).
16. M. Krofta and L. K. Wang. *Flotation Engineering*, Technical manual no. Lenox/1-06-2000/368, Lenox Institute of Water Technology, Lenox, MA (2000).
17. M. Krofta, D. Guss, and L. K. Wang. Development of low-cost flotation technology and systems for wastewater treatment. *Proceedings of 42nd Industrial Waste Conference*, May 12–14, 1987, Purdue University, Lewis Publishers, Chelsea, MI, pp. 185–195 (1988).
18. M. Krofta and L. K. Wang. Design of dissolved air flotation systems for industrial pretreatment and municipal wastewater treatment—Design and energy considerations. American Institute of Chemical Engineers National Conference, Houston, TX, 30 p. (NTIS-PB83-232868) (1983).
19. M. Krofta and L. K. Wang. Flotation technology and secondary clarification. *Tappi Journal*, Vol. 69, No. 4, pp. 92–96 (1987).

20. M. Krofta, D. Miskovic, D. Burgess, and E. Fahey. The investigation of the advanced treatment of municipal wastewater by modular flotation-filtration systems and reuse for irrigation. *Water Science and Technology*, Vol. 33, No. 10–11, pp. 171–179 (1996).
21. N. K. Shammas. Physicochemically-enhanced pollutants separation in wastewater treatment, *Proceedings of International Conference: Rehabilitation and Development of Civil Engineering Infrastructure Systems: Upgrading of Water and Wastewater Treatment Facilities*, The American University of Beirut and University of Michigan, Beirut, Lebanon, June 9–11 (1997).
22. N. K. Shammas and M. Krofta. A compact flotation: Filtration tertiary treatment unit for wastewater reuse, Water reuse symposium, AWWA, Dallas, TX, pp. 97–109, February 27–March 2 (1994).
23. L. K. Wang, N. K. Shammas, M. S. Holtorff, and D. Khaitan. Treatment of domestic sewage and combined sewer overflow by dissolved air flotation and filtration, *International Conference of Water, Asia 2000*, India, September (2000).
24. M. Krofta, L. K. Wang, and D. Khaitan. Treatment of domestic sewage and combined sewer overflow by dissolved air flotation and filtration, Technical report no. Lenox/July-23-2000/388, July 23 (2000).
25. M. S. Holtorff. Municipal waste treatment by innovative physical-chemical and biological processes, Master Thesis, Lenox Institute of Water Technology, Lenox, MA, 113 p. (2000).
26. L. K. Wang, M. H. S. Wang, and J. C. Wang. *Design, Operation and Maintenance of the Nation's Largest Physicochemical Waste Treatment Plant*, Volume I, Lenox Institute for Water Technology, Technical report no. LIR/03-87/248, 183 p. (1987).
27. L. K. Wang, W. J. Mahoney, and M. H. S. Wang. Treatment of storm runoff for Imperial Oil Company. *Water Treatment*, Vol. 9, pp. 223–233 (1994).
28. J. R. Bratby. Treatment of raw wastewater overflows by dissolved air flotation. *Journal of Water Pollution Control Federation*, Vol. 54, pp. 1558–1565 (1982).
29. L. K. Wang, M. H. S. Wang, and W. J. Mahoney. Management of waste sludge for Imperial Oil Company. *Proceedings of the 44th Industrial Waste Conference*, Lafayette, IN, pp. 667–673 (1990).
30. APHA, AWWA and WEF. *Standard Methods for the Examination of Water and Wastewater*, 21st Edition, American Public Health Association, American Water Works Association and Water Environment Federation, Washington, DC, 1368 p. (2005).
31. US EPA. Costs of urban stormwater control, EPA-600/R-02/021. US Environmental Protection Agency, Cincinnati, OH, January (2002).
32. US EPA. Manual: Combined sewer overflow control, EPA-625/R-93-0007. US Environmental Protection Agency, Washington, DC (1993).
33. US ACE. Yearly average cost index for utilities. In: *Civil Works Construction Cost Index System Manual*, 110-2-1304, U.S. Army Corps of Engineers, Washington, DC, p. 44. www.nww.usace.army.mil/cost (2015).
34. B. Urbonas. An assessment of stormwater BMP technology. In: *Innovative Urban Wet-Weather Flow Management Systems*, Chapter 7, J. P. Heaney, R. Pitt, and R. Field (Eds). United States Environmental Protection Agency, Cincinnati, OH, 1999. www.epa.gov/ednnrmrl/publish/book/epa-600-r-99-029/achap07.pdf (2015).
35. B. K. Ferguson. *Introduction to Stormwater: Concept, Purpose, Design*, John Wiley & Sons, New York (1998).
36. G. K. Young, S. Stein, P. Cole, T. Kammer, F. Graziano, and F. Bank. Evaluation and management of highway runoff water quality, Technical report. The Federal Highway Administration, Washington, DC (1996).
37. C. Wiegand, T. Schueler, W. Chittendren, and W. Jellick. Cost of urban runoff controls in urban runoff quality. In: *Impact and Quality Enhancement Technology*, B. Urbonas and L. A. Roesner (Eds). American Society of Civil Engineers, Reston, VA (1986).
38. T. R. Schueler. *Controlling Urban Runoff: A Practical Manual for Planning and Designing Urban BMPs*. Department of Environmental Programs, Metropolitan Washington Council of Governments, Washington, DC (1987).
39. T. R. Schueler. Developments in sand filter technology to improve stormwater runoff quality. *Water Protection Techniques*, Vol. 1, No. 2, pp. 47–54 (1994).
40. T. R. Schueler, P. Kumble, and M. Heraty. A current assessment of urban best management practices: Techniques for reducing nonpoint source pollution in the coastal zone. Anacostia Research Team, Metropolitan Washington Council of Governments, Washington, DC (1992).
41. L. K. Wang, M. H. S Wang, and J. C. Wang. *Design, Operation and Maintenance of the Nation's Largest Physicochemical Waste Treatment Plant*, Volume II, Lenox Institute of Water Technology, Technical report no. LIR/03-87/249, 161 p. (1987).

42. L. K. Wang, M. H. S Wang, and J. C. Wang. *Design, Operation and Maintenance of the Nation's Largest Physicochemical Waste Treatment Plant*, Volume III, Lenox Institute of Water Technology, Technical report no. LIR/03-87/250, 227 p. (1987).
43. US EPA. Combined sewer overflows, guidance for funding options. EPA 832-B-95-007, US Environmental Protection Agency, Washington, DC, August (1995).
44. US EPA. Combined sewer overflow O&M fact sheet proper operation and maintenance. EPA 832-F-99-039, US Environmental Protection Agency, Washington, DC, September (1999).
45. B. Pascual, B. Tanselm and R, Shalewitz. Economic sensitivity of the dissolved air flotation process with respect to the operational variables. *Proceedings of 49th Industrial Waste Conference*, Lewis Publishers, Chelsea, MI (1994).
46. US EPA. The O & M in CMOM: Operation & maintenance: A reference guide for utility operators. Capacity, Management, Operations, and Maintenance (CMOM) Regulations. US EPA Web Site for Management of O & M. www.cmom.net/CMOMGuide_OandM.htm (2007).
47. C. Yapijakis, R. L. Trotta, C. C. Chang, and L. K. Wang. Stormwater management and treatment. In: *Handbook of Industrial and Hazardous Wastes Treatment*, L. K. Wang, Y. T. Hung, H. H. Lo, and C. Yapijakis (Eds). Marcel Dekker, New York, pp. 873–922 (2004).
48. L. K. Wang, D. A. Vaccari, Y. Li, and N. K. Shammas. Chemical precipitation. In: *Physicochemical Treatment Processes*, L. K. Wang, Y. T. Hung, and N. K. Shammas (Eds). Humana Press, Totowa, NJ, pp. 141–198 (2005).
49. L. K. Wang, E. M. Fahey, and Z. Wu. Dissolved air flotation. In: *Physicochemical Treatment Processes*, L. K. Wang, Y. T. Hung, and N. K. Shammas (Eds). Humana Press, Totowa, NJ, pp. 431–500 (2005).
50. Y. T. Hung, H. H. Lo, L. K. Wang, J. R. Taricska, and K. H. Li. Granular activated carbon adsorption. In: *Physicochemical Treatment Processes*, L. K. Wang, Y. T. Hung, and N. K. Shammas (Eds). Humana Press, Totowa, NJ, pp. 573–634 (2005).
51. J. P. Chen, L. Yang, L. K. Wang, and B. Zhang. Ultraviolet radiation for disinfection. In: *Advanced Physicochemical Treatment Processes*, L. K. Wang, Y. T. Hung, and N. K. Shammas (Eds). Humana Press, Totowa, NJ, pp. 317–366 (2006).
52. P. Kajitvicchyanukul, Y. T. Hung, and L. K. Wang. Oil water separation. In: *Advanced Physicochemical Treatment Processes*, L. K. Wang, Y. T. Hung, and N. K. Shammas (Eds). Humana Press, Totowa, NJ, pp. 521–548 (2006).
53. G. F. Bennett and N. K. Shammas. Separation of oil from wastewater by air flotation. In: *Flotation Technology*, L. K. Wang, N. K. Shammas, W. A. Selke, and D. B. Aulenbach (Eds). Humana Press, Totowa, NJ, pp. 85–120 (2010).
54. L. K. Wang and N. K. Shammas. Ozone-oxygen oxidation flotation. In: *Flotation Technology*, L. K. Wang, N. K. Shammas, W. A. Selke, and D. B. Aulenbach (Eds). Humana Press, Totowa, NJ, pp. 269–326 (2010).
55. N. K. Shammas and G. F. Bennett. Principles of air flotation technologies. In: *Flotation Technology*, L. K. Wang, N. K. Shammas, W. A. Selke, and D. B. Aulenbach (Eds). Humana Press, Totowa, NJ, pp. 1–48 (2010).
56. L. K. Wang, M. H. S. Wang, N. K. Shammas, and M. Krofta. Innovative and cost-effective flotation technologies for municipal and industrial wastes treatment. In: *Handbook of Environment and Waste Management: Air and Water Pollution Control*, Y. T. Hung, L. K. Wang, and N. K. Shammas (Eds). World Scientific, Hackensack, NJ, pp. 1151–1176 (2012).
57. N. K. Shammas and L. K. Wang. *Water Engineering: Hydraulics, Distribution and Treatment*. John Wiley & Sons, Hoboken, NJ, 806 p. (2016).
58. US ACE. Civil works construction cost index system manual, 110-2-1304. US Army Corps of Engineers. Washington, DC, p. 44 (2015).
59. N. K. Shammas. Wastwater renovation by flotation for water pollution control. In: *Advances in Water Resources Management*, L. K. Wang, C. T. Yang, and M. H. S. Wang (Eds). Springer, New York, pp. 403–421 (2016).
60. Portland City. Stormwater management manual, Portland, OR. www.portlandoregon.gov/bes/64040 (2015).
61. Seattle City. Stormwater code and manual, City of Seattle, Department of Construction and Inspections, Washington, DC. www.seattle.gov/dpd/codesrules/codes/stormwater/ (2016).

4 Waste Treatment and Management in Chlor-Alkali Industries

Hamidi Abdul Aziz, Miskiah Fadzilah Ghazali, and Mohd. Suffian Yusoff
Universiti Sains Malaysia

Yung-Tse Hung
Cleveland State University

CONTENTS

4.1	Introduction	100
	4.1.1 Industry Description	100
	4.1.2 End Use of Chlor-Alkali Products	101
4.2	Chlor-Alkali Manufacturing Process	102
	4.2.1 Diaphragm Cell Process	104
	4.2.2 Membrane Cell Process	104
	4.2.3 Mercury Cell Process	105
	4.2.4 Chlorine Processing	108
	4.2.4.1 The U.S. Chlorine Industry	108
	4.2.5 Hydrogen Processing	108
	4.2.6 Caustic Soda Processing	112
	4.2.7 Brine Processing	112
	4.2.8 Sodium Hypochlorite Manufacturing Process	113
	4.2.9 Sodium Chlorate Manufacturing Process	113
4.3	Energy Requirements for the Chlor-Alkali Industry	114
4.4	Waste Characterization	116
	4.4.1 Solid Wastes	116
	4.4.2 Liquid Effluents	123
	4.4.3 Air Emission	126
4.5	Contaminants of Concern	127
	4.5.1 Mercury	127
	4.5.2 Asbestos	128
	4.5.3 Lead	128
4.6	Environmental Impacts	128
	4.6.1 Mercury	128
	4.6.2 Sodium Chloride	130
4.7	Occupational Health and Safety Issues	130
	4.7.1 Chlorine	130
	4.7.2 Mercury	130
	4.7.3 Asbestos	130
	4.7.4 Caustic Soda	131

4.8	Pollution Prevention and Abatement	131
	4.8.1 Material Substitution	131
	4.8.2 Waste Segregation	132
	4.8.3 Process Related Measures	132
	4.8.4 Mercury Recovery	132
4.9	Treatment Technology	133
	4.9.1 Solid Waste	133
	4.9.2 Air Emission	133
	4.9.3 Liquid Effluents	133
	4.9.3.1 Mercury Cell Process	133
	4.9.3.2 Diaphragm Cell Plant	134
	4.9.4 BAT	134
	4.9.5 Disposal Option and Alternative Treatment Technologies	135
4.10	Capital and Operation and Maintenance Costs	136
	4.10.1 Olin Corporation	136
	4.10.2 Operation and Maintenance Costs	138
4.11	Monitoring and Reporting	138
4.12	Global Overview of Discharge Requirements	138
	4.12.1 Liquid Effluents	138
	4.12.2 Air Emissions	140
4.13	Conclusion	142
References		142

Abstract

The chlor-alkali industry produces chlorine (Cl_2) and caustic soda from brines, by electrolysis of a salt solution. The world's chlorine capacity at the beginning of 1988 was approximately 42.3 million metric tons per year and the corresponding caustic capacity was approximately 46.1 million metric tons per year. Three main technologies applied for chlor-alkali production are mercury cell, diaphragm cell, and membrane cell process. Each process has a different method for producing chlorine at the anode, and producing caustic soda and hydrogen, directly or indirectly, at the cathode. This chapter describes the manufacturing processes, energy requirement, waste characterization, environmental impacts, pollution prevention and abatement, treatment technologies, costs, monitoring, global overview, and standard requirement of air emission and effluents discharge of the chlor-alkali industry.

4.1 INTRODUCTION

The U.S. chemical industry is the largest in the world and responsible for about 11% of U.S. industrial production measured as value [1]. One of the most important industries is the chlor-alkali industry. The chlor-alkali process produces chlorine, caustic soda, and hydrogen through the electrolysis of sodium chloride solution. The corresponding industry is one of the most important segments of the chemical industry, the products of which are used in over 50% of all industrial chemical processes [2]. Chlorine and sodium hydroxide are among the top ten chemicals produced in the world and are involved in the manufacturing of a wide variety of products used in day-to-day life. These include pharmaceuticals, detergents, deodorants, disinfectants, herbicides, pesticides, and plastics [3].

4.1.1 INDUSTRY DESCRIPTION

Since 1950, chlorine has become increasingly important as a raw material for synthetic organic compounds. Chlorine is one of the more abundant chemicals produced by industry and has varied

TABLE 4.1
Estimated Production of Caustic Soda between 1980 and 2000

Year	Total Production (million t/year)	Mercury (%)	Diaphragm (%)	Membrane (%)
1980	44.4	45	53	2
1990	46.5	39	45	16
2000	55.5	15	35	50

Source: U.S. EPA; Ayres, R., *J. Ind. Ecol.*, 1(1), 81–94, 1997.

industrial uses. Since 1982, the chlor-alkali industry has increased production in Canada and Western Europe, but it has declined in the United States and Japan due to increasing concerns about the toxicity of chlorinated end-use materials and chlorine pollution from the wastewater. Table 4.1 shows the estimated production of caustic soda in the world during 1980–2000, using various processes.

Annual production from facilities in the United States was 9.9 million megagrams (10.9 million tons) in 1990 after peaking at 10.4 million megagrams (11.4 million tons) in 1989. In 1991, about 52 chlor-alkali plants were in operation in 23 states around the country. Louisiana and Texas have the largest number of plants operating within their borders [4,5].

The chlor-alkali industry has been growing at a slow pace over the past 10 years, and this rate is expected to continue in the early years of the new century. Chlorine and sodium hydroxide are coproducts, and the demand for one will highly influence the demand for the other. Over the past several decades, market forces have switched between chlorine and sodium hydroxide a number of times. Chlorine demand drives the chlor-alkali industry, but the demand is cyclic, with chlorine and caustic soda out of phase in the marketplace. When caustic soda reaches a high level of demand, the direction of product flow is dependent upon Asian and European economies and the foreign exchange rate. Foreign producers may often export caustic soda to the United States to keep chlorine production high, which may have an impact on both market and production in this country.

There are three basic processes of manufacturing chlorine and caustic soda that include mercury cell, diaphragm cell, and membrane cell [6,7]. The diaphragm cell process and the mercury cell process were both developed in the late 1800s, while the membrane cell process was developed much more recently in 1970. The membrane cell is the most modern, economic, and environment-friendly process. The mercury cell process and diaphragm cell process release hazardous waste containing mercury and asbestos, respectively [6,7].

4.1.2 END USE OF CHLOR-ALKALI PRODUCTS

The caustics chain begins with sodium chloride (NaCl) and forms the basis for what is often referred to as the chlor-alkali industry. Major products of the chlor-alkali industry, as stated earlier, include chlorine, sodium hydroxide (caustic soda), soda ash (sodium carbonate), sodium bicarbonate, potassium hydroxide, and potassium carbonate. Of these products, chlorine, sodium hydroxide, and soda ash account for the largest share of shipments from the chlor-alkali industry. These products are also very important economically, as they are in the eighth, ninth, and tenth place in production in the United States [8]. Figure 4.1 presents the chain of chlor-alkali production.

The first use of chlorine for disinfection dates back to 1823, when it was used in hospitals. Chlorine water was employed in obstetric wards to prevent puerperal fever in 1826, and fumigation with chlorine was practiced during the great European cholera epidemic. Between 1920 and 1940, several new applications for chlorine were developed, for example, in the manufacture of ethylene glycol, chlorinated solvents, and vinyl chloride. About 70% of the chlorine produced in the United States is used to manufacture organic chemicals such as vinyl chloride monomer, ethylene dichloride, glycerin,

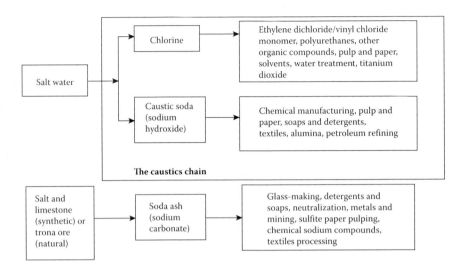

FIGURE 4.1 Chlor-alkali product chain. (From U.S. EPA; Nunes, S.P., Peinemann, K.-V., *Membrane Technology in the Chemical Industry*, Wiley-VCH, Weinheim, Germany, 2006.)

chlorinated solvents, and glycols (Figure 4.2). About 40% is used for the production of vinyl chloride, an important building block for polyvinyl chloride (PVC) and a number of petrochemicals. Presently, the primary uses of chlorine are in the pulp and paper manufacturing operations for bleaching to produce a high-quality whitened material and in water treatment operations as a disinfectant. About 15% of chlorine is consumed in the processing of pulp and paper, and the rest is used for manufacturing inorganic chemicals, disinfection of water, and production of hypochlorite [8].

The end uses of caustic (sodium hydroxide) are diverse compared with the uses of chlorine. Its primary application is in neutralization reactions and in the formation of anionic species such as aluminates and zincates. About 30% of the sodium hydroxide produced is used by the organic chemical industry and about 20% is consumed by the inorganic chemical industry for neutralization and off-gas scrubbing, and as an input in the production of various chemical products such as alumina, propylene oxide, polycarbonate resin, epoxies, synthetic fibers, soaps, detergents, rayon, and cellophane. Another 20% of sodium hydroxide production is used by the pulp and paper industry for pulping wood chips and for other processes (Figure 4.3). Sodium hydroxide is also used to manufacture soap and cleaning products, and as a drilling fluid for oil and gas extraction.

Soda ash is used primarily by the glass industry as a flux to reduce the melting point of sand. It is also a raw material in the manufacture of sodium phosphates and sodium silicates, which are important components of domestic and industrial cleaners. It is also used in both the refining and the smelting stages in the production of metals, in sulfite paper pulping processes, and in textiles processing. Soda ash is also an intermediate in the production of sodium compounds, including phosphates, silicates, and sulfites.

4.2 CHLOR-ALKALI MANUFACTURING PROCESS

A chlor-alkali process implies the electrolysis of common salt or sodium chloride. Generally, there are three basic processes to manufacture chlorine and caustic soda from brine: the membrane cell, the diaphragm cell, and the mercury cell. About 59 million metric tons of chlorine, which is about 84% of the total world capacity, was produced electrolytically using diaphragm and membrane cells, while about 13% was produced using mercury cells in 2006 (Figure 4.4). In each process, a salt solution is electrolyzed by the action of direct electric current, which converts chloride ions to elemental chlorine [9]. The overall process reaction is as follows:

$$2NaCl + 2H_2O \rightarrow Cl_2 + H_2 + 2NaOH \tag{4.1}$$

Waste Treatment and Management in Chlor-Alkali Industries

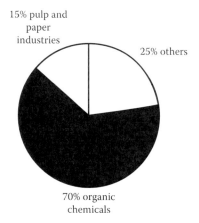

FIGURE 4.2 Chlorine end use.

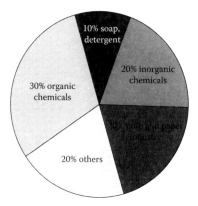

FIGURE 4.3 Sodium hydroxide end use.

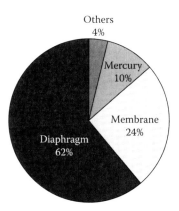

FIGURE 4.4 Chlorine cell technology in the United States.

This process represents a different method of chlorine production at the anode and the caustic soda (NaOH) and hydrogen (H_2) production at the cathode directly or indirectly. The membrane cell process has economic and environmental advantages and it is the most modern process, while the other two processes generate hazardous waste containing mercury or asbestos. However, all three basic cell technologies generate chlorine at the anode and hydrogen along with sodium hydroxide in the cathode compartment (or in a separate reactor for mercury cells).

The distinguishing difference between the technologies lies in the manner by which the anolyte and the catholyte streams are prevented from mixing with each other. Separation is achieved in a diaphragm cell by a separator and in a membrane cell by an ion exchange membrane. In mercury cells, the cathode itself acts as a separator by forming an alloy of sodium and mercury (sodium amalgam), which subsequently reacts with water to form sodium hydroxide and hydrogen in a separate reactor [4–6,9]. The three basic processes are described in the following sections [4,9,10].

4.2.1 Diaphragm Cell Process

Figure 4.5 shows the process flow diagram of the diaphragm cell process. This process was developed in the late 1800s. In the diaphragm process, there are two compartments, which are anode and cathode separated by a permeable diaphragm, often made of asbestos fiber (Figure 4.6). The chlorine is produced at the anode, while hydrogen and hydroxyl ions are produced at the cathode. Positively charged sodium ions (cations) react with negatively charged hydroxyl ions (anions) to form a caustic solution. The conversion of sodium chloride is approximately 50% per pass. Polymer material is replaced for asbestos for the diaphragm in new cell designs [4,9,10].

Chlorine gas is processed in the a similar way as in the mercury cell process. Only the condensate from cooling the chlorine does not contain mercury, while other processes such as chlorine cooling, drying, compression, and liquefaction generate the same residuals. Hence, the condensate can be sent to disposal after being dechlorinated rather than being recycled to the brine system.

The hydrogen gas can be either marketed or used as a fuel after removing the water vapor through a venting or cooling process. Although venting of hydrogen is practiced by some companies, it is not a safe process because this gas has high flammability. Therefore, this practice is not recommended. The caustic solution that is produced has a concentration of about 10%–12% sodium hydroxide and a sodium chloride content as high as 18%. Normally, the caustic solution is filtered to remove the impurities and then evaporated in a multiple-effect evaporator to 50% of sodium hydroxide.

The vapor resulting from the last of the evaporators is condensed in barometric condensers by contact with cooling water or in surface condensers using noncontact cooling water. Sodium chloride remains as a solid salt and is sent to the brine system. The salt separated from the caustic brine is recycled to saturate the dilute brine. Production of 50% caustic contains about 1% sodium chloride. For certain applications such as rayon production, purification of caustic is needed. The small amounts of impurities can be effectively removed using either extraction or adsorption techniques [1, 7, 11].

4.2.2 Membrane Cell Process

The membrane cell process was developed in the 1970s and quickly gained acceptance. Membrane cells are acknowledged as most efficient for chlor-alkali [9]. Figure 4.7 shows the simplified block diagram for the membrane cell process. In a membrane cell, a perfluoropolymer membrane containing cation exchange groups separates the anode and cathode compartments. This separator selectively transmits sodium ions, but suppresses the migration of hydroxyl ions from the catholyte to the anolyte.

The saturated brine is fed into the anode compartment where chlorine is liberated at the anode; the sodium ions migrate to the cathodic compartment along with some water. Hydrogen ionizes at the cathode compartment, and caustic soda will be produced as the hydroxyl ions combine with the sodium ions [1,11]. Because of the corrosive nature of the chlorine produced, the anode has to be made from a nonreactive metal such as titanium, whereas the cathode can be made from steel. Figure 4.8 illustrates the schematic of a membrane cell. About 50% of the sodium chloride is converted in the cell. The depleted brine is dechlorinated and returned to the brine purification system.

Waste Treatment and Management in Chlor-Alkali Industries

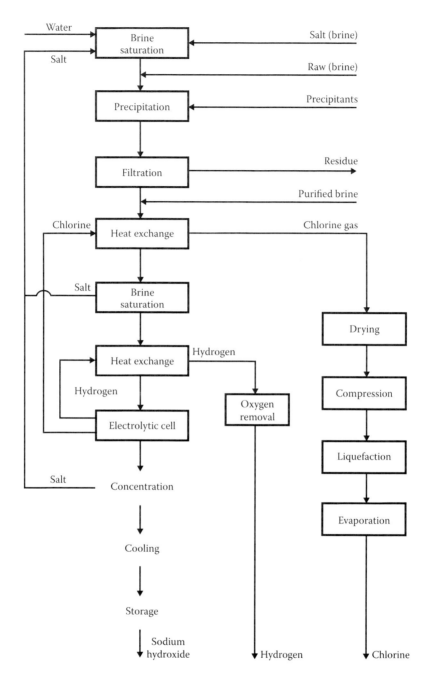

FIGURE 4.5 Diaphragm cell process flow. (From U.S. EPA; Multilateral Investment Guarantee Agency, Environmental guidelines for chlor-alkali industry, www.miga.org/documents/chloralkali.pdf.)

4.2.3 MERCURY CELL PROCESS

The mercury cell process was developed about the same time as the diaphragm cell process in the late 1800s. This process has been used extensively until the toxicological effects of mercury were discovered in the 1970s. Figure 4.9 shows the simplified basic layout and the waste streams of this process. The mercury cell has two sections—the decomposer or denuder and the electrolyzer. The electrolyzer is an elongated steel bar that inclines slightly from the horizontal.

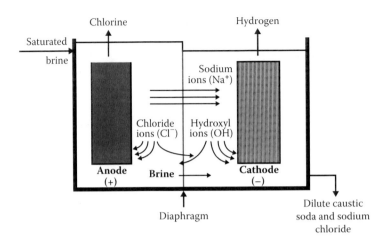

FIGURE 4.6 Schematic of a diaphragm cell.

This cell has a steel bottom with rubber-coated steel sides, as well as end boxes for brine and mercury feed and exit streams with a flexible rubber or rubber-coated steel cover. Figure 4.10 illustrates the schematic of a mercury cell. Mercury will flow through this cell and act as the cathode, while the brine flows on top of the mercury. Parallel activated titanium anode plates are suspended from the cover of the cell. The current flowing through the cell decomposes the brine, producing chlorine at the anode and sodium metal at the cathode. The sodium combines with mercury to form an amalgam. The amalgam flows from the electrolyzer to the decomposer [12].

In the United States, there are currently eight chlor-alkali plants still using mercury cell technology. A plant in Louisiana is expected to convert to nonmercury technology in 2007, and an Alabama plant is expected to close in 2008. The eight plants are located in seven states in the South and Midwest [13].

The sodium–mercury amalgam reacts with deionized water in the composer to form hydrogen and caustic soda in the presence of a catalyst. Graphite is the most common catalyst that is used for this process. The catalyst will be activated by oxides of iron, nickel, or cobalt, or by carbides of molybdenum or tungsten. A major part of mercury is removed in an initial cooling unit using water as a coolant and returned to the electrolyzer. The hydrogen gas is cooled by refrigeration for removing water vapor and mercury and is cooled further for additional removal of mercury prior to sale or use as fuel.

The impurities in the solution can be removed or reduced with the addition of certain chemicals and filtration processes. In most cases, the caustic solution then goes to storage or is evaporated if a more concentrated product is required. Generally, the caustic soda solution that flows out from the decomposer has a concentration of 50% sodium hydroxide.

The chlorine liberated at the anode is cooled to remove water, sodium chloride, and other impurities, including mercury. Normally, the condensate will be steam-stripped and returned to the brine system. After the cooling process, the chlorine gas is further dried by scrubbing with sulfuric acid. Sulfuric acid can be used until its concentration is 50%–70%. Then this diluted acid is regenerated for sale, reuse, or pH control. The dry chlorine gas is compressed and liquefied. The liquefying procedure results in a mixture residual of noncondensable gases, or tail gases, which are usually scrubbed with caustic soda or lime. The scrubbing process generates a hypochlorite solution, which is decomposed and used on site or sold.

About 12%–16% of the sodium chloride is converted in the cells. The brine that has been used is dechlorinated and then recycled using the brine purification process. The dechlorination process is

Waste Treatment and Management in Chlor-Alkali Industries

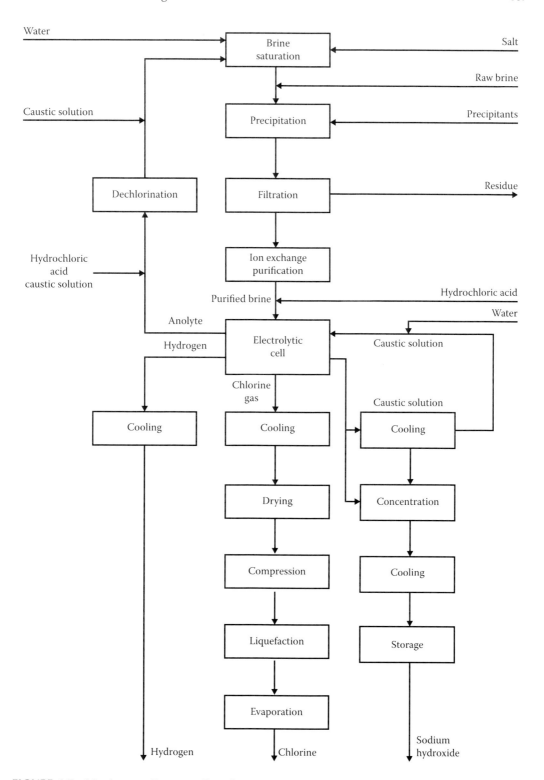

FIGURE 4.7 Membrane cell process flow diagram. (From U.S. EPA; Multilateral Investment Guarantee Agency, Environmental guidelines for chlor-alkali industry, www.miga.org/documents/chloralkali.pdf.)

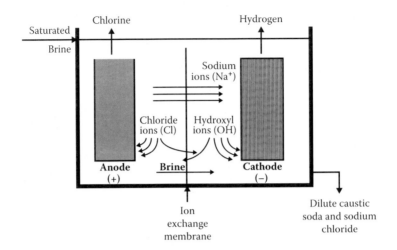

FIGURE 4.8 Schematic of a membrane cell.

carried out by acidifying the brine to a pH 2–2.5 and passing it down a packed column or spraying it into a vacuum vessel. The chlorine in the vacuum system that has been removed is combined with the chlorine stream [3–5].

4.2.4 Chlorine Processing

The chlorine gas from the anode compartment contains moisture, by-product oxygen, and some back-migrated hydrogen. In addition, if the brine is alkaline, it will contain carbon dioxide and some oxygen and nitrogen from the air leakage through the process or pipelines. Removal of the water droplets and the particulates of salt can be achieved by cooling the chlorine to 16°C (60F) and passing them through demisters. After the cooling process, the gas will go to a sulfuric acid tower, which is operated in series. Normally, removal of the moisture is done in three towers. Then, the dried chlorine goes via demisters before it is compressed and liquefied at low temperatures. Snift gas, which is a noncondensate gas, can be used to produce hydrochloric or hypochlorite acid. In the form of hydrochloric acid, the snift gas can be converted to hypochlorite acid by neutralizing it with caustic soda or lime. The hypochlorite is either sold as bleach or decomposed to form salt and oxygen.

4.2.4.1 The U.S. Chlorine Industry

Table 4.2 shows the data on chlorine production by plant and cell type for the United States in 1994. From the table, it can be seen that about 24 companies are engaged in chlorine production with a total cell capacity of 11,525 kt. It also shows that the largest companies include Dow, Occidental, PPG Industries, and Olin with a share of total capacity of 27%, 25%, 13%, and 7% respectively. The vast majority (83%) of production took place in the South, where companies are able to take advantage of low electricity prices and low labor costs. Total capacity utilization in 1994 was 95%. A total capacity of 2.1 million ton/year is expected to come online by 2000 in the United States. Figure 4.11 indicates the age distribution of plants. Most of the plants are 20–25 years old and some are considerably older [1].

4.2.5 Hydrogen Processing

The hydrogen gas from chlor-alkali cells is usually used for the production of hydrochloric acid or as a fuel to produce steam. First, hydrogen from the mercury cells is cooled to take out the mercury.

Waste Treatment and Management in Chlor-Alkali Industries

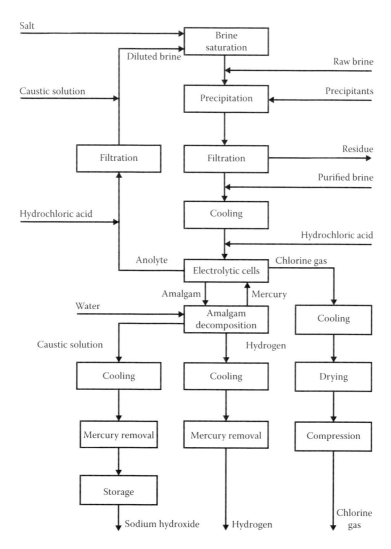

FIGURE 4.9 Mercury cell process flow diagram. (From U.S. EPA; Multilateral Investment Guarantee Agency, Environmental guidelines for chlor-alkali industry, www.miga.org/documents/chloralkali.pdf.)

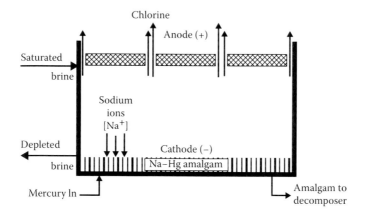

FIGURE 4.10 Schematic of a mercury cell.

TABLE 4.2
Chlorine Plants in the United States (Situation at the End of 1997)

Company	City	State	1994 Capacity (Kt/year)[a]	Age of Process in 1994[b,c]	Process[d]
Ashta Chemicals	Ashtabula	OH	36	31	KCl electrolysis; mercury cell; KOH by-product
BF Goodrich	Calvert City	KY	109	28	Salt; mercury cell; diaphragm
Dow	Freeport	TX	041	54	MgCl containing brines; diaphragm and by-product of magnesium metal production
	Plaquemine	LA	1075	36	Brine; diaphragm
DuPont	Niagara Falls	NY	77	96	Downs; by-product of metallic sodium production
Elf Atochem	Portland	OR	169	47	Salt; diaphragm
Formosa Plastics	Baton Rouge	LA	180	13	Brine; diaphragm (upgraded 1981)
	Point Comfort	TX	505	0	Membrane
Fort Howard	Green Bay	WI	8	26	Rocksalt; diaphragm
	Muskogee	OK	5	9	Rocksalt; membrane (upgraded in 1981)
	Rincon	GA	6	4	
GE	Burkville	AL	24	7	Membrane
	Mount Vernon	IN	50	18	Captive brine; diaphragm
Georgia Gulf Corp.	Plaquemine	LA	410	19	Captive brine; diaphragm
	Bellingham	WA	82	29	Salt; mercury cell
HoltraChem	Acme	NC	48	31	Salt; mercury cell
	Orrington	ME	73	27	Salt; mercury cell
laRoche Holding Inc.	Gramercy	LA	181	36	Brine; diaphragm
Miles Inc.	Baytown	TX	82	22	By-product HCl; HCl electrolysis
Niachlor Inc.	Niagara Falls	NY	218	7	Brine; membrane (upgraded in 1981)
Occidental	Convent	LA	279	13	Captive; salt dome; diaphragm
	Corpus Christi	TX	417	20	Brine; diaphragm
	Deer Park	TX	347	56	Captive salt dome; mercury cell; diaphragm
	Delaware City	DE	126	29	Salt mercury cell
	La Porte	TX	480	20	Captive salt dome; diaphragm
	Mobile	AL	41	3	Mercury cell; KOH is produced; membrane (upgraded in 1981)
	Muscle Shoals	AL	132	42	Mercury; KOH is produced
	Niagara Falls	NY	293	20	Captive salt dome; diaphragm (upgraded in 1974)

(*Continued*)

TABLE 4.2 (*Continued*)
Chlorine Plants in the United States (Situation at the End of 1997)

Company	City	State	1994 Capacity (Kt/year)[a]	Age of Process in 1994[b,c]	Process[d]
	Tacoma	WA	195	6	Rock salt; diaphragm; membrane (upgraded in 1975); membrane (upgraded in 1986)
	Taft	LA	581	19/8	Diaphragm/membrane
Olin Corporation	Augustus	GA	102	29	Salt; mercury cell
	Charleston	TN	230	32	Rock salt; mercury cell
	McIntosh	AL	365	17	Brine diaphragm (upgraded 1977)
	Niagara Falls	NY	82	34	Rock salt; mercury cell (upgraded 1960)
Oregon Metallurgical Corp.	Albany	OR	2	23	Magnesium chloride; by-product of metallic magnesium
Pioneer Chlor Alkali Co.	Henderson	NV	104	18	Salt; diaphragm (upgraded in 1976)
	St. Gabriel	LA	160	24	Salt; mercury cell
PPG Industries	Lake Charles	LA	1126	25/17	Brine; mercury cell (upgraded in 1969); diaphragm (upgraded in 1977)
			356	36/10	Brine; mercury cell (upgraded in 1958); diaphragm (upgraded in 1984)
Renco Group (Magnesium Corp. of America)	Rowley	UT	14	17	Brine; by-product of magnesium metal production
Vicksburg Chemical Co.	Vicksburg	MS	33	32	By-product of production of potassium nitrate from KCl
Vulcan Materials Co.	Geismar	LA	243	18	Brine; diaphragm
	Port Edwards	WI	65	27	Salt; mercury cell: KOH is also produced
	Wichita	KS	239	19/11	Brine; diaphragm (upgraded in 1975); membrane (upgraded in 1984)
Weyerhaeuser	Longview	WA	136	19	Brine; diaphragm (upgraded in 1975)
Total			11525		
Total production			10973[f]	29[e]	

Source: U.S. EPA; U.S. Environmental Protection Agency. Sector notebook for inorganic chemicals, Washington, DC, 1995a.

[a] Capacity numbers taken from Directory of Chemical Producers (DCP), SRI International.
[b] Time since last major upgrade.
[c] Cell type data and year of plant upgrade taken from Chlorine Institute Pamphlet 10 (1994).
[d] Process data taken from DCP.
[e] Capacity weighted average age.
[f] Chemical Manufacturers Association.

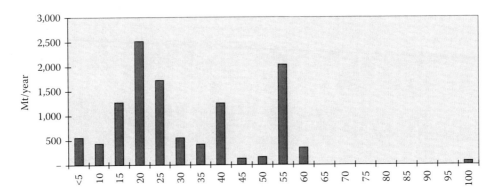

FIGURE 4.11 Age of U.S. chlorine plants (From U.S. EPA; Worrell, E., et al., *Energy Use and Energy Intensity of the U.S. Chemical Industry*, University of California, Berkeley, CA, 2000.)

After that, it is returned to the cell. Occasionally, it will go through a secondary treatment to remove trace levels of mercury in the hydrogen via molecular sieve columns. Then the hydrogen gas is normally compressed. The heat value in the hydrogen cell gas can be recovered in a heat exchanger via heating the brine feed to the cells.

4.2.6 Caustic Soda Processing

Caustic soda is marketed as 50%, 73%, or anhydrous beads or flakes. The mercury cell can produce 50% and 73% caustic directly. The caustic from the decomposer is cooled and passed once or twice through an activated carbon filter to reduce the mercury levels. After filtration, the mercury concentration in the caustic soda is lowered to parts-per-million (ppm) levels.

The mercury cell caustic soda has a few ppm salt and <5 ppm sodium chlorate. The mercury cell caustic is the highest purity caustic that can be made electrolytically if trace concentrations of mercury are tolerable in the end use of the caustic. The membrane cell caustic is concentrated in a multiple-effect falling film evaporator, which increases the caustic soda concentration to 50% with a high steam economy. Caustic soda from membrane cells generally has 30 ppm sodium chloride and 5–10 ppm sodium chlorate. The catholyte from the diaphragm cells contains ~12% sodium hydroxide, ~14% sodium chloride, 0.25%–0.3% sodium sulfate, and 100–500 ppm sodium chlorate. The catholyte is evaporated in a multi-effect evaporator. Most of the salt from the catholyte will precipitate during the concentration of the caustic soda to 50% sodium hydroxide.

4.2.7 Brine Processing

Sodium chloride exists in the form of solid salt that can be mined by excavation or by evaporating seawater. It is also available as a liquid by solution mining the salt domes. Salt has different concentrations of impurities. These impurities should be removed to operate the electrolytic cells at a high current efficiency. Calcium, magnesium, and sulfates are the major impurities. The other minor impurities include barium, strontium, manganese, aluminum, silica, iron, vanadium, chromium, molybdenum, and titanium, which are undesirable depending on the type of chlor-alkali process selected.

The solution-mined brine or the solid salt dissolved in a salt dissolver is treated in a reactor with sodium carbonate and caustic soda to precipitate calcium carbonate and magnesium hydroxide. These precipitates are settled in a settler. The underflow carries the solid slurry, which is pumped to a filter to remove it as sludge; sometimes, it is disposed of along with the rest of the liquid effluent from the plant. The calcium carbonate precipitates are heavy, and drag with it the hydroxides of aluminum, magnesium, strontium, etc. The overflow from the settler, which carries 10–50 ppm (parts per million) of suspended solids, is filtered.

In all the cell processes, the filtered brine is heated and passed through a bed of salt in a saturator in order to increase the salt concentration before feeding it to the electrolyzers. In some plants, the brine feed is acidified to improve the cell current efficiency. The acidification reduces the alkalinity, which would otherwise react with the chlorine in the anolyte compartment forming chlorate.

4.2.8 Sodium Hypochlorite Manufacturing Process

Another useful product generated by the electrolysis of weak brine is sodium hypochlorite, otherwise known as "bleach." Sodium hypochlorite cells generally do not require saturated brine, but can utilize weak brine or even seawater. Bleach is produced on-site for disinfection of drinking water and wastewater. The cells employed for this purpose are the same as those used for chlorate manufacture, that is, they consist of an anode and a cathode without a separator or a diaphragm. The anodic and the cathodic reactions are the same as in chlor-alkali and chlorate cells, the difference being the pH of the electrolyte, which is maintained in the range of 10–12. The electrolytically generated chlorine reacts with sodium hydroxide to form sodium hypochlorite. However, the hypochlorite ion, formed in bulk, is easily reduced at the cathode to re-form chloride. Therefore, only dilute solutions of bleach can be produced in the cell. Hypochlorite can also react further to form chlorate, but this can be minimized by keeping the solution basic and the temperature low, which is close to room temperature.

4.2.9 Sodium Chlorate Manufacturing Process

One of the energy intensive electrolytic industries is the production of sodium chlorate by the electrolysis of sodium chloride solutions in an electrolytic cell without a separator. The products of the electrode reactions, the chlorine and the caustic, are allowed to intermix and react, producing sodium chlorate as the final product.

The major raw material is sodium chloride, either very pure, such as solar rock salt, or partially purified evaporated salt. The salt is stored and dissolved in lixiviators to produce a saturated sodium chloride solution. This solution is purified by removing calcium, magnesium, fluoride, sulfate, and iron as insoluble compounds, through the addition of sodium carbonate, sodium hydroxide, sodium phosphate, and barium chloride.

The impurities or precipitates are removed in a pressure leaf filter, with diatomaceous earth as a filter precoat and filter aid. This filter cake, containing approximately 35% water, is the only solid waste stream from the process. A polishing filtration stage and an ion exchange system follow pressure leaf filtration.

The electrochemistry and chemistry of chlorate formation dictate that an efficient and economical cell should embody several distinct zones. In the electrolysis zone, the electrolytic reactions take place along with the hydrolysis of chlorine. Since the chemical chlorate formation proceeds very slowly, a relatively large volume of chemical reaction zone is needed. A cooling zone is also required to remove the excess heat generated from the reaction and control the operating temperature. The cooling zone may be located within the chemical reactor or in an external heat exchanger. Hydrogen gas generated at the cathode must be released from the cell liquor. This hydrogen release takes place in an electrolysis cell, a separate vessel, or a chemical reactor.

A continuous stream of cell liquor flows from the electrolysis system to the "hypo removal" system, where the sodium hypochlorite concentration is reduced to low levels simply by heating the cell liquor to about 185–200 F (85°C–95°C) under careful pH control. Final traces of hypochlorite can be completely removed by treatment with a reducing agent such as sodium sulfite or hydrogen peroxide.

Sodium chlorate is usually recovered from cell liquors by concentration, followed by cooling to facilitate crystallization. Hot cell liquor, following hypo removal, is fed continuously into the circulation leg of a draft tube baffle evaporator/crystallizer. Crystal slurry is withdrawn from the bottom of the crystallizer section. The crystals are separated from the mother liquor and washed with water in a pusher centrifuge. They are thoroughly washed to remove sodium dichromate, which is an additive to the cell solution to increase current efficiency, from the chlorate crystals. Sodium dichromate contains chromium in the hexavalent state, which is a recognized human carcinogen. A white sodium chlorate crystal, containing about 1%–1.5% moisture, is obtained from the centrifuge. Mother liquor from the centrifuge is mixed with fresh purified brine and recycled to the electrolytic cells.

Approximately 98% of the sodium chlorate capacity in North America is produced directly in sodium chlorate cells. The remaining 2% is produced "chemically" by the reaction of chlorine and caustic.

4.3 ENERGY REQUIREMENTS FOR THE CHLOR-ALKALI INDUSTRY

Production of chlorine and caustic soda is the third most important product from the energy perspective. Chlorine is produced through electrolysis of a salt solution. Chlorine production is the main electricity-consuming process in the chemical industry. Table 4.3 identifies estimated final energy consumption and carbon emissions in 1994 for chlorine/caustic production. The U.S. Manufacturing Energy Consumption Survey reports that the total final energy consumption in 1994 for the chlor-alkali industry was 136 PJ, with a net electricity consumption of 49 PJ [1].

Electricity fuels the electrolysis process and represents the primary energy source. The amount of electricity required depends on the design of the cell, the operating current, concentration of electrolytes, temperature, and pressure. The values showed in Table 4.4 represent an average energy consumption for the various cell types. Energy in the form of fuels or steam is used primarily for evaporation of the sodium hydroxide solution to a useable state. Some fuels are also consumed in the production and purification of brine feedstock before it is sent to the electrolysis cell.

Among the three types of chlorine cells, the mercury cell is the most energy intensive, with electricity requirements of nearly 3600 kWh per metric ton of chlorine. The membrane cell is the least energy intensive in terms of both steam and electricity requirements. Steam requirements are less than half those of the diaphragm or mercury cell. Electricity requirement for the membrane cell is in the range of 2800 kWh per metric ton of chlorine. The diaphragm cell is intermediate between these energy consumption ranges.

Overall, the electrical energy requirements for chlorine electrolysis cells are high, accounting for nearly 130 trillion Btu annually. When losses due to transmission and generation of electricity are considered, they reach nearly 400 trillion Btu annually. Thus, efficient operation of the cell is critical for optimized energy use and cost-effective production. Sources of energy losses in chlorine cells include anode or cathode overvoltage, too large a drop across the diaphragm, oxygen evolution at the anode, and failure to recover heat and energy from hydrogen, chlorine, and cell liquor streams. A key consideration in membrane processes is the purity of the brine. Using very pure brine at an optimum flow rate minimizes blockage through the membrane and allows sodium to penetrate freely. Brine purity is also important in mercury cells. Impurities tend to increase hydrogen by-product and reduce current efficiency. Another issue is brine flow rate. High flow rates will increase the cell temperature and electrical conductivity of the medium. When Brine rates are too low, temperatures and high cell voltages are created that are higher than the most efficient voltage (3.1–3.7 V).

At present, chlorine production in the United States is dominated by the diaphragm cell; the use of the more energy-intensive mercury cell continues to decline. In Europe, however, production is still dominated by the use of the mercury cell, although European manufacturers have committed to build no new mercury cells in the future [14].

TABLE 4.3
Energy Use and Carbon Dioxide Emissions in the Chlorine Industry

Process Stage	Final Electricity Intensity[a] (GJ/t Cl$_2$)	Fuel Intensity[a] (GJ/t Cl$_2$)	Final Energy Intensity[a] (GJ/t Cl$_2$)	Final Electricity Use[b] (PJ)	Final Fuel Use[b] (PJ)	Final Energy Use[b] (PJ)	Carbon Emission[c] (KtC)
Rectifier	0.28	0.00	0.28	3.1	0.0	3.1	132
Brine preparation	0	0.02	0.02	0.0	0.2	0.2	10
Cell use	13.63	0.00	13.63	149.6	0.0	149.6	6365
NaOH concentration	0	3.42	3.42	0.0	37.6	37.6	435[d]
NaOH cooling	0.27	0.00	0.27	2.9	0.0	2.9	125
Hydrogen cooling/drying	0.58	0.00	0.58	6.4	0.0	6.4	272
Chlorine cooling/drying	0.39	0.00	0.39	4.3	0.0	4.3	181
Chlorine compression	0.63	0.00	0.63	6.9	0.0	6.9	295
Total	15.78	3.45	19.23	173.20	37.81	211.02	7817
Hydrogen heating value			−3.35[e]			−36.8[e]	—

Source: U.S. EPA; Ayres, R., *J. Ind. Ecol.*, 1(1), 81–94, 1997.

[a] 1994 Industry final energy intensity based on a calculation of weighted energy use by cell type and plant age. Cells and intensities for a modern 1989 plant were: membrane (9.8 GJ/t electricity, 0.5 GJ/t fuel), diaphragm (11.3 GJ/t electricity, 2.6 GJ/t fuel), mercury (12.3 GJ/t electricity, 0.1 GJ/t fuel) (Doesburg 1994). Shares of cell type for 1994 are: membrane 8%, diaphragm 75%, mercury 15%. These are based on the study by Kirk-Othmer (1994) and currently exclude plants that manufacture chlorine from sources other than sodium chloride (e.g., magnesium chloride). Using the 1989 efficiencies for each cell type, and an assumed 10% efficiency improvement over 20 years, we have calculated efficiency for each plant and taken a production weighted average.

[b] Industry total energy use based on a production of 11.05 Mt of chlorine in 1994. Electricity production from cogeneration is currently not included in the calculation due to lack of reliable data.

[c] Carbon emissions are based on the following factors from EIA, 1995. Electricity 42.6 KtC/PJ, Oil 20.4 KtC/PJ.

[d] This value assumes that half the hydrogen produced is used to concentrate the caustic. The rest is sold or used in other production processes.

[e] This is the fuel value of hydrogen. It is listed as a negative because it represents energy that can be used to fuel other processes.

TABLE 4.4
Estimated Energy Use in Manufacture of Chlorine/Sodium Hydroxide (Caustic Soda) Co-products—1997

Energy	Specific Energy[d] (Btu/lb of Cl_2)	Average Specific Energy[d] (Btu/lb of Cl_2)	Total Industry Use[e] (10^{-12} Btu)
Electricity[a]	4353–5561	4957	128.8
Fuel oil and LPG[b]	62–112	87	2.3
Natural gas	1601–2880	2240	58.3
Coal and coke	208–374	291	7.6
Others[c]	208–374	291	7.6
Net processing energy	6432–9301	7867	204.5
Electricity losses	9039–11547	10293	267.6
Energy export	0	0	0.0
Total process energy	15471–20848	18160	472.2

Source: U.S. EPA; Worrell, E., et al., *Energy Use and Energy Intensity of the U.S. Chemical Industry*, University of California, Berkeley, CA, 2000; IND CHEM 1990; Brown 1996; Orica 1999; Ayres 1999.

[a] Does not include losses incurred during the generation and transmission of electricity. Conversion factor: 3412 Btu/kWh. Based on the range of cell currents for diaphragm, mercury, and membrane cells. Includes electricity for brine purification.

[b] Includes ethane, ethylene, propane, propylene, normal butane, butylenes, and mixtures of these gases.

[c] Includes net purchased steam and any other energy source not listed (e.g., renewables).

[d] Steam/fuel use is estimated based on current distribution of fuels in chemical plants [8]. Values are based on the published fuel use and electricity requirements for diaphragm, mercury, and membrane cell technologies. Includes steam for brine purification and caustic evaporation.

[e] Calculated by multiplying average energy use (Btu/lb) by 1997 production values for chlorine (26 billion lbs) [8]. Note that 1.12–1.43 lbs of sodium hydroxide are produced for every lb of chlorine.

4.4 WASTE CHARACTERIZATION

In this section, we will discuss the characterization of wastes generated in the chlor-alkali plants. Wastes that are generated from this plant are solid, liquid, and gaseous. The process flowcharts for wastes generated in the mercury cell, diaphragm cell, and membrane cell processes are shown in Figures 4.12 through 4.14. Tables 4.5 and 4.6 describe the waste streams in more detail.

4.4.1 SOLID WASTES

Solid wastes that are generated in chlor-alkali plants are produced from the diaphragm cell plants and the membrane cell plants. The major waste stream from the processes consists of brine mud, which is the sludge that is produced from the brine purification step. Brine muds are the main sources of solid wastes in chlorine plants. The quantity of brine muds generated is in the range 10–45 kg per metric ton of chlorine. This varies with the purity of the salt used to produce the brine. For example, pre-purified salt will generate about 0.7–6.0 kg per 1000 kg of chlorine produced.

This sludge may contain a variety of compounds, typically magnesium, calcium, iron, and other metal hydroxides. It depends on the source and purity of the brines. The sludge or brine mud containing these impurities must be disposed of in a landfill [4,6,7,15]. If a mercury process is being used, the brine mud may contain trace levels of mercury. Mercury concentration in brine muds from mercury cell plants varies from 13 to 1000 ppm, with an average concentration of 200 ppm. In this case, the sludge has to be treated with sodium sulfide to form mercury sulfide, which is an insoluble

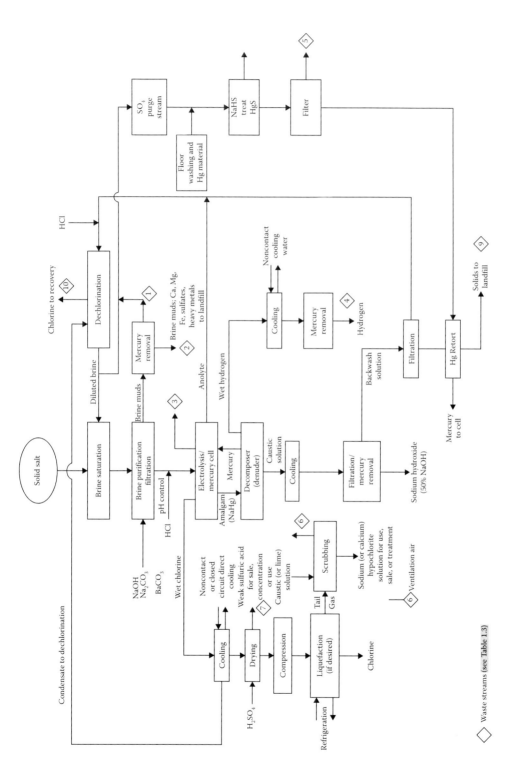

FIGURE 4.12 Process flow and waste stream of mercury cell. (From World Bank and United Nations Industrial Development Organization. Industrial pollution prevention and abatement: Chlor-alkali industry. Pre-publication Draft, Environment Department, Washington, DC, 1994.)

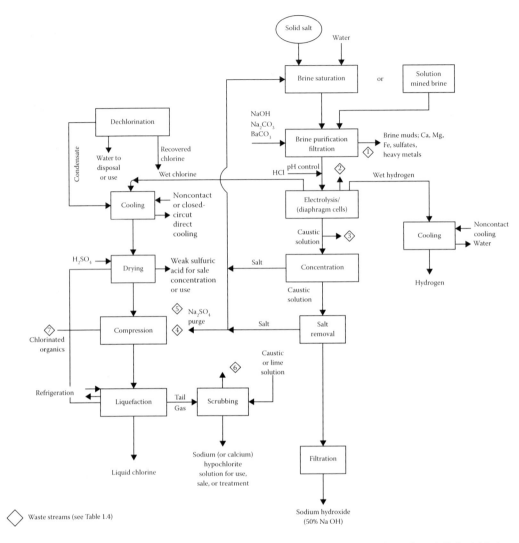

FIGURE 4.13 Process flow diagram for diaphragm cell process. (From World Bank and United Nations Industrial Development Organization. Industrial pollution prevention and abatement: Chlor-alkali industry. Pre-publication Draft, Environment Department, Washington, DC, 1994.)

compound. The sludge is further treated by casting it into concrete blocks, which are treated for leachibility and sent to a controlled landfill.

The worldwide activity of chlor-alkali plants that use elemental mercury (Hg) for the production of chlorine and caustic soda was estimated to be responsible for the emission of 9.5 tons Hg in 1998. In Europe, this amount has now significantly decreased due to the recent legislation that imposes the use of alternative and less polluting processes. Although the use of elemental Hg as cathode for chlor-alkali plants has been forbidden in Europe since 2008, some facilities benefit from specific authorizations and are still in operation [16].

Diaphragm and membrane cell processes generate solid wastes from scrapping of cell parts including cell covers, piping, used membranes, cathodes, and anodes. The cell parts that are discarded are landfilled on-site or shipped off-site to a third party recovery facility. The cathodes may be refurbished and reused, particularly those made of nickel, but it depends on the cell technology. The anodes for diaphragm cells are refurbished and re-coated with RuO_2/TiO_2

Waste Treatment and Management in Chlor-Alkali Industries

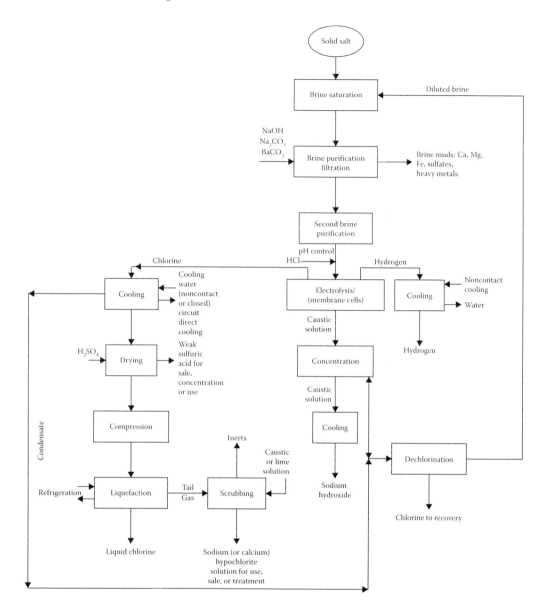

FIGURE 4.14 Process flow diagram for membrane cell process. (From U.S. EPA; Electrochemistry Encyclopedia, Brine Electrolysis, http://electrochem.Cwru.edu/ed/encycl/art-b01-brine.htm.)

and returned to service. A significant quantity of asbestos is present in the water released from a diaphragm cell plant, which originates from wash down and cell repair or cleaning. Asbestos from cell wash operations and precipitated solids from metal treatment generate a solid waste of 0.83 kilogram per ton (kg/t) of chlorine.

Solid wastes from mercury cell plants include spent graphite from decomposer cells and spent caustic filtration cartridges from the filtration of caustic soda solution. Mercury cell plants are also expected to produce some mercury that is present in the solid waste, which is discharged by this plant. Mercury is recovered from these wastes where possible, and the remainder is disposed of in secured landfills to prevent migration of mercury, which can cause damage to

TABLE 4.5
Waste Streams from a Modern Mercury Cell Process

Stream	Source and Comments	Components	Quantity (kg/1000 kg Cl$_2$)	Normal Deposition
1	Precipitated impurities from brine contain adsorbed soluble mercury removed by water washing	Complex and ionic mercury in aqueous solution	n.a.	Wash water sent to sulfide precipitation system
2	Waste sludge from brine purification	CaCO$_3$, Mg(OH)$_2$, heavy metal hydroxides Mercury (ionic)[a] Water	17 0.5 × 10^{-3} 55	To lagoon
3	Rubber cell linings, decomposer graphite packing, etc., from cell rebuilding operations	Graphite, rubber, steel, some adsorbed mercury	3	Mercury-containing materials retorted to distill out mercury for recycling; residue and other materials to landfill
4	Hydrogen from decomposer contains mercury vapor at equilibrium pressure of mercury temperature	Mercury vapor[a]	6.4 × 10^{-7}	Hydrogen refrigerated to condense and recycle mercury, then passed through carbon filter process for mercury recovery
5	Purge stream of brine plus wash water from purification sludge washing, and housekeeping water after sulfide treatment for mercury removal	Salt Water Miscellaneous dirt Mercury at (20 parts per billion)[a]	80 800 0.5 2 × 10^{-5}	Discharge to sewer or receiving water
6	Cell house air changed, to keep Hg ≤50 µg/m^3	Air Mercury vapor[a]	2.5 × 10^{-3}	Discharge to atmosphere
7	Diluted sulfuric acid from chlorine drying towers	H$_2$SO$_4$ (93%–70%)	12	Sold or used for reconcentration
8	Final vent scrubbers after normal removal of chlorine from tail gas and other vents. Dilute caustic used as an absorbent.	NaOCl NaHCO$_3$ Water (approx.)	2 Negligible 40	To lagoon or sewer
9	Filter aid from caustic filter and carbon particle removal	Activated carbon (containing mercury and graphite particles)	<1	Treated for mercury recovery and recycle, then sent to landfill
10	Vent from air blowing of recycled brine for dechlorination	Air Chlorine	 0.45	In-plant water chlorination, caustic scrubbing or vent to atmosphere

Source: U.S. EPA; Guédron, S., et al., *Sci. Total Environ.*, 445–446, 356–364, 2013.
n.a., data not available.
[a] Amount of mercury shown is after treatment.

TABLE 4.6
Waste Streams from a Diaphragm Process

Stream	Source and Comments	Components	Quantity (kg/1000 kg Cl$_2$)	Normal Deposition
1	Sludge from brine purification, amount and nature depends upon salt purity	Mg(OH)$_2$, CaCO$_3$, iron and other metal hydroxides	15	To lagoon
		Filter aid	0.20	
		NaCl	3.00	
		Water	45.00	
2	Materials from rebuilding of electrolytic cells	Concrete and graphite rubble	3.00	To landfill
		Asbestos fibers	0.40	
3	Housekeeping operations, equipment cleaning	NaCl	50.00	To lagoon
		NaOH	5.00	
		Lead	0.04	
		Copper	<0.01	
		Water (approx.)	600.00	
4	Recycle salts from evaporators purified to remove sodium sulfate	Na$_2$SO$_4$	10.00	To lagoon, or neutralized and discharged into receiving water, where allowable
		NaOH	20.00	
		NaCl	50.00	
		Water (approx)	150.00	
5	Diluted sulfuric acid from chlorine drying towers	H$_2$SO$_4$ (93%–70%)	10–35	Sold for use or for re-concentration
			10.00	
6	Final vent scrubber after normal removal of chlorine from tail gas and other vents. Dilute caustic soda or carbon tetrachloride used as absorbent	NaOCl	1.00	To lagoon or sewer
		NaHCO$_3$	3.00	
		Water (approx.)	40.00	
7	Product of chlorine purification	Chlorinated organics of unspecified nature	0.45	Discharged to drums for disposal by incineration or in landfill

Source: U.S. EPA; Multilateral Investment Guarantee Agency, Environmental guidelines for chlor-alkali industry, www.miga.org/documents/chloralkali.pdf.

the environment. Mercury cell brine mud may also contain mercury in elemental form or as mercuric chloride. These muds are considered hazardous and must be disposed of in an RCRA Subtitle C landfill after being treated with sodium sulfide, which creates an insoluble sulfide compound. Wastewater treatment that is conducted in mercury cell plants generates 1.4 kg of sludge per metric ton of chlorine produced. Mercury content in these wastes is 0.2%–25%, with an average concentration of 0.42% mercury [7].

Table 4.7 summarizes the effluents, solid wastes, and hazardous wastes from chlor-alkali production. Specific waste from chlor-alkali manufacture that is listed by the U.S. Environmental Protection Agency (U.S. EPA) as hazardous is shown in Table 4.8. From the table, it can be seen that most of the hazardous wastes are generated from the mercury cell process.

The caustic filter wash down and the cell room wastes also produce solid wastes. Concentration of solids in the filter backwashing can vary from 2% to 20% and the volume in the range from 0.04 to 1.5 m^3/t of chlorine produced. The dissolving and clarification steps in the production of

TABLE 4.7
Summary of Effluents and Solid and Hazardous Waste Streams from Chlorine/Sodium Hydroxide Production.

Source	Diaphragm/Membrane Cells	Mercury Cells	Caustic Evaporation	Chlorine/Hydrogen Processing	Brine Purification
Wastewater	Ion exchange wash water, cell wash water, brine purge	Brine pumps, cell wash, sumps (small amounts of mercury)	5 t water/t 50% caustic soda solution	79% sulfuric acid solution (6–35 kg/1000 kg chlorine)	
Solid wastes	Scrapped cell parts (cell covers, piping, used diaphragm, used membranes, and anodes)	Spent caustic filtration cartridges from the filtration of caustic soda solution, spent graphite from decomposer		Spent activated carbon	Brine mud (0.7–30 kg/1000 kg chlorine)
Hazardous wastes	Chlorinated hydrocarbon waste from the purification step of the diaphragm cell process (K073)	Spilled mercury from sumps and mercury cell "butters," wastewater treatment sludge (K106)			Brine mud containing mercury (K071)

Source: U.S. EPA; Nunes, S.P., Peinemann, K.-V., *Membrane Technology in the Chemical Industry*, Wiley-VCH, Weinheim, Germany, 2006.

TABLE 4.8
Hazardous Waste from Chlor-Alkali Manufacture

Waste Classification	Description	Hazardous Constituents
K071	Brine purification mud from the mercury cell process for chlorine production, where separately prepurified brine is not used	Mercury
K073	Chlorinated hydrocarbon waste from the purification step of the diaphragm cell process	Chloroform, carbon tetrachloride, hexachloroethane, trichloroethane, tetrachloroethylene, 1,12,2- tetrachloethylene
K106	Wastewater treatment sludge from the mercury cell process for chlorine production	Mercury

Source: U.S. EPA; Nunes, S.P., Peinemann, K.-V., *Membrane Technology in the Chemical Industry*, Wiley-VCH, Weinheim, Germany, 2006.

sodium carbonate also create a waste sludge containing nonhazardous impurities, such as salts and minerals. This sludge is disposed of in landfill.

4.4.2 Liquid Effluents

Liquid effluents are different from plant to plant, and this depends on various factors such as by-product recovery, recycling practices, housekeeping procedures, and contact versus noncontact water usage. Acid and caustic wastewaters are generated in both the process and the materials recovery stages. Table 4.9 presents wastewater quantities of various plant sources in mercury, diaphragm, and membrane cell processes.

TABLE 4.9
Average Wastewater Flow (Cubic Meters per Cubic Ton of Chlorine Produced Using Metal Anodes)

Source	Diaphragm Plant	Mercury Plant	Membrane Plant
Cell room wastes and cell wash[a]	0.02–0.67	0.01–1.5	0.02–0.67[b]
Chlorine condensate[c]	0.16–0.90	0.01 (1 plant)	n.a.
Spent sulfuric acid	0.01	n.a.	0.02
Tail gas scrubber	0.10–0.29	0.04–0.58	n.a.
Caustic filter wash	0.42	n.a.	Recycled to system
Caustic cooling blow down	0.82–0.89	0.01–0.45	n.a
Brine mud	0.04–1.5	n.a.	0.01–0.15
Total wastewater	2.4	2.1(0.36–6.3)	2.5

Source: Energetics Incorporated. *Energy and Environmental Profile of the U.S. Chemical Industry*. U.S. Department of Energy, Washington, DC, 2000.

[a] Cell washing is not required during normal operation. Effluents are generated during cell washing only if the electrolyzer is disassembled.
[b] Similar to diaphragm cell (depends on operating practices).
[c] Chlorine condensate is dechlorinated and recycled to the system.
n.a., data not available.

Each process needs a brine purification stage that produces brine mud wastes, which are filtered or settled in lagoons. The overflow or filtrate will be discharged or recycled to the brines system. The average filtrate volume was about 0.42 cubic per metric ton (m^3/t) of chlorine produced for the diaphragm cell process.

The volume of cell room liquid wastes depends on housekeeping practices and varies widely from plant to plant. Liquid wastes from the cell room include leaks, spills, area wash down, and cell wastewater. In the waste streams of diaphragm and mercury cell plants, the treatment of cell room wastes is the major requirement because of the presence of asbestos and mercury. However, the wastes from older plants may also contain lead and chlorinated organics because of the application of graphite anodes rather than activated titanium anodes. The volume of wastewater varies widely in each process: the volume of wastewater from a mercury cell plant may range from 0.01 to 1.5 m^3/t of chlorine produced; in a diaphragm cell plant using metal anodes, the average wastewater volume is about 0.38 m^3/t of chlorine produced. It is different when using graphite anodes where the average wastewater flow is about 1.2 m^3/t of chlorine produced. Only membrane cell room wastes do not contain any unique contaminants of concern.

Generally, condensate from the indirect cooling of gas cell gas is contaminated with chlorine. The chlorine is removed and/or recovered from the liquid stream before the condensate is discharged or recycled. Average wastewater flow in diaphragm cell plants that use graphite anodes is 0.78 m^3/t of chlorine, while for metal anodes plants it is 0.49 m^3/t of chlorine produced.

During chlorine gas processing, water vapor is removed by scrubbing with concentrated sulfuric acid. Between 6 and 35 kg of 75% sulfuric acid wastewater is generated per 1000 kg of chlorine produced [15]. Most of this wastewater is shipped off-site for processing into concentrated sulfuric acid or for use in other processes. The remainder is used for pH control or discharged to water treatment facilities for disposal [15].

The liquefaction stage will produce tail gas containing uncondensated chlorine gas with some air and others gases. This tail gas will be scrubbed with sodium/calcium hydroxide to form sodium/calcium hypochlorite solution. The scrubber tail gas that is released from mercury cell plants is in the range of 0.04–0.58 m^3/t of chlorine produced. In diaphragm cell plants, the average discharge from the tail gas scrubber for metal and anode plants is 0.17 and 0.11 m^3/t of chlorine, respectively. Additional hypochlorite solution is produced when the equipment is purged for maintenance. The hypochlorite can be sold, decomposed, or used on-site in other processes.

The clarifying process of caustic products by backwashing filters may produce significant quantities of wastewater, which are wholly or partially recycled to the process. In addition to salt, small amounts of mercury or asbestos fibers may be present in the waste from the mercury cell and diaphragm cell processes. The filtered solids either may be disposed of on land after treatment or can be reprocessed for recovery of mercury in the case of mercury cell plant.

The caustic solutions that are produced from the electrolyzer in diaphragm and membrane cell plants are normally concentrated by evaporation to achieve 50% product. Caustic evaporation where the sodium hydroxide solution is concentrated to a 50% or 70% solution evaporates about 5 t of water per ton of 50% caustic soda produced. A significant amount of wastewater is generated when vapors are cooled by direct contact cooling water on a once-through basis. The discharge of wastewater can be reduced greatly by recirculation of the cooling water, but it needs a cooling step and a blow down discharge.

In diaphragm cell plants, sodium chloride will precipitate out during the concentration of the caustic soda solution. It will contain about 15% caustic soda solution and a relatively high salt content of 15%–17%. The salt is removed and washed with water to remove sodium sulfate. If sodium sulfate is not removed during the brine purification process, salt recovered from the evaporators may be recrystallized to avoid buildup of sulfate in the brine. A portion of water from this process will be recycled and the remainder is purged to prevent the buildup of sulfate. If the salt is recrystallized, the wastewater may contain sodium sulfate. This purge stream is the major source of wastewater from diaphragm cell plants. The stream can be controlled by treating it with calcium chloride ($CaCl_2$), but it depends on the sulfate content of the feed salt to generate calcium sulfate ($CaSO_4$). The calcium sulfate is settled and filtered together with brine sludge and is disposed of as solid waste. This is the treatment that may assist in salt recovery and converts liquid waste into solid waste.

Wastewater from membrane processes contains caustic soda solution, but is virtually free of salt or sodium sulfates. Wastewater from caustic soda processing is usually neutralized with hydrochloric acid and discharged to receiving ponds and lagoons.

Mercury is the main pollutant of concern in wastewater that is released by mercury cell plants. Asbestos is the separator material that is used in the diaphragm cell plants for separating the anode from the cathode and is the major pollutant in the diaphragm cell process wastewater.

The average amount of liquid effluent flow from mercury cell plants is $2.1\,m^3/t$ of the chlorine produced, totally. But this does not include brine mud water, which is reused instead of being discharged. Hence, it does not affect the total flow. Wastewater streams are generated from mercury cells during the chlorine drying process, brine purge, and from other sources. Mercury is present in the brine purge and other sources such as floor sumps and cell wash water in small amounts. This mercury is generally present in concentrations ranging from 0 to 20 ppm, and is precipitated out using sodium hydrosulfide to form mercuric sulfide. Mercuric sulfide is removed through filtration before the water is discharged [15]. In old installations, good operations can restrict total mercury emissions to <10 grams per ton (g/t) chlorine, while in the new plants the total measured emissions can be <3 g/t.

Ion-exchange wash water from membrane cell processes normally contains dilute hydrochloric acid with small amounts of dissolved calcium, magnesium, and aluminum chloride. This wastewater is usually treated along with other acidic wastewater by neutralization.

Sodium carbonate manufacture creates significant volumes of wastewater that must be treated before the discharging and recycling process. The wastewater may contain both mineral and salt impurities [18]. The limitation for toxic or hazardous compounds contained in these wastewaters are given by the U.S. EPA in 40 CFR, Chapter I, Part 415, which was originally promulgated in 1974 and has been revised several times since then.

Pollution parameters that are important in chlor-alkali plants are total suspended solids (TSS) and hydrogen ion concentration (pH). The specific limitations for restricted compounds and TSS are given in Tables 4.10 through 4.12. Table 4.12 provides best practical control technology (BPT) limitations only for the production of sodium chloride by solution brine mining. In this process, water is pumped into a salt deposit, and the saturated salt solution is removed that is about one-third of the salt.

TABLE 4.10
Effluent Pretreatment Standards: Mercury Cell

Effluents	BPT Standards: Average of daily Values for 30 Consecutive Days (lb/1000 lb Product)	BAT Standards: Average of daily Values for 30 Consecutive Days (lb/1000 lb Product)	NSPS Standards: Average of Daily Values for 30 Consecutive Days (lb/1000 lb Product)
TSS	0.32	—	0.32
Mercury	0.00014	0.00010	0.00010
Total residual chlorine	—	0.0019	0.0019
pH	6–9	6–9	6–9

Source: Energetics Incorporated. *Energy and Environmental Profile of the U.S. Chemical Industry.* U.S. Department of Energy, Washington, DC, 2000.

Note: 40 CFR Chapter 1, Part 415, Inorganic Chemicals Manufacturing Point Source Category, Subpart F.

TABLE 4.11
Effluent Pretreatment Standards: Diaphragm Cells

Effluents	BPT Standards: Average of Daily Values for 30 Consecutive Days (lb/1000 lb Product)	BAT Standards: Average of Daily Values for 30 Consecutive Days (lb/1000 lb Product)	NSPS Standards: Average of daily Values for 30 Consecutive Days (lb/1000 lb Product)
TSS	0.51	—	0.51
Copper	0.0070	0.0049	0.0019
Lead	0.010	0.0024	0.0019
Nickel	0.0056	0.0037	—
Total residual chlorine	—	0.0079	0.0079
pH	6–9	6–9	6–9

Source: Energetics Incorporated. *Energy and Environmental Profile of the U.S. Chemical Industry.* U.S. Department of Energy, Washington, DC, 2000.

TABLE 4.12
Effluent Pretreatment Standards: Production of Sodium Chloride by Solution Brine Mining

Effluent	BAT Standards: Average of Daily Values for 30 Consecutive Days (lb/1000 lb Product)
TSS	0.17
pH	6–9

Source: Energetics Incorporated. *Energy and Environmental Profile of the U.S. Chemical Industry.* U.S. Department of Energy, Washington, DC, 2000.

Small amounts of metals such as arsenic, antimony, cadmium, chromium, copper, lead, nickel, silver, thallium, and zinc may be present in the wastewater. The main sources of these metals are the raw salt or brine and the products of corrosion reactions between chlorine and the materials in the process equipment. Treatment steps of mercury such as sulfide precipitation followed by filtration can also reduce the level of these toxic metals to achieve standard limits. Precipitation of the toxic metals in diaphragm and membrane process cell plants can be accomplished by the addition of soda ash.

4.4.3 AIR EMISSION

Air emissions that are produced by the chlor-alkali plants include fugitive chlorine gas, carbon dioxide, carbon monoxide, and hydrogen [17]. Gaseous emissions of chlorine reaching the environment can be produced from fugitive emissions, plant start-up and shutdown, and abnormal operating conditions. Fugitive emissions arise from cell scrubbers and vents throughout the system. While individual leaks may be minor, the combination of fugitive emissions from various sources can be substantial. In 1995, nearly 3 million pounds of fugitive chlorine and point source emissions were reportedly released by the inorganic chemical industry [18]. These emissions can be controlled through leak-resistant equipment modifications, source reduction, and programs to monitor such leaks. A caustic scrubber system should be installed to handle gas emissions from sources such as vents on returned tank cars, cylinders, storage tanks, and process transfer tanks during handling and loading of liquid chlorine.

TABLE 4.13
Mercury Emissions from Chlorine/Caustic Soda Manufacture

Source	Mercury Gas (lb/t of Chlorine Produced)	Mercury Gas (g/metric t of Chlorine Produced)
Uncontrolled hydrogen vents	0.0033	0.17
Controlled hydrogen vents	0.0012	0.6
End box	0.010	5.0

Source: World Bank and United Nations Industrial Development Organization. Industrial pollution prevention and abatement: Chlor-alkali industry. Pre-publication Draft, Environment Department, Washington, DC, 1994.

Chlorine is a highly toxic gas, and strict precautions are necessary to minimize risk to employees and possible release during its handling [19]. Diaphragm cells and membrane cells release chlorine as fugitive emissions from the cell itself and in process tail gases, which are wet-scrubbed with soda ash or caustic soda to remove chlorine. The spent caustic solution from this wash is neutralized and then discharged to water treatment facilities [5,15]. Mercury cells release small amounts of mercury vapor and chlorine gas from the cell itself. Process tail gases from chlorine processing, caustic soda processing, and hydrogen processing also release small amounts of emissions. Mercury is removed from the hydrogen gas stream by cooling, followed by absorption with activated carbon [5]. Table 4.13 lists emission factors for mercury emissions from chlor-alkali production [5]. Mercury emissions from chlorine production have dramatically declined in response to tighter regulations, better housekeeping, and technology improvements. However, there are considerable variations in the estimated mercury emissions from chlorine manufacture [14].

Brine preparation and caustic evaporation processes release emissions through the combustion of fuel in process heaters and in boilers that produce steam. When operating in an optimum condition and burning cleaner fuels such as natural gas and refinery gas, these heating units create relatively low emissions of SO_x, NO_x, CO, particulates, and volatile hydrocarbon emissions. Production of sodium carbonate also release particulate emissions that are created from ore calciners, soda ash coolers and dryers, screening and transportation, product handling and shipping, and ore crushing. Combustion products such as SO_x, NO_x, CO, particulates, and volatile hydrocarbons are emitted from direct-fired process heating units.

More than 88% of emissions from the use of sodium carbonate were associated with flue gas desulfurization. In 1997, the reported annual emissions of carbon dioxide from the manufacture of sodium carbonate were 1.08 million metric tons and emissions from the use of sodium carbonate were 0.86 million metric tons [20]. Emissions of particulates from calciners and dryers are most often controlled by venture scrubbers, electrostatic precipitators, or cyclones. Exiting gases have a high content of moisture that makes them difficult to use bag houses as filters. Control of particulates from ore and product handling systems, however, is often accomplished by bag house filters or venture scrubbers. These processes are cost-effective as they permit capture and recovery of valuable products [4,6,7].

4.5 CONTAMINANTS OF CONCERN

4.5.1 Mercury

Mercury is an atmospheric pollutant with a complex biogeochemical cycle. Atmospheric cycling includes chemical oxidation/reduction in both gaseous and aqueous phases, and deposition to and re-emission from natural surfaces in addition to emissions from both natural and anthropogenic sources. The mercury (Hg) consumption by the chlor-alkali sector was found to be much greater than that understood to be released or transferred by the industry in its reporting to the U.S. EPA's Toxics Release Inventory, in 1997. At that time, the Chlorine Institute, an association of the chlor-alkali industry concerned with safety, health, and environmental matters, launched a voluntary program to reduce mercury consumption from the 14 plants located in the United States by at least 50% within 20 years. Mercury and its compounds are highly toxic to humans, ecosystems, and wildlife. High doses can be fatal, but even relatively low doses can damage the nervous system [20].

Mercury can accumulate in the food chain and cause various health problems. One of them is neurological problems that occur when the concentration of mercury in the brain reaches 20 ppm. Symptoms of long excessive exposure to mercury include numbness of the lips, hands, and feet; lack of muscular coordination; speech, vision, hearing disorders, and emotional disorders. The offspring of a pregnant woman who consumed mercury may be mentally retarded or have cerebral palsy [4].

4.5.2 Asbestos

Asbestos fiber is one of the materials that is used as a separator in diaphragm cell processes. The asbestos fiber content in wastewater is a matter of concern. It is also of concern for the cell room workers who may be exposed to airborne fibers. In the United States, the maximum asbestos contamination in drinking water is 7×10^6 fibers per liter (fibers/L) for fiber >10 μm in size. The U.S. EPA limits the asbestos contamination to <30,000 fibers/L for water supporting aquatic life [4].

4.5.3 Lead

Graphite anodes that are used in the older plants can cause elevated levels of lead and chlorinated organics in wastewater. However, almost all modern plants use the more efficient activated titanium anodes; new plants do not use graphite anodes [4].

4.6 ENVIRONMENTAL IMPACTS

4.6.1 Mercury

Wastewater from chlor-alkali plants were discharged into nearby lakes or rivers before the 1970s. Since mercury is insoluble in water, the heavy mercury was expected to settle at the bottom of rivers and remain harmless. In the environment, mercury is transformed through complex biogeochemical interactions that affect environmental and biological forms and concentrations. Some mercury compounds are more easily absorbed by living organisms than elemental mercury. When atmospheric mercury falls on earth, it may be altered by bacterial or chemical action into an organic form known as methyl mercury. Methyl mercury is a compound with a tendency to bioaccumulate [4].

Methyl mercury is much more toxic than the original metal molecules that drift in the air, and has the ability to migrate through cell membranes and bioaccumulate in living tissue. Bioaccumulation is the process by which a substance builds up in a living organism from the surrounding air or water, or through the consumption of contaminated food. Bioaccumulation will vary for different species and will depend on emission sources as well as local factors like water chemistry and temperature. Figure 4.15 illustrates the concept of accumulated methyl mercury by the dots.

The bioaccumulation of methyl mercury in natural ecosystems is an environmental concern because it inflicts increasing levels of harm on species higher up the food chain. This occurs through a process known as "bio-magnification." Small organisms that feed on the mercury in sediments are eaten by small fish, which in turn are eaten by larger fish. Subsequently, fish at the top of the food chain in a contaminated area contain mercury levels greater than those considered safe for human consumption. Elevated methyl mercury levels may lead to a decline in affected wildlife populations

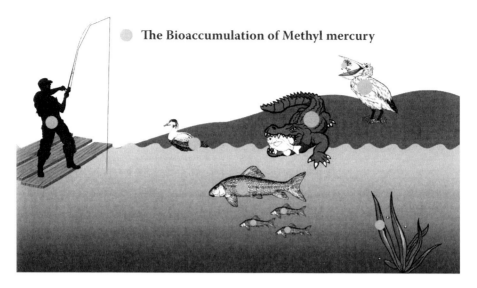

FIGURE 4.15 The bioaccumulation of methyl mercury in organisms.

TABLE 4.14
EC Standards for Mercury in Aquatic Environments

Pollutant	Inland Surface Water	Estuary Water	Territorical Sea and Coastal Water
Maximum mercury concentration in micrograms per liter (arithmetic mean over 1 year)	1.0	0.5	0.3

Source: U.S. EPA; Guédron, S., et al., Sci. Total Environ., 445–446, 356–364, 2013.

---

and may affect human health when people consume significant quantities of fish or other contaminated foods. The most infamous case of this impact occurred in Minamata, Japan, where local residents consumed fish with toxic levels of methyl mercury originating from an industrial sewer discharge, leading to the deaths of more than 1000 people. This type of exposure has now come to be known as Minamata disease.

Although chlor-alkali plants have reduced the amount of mercury discharge greatly, the pollution problem has not disappeared because the mercury discharged to surface water prior to the statutory regulations is still present. The quality objectives adopted by the European Community (EC) for mercury pollution in the aquatic environment is presented in Table 4.14.

Chaleur Bay in northeastern New Brunswick, Canada, has received Hg inputs from a variety of anthropogenic sources. Former industries in the area included a chlor-alkali plant, pulp and paper mill (both closed in 2008), and a thermal generating station (closed in 2011), all known for Hg releases in the environment. The largest industrial use of Hg in the 20th century was the chlor-alkali process, which used electrolysis (Hg being the anode) for separating chlorine (for bleaching in the pulp and paper industry) and sodium (to make caustic soda) from brine. The chlor-alkali plant located in Dalhousie, New Brunswick, used Hg cell technology to produce chlorine and caustic soda. Production began in 1963, with treated effluents discharged into Chaleur Bay. Contamination sources from the chlor-alkali process were effluents and atmospheric emissions, reportedly releasing 1.5 and 45.6 kg of Hg, respectively, in 2002 [21], making it one of the highest local point sources of Hg in the region during operations [22].

4.6.2 SODIUM CHLORIDE

Discharge of salts into fresh water has created problems in some areas, for example, along the Rhine River. The salt has built up to such levels that the water is no longer usable as drinking water. Once the salt enters freshwater, it can build up to concentration levels that further affect aquatic plants and other organisms. While the major effect on public drinking water supplies for humans is merely an alteration of taste, high concentrations of sodium in drinking water can lead to increased dietary intake and possibly hypertension.

4.7 OCCUPATIONAL HEALTH AND SAFETY ISSUES

4.7.1 CHLORINE

The major sources of fugitive air emission of chlorine are vents, seals, and transfer operations. Chlorine is a highly toxic gas, and strict precautions are needed to minimize risk to workers and possible release during its handling. It has a penetrating odor and a yellow-green color when released in the air; it is a respiratory irritant.

The chlorine gas olfactory threshold is in the range 0.2–3.5 milliliters per cubic meter (mL/m^3), although lengthened exposure appears to raise the olfactory threshold. A range of concentrations from 3 to 5 mL/m^3 can be tolerated for up 30 min without any subjective feeling of malaise. At exposures of between 5 and 8 mL/m^3, mild irritation of the upper respiratory tract occurs. Concentrations of 15 mL/m^3 and higher can cause running of the eyes and coughing. If the concentration is above 30 mL/m^3, it will cause nausea, vomiting, oppressive feeling, shortness of breath, and fits of coughing, sometimes leading to bronchial spasms. It can lead to the development of toxic tracheobronchitis when somebody is exposed to concentrations in the range of 40–60 mL/m^3. Pulmonary edema may occur after a latent period of several hours due to alveolar membrane destruction, indicated by increased shortness of breath, restlessness, and cyanosis. Pneumonia may develop because of the infection of the injured pulmonary tissue and can lead to death.

Workers in chlorine plants should be protected by providing them with escape-type respirators. If the plants have high concentration of chlorine, filter masks are inadequate to protect the workers. Anybody entering the area with high concentration of chlorine should be equipped with self-contained breathing apparatus and full protective clothing suitable for dealing with liquid chlorine. Actual wind direction indicators should be located near the chlorine installation [4,6,7].

4.7.2 MERCURY

Employees who work in the cell mercury area of mercury cell plants should go for regular health checks. The maximum allowable concentration or threshold limit value (TLV) of mercury in the atmosphere is in range of 0.025–0.100 mg/m^3 for Western Europe and the United States.

A "Mercury Code of Practice" was prepared by the Western European chlorine manufacturers in the Bureau of International Technique Chlore (BITC). It presented the recommendations for dealing with mercury, including protective measures and medical tests. The U.S. EPA has established 18 rules relating to cleanliness of the cell room.

4.7.3 ASBESTOS

Asbestos is a toxic material that can cause medical problems including lung cancer, asbestosis, pleural thickening, and mesothelioma. As a result, in 2007, a bill was adopted to ban most uses of asbestos in the United States. In the United States, the TLV of asbestos is 0.2 fibers per cubic centimeter (crocidolite, >5 μm diameter), 0.5 fibers/cm^3 (amosite, >5 μm diameter), and 2 fibers/cm^3 (chrysolite and other forms, >5 μm diameter). Chrysolite fibers are used in diaphragm cells. Exhaust extract ventilation equipment and protecting cloth should be provided to control asbestos in the workplace [4].

Regulations governing asbestos have been issued by several agencies, including the U.S. EPA and the Occupational Safety and Health Administration (OSHA).

4.7.4 Caustic Soda

Caustic soda does not cause immediate pain when it comes in contact with the skin. However, it will cause damage and should be washed away with water. It is particularly damaging when it is hot and concentrated or when it comes in contact with eyes. Caustic soda solution is colorless, clear, and odorless. Workers should be provided with proper protective clothing and equipment when handling caustic soda. They must wear chemical safety goggles or face shields, and should have immediate access to showers and eye washers [4].

4.8 POLLUTION PREVENTION AND ABATEMENT

The following are several practices recommended by the U.S. EPA in order to reduce toxic contaminant flow from chlor-alkali plants to the environment:

- Store precipitated wastes in a lined pond or dispose of them after treatment in a secure landfill.
- Remove chlorine from air emissions resulting from abnormal operating conditions, from plant startup and shutdown, or from vents on returned tank cars, cylinders, storage tanks, and process transfer tanks during handling and loading of liquid.
- Divert and contain storm runoff from plant areas. Collected runoff can then be sent to the wastewater treatment systems.
- Pave the brine treatment area and the cell room areas with smooth, sloping floors to facilitate recovery of leaks, spills, and wash water. Mercury cell room floors should use fiberglass gratings.

The following pollution prevention measures should be considered:

- Use metal rather than graphite anodes to reduce lead and chlorinated organics.
- Re-saturate brine in closed vessels to reduce the generation of salt sprays.
- Use noncontact condensers to reduce the amount of process wastewater.
- Scrub chlorine tail gases to reduce chlorine discharges and to produce hypochlorite.
- Recycle condensates and waste process water to the brine system, if possible.
- Recycle brine wastes, if possible.

For the chlor-alkali industry, an emergency preparedness and response plan is required for potential uncontrolled chlorine and other releases. Carbon tetrachloride is sometimes used to scrub nitrogen trichloride (formed in the process) and to maintain its levels below 4% to avoid explosion. Substitutes for carbon tetrachloride may have to be used, as the use of carbon tetrachloride may be banned in the near future [4,6,7].

4.8.1 Material Substitution

Recently, most of the chlor-alkali plants have started using the more powerful and efficient metal anodes rather than graphite anodes. This change may reduce the potential pollutant load of lead and chlorinated organics. Modern diaphragm cell chlor-alkali plants are able to meet regulations limiting exposure to asbestos fibers. However, the increased cost for disposal of used asbestos diaphragm has prompted the industry to search for alternatives. Fluorocarbon polymer sheets and inorganic ceramic-like fibers bonded with a fluorocarbon polymer are the possible replacements for the asbestos diaphragm. Diaphragm cells can be retrofitted as membrane cells by the installation of an ion exchange membrane between the anodes and cathodes in place of the asbestos diaphragm.

4.8.2 Waste Segregation

Process water and cooling water should be disposed of separately, while the condensate should be recycled to the brine system when possible. In a mercury cell plant, mercury-containing wastes can be treated for mercury recovery before being combined with other process wastewater.

4.8.3 Process Related Measures

Reducing of the wastewater amount can be achieved by changing from contact to noncontact cooling of vapors that come from the caustic soda concentration, or by recirculating barometric condenser water. Demisters or similar control devices can be installed to reduce the salt and caustic carryover in the vapors, if condenser water is too expensive or not feasible.

The amount of wastewater from the system can be reduced by indirect cooling or direct cooling (preferably) of chlorine with recycling of the cooling water. The direct contact method is preferable because it tends to remove much of the salt carryover and other impurities from the chlorine. The chlorine can be recovered from the condensate prior to being recycled to the brine system or sent for disposal.

Good management and housekeeping procedures also may eliminate wastewater discharge. Condensates and process waters should be sent to the brine system for recycling if the water balance permits. A closed brine circulation system is particularly significant for the mercury cell process in reducing the waste volume requiring mercury treatment. Since the amount of mercury-contaminated sludge depends on the purity of salt used, where feasible, higher purity salt such as evaporated salt is preferred to minimize the quantities of brine mud produced. In older plants, the depleted brine from the cells was sprayed into the open salt storage vessels or pits to be saturated. Modern saturators are used to reduce the environmental pollution caused by the salt spray or mist, which includes mercury from a mercury cell process.

Useful or saleable by-products such as sodium hypochlorite solution can be recovered from chlorine tail gases. The use of high pressure and refrigeration for chlorine recovery reduces the chlorine content of tail gases. The tail gas should be scrubbed with caustic soda before venting it to the atmosphere. It will produce a hypochlorite solution that can be sold, used on-site, or catalytically decomposed to regenerate and recover the sodium chloride salt. Decomposition of the hypochlorite can be done by thermal and chemical treatment. In a mercury cell plant, mercury can be recovered from the wastewater and solid wastes.

4.8.4 Mercury Recovery

Removal of mercury from the waste gases can be achieved by cooling the gases to condense the mercury vapor and/or by filtration or adsorption techniques. All equipment, products, waste gases, and other waste materials that come into contact with mercury are slightly contaminated with it. Mercury sieved cabs have been installed on cell end boxes to reduce the mercury content in the air vented from the cells. Implementation of this technique not only can clean the air but it may reduce the mercury in the plants area runoff as well.

Mercury that is present in the chlorine products must be removed. The chlorine-cooling process can remove the mercury. The mercury-containing condensate may feedback to the brine system. The cooled and dried chlorine gas contains minute traces of mercury in the range of 0.001–0.1 mg/kg.

Mercury from hydrogen gas can be removed by the cooling process. Normally, it is accomplished by adsorption on an activated carbon impregnated with sulfur or sulfuric acid, which leaves a mercury concentration of 0.002–0.015 milligrams per cubic meter (mg/m^3) in the hydrogen. Adsorption on copper/aluminum oxide or silver/zinc oxide can be used to achieve mercury concentration of <0.001 mg/m^3. Cooling or cooling and compressing the hydrogen can also reduce the mercury content. This is achieved by adding chlorine to form mercurous chloride, which can be collected on rock salt in a packed column, or by washing the hydrogen with solutions containing active chlorine. Centrifugation or filtration in filters precoated with charcoal may reduce the mercury content in sodium hydroxide solutions [23].

Mercury-containing residues include brine filter slurry, discarded cell components, residue from the purification of products, spent decomposer catalyst, residue from the purification of products, waste material from rinsing media, adsorption materials, and ion exchange media. Recovery of mercury from these materials can be done by distillation in closed retorts. After distillation, the residue must be disposed of at special areas [4,13].

4.9 TREATMENT TECHNOLOGY

4.9.1 Solid Waste

Brine mud and wastewater treatment sludge release the largest amount of solid wastes in chlor-alkali plants. Steps must be taken to reduce chloride losses in the sludge. Solid waste may contain toxic heavy metals, and should be stabilized.

Mercury in the mercury cell plants should be removed from the sludge. The dehydrated brine muds can be disposed of at a designated secure location. The U.S. EPA standard for K071 sludge presents the best demonstrated available technology (BDAT) for removal of mercury from brine muds. K071 sludges are brine purification muds from the mercury cell process in which prepurified brine is not used. The BDAT process includes the following:

1. Making alkaline hypochlorite to oxidize the mercury in the brine sludge soluble by leaching with acid.
2. Using an alkaline hypochlorite to oxidize the mercury to a highly soluble mercuric chloride.
3. Washing with hydrochloric acid and water during a filtration step.

The filtrate contains the solubilized mercury, which is then precipitated out as mercury sulfide sludge. The sulfide sludge is also filtered and/or dewatered. The aqueous residual from this process is classified as K071 wastewater, while the sulfide sludge is classified as K071 nonwastewater. Aqueous residual must not contain more than 0.031 milligrams per liter (mg/L) mercury for any grab sample and sulfide sludge must have no more than a maximum toxic characteristic leaching procedure (TCLP) leachate concentration of 0.025 mg/L mercury for any single grab sample.

4.9.2 Air Emission

Airborne emission can be kept within required air quality limitations through the use of cyclones, scrubbers, trippers, and other methods. The use of stabilized anodes instead of graphite anodes has greatly reduced and even eliminated the production of carbon monoxide and carbon dioxide. In many instances, the gases can be recovered and reused or marketed as saleable products.

4.9.3 Liquid Effluents

4.9.3.1 Mercury Cell Process

Sulfide precipitation followed by pressure filtration is the BPT treatment of mercury-bearing wastewaters. This treatment will also reduce other heavy metals in the wastewater. Stage of treatment includes recycling of the brine mud overflow filtrate back to the process, and the settling and the storage of brine muds. Mercury-bearing sludge can be disposed of at landfills or can be sent for mercury recovery. Removal of residual mercury and metal sulfides that are contained in the filtered effluent from these treatments can be achieved by passing through a granular activated carbon.

Best available technique (BAT) limitations are on the basis of the BPT requirements that have been described earlier, with the addition of the dechlorination of the wastewater to protect aquatic life. Chlorine can be removed by treatment with bisulfate or sulfur dioxide. Best available technology specific to mercury cell plants is considered to be membrane cell technology. During the remaining life of mercury cell plants, all possible measures should be taken to protect the environment as a whole. The best performing mercury cell plants encounter mercury losses to air, water, and with products in the range of 0.2–0.5 g Hg/t of chlorine capacity as a yearly average. The majority of mercury losses are in the various wastes from the process. Measures should be taken to minimize current and future mercury emissions from handling, storage, treatment, and disposal of mercury-contaminated wastes. Decommissioning of mercury cell plants should be carried out in a way that it prevents environmental impact during and after the shutdown process and safeguards human health [4,6,7].

4.9.3.2 Diaphragm Cell Plant

U.S. EPA has set the standard for the diaphragm cell process on the basis of plants using graphite anodes, because of the higher pollutant load. The BPT treatment standard includes treating the asbestos-contaminated cell room wastes with flocculating agent and a filter aid prior to filtration. The filtrate is sent to a holding tank and the solids are removed to a landfill where it is combined with other process waste sources containing treatable levels of lead and other toxic metals. Alkaline precipitation of the toxic metals is accomplished by the addition of soda ash. The solids are removed by settling while the filtrate is combined with other process waste streams (e.g., chlorine condensate, caustic filter backwash, tail gas scrubber water, and barometric condenser blow down waters) contaminated with toxic metals at levels usually below the limits of treatability and removed by alkaline precipitation. Plants that use metal anodes are expected to have lower levels of toxic metal emissions and may not require alkaline precipitation to meet the proposed BPT limitations. The combined flow will pass through a polishing pond before discharging for additional purification. The brine mud will be collected in lagoons, and the effluents are recycled to the process for every level of treatment. Although specific limits are not set, this level of treatment can reduce the total asbestos discharges.

U.S. EPA set the new source performance standards (NSPS) for diaphragm cell processes on the basis of the BAT technology for alkaline precipitation, filtration, and dechlorination, and on the performance achieved when leads are excluded from the cell construction. Dual media filtration of the combined effluent from BPT treatment is included as a part of BAT treatment. BAT treatment also includes dechlorination of the wastewaters with sulfur dioxide or bisulfate [4,6,7].

4.9.4 BAT

BAT for the production of chlor-alkali is considered to be membrane technology. Nonasbestos diaphragm technology can also be considered as BAT. The total energy use associated with BAT for producing chlorine gas and 50% caustic soda is <3000 kWh (AC) per ton of chlorine when chlorine liquefaction is excluded and <3200 kWh (AC) per ton of chlorine when chlorine liquefaction and evaporation are included.

BATs for the manufacture of chlor-alkali for all cell plants include the following measures:

1. Use of management systems to reduce the environmental, health, and safety risks of operating a chlor-alkali plant. The risk level should tend to zero. The management systems will include:
    a. Training of personnel.
    b. Identification and evaluation of major hazards.

Waste Treatment and Management in Chlor-Alkali Industries 135

   c. Instructions for safe operation.
   d. Planning for emergencies and recording of accidents and near misses.
   e. Continuous improvement including feedback and learning from experience.
2. A chlorine destruction unit designed to be able to absorb the full cell-room production in the event of a process upset until the plant can be shut down. The chlorine absorption unit prevents emissions of chlorine gas in the event of emergencies and/or irregular plant operation. The absorption unit should be designed to lower the chlorine content in the emitted gas to <5 mg/m^3 in the worst case scenario. All chlorine-containing waste gas streams should be directed to the chlorine absorption unit. The chlorine emission level to air associated with BAT during normal operation is <1 mg/m^3 in the case of partial liquefaction and <3 mg/m^3 in the case of total liquefaction. No systematic discharge of hypochlorite to water should take place from the chlorine destruction unit.
3. Minimizing consumption/avoiding discharge of sulfuric acid by means of one or more of the following options or equivalent systems:
   a. On-site re-concentration in closed loop evaporators.
   b. Using the spent acid to control pH in process and wastewater streams.
   c. Selling the spent acid to a user who accepts this quality of acid.
   d. Returning the spent acid to a sulfuric acid manufacturer for re-concentration. If the sulfuric acid is re-concentrated on-site in closed loop evaporators, the consumption can be reduced to 0.1 kg of acid per ton of chlorine produced.
4. Minimizing the discharge of free oxidants to water by applying the following methods:
   a. Fixed-bed catalytic reduction.
   b. Chemical reduction.
   c. Any other method with equally efficient performance. The emission level of free oxidants to water associated with BAT is <10 mg/L. The overall environmental impact should be considered when the destruction method is chosen.
5. Use of carbon tetrachloride-free chlorine liquefaction and purification processes.
6. Hydrogen should be used as a chemical or as a fuel to conserve resources.

4.9.5 Disposal Option and Alternative Treatment Technologies

Solid wastes that contain mercury are normally disposed of in a landfill. The concentration of mercury in such wastes is regulated in some places. In Japan, muds and sludge containing <0.005 mg/kg mercury can be disposed of in landfills. For higher mercury contents, the muds and sludge should be mixed with concrete before disposal. The concrete block must be embedded in larger blocks if the leachate exceeds 0.005 mg/kg. Muds and sludge also can be dumped at sea, but according to the regulations set by the Ocean Contaminants Law. If the mercury content is below 0.005 mg/kg, the muds and sludge can be disposed of at depths of 3500 m or more. For mercury content between 0.005 and 2.0 mg/kg, the waste must be mixed with concrete and dumped in areas with low currents and few fish. The mercury content in the leachate must be <0.005 mg/kg. Muds and sludge with mercury content above 2 mg/kg must be treated or incinerated.

In the United Kingdom, the nonalkyl mercury content of wastes before disposal to landfill is not to exceed 2% by weight. If the mercury content exceeds 4 mg/kg, additional safeguards are required.

In Europe, many facilities use ion exchange for treatment of wastewater generated from the mercury chlor-alkali process. There are also several facilities that use activated carbon as an adsorbent. The spent carbon or ion exchange resin can be retorted or incinerated to recover mercury. One plant precipitates mercury with hydrazine mercury. Reduction of mercury to the metallic state by borohydride, sulfide, hypophosphite, and iron followed by filtration and carbon adsorption has also been patented. Stabilization treatment, which typically binds metals into a solid form that is more resistant to leaching than are metals in the untreated wastes, is potentially applicable for treatment of mercury cell wastewater treatment sludge (classified as K106 by the U.S. EPA).

4.10 CAPITAL AND OPERATION AND MAINTENANCE COSTS

Costs of energy are the most important expenses for a chlor-alkali plant. These costs are varying in every plant location as do construction costs. Only relative cost comparisons are possible. If the electrical energy is relatively expensive in an area, the energy consumption can be reduced at the expense of higher capital investment by using a greater number of cells. Generally, the diaphragm process needs the biggest capital investment for a medium-sized plant which produces 100,000 t/year of chlorine. This is because of the high cost of caustic soda evaporation. The least expensive of the three processes is membrane cell process. The capital cost that is required for this process is approximately 80% of the investment for the diaphragm cell plant. The mercury cell process investment is 90%–95% than that of the diaphragm process. The equipment required to prevent emissions of mercury into the environment accounts for 10%–15% of the total capital investment. Table 4.15 presents the comparison of each process.

The caustic product from the diaphragm cell process contains more impurities, especially salt, than that obtained from the mercury and membrane cell processes. Hence, additional capital should be spent to obtain a low-salt caustic product from this process.

4.10.1 OLIN CORPORATION

Olin Corporation is a Virginia corporation, incorporated in 1892, having its principal executive offices in Clayton, MO. They have been involved in the U.S. chlor-alkali industry for more than 100 years and are a major participant in the North American chlor-alkali market. Chlorine and caustic soda are co-produced commercially by the electrolysis of salt. As reported in the Form 10-K Annual Report, electricity and salt are the major raw materials purchased for their chlor-alkali products segment. Raw materials represent approximately 50% of the total cost of producing an ECU (an electrochemical unit). Electricity is the single largest raw material component in the production

TABLE 4.15
Comparison of Processes

Comparison Item	Mercury	Diaphragm	Membrane
Raw materials	More expensive, solid salt preferred	Cheaper brine acceptable	More expensive solid salt preferred
Electrolytic cell	Most expensive, but requires least amount of surface area; requires mercury makeup	Requires replacement of asbestos diaphragm	Requires replacement of membranes
Brine system	Requires larger system because of lower prepass conversion	Least expensive: no sulfate precipitation required; no storage required if brine is used	Added expense: second purification necessary
Caustic concentration to 50%	Not required	Requires nickel-plated multiple-effect evaporators	Requires stainless steel evaporators in the third step.
Pollution control	Most expensive: requires mercury pollution prevention	Requires asbestos disposal	Least expensive: lowest environmental risk
Product quality	Acceptable	Extra expense for caustic purification	Acceptable

Source: World Bank and United Nations Industrial Development Organization. Industrial pollution prevention and abatement: Chlor-alkali industry. Pre-publication Draft, Environment Department, Washington, DC, 1994.

of chlor-alkali products. During the past few years, the corporation experienced an increase in the cost of electricity from the suppliers primarily due to an increase in the energy cost and regulatory requirements. They are supplied by utilities that primarily utilize coal, hydroelectric, natural gas, and nuclear power. The commodity nature of this industry places an added emphasis on cost management, and they have managed their manufacturing costs in a manner that has made them one of the lowest cost producers in the industry [20]. Table 4.16 lists the products of Olin Corporation.

TABLE 4.16
List of Products Manufactured by Olin Corporation

Products and Services	Major End Uses	Plants and Facilities	Major Raw Materials and Components for Products/Services
Chlorine/caustic soda	Pulp and paper processing, chemical manufacturing, water purification, manufacture of vinyl chloride, bleach, swimming pool chemicals, and urethane chemicals	Augusta, GA; Becancour, Quebec; Charleston, TN; Dalhousie, New Brunswick; Henderson, NV; McIntosh, AL; Niagara Falls, NY; St. Gabriel, LA	Salt, electricity
Sodium hypochlorite (bleach)	Household cleaners, laundry bleaching, swimming pool sanitizers, semiconductors, water treatment, textiles, pulp and paper, and food processing	Augusta, GA; Becancour, Quebec; Charleston, TN; Dalhousie, New Brunswick; Henderson, NV; McIntosh, AL; Niagara Falls, NY; Santa Fe Springs, CA; Tacoma, WA; Tracy, CA	Chlorine, caustic soda
Hydrochloric acid	Steel, oil and gas, plastics, organic chemical synthesis, water and wastewater treatment, brine treatment, artificial sweeteners, pharmaceuticals, food processing, and ore and mineral processing	Augusta, GA; Becancour, Quebec; Charleston, TN; Henderson, NV; McIntosh, AL; Niagara Falls, NY	Chlorine, hydrogen
Potassium hydroxide	Fertilizer manufacturing, soaps, detergents and cleaners, battery manufacturing, food processing chemicals, and deicers	Charleston, TN	Potassium chloride, electricity
Hydrogen	Fuel source, hydrogen peroxide, and hydrochloric acid	Augusta, GA; Becancour, Quebec; Charleston, TN; McIntosh, AL; Niagara Falls, NY; St. Gabriel, LA	Salt, electricity
Sodium chlorate	Pulp and paper manufacturing	Dalhousie, New Brunswick	Sodium chloride, electricity
Sodium hydrosulfite	Paper, textile, and clay bleaching	Charleston, TN	Caustic soda, sulfur dioxide

Source: U.S. EPA; U.S. Environmental Protection Agency, Sector notebook for inorganic chemicals, Washington, DC, 1995a.

4.10.2 OPERATION AND MAINTENANCE COSTS

Energy consumption results in the greatest difference in operating costs. The mercury cell process has the greatest requirement for electricity, but overall energy consumption is less than that of the diaphragm cell process that requires a significant amount of steam to bring the caustic product to 50% concentration. Membrane process is the lowest energy-consuming process.

The pollution control cost for a modern mercury cell plant was $7.30/t of chlorine as mentioned by a U.S. EPA document (1976). The cost for membrane and diaphragm plants (with stabilized anodes) is lower than the mercury plants, which are $1.00 and $1.10, respectively. All costs escalated in 1990 using "The Statistical Abstract of the United States" (1992). Hence, assuming the difference between pollution control cost is primarily for mercury recovery, the cost of mercury recovery can be estimated at around $6.20/t of chlorine produced (in 1990 U.S. dollars).

4.11 MONITORING AND REPORTING

Chlor-alkali plants should be monitored continuously for their air emissions, and liquids and solid wastes discharged. Daily monitoring is recommended for parameters other than pH, especially for the effluents from the diaphragm process. Samples should be taken at least once every 8 or 12 h shift. This is in order to make sure that the applicable 24 h and 30 days average for TSS, mercury, copper, lead, nickel, and residual chlorine are not exceeded. The pH of the liquid effluents should be monitored continuously, and where appropriate, monitoring of chlorinated organics and asbestos level should also be done to the effluents. Chlorine monitors should be strategically located within the plant to detect chlorine releases or leaks on a continuous basis [20].

In case of mercury emission, all potential sources of the emissions should be monitored daily to make sure they do not exceed the standard limits. The main mercury sources are from the cell room ventilation air, the hydrogen by-product stream, the cell end-box ventilation stream, gases exhausted from the retort, and gases exhausted from tanks. Monitoring of the solid waste also should be done for chlorinated organics and mercury to ensure proper treatment and disposal. Monitoring wells, strategically located around the burial sites of mercury contaminated wastes, should be sampled at least once a year to make sure that mercury is not leaching into the groundwater [18]. Table 4.17 shows the monitoring techniques for mercury.

During start-up and upset conditions, frequent sampling may be required. Once a record of consistent performance has been established, sampling for the parameters listed earlier should be as follows. Analyzing and reviewing of monitoring data should be done at regular intervals, and it must be compared with the operating standards. After that, if any necessary correction is required, action should be taken. Monitoring results report should be kept in an acceptable format. These should be reported to the responsible authorities and relevant parties, as required, and provided to MIGA if requested [4,6,7].

4.12 GLOBAL OVERVIEW OF DISCHARGE REQUIREMENTS

4.12.1 LIQUID EFFLUENTS

Table 4.18 presents liquid effluent standards for the chlor-alkali industry from selected countries. The U.S. limits for new sources are given in Table 4.19. The U.S. standards are for discharge into navigable water and must be satisfied by the existing plants using BAT [24].

From Table 4.18, it can be seen that the mercury standard for industrial discharge in Canada is 0.00250 kg/t of chlorine produced, while in Japan it is 0.005 mg/L. The Canadians restrict chlorine discharge based on the reference production rate of the plant. The mercury standards in Japan are

TABLE 4.17
Mercury Monitoring Techniques

Potential Source	Method	Principal of Detection	Typical Equipment
Process equipment, noncombustible gas or vapor	1. Portable mercury vapor analyzer–UV absorption detector	A sample of gas is drawn through a detection cell where UV light at 253.7 nm is directed perpendicularly through the sample toward a photo detector. Mercury absorbs the incident light in proportion to its concentration in the stream	MV2
	2. Portable mercury vapor analyzer–gold film amalgamation detector	A sample of gas is drawn through a detection cell containing a gold film detector. Mercury amalgamates with the gold film, changing the resistance of the detector in proportion to the mercury concentration in the air sample	Jerome Model 411 Note: This model is no longer being manufactured
	3. Portable short-wave UV light, fluorescent background–visual indication	UV light is directed toward a fluorescent background positioned behind a suspected source of mercury emissions. Mercury vapor absorbs the UV light projecting a dark shadow image on the fluorescent background	Portable short-wave UV light source and any portable fluorescent background.
Hydrogen gas-piping and associated equipment	Methods 1–3	As for Methods 1–3	As for Methods 1–3
	4. Portable combustible gas meter	Since mercury is likely to be present in significant concentrations in hydrogen from the decomposer until treatment, detection of hydrogen as a combustible gas is a surrogate for mercury vapor.	Any standard portable combustible gas meter.
Area monitoring	Methods 1, 2 or 5. Permanganate impingement	A known volume of gas sample is adsorbed in $KMnO_4$ solution. Mercury in the solution is determined using CVAA and the concentration of mercury in the gas sample is calculated.	SKC Model 224-PCXR4 gas sample pump or equivalent and LKS Ultrascan Model 709 low calibrator or equivalent and CVAA laboratory equipment

Source: U.S. EPA; U.S. Department of Energy, Energy Information Administration, Emissions of greenhouse gases in the United States 1997, DOE/EIA-0573(97), Washington, DC, 1998.

TABLE 4.18
Selected Liquid Effluent Standards

	Maximum kg/t Chlorine	
	24-h Average	30-Day Average

United States[a]

Mercury Cell Process

Mercury	0.00023	0.00010
Residual chlorine	0.0032	0.0019
pH	6–9	6–9

Diaphragm Cell Process

Copper	0.012	0.0049
Lead	0.0059	0.0024
Nickel	0.0097	0.0037
Residual chlorine	0.013	0.0079
pH	6–9[b]	6–9[b]

Canada[b]

All processes	
Mercury	0.00250 kg/t chlorine

Japan[c] — *Maximum levels*

Mercury	0.005 mg/L
pH (freshwater)	5.8–8.6

European Community[d] — *Monthly Average Limit for Mercury*[e]

Recycled brine and lost brine	0.05 mg/L (from plant site)
Recycled brine	0.5 g/t chlorine (from plant site)
	1.0 g/t chlorine (from plant site)
Lost brine	5.0 g/t chlorine (from plant site)

Source: World Bank and United Nations Industrial Development Organization. Industrial pollution prevention and abatement: Chlor-alkali industry. Pre-publication Draft, Environment Department, Washington, DC, 1994.

[a] 40 CFR (The U.S. Code of Federal Regulations) Part 415 subpart F (1992).
[b] Chlor-alkali mercury liquid effluent regulations. Canada Gazette, Part II, III(7): 1666 (March 28, 1977).
[c] Environmental legislation of Japan. 1993. Japan: Chu-Oh Hoki Publishing Company.
[d] European Economic Community Council Directive 32/176/EEC, March 22, 1982.
[e] Daily average limits are four times the corresponding monthly average limits.

the joint responsibility of the Environmental Agency and the Ministry of International Trade and Industry (MITI). MITI has banned the use of the mercury cell process in Japan.

4.12.2 Air Emissions

Air emission standards of selected countries for mercury emissions from chlor-alkali plants using the mercury cell process are given in Table 4.20.

TABLE 4.19
New Source Performance Standards (U.S. EPA) for Mercury and Diaphragm Cell Processes

	Maximum Kilograms per Metric Ton of Chlorine	
Source	24-h Average	30-Day Average
Mercury Cell Process		
TSS	0.64	
Mercury	0.00023	0.00010
Total residual chlorine	0.0032	0.0019
pH	6–9	6–9
Diaphragm Cell Process		
TSS	1.1	0.51
Lead	0.0047	0.0019
Total residual chlorine	0.013	0.0079
pH	6–9	6–9

Source: World Bank and United Nations Industrial Development Organization. Industrial pollution prevention and abatement: Chlor-alkali industry. Pre-publication Draft, Environment Department, Washington, DC, 1994.

Note: 40 CFR (U.S. Code of Federal Regulations) Part 415 Subpart 5, 1992.

TABLE 4.20
Selected Air Emission Standards for Mercury

United States[a]

Mercury in
- Waste air and hydrogen — 1 kg/day/facility
- Ventilation air — 1.3 kg/day/facility

Canada[b] — Maximum (g/day/t of Chlorine)

Mercury in
- Cell room ventilation air — 5
- Hydrogen stream from denuders — 0.1
- Ventilation air from end boxes — 0.1
- Air exhausted from retorts — 0.1
- Total mercury from all sources — <1.68 kg/day/facility

Sweden[c]

Mercury in
- Ventilation air — 1 g/t chlorine
- Hydrogen stream — 0.5 g/t chlorine

Source: World Bank and United Nations Industrial Development Organization Industrial pollution prevention and abatement: Chlor-alkali industry. Pre-publication Draft, Environment Department, Washington, DC, 1994.

[a] 40 CFR (U.S. Code of Federal Regulations) Part 61 Subpart E.
[b] Chlor-alkali Mercury National Emission Standard Regulation. Canada Gazette. Part II, III(14): 2985–2290 (1997, as revised July 1, 1978).
[c] Sweden, States Naturvarsadek, 1974, Anvisningar fur knotrol av miljoralrig verksamket vid klor-alkalifabriker enlight amalgammetoeden, Vallingby, Almanna Forlaget. In 1983, the National Environment Protection Board of Sweden expected all existing plants to convert to non-mercury technology within 5–7 years.

4.13 CONCLUSION

The membrane cell should be used in all new chlor-alkali plants. The membrane process has both economic and environmental advantages compared with the other two processes. This is because the membrane cell process needs the least amount of electricity and has the lowest overall energy requirements. The capital investment for new membrane cell plants is also less than that required for the mercury and diaphragm cell plants.

Discharge from the mercury cell plants should be brought in line with the emissions requirements in the interim. The use of mercury cell plants should be phased out.

The membrane cell plants also have the added advantage of posing the least pollution problems. Precautions must be taken in the mercury cell process to prevent mercury emissions, and in diaphragm plants to prevent asbestos, which may be a potential pollutant.

REFERENCES

1. Worrell, E., Phylipsen, D., Einstein, D., Martin, N. 2000. *Energy Use and Energy Intensity of the U.S. Chemical Industry.* University of California, Berkeley, CA.
2. Wang, X., Teichgraeber, H., Palazoglu, A., El-Farra, N.H. 2014. An economic receding horizon optimization approach for energy management in the chlor-alkali process with hybrid renewable energy generation. *Journal of Process Control*, 24, 1318–1327.
3. Bommaraju, T.V., Orosz, P.J., Sokol, E.A. 2007. Brine electrolysis. *Electrochemistry Encyclopedia.* Case Western Reserve Univ. http://electrochem.cwru.edu/ed/encycl/art-b01-brine.htm (Accessed June 15, 2008).
4. World Bank and United Nations Industrial Development Organization. 1994. Industrial pollution prevention and abatement: Chlor-alkali industry. Pre-publication Draft, Environment Department, Washington, DC.
5. Pacific Environmental Services, Inc., for the U.S. Environmental Protection Agency. 1992. Background report: AP-42 Section 5.5. Chlor-Alkali Industry, Washington, DC.
6. Multilateral Investment Guaranty Agency. 1997. Environmental guidelines for chlor-alkali industry. www.miga.org/documents/chloralkali.pdf (Accessed June 10, 2008).
7. World Bank Group. 1998. *Pollution Prevention and Abatement: Chlor-Alkali Plants.* Environment Department, Washington, DC.
8. Chemical Manufacturers Association. 1998. *U.S. Chemical Industry Statistical Handbook 1998.* Hemisphere Publishing Corporation, New York.
9. Nunes, S.P., Peinemann, K.-V. 2006. *Membrane Technology in the Chemical Industry.* Wiley-VCH, Weinheim, Germany.
10. Wikipedia Encyclopedia. 2017. Chlor-alkali process. Last modified on March 24, 2017. http://en.wikipedia.org/wiki/Chloralkali_process (Accessed June 15, 2008).
11. U.S. Environmental Protection Agency. 1995b. Sector notebook for inorganic chemicals. Seth Heminway, Coordinator, Sector Notebook Projec US EPA Office of Compliance 401 M St., SW (2223-A), Washington, DC.
12. Kinsey, J.S. 2002. Characterization of mercury emissions at chlor-alkali plant, Volume 1, Report and Appendices A-E. Office of Research and Development, U.S. Environmental Protection Agency, Durham, NC.
13. U.S. Environmental Protection Agency. 2007. Background paper for stakeholder panel to address options for managing U.S. non-federal supplies of commodity grade mercury. Washington, DC.
14. Ayres, R. 1997. The life cycle of chlorine, Part I. *Journal of Industrial Ecology*, 1(1): 81–94.
15. U.S. Environmental Protection Agency. 1995a. Sector notebook for inorganic chemicals. Seth Heminway, Coordinator, Sector Notebook Projec US EPA Office of Compliance 401 M St., SW (2223-A), Washington, DC.
16. Guédron, S., Grangeon, S., Jouravel, G., Charlet, L., Sarret, G. 2013. Atmospheric mercury incorporation in soils of an area impacted by a chlor-alkali plant (Grenoble, France): Contribution of canopy uptake. *Science of the Total Environment*, 445–446, 356–364.
17. Pacific Environmental Services, Inc., for the U.S. Environmental Protection Agency. 1997. Background report: AP-42, Section 5.16, Sodium Carbonate Production. Washington, DC.

18. U.S. Environmental Protection Agency. 1998. Sector notebook data refresh 1997. Office of Compliance, Office of Enforcement and Compliance Assurance, U.S. Environmental Protection Agency, Washington, DC.
19. U.S. Department of Energy. Energy Information Administration. 1998. Emissions of greenhouse gases in the United States 1997, DOE/EIA-0573(97), Washington, DC.
20. Kinsey, J.S., Anscombe, F.R., Lindberg, S.E., Southworth, G.R. 2004. Characterization of the fugitive mercury emissions at a chlor-alkali plant: Overall study design. *Atmospheric Environment*, 38, 633–641.
21. United States Securities and Exchange Commission. 2007. Annual report, form 10-K. Washington, DC.
22. The Chlorine Institute, Inc. 2001. Guidelines for mercury cell chlor-alkali plants emissions control: Practices and techniques. The Chlorine Institute, Inc., Washington, DC.
23. EC. National Pollutant Release Inventory (NPRI). 2014. 2002 facility substance summary. www.ec.gc.ca/npri (February, 2014).
24. Walker, T.R. 2016. Mercury concentrations in marine sediments near a former mercury cell chlor-alkali plant in Eastern Canada. *Marine Pollution Bulletin*, 107(1), 398–401.
25. Energetics Incorporated. 2000. *Energy and Environmental Profile of the U.S. Chemical Industry*. U.S. Department of Energy, Washington, DC.
26. Pacific Environmental Services, Inc., for the U.S. Environmental Protection Agency. 1992. Background Report: AP-42 Section 5.5, Chlor-Alkali Industry. Washington, DC.
27. Carol, A.M., Brooks, M.D., Virginia, A.P. 2000. The Chlorine Industry: A Profile. Draft Report. Research Triangle Institute Center for Economics Research, Research Triangle Park, NC.
28. World Chlorine Council. 2002. *The World Chlorine Council and Sustainable Development*. Grafiks Marketing and Communications, Sarina, ON, Canada.
29. Wikipedia Encyclopedia. Mercury in Fish. https://en.wikipedia.org/wiki/Mercury_in_fish (Accessed June 23, 2016).
30. Doesburg, R. van. 1994. Energiebesparing in de Nederlandse Chloor-Alkali Industrie (Energy Savings I the Netherlands' Alkalies and Chlorine Industry, in Dutch), Dept. of Science, Technology & Society, Utrecht University, The Netherlands.
31. IND CHEM 1990 Stocchi, E. 1990. Industrial chemistry, Volume I. Ellis Horwood, New York.
32. Brown, H.L., Hamel, B.B., Hedman, B.A., et al. 1996. *Energy Analysis of 108 Industrial Processes*. Fairmont Press, Lilburn, Georgia.
33. Ayres, R. 1997. The life cycle of chlorine, Part I. *Journal of Industrial Ecology*, 1(1), 81–94.
34. Orica Australia Limited. 1997–1999. Orica chemical fact sheets. http://www.orica.com.au/resource/chemfact/ (Accessed January–April 1999).

5 Dissolved Air Flotation (DAF) for Wastewater Treatment

Puganeshwary Palaniandy, Hj. Mohd Nordin Adlan, Hamidi Abdul Aziz, and Mohamad Fared Murshed
Universiti Sains Malaysia

Yung-Tse Hung
Cleveland State University

CONTENTS

5.1 Introduction	146
5.2 Types of Flotation	146
5.2.1 Electroflotation	146
5.2.2 Dispersed Air Flotation	147
5.2.3 Dissolved Air Flotation	147
5.3 Process Description of DAF	148
5.4 Theory of DAF	150
5.4.1 Bubble Formation	151
5.4.2 Bubble–Particle Attachment	151
5.4.3 Flotation of Bubble–Particle Agglomerate	151
5.4.4 Kinetics of Flotation	151
5.4.5 Solubility of Air	157
5.4.6 Bubble Generation	159
5.4.7 Collision	160
5.4.8 Interception and Diffusion	163
5.4.9 Tank Design	166
5.5 Advantages of DAF Application in Wastewater Treatment	170
5.6 Application of DAF Process in Wastewater Treatment	170
5.7 Application of DAF Process in Landfill Leachate Treatment	173
List of Nomenclature	176
References	178

Abstract

Flotation system consists of four major components—air supply, pressurizing pump, retention tank, and flotation chamber. The theory of dissolved air flotation (DAF) process is to separate suspended particles from liquids by bringing the particles to the surface of the liquid. In most cases, DAF is an alternative process to sedimentation and offers several advantages, including better final water quality, rapid startup, higher rates of operation, and thicker sludge. Additionally, DAF systems need less space compared with normal clarifiers, and due to their modular components, they allow easy installation and setup. This chapter covers types of flotation, process description of DAF, theory of DAF, advantages of DAF application in wastewater treatment, application of DAF process in wastewater treatment, and application of DAF in landfill leachate treatment.

5.1 INTRODUCTION

There are various types of flotation process available for different applications. The technology has been applied in many industries such as in mineral processing [1,2], wastewater clarification [3–6], artificial recharge [7], and potable water treatment [8,9]. Basically, flotation is a process of separating solids from a body of liquid by using air bubbles.

Flotation has been used in the mining and chemical processing industries for over 100 years [10]. However, the history of flotation goes back even earlier. The ancient Greeks used this process over 2000 years ago to separate minerals from gangue [11]. The development of the process to its current modern practices took several years. According to Kitchener, Haynes was able to separate minerals using oil in 1860 [12]. His method was patented. In 1905, Salman, Picard, and Ballot developed a process to separate sulfate grains from water by adding air bubbles and a small amount of oil to enhance the process. This was called "froth flotation." In 1910, T. Hoover developed the first flotation machine, which was not much different from the current equipment. A few years later, in 1914, Callow introduced a new process called "foam flotation" [12]. This process involved the introduction of air bubbles through submerged porous media. In fact, froth and foam flotation processes are generally known as dispersed air flotation processes and are used widely in the mineral processing industry. The development of the electrolytic flotation process can be traced back to 1904. The process was suggested by Elmore brothers who showed that electrolysis could produce bubbles for flotation. It was not used commercially at that time.

Dissolved air flotation was patented in 1924 by Niels Peterson and Carl Sveen in Scandinavia [13]. It was initially used to recover fibers and white water in the paper industry. The use of dissolved air flotation (DAF) in the treatment of wastewater and potable water began in the late 1960s. Edzwald and Walsh reported that DAF has been used for water clarification in Europe especially in the Scandinavian countries for more than 20 years, [10]. Heinanen, in his survey on the use of flotation in Finland, indicated that the first DAF plant for potable water clarification was constructed in 1965, and that by 1988, there were 34 plants in operation [14]. However the first application of flotation for a water reclamation plant was introduced in the early 1960s in South Africa [15]. In the United Kingdom, the first full-scale water treatment plant using this process was commissioned in 1976 at the Glendye Treatment Works of the Grampian Regional Council, Scotland [16]. Experiments carried out by researchers at the Water Research Centre showed that flotation is a more rapid method of solid–liquid separation than sedimentation [17].

5.2 TYPES OF FLOTATION

The basic idea in the flotation process is the solid–liquid separation process using bubbles. The bubbles are produced using any gas that is not highly soluble in liquid. However, in actual practice, air is the gas most commonly used because it is easily accessible, safe to use, and less expensive. The method of producing bubbles gives rise to different types of flotation processes, namely, electrolytic flotation, dispersed air flotation, and dissolved air flotation [18,19].

5.2.1 ELECTROFLOTATION

Electrolytic flotation is also known as electroflotation. The bubbles are produced by passing a direct current between two electrodes and by generating oxygen and hydrogen in a diluted aqueous solution. Bubbles produced from electrolytic flotation are smaller compared with those produced from dispersed air flotation and DAF. Thus, this process is favorable for the removal of low-density fragile flocs. This process is suitable for effluent treatment [20], sludge thickening, and water treatment installations of $10–20\,m^3/h$. Figure 5.1 shows a typical arrangement of an electroflotation tank.

Dissolved Air Flotation (DAF) for Wastewater Treatment

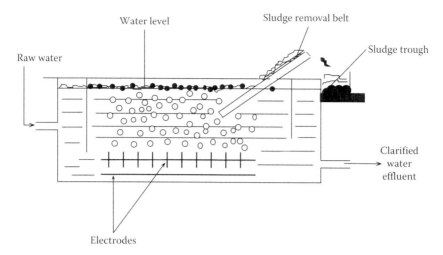

FIGURE 5.1 Electroflotation tank. (From Adlan, M.N., A study of dissolved air flotation tank design variables and separation zone performance. PhD thesis, University of Newcastle Upon Tyne, 1998.)

5.2.2 DISPERSED AIR FLOTATION

Dispersed air flotation has two different systems to generate bubbles, namely, foam flotation and froth flotation. In the foam flotation system, bubbles are generated by forcing air through a porous media made of ceramic, plastic, or sintered metal [21]. Figure 5.2 shows a typical arrangement for bubble generation through a medium or diffuser.

In the froth flotation system (shown in Figure 5.3), a high-speed impeller or turbine blade rotating in the solution is used to produce air bubbles.

Dispersed air flotation normally produces large air bubbles measuring >1 mm in diameter [22]. It is used mainly for the separation of minerals and removal of hydrophobic materials such as fat emulsions in selected wastewater treatment. This process was assessed for potable water treatment but was found unsuitable [23].

5.2.3 DISSOLVED AIR FLOTATION

The bubbles in DAF are produced by the reduction in pressure of a water stream saturated with air. The three types of DAF are vacuum flotation, microflotation, and pressure flotation, of which

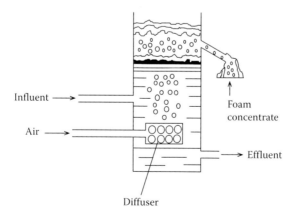

FIGURE 5.2 Foam flotation. (From Adlan, M.N., A study of dissolved air flotation tank design variables and separation zone performance. PhD thesis, University of Newcastle Upon Tyne, 1998.)

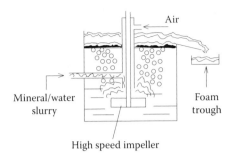

FIGURE 5.3 Froth flotation. (From Adlan, M.N., A study of dissolved air flotation tank design variables and separation zone performance. PhD thesis, University of Newcastle Upon Tyne, 1998.)

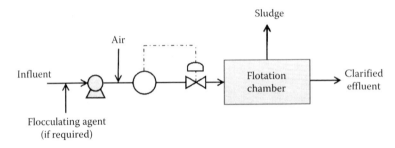

FIGURE 5.4 Full flow pressure operational modes. (From Adlan, M.N., A study of dissolved air flotation tank design variables and separation zone performance. PhD thesis, University of Newcastle Upon Tyne, 1998.)

the pressure flotation is the most important and widely used in water and wastewater treatments. In the pressure flotation, the air is dissolved in water under high pressure and released at atmospheric pressure through a needle valve or nozzle, which produces small air bubbles.

There are three kinds of pressures dissolved in air flotation that can be used: full-flow, spilt-flow, and recycle flow pressure flotation [24,25].

- Full flow: The entire influent is pressurized and then released in the flotation tank, where the bubbles are formed. This type of flotation is used for influents that do not need flocculation but require large volumes of air bubbles (Figure 5.4).
- Spilt flow: A part of the influent is pressurized, and the remaining flows directly into the flocculation or flotation tank. This type of flow is cost-effective compared with full-flow pressure flotation, because the saturator and the feed pump handle only a portion of the total flow, thus requiring a smaller saturator and feed pump. However, split flow provides less air in the system. As a result, it has to be operated at high pressure to provide the same amount of air. This type of flow is used for influents containing suspended particles susceptible to the shearing effects of a pressure pump. It is also suitable for influents containing suspended particles at low concentration, due to low air requirement (Figure 5.5).
- Recycle flow: The influent flows into the flocculation or flotation tank if flocculation process is not required. A portion of the treated influent is recycled, pressurized, saturated with air, and released to the flotation tank. This type of flow is applied to influents that need coagulation and flocculation. It is a common type of flow, and it is used more often than other types (Figure 5.6).

5.3 PROCESS DESCRIPTION OF DAF

The DAF system consists of air and water supply, saturator, and flotation chamber or tank. There are two types of flotation tanks: circular tank and rectangular tank. Wastewater treatment and sludge

Dissolved Air Flotation (DAF) for Wastewater Treatment

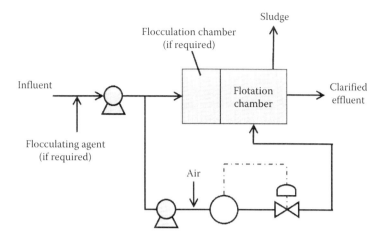

FIGURE 5.5 Split flow operational modes. (From Adlan, M.N., A study of dissolved air flotation tank design variables and separation zone performance. PhD thesis, University of Newcastle Upon Tyne, 1998.)

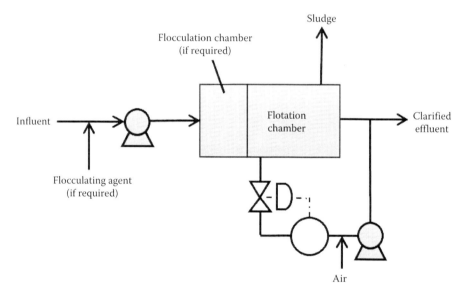

FIGURE 5.6 Recycle flow operational mode. (From Adlan, M.N., A study of dissolved air flotation tank design variables and separation zone performance. PhD thesis, University of Newcastle Upon Tyne, 1998.)

thickening mostly use circular tanks in the flotation process, which is carried out in small-size flotation plants, and require no pre-flocculation prior to flotation. However, there are large flotation plants that use circular flotation tanks and include the flocculation process. Here, the flocculation and flotation processes are contained within the same circular tank to achieve an even distribution of bubbles and particles/flocs attachment. In contrast, the advantages of using rectangular flotation tanks are its simple design that makes the introduction of flocculated water into, and removal of the floated sludge from, the tank easier; simple dimensional scale-up; and the smaller area than the circular tank.

In rectangular tanks, an inclined baffle is fixed (60° to the horizontal or at 90°) between the contact zone and the separation zone. The baffle is fixed to elevate the bubble–floc agglomerates toward the surface. At the same time, the baffle also reduces the turbulence condition created by water/wastewater entering the separation zone. If the water/wastewater enters the separation zone at high velocity, it would create a turbulence condition, which would disturb the floated sludge layer

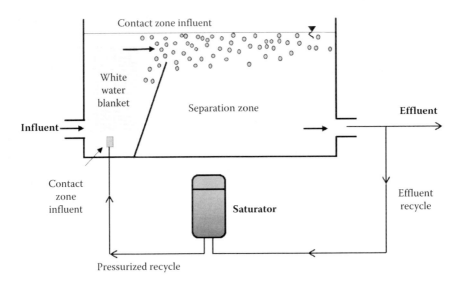

FIGURE 5.7 Rectangular flotation tank with recycle flow system. (From Murshed, M.F., Removal of turbidity, suspended solid and aluminum using DAF pilot plant. MSc thesis, Universiti Sains Malaysia, 2007.)

that accumulates continuously on the surface of the flotation tank. Figure 5.7 shows a typical setup of a recycle flow DAF system with a rectangular flotation tank.

As can be seen in Figure 5.7, the flotation tank is divided into two zones: the front zone, which is the contact zone or reaction zone, and the separation zone. A baffle is fixed between the contact zone and the separation zone. The contact zone is designed to form bubble–floc agglomerates. Small air bubbles are introduced into the contact zone. To obtain the preferred bubble size, air is dissolved in a saturator under pressure in the range of 400–600 kPa. Thus, the pressure and the recycle flow control the total amount of air introduced into the contact zone [26]. Once the pressurized water is released into the flotation tank at atmospheric pressure, bubbles are produced. In DAF, small bubbles are required to achieve a good solid–liquid separation. Bubble size in the range of 50–100 μm is the most suitable for the DAF process. If the bubbles are larger than this range, they may create turbulence in the flotation tank and, at the same time, decrease the surface area of the bubble–particle attachment. Once the bubbles and particles come into contact through adhesion, trapping, or absorption process in the reaction zone, the bubble–floc aggregates move to the separation zone. Here, the bubble–floc aggregates will rise steadily to the surface of the flotation tank, while the treated water/wastewater will be withdrawn from the bottom of the tank. Later, the bubble–floc aggregates move to the separation zone. Here, the floc rises to the surface and floats as a thick layer of sludge. The rising velocity of the bubble–floc aggregates can be estimated using Stokes' law [27]. The aggregates that do not reach the surface are swept out with the clarified water.

5.4 THEORY OF DAF

To achieve a good solid–liquid separation using DAF for influents containing particles and natural color, coagulation or flocculation is necessary prior to the introduction of microbubbles to form bubble–floc aggregates [19]. The main idea in DAF is to float the particles having specific gravity more or less equal to the specific gravity of water. This should be carried out using a low-density gas bubble, usually air. The air bubbles adhere with the particles and reduce the specific gravity to <1.0, aggregate the particles, and float them to the surface of the flotation tank [28]. In DAF, there are three main processes for removal of particles: bubble formation, bubble–particle attachment, and flotation of the bubble–particle agglomerate [29].

Dissolved Air Flotation (DAF) for Wastewater Treatment

5.4.1 Bubble Formation

There are two processes involved in bubble formation. First, the nucleation process, which initiates as soon as the pressurized water with air is released through the nozzle. The second step occurs when the excess air in the saturated water is transferred to the flotation tank in the form of gas. In this step, the bubbles begin to increase in size due to coalescence and decrease in hydrostatic pressure as they rise through the flotation tank. However, during this step, the air volume remains constant.

5.4.2 Bubble–Particle Attachment

There are three types of bubble–particle attachment mechanisms:

1. Precipitation or collisions.
2. Bubbles trapped in a floc structure as the rise through the liquid medium.
3. Bubbles absorbed in a floc structure as the floc is formed.

The adhesion or collision, trapping, and absorption are sequence mechanisms in the bubble–particle attachment (Figure 5.8a through c). The particle must be destabilized to obtain good agglomeration between particle and bubble. However, the particle has to fulfil two most significant conditions: neutral charge and hydrophobic surface. Under these conditions, attachment between particle and bubble is very strong, resulting in successful flotation. Influent required coagulation process, a proper dosing and pH value resulting in low particle charge and the formation of hydrophobic particle surface.

5.4.3 Flotation of Bubble–Particle Agglomerate

In the DAF process, the particles float due to bubbles that reduce the density of the bubble–particle agglomerates. As long as bubble–particle agglomerates have lesser density than water ($1.00\,g/cm^3$), they will rise and float to the surface. If the particles are small, fewer bubbles are required to decrease the density compared with big particles that require more bubbles. The bubble–particle agglomerates should rise to the surface of the flotation tank. The agglomerates that do not reach the surface are swept out with the clarified water. The rising velocity of the bubble–particle can be estimated using Stokes' law.

The three main theories in DAF show that there are some factors affecting the DAF system that should be taken into consideration before utilizing the DAF system in water or wastewater treatment. Therefore, the system operation and all factors should be accounted for in designing and utilizing the DAF system.

5.4.4 Kinetics of Flotation

Harper indicated that experiments seldom agree with the prediction that a bubble rising in a Newtonian liquid can be treated in isolation, unless great care is taken to remove impurities [31]. A bubble with constant surface tension rising under gravity will rise steadily if:

1. Its motion is stable relative to random small disturbances, and
2. The time taken to approach close to terminal velocity is much less than the time required for the bubble to change its size significantly.

At a low Reynolds number, the retarding or drag force is parallel and opposite to the terminal velocity with a magnitude of

$$D = 6\pi a \mu U, \tag{5.1}$$

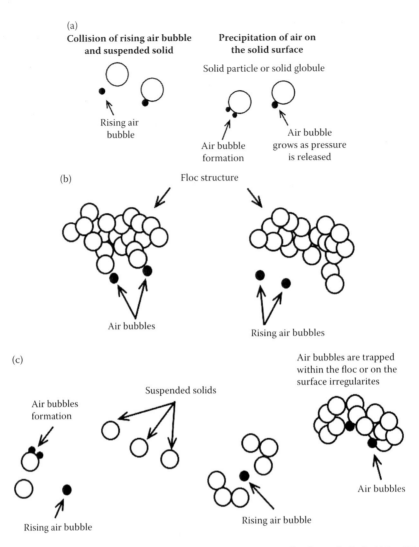

FIGURE 5.8 Bubble–particle attachment mechanism (a)–(c). (a) Adhesion of air bubble, (b) rising air attaches to floc structure, and (c) entrapment within floc structure during formation. (From Adlan, M.N., A study of dissolved air flotation tank design variables and separation zone performance. PhD thesis, University of Newcastle Upon Tyne, 1998.)

where
 D = drag force (kN)
 a = radius of bubble (m)
 μ = dynamic viscosity (kg/m/s)
 U = terminal velocity (m/s)

Equation 5.1 was obtained by Stokes for slow motion of a sphere in viscous fluid [32,33]. The expression is usually known as Stokes' law for the resistance to a moving sphere [34]. The derivation of Stokes' law is based on the assumption that the motion of the spherical particle is extremely slow and that the liquid medium boundary is at an infinite distance from the particle and also is of a large volume compared with the dimensions of the particle [35]. Clift et al. showed that bubbles are closely approximated by spheres if the interfacial tension and/or viscous forces are much more important than inertia forces and the term "spherical" can be used if the ratio of minor axis to major axis lies within 10% of unity [36].

Dissolved Air Flotation (DAF) for Wastewater Treatment

When a solid sphere falls vertically in a liquid, the viscous liquid produces a terminal velocity U. By equating the weight of the sphere to the sum of the up thrust and the drag [37], the following equation is obtained:

$$\frac{4}{3}\pi a^3 \sigma g = \frac{4}{3}\pi a^3 \rho g + 6\pi a \mu U$$

$$U = \frac{2}{9}(\sigma - \rho)a^2 \frac{g}{\mu}, \quad (5.2)$$

where
a = radius of sphere (m)
σ = density of sphere (kg/m³)
ρ = density of liquid (kg/m³)

Packham and Richards indicated that the alum sludge from a DAF water treatment plant rose at a rate of 20–35 mm/s [38]. The rising velocity is far greater than the settling velocity of an aluminum or iron floc encountered in a water treatment works (i.e., normally <0.5 mm/s). Basing their judgment on the fact that the rising rate of the floc in flotation was far greater than the settling rate, Packham and Richards considered the rate of separation of suspended matter in the flotation process from the viewpoint of Stokes' equation governing the motion of a sphere through a viscous medium and thus showed that Equation 5.2 was appropriate to describe the rising rate of the particle in the flotation process. Packham and Richards, in reviewing Equation 5.2, were of the opinion that if the size of the suspended matter is increased, a higher separation rate may be achieved. This is because the rate of separation is directly proportional to the square of the radius of the particles, the difference in the densities of liquid and the suspended particles, and inversely proportional to the liquid viscosity, as shown by Equation 5.2.

Research carried out in Russia [39] showed that at small Reynolds numbers, gas bubbles moved like solid spheres. Theoretical values of bubble rise velocity in water were not in agreement with much experimental data. For a gas bubble, which is assumed to behave like a solid, the surface can sustain a finite shear stress, the tangential velocity of the surface is everywhere zero relative to the center of the bubble, and the conventional Stokes' solution applies. According to Jameson, a force balance equation will result [40]:

$$6\pi \mu U a = \frac{4}{3}\pi a^3 (\rho - \rho_g) g. \quad (5.3)$$

When the density of gas ρ_g is negligible compared with the density of liquid ρ, the terminal velocity is given by

$$U = \frac{2\rho g a^2}{9\mu}. \quad (5.4)$$

Equation 5.4 shows that the rise velocity of a bubble is controlled by the size of the bubble and the viscosity of the fluid. If the radius of the bubble is increased, the rise velocity will be increased. The kinematic viscosity is affected by the density and the temperature of the fluid. An increase in temperature will result in the decrease in viscosity, and hence an increase in rising velocity of the bubble. Shannon and Buisson indicated that bubble rise rates at 80°C increased three times compared with those at 20°C [41]. Force balance is presented in terms of drag coefficient C_D by Harper [31] as follows:

$$C_D = \frac{\text{Force on bubble}}{1/2 \rho U^2 \pi a^2} = \frac{4/3 \pi \rho g a^3}{1/2 \rho U^2 a^2} = \frac{4gd}{3U^2}. \quad (5.5)$$

This coefficient is the force per unit cross-sectional area, made dimensionless by the dynamic pressure $1/2\rho U^2$. Substituting Equation 5.4 in 5.5 yields:

$$C_D = \frac{24}{Re}, \qquad (5.6)$$

where Re is the Reynolds number. Equation 5.6 is used for viscous resistance at low Reynolds numbers, $Re < 0.5$ [42]. The same equation was used by Vrablik to determine the maximum bubble size of 130 μm for a complete viscous flow [43]. He indicated that the maximum value of the Reynolds number for laminar or viscous flow is 1.13. The relationship governing bubble size, laminar flow, bubble rising velocity (as per Equation 5.4), and temperature has been established. This relationship is shown in Table 5.1. The optimum bubble rising velocity for the DAF system is about 300 mm/min [44]. The rising velocity should not be <125 mm/min or >500 mm/min.

Experimental work by Fukushi et al. showed that Equation 5.4 cannot be used to describe bubble rise velocity [44]. This is due to the turbulent environment that occurs in the mixing zone (reaction zone) of the DAF tank. They suggested that the following equation is more appropriate and that it agreed with their experimental results:

$$U = \frac{\rho g a^2}{3\mu}. \qquad (5.7)$$

The properties of air bubbles produced in the DAF process developed by Fukushi et al. were compared with those developed by Edzwald et al. [45,46]. There are many differences between the models. These are shown in Table 5.2. The model developed in 1985 by Fukushi et al. was based

TABLE 5.1
Relationship between Bubble Size, Rise Velocity, Temperature, and Laminar Flow

Bubble Size (μm)	Rise Velocity (m/h) Above Which Turbulent Flow Exists[a] 4°C	20°C	Terminal Rise Velocity (m/h) Based on Stokes' Law 4°C	20°C
10	565	360	0.125	0.196
20	283	180	0.499	0.783
30	188	120	1.12	1.76
40	141	90	2.00	3.13
50	113	72	3.12	4.89
80	70.7	45	7.99	12.5
110	51.4	32.7	15.1	23.7
120	47.1	30	18.0	28.2
130	43.5	27.7	21.1	33.1[b]
140	40.4	25.7	24.5	38.3[b]
160	35.3	22.5	31.9	50.1[b]
170	33.2	21.2	36.1[b]	56.5[b]

Source: Malley, J.P. Jr., A fundamental study of dissolved air flotation for treatment of low turbidity waters containing natural organic matter. Unpublished Ph.D. dissertation, University of Massachusetts, 1988.

[a] Based on a critical Reynold's number of 1.0 for the upper limit of laminar flow.
[b] Indicates that the terminal rise velocity will result in turbulent flow.

TABLE 5.2
Models Developed for the Dissolved Air Flotation Process

	Fukushi et al. [47]	Edzwald et al. [46]
Generated Air Bubbles		
Size range d_a (μm)	10–120 (average 60)	10–100 (average 40)
Rise velocity (cm/s)	$gd_a^2/12\nu$	$gd_a^2/18\nu$
Zeta potential (mV)	−150 at pH 7	Not measured
Pressure P (kPa)	392	345–585
Recycle ratio r	0.1	0.08
Concentration n_a (cm^{-3})	10^4–10^5	10^4–10^5
Produced Flocs		
Size range d_f (μm)	10^0–10^3	10^0–10^2 (10–30 μm is best)
Density ρ_f (g/cm^3)	Floc density function	1.01 (assumed)
Suitable mobility (μm/s/V/cm)	0 to +1 (clay floc) −1 to +1 (colour floc)	0.5 or less
Bubble–Floc Collision and Attachment		
Collision model	Population balance model	Single collector collision model
Flow regime	Turbulent flow	Laminar flow
Mechanism	Locally isotropic turbulence, viscous subrange diffusion	Brownian diffusion, interception, gravity settling
Attachment mechanism	Electrical charge interactions (coverage of precipitated coagulant on a floc surface)	Electrical charge interaction, water layer at floc surface
Rise velocity of agglomerate (cm/s)	0.1–2.6 (observed)	About 0.3 (nearly equal to bubble rise velocity)

Source: Fukushi, K., et al., *Water Sci. Technol.*, 31, 37–47, 1995.
g, gravity; d_a, diameter of bubble; ν, kinematic viscosity; d_f, diameter of floc; ρ_f, density of floc.

---

on the population balance model of bubbles and flocs in a turbulent flow environment Population Balance Model (PBM model) [47]. However, the model developed by Edzwald was derived from a single collision theory in a laminar flow condition Single Collision Theory (SCT model). In the SCC model, collision occurs due to Brownian diffusion, interception, and gravity settling. Fukushi et al. indicated that Brownian diffusion and gravity settling cannot be dominant for a normal floc (10–1000 μm) and bubble size range in flotation [45]. Interception also cannot be dominant because, in practice, the mixing zone is apparently in a turbulent flow where certain energy dissipation occurs.

In fact, a literature survey indicates that Equation 5.7 was initially suggested by V. G. Levich in 1962. Levich showed that Equation 5.7 is applicable for small Reynolds numbers, $Re \ll 1$ and when the following inequality holds [48]:

$$\frac{ga^3}{3\nu^2} \ll 1, \tag{5.8}$$

where $\nu = \mu/\rho$ (i.e., kinematic viscosity equals dynamic viscosity divided by density of liquid). If the medium is water, the size of moving bubbles will be $a \ll 2 \times 10^{-2}$ cm. Levich also indicated that the theoretical value of the drag coefficient for a gas bubble in water is equal to $8/Re$ (i.e., one and one-half times smaller than for a solid sphere). This value is not in agreement with Equation 5.6. However, Levich indicated that Allen's experimental results with small Reynolds numbers completely disagreed with the theory and led to values for the drag coefficient that coincided exactly with the drag on a solid sphere [48].

A mathematical equation for solid/liquid separation was developed by Howe limited to flotation of discrete particles without the interference of surface-active forming agents. It was derived from a differential equation of motion, which was expanded to give solutions for the rising velocity of a particle with changes in the applied rising force, particle diameter, liquid viscosity, and particle density [49]. The equation is as follows:

$$R = 1 - e^{-\left(\frac{V_r}{Q/A_h}\right)}, \quad (5.9)$$

where
R = ratio of total removal of solid concentration after flotation to the inflow solid concentration
  $= 1 - C_o/C_i$
C_o = the effluent suspended solids
C_i = the influent suspended solids
V_r = the rising velocity of a single particle/air bubble (m/s)
Q = the flow applied to the flotation unit (m³/s)
A_h = the horizontal area of the unit (m²)

Equation 5.9 is limited to discrete particles without the interference of surface-active forming agents. Karamanev, in his article on the rise of bubbles in quiescent liquid, showed that equations based on the model of bubble with internal circulation often fail to describe the real systems adequately [50]. This is because even highly purified liquids (such as triple distilled water) contain enough surface-active components to affect internal bubble recirculation. Recirculation is normally due to the presence of surface-active substances, and the resulting variable surface tension leads to a change in boundary conditions of the bubble [48]:

$$U = 25V^{1/6}, \quad (5.10)$$

where V is the volume of the bubble. However, this equation works only for large, spherical cap-shaped bubbles. The drag coefficient C_D of the gas bubble calculated on the basis of equivalent sphere diameter by most authors was found to have a large deviation of C_D as a function of the Reynolds number when different liquids are used. The assumption made by most authors that free-falling heavy spheres behave exactly like free-rising solid spheres is found to be incorrect, especially for particles with densities <0.3 gm/cm³ and $Re > 130$ rising in water. Karamanev suggested the following equation based on the balance of forces acting on a rising bubble [50]:

$$\frac{1}{2} C_D S \rho U^2 = \Delta \rho g V, \quad (5.11)$$

where ρ is the liquid density, $\Delta\rho$ is the difference of density between liquid and gas, and S is the area of the bubble. To obtain C_D based on real bubble geometry, the area S should be determined from the diameter projected on the horizontal plane circle, d_h; $S = \pi d_h^2/4$. Then, the volume of the bubble is calculated using the equivalent diameter, $V = \pi d_e^3/6$. These values are substituted in Equation 5.11 and become

$$C_D = \frac{4g\Delta\rho d_e^3}{3\rho d_h^2 U^2}. \quad (5.12)$$

Equation 5.11 can be written in terms of U:

$$U = \left(\frac{8gV}{\pi C_D d_h^2}\right)^{1/2}. \tag{5.13}$$

By substituting from Equation 5.12,

$$U = \left(\frac{8g}{6^{2/3}\pi^{1/3}C_D}\right)^{1/2} V^{1/6} \frac{d_e}{d_h}. \tag{5.14}$$

For $Re < 130$, Karamanev suggested that C_D can be calculated using the following equation:

$$C_D = \frac{24\left(1+0.173\,Re^{0.657}\right)}{Re} + \frac{0.413}{1+16300\,Re^{-1.09}}. \tag{5.15}$$

For spherical bubbles at $Re < 1$, $C_D = 24/Re$ and $d_e/d_h = 1 = 1$ and Equation 5.12 transforms to Stokes' equation.

5.4.5 Solubility of Air

In flotation, the quantities of air used are normally expressed in terms of volume of air supplied per volume of water treated [10]. Henry's law is used when treating saturated water as a dilute solution of air in water [43]. It must be remembered that Henry's law was originally based on his experiment with N_2, O_2, N_2O, H_2S, CO_2 and water at only one temperature. The concept that the law could be used for general application is unfounded [51]. However, experimental work on wastewater with dissolved solids up to 1000 mg/L with pressures up to 500 kPa showed that Henry's law constant could be used to calculate the mass of dissolved air [52]. For ideal dilute solutions where the solute obeys Henry's law but not Raoult's law, and the solvent obeys Raoult's law, then the use of Henry's law is applicable [53,54].

$$PB = x_B K_B, \tag{5.16}$$

where PB is the vapour pressure, x_B is the mole fraction of the solute, and K_B is constant. Based on Henry's law, Edzwald and Walsh suggested the following:

$$C_S = f\frac{p}{k}, \tag{5.17}$$

where C_S is the concentration of air in the saturated liquid, p is the absolute pressure, k is the Henry's law constant, and f is the efficiency factor, which is about 70% for unpacked saturators and up to 90% for packed systems. Values of k at 0°C and 25°C are 2.72 and 4.53 kPa/mg/L, respectively. However, others indicated that Henry's law is not strictly applicable when treating saturated water [55,56]. The equation has to be modified with an exponent m on the pressure p as follows [56]:

$$C_S = \frac{p^m}{k}. \tag{5.18}$$

Klassen and Mokrousov, in their review on the solubility of gases in water, were of the opinion that the solubility of gases depends on the partial pressure, temperature, and concentration of other substances in the solution [57]. If the partial pressure is increased, then the solubility of gas will be increased. However, if the concentration of soluble substances in water is increased, gas solubility will be decreased as definite quantities of water molecules complex in the form of hydrated ions. Edwards added that if the total pressure is <507 kPa (5 atm.), the solubility for a particular partial pressure of solute gas is normally independent of the total pressure of the system [58]. In its relationship to temperature, the solubility of a gas will be decreased when the temperature is increased [43,59]. This is as illustrated in Figure 5.9. In the case of distilled water, when the temperature is increased from 0°C to 30°C, the solubility of air is reduced by 45%. Liquid solubility of the gases varies as shown in Table 5.3.

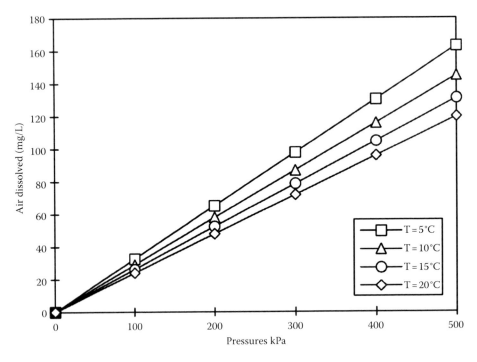

FIGURE 5.9 Solubility of air in water. (From Adlan, M.N., A study of dissolved air flotation tank design variables and separation zone performance. PhD thesis, University of Newcastle Upon Tyne, 1998.)

TABLE 5.3
Solubility of Various Gases at 20°C and 760 mm Hg

Type of Gas	Cubic cm Gas/Cubic cm Water	Gram of Gas/100 gm of Water
Nitrogen	0.015	0.0019
Oxygen	0.031	0.0043
Hydrogen	0.018	0.00016
Carbon dioxide	0.88	0.17
Carbon monoxide	0.023	0.0028
Air	—	1.87
Hydrogen sulfide	2.58	0.38
Sulfur dioxide	39.4	11.28

Source: Vrablik, E.R., *Proceedings of the 14th Industrial Waste Conference*, Purdue University, West Lafayette, IN, 743–779, 1959.

Bratby and Marais in their studies on saturator performance indicated that it would be difficult to achieve full saturation at a saturator pressure <350 kPa [60]. From an economic point of view, the efficiency of the saturator system was important. They found that by using a packed system of 0.5 m depth with Raschig rings of 25 mm diameter, full saturation was achieved at saturator pressures beyond 250 kPa for a surface loading up to 2500 m/day. A similar level of saturation was found by Zabel and Hyde using a packed saturator of 0.75 m depth with 25 mm Berl saddles [23].

For design purposes, Bratby and Marais suggested that at a temperature of 20°C with a pressure of 3 atm., the concentration of air precipitated on reducing the pressure to atmospheric pressure is given by the following equation [60]:

$$a_P = 19.5\, p\,(\text{mg/L}), \tag{5.19}$$

where p is the saturator pressure in atmospheres.

5.4.6 Bubble Generation

Rykaart and Haarhoff indicated that the geometrical design and operating conditions of the injection nozzles were important determining factors for bubble size [61]. They reported that saturator pressure does not have a consistent effect on nozzle efficiency. There were contradicting claims regarding whether a higher pressure produces smaller bubbles [55,62] or bigger bubbles [52,63]. But Jones and Hall reported that there was no significant relationship between pressure variation and bubble size [64].

Studies by Bratby and Marais [65] showed that the shape and roughness of the valve, the degree of turbulence and dilution of saturator feed downstream of the valve, and the concentration of particulate nuclei in the dilution water had a negligible effect on the precipitation of air from a solution (i.e., mass of air precipitated to unit volume of saturator feed). However, these findings were contradicted by those reported by others [55,61] in terms of the shapes and roughness of the valves. Rykaart and Haarhoff showed that at a saturator pressure of 500 kPa, a nozzle with a bend in its channel produced a bubble size of 49.4 µm (median diameter) compared with a nozzle with a tapering outlet that produced 29.5 µm [61]. When the saturator pressure was reduced, the bubble sizes were reduced.

For a continuous-flow DAF plant, Edzwald and Walsh predicted that the concentration of air released in the tank (C_r) would be as follows:

$$C_r = \left[\left(\frac{C_S - C_a}{1 + R_r}\right)\right] R_r - K, \tag{5.20}$$

where C_a is the concentration of air that remains in solution at atmospheric pressure, R_r is the recycle ratio, which is equal to the recycle flow rate divided by the influent flow rate, and K is the influent saturation factor defined as $(C_S - C_a)$, where C_o is the concentration of air in the influent water. In most cases, C_o is saturated, and this means $K = 0$. To find the bubble volume concentration (ϕ_b), Edzwald and Walsh suggested that C_r be divided by the saturated density of air ρ_{sat} as shown in the following equation,

$$\phi_b = C_r / \rho_{sat} \tag{5.21}$$

To get the generated air volume at the same temperature under atmospheric pressure, Takahashi et al. used the following equation by considering air as an ideal gas [55]:

$$V_A = \left(\frac{\rho_w}{M_w}\right)\left(\frac{P_A - P_O}{P_O}\right)\frac{RT}{H_E}, \tag{5.22}$$

where ρ_w is the density of water in gm/cm³, P_A is the dissolved pressure in dyne/cm², P_O is the atmospheric pressure in dyne/cm², R is the gas constant in erg/k · mol, T is the absolute temperature in Kelvin, M_w is the molecular weight of water in g/g·mol, and H_E is Henry's law constant in dyne/cm².

By assuming that all the dissolved air in water changes into bubbles, the theoretical generated flow rate can be obtained using Equation 5.18. The experimental results by Takahashi et al. showed that the generated air flow rate increased with an increase in dissolved pressure and also with an increase in liquid flow rate [55]. To obtain the volume of air occupied by a single spherical bubble, V_b, Takahashi et al. suggested the following equation:

$$V_b = \left(\frac{P_O + \rho_w gh + 4\sigma/d_{av}}{P_O}\right)\frac{\pi}{6}d_{av}^3, \qquad (5.23)$$

where h is the depth from the liquid surface to the bubble in cm, σ is the surface tension in dyne/cm², and d_{av} is the volumetric mean diameter of the bubble in cm. According to the authors, the effect of liquid depth is negligible; thus, the measurement of the bubble diameter was carried out at the top of the flotation tank. The number of bubbles generated per cm³ of water could be obtained from the following equation:

$$N_b = \frac{G/Q}{V_b}, \qquad (5.24)$$

where G is the volumetric flow rate of air generated under decrease in pressure (cm³/s), and Q is the volumetric flow rate of liquid (cm³/s). Their experimental results showed that by increasing the dissolved pressure and liquid flow rate, the number of bubbles will be increased. The geometry of the nozzle also affects the bubble size. By using a needle valve, Takahashi et al. obtained the following equation:

$$N_b = 1 \times 10^4 \left(\frac{P_A - P_O}{P_O}\right)^2 Q. \qquad (5.25)$$

The calculated and experimental values of the number of bubbles were compared, and the results were claimed to be remarkably in agreement with the equation used.

For an efficient solid–liquid separation process, small bubbles are needed [66–68]. Bubble sizes in the range of 20–80 µm are capable of good attachment to floc particles. Larger bubbles will create a hydraulic disturbance along their rising path toward the surface and a decrease in the surface area. For example, a 2 mm bubble contains the same amount of air as 64,000 bubbles of 50 µm in size. Collin and Jameson reported that the optimum bubble size in the microflotation process is approximately 50 µm [67].

5.4.7 Collision

Reay and Ratcliff, in their study of dispersed air flotation, defined the collection efficiency of a bubble as the fraction of particles in the bubble's path that are picked up by the bubble [69]. Particles of about 3 µm diameter or larger will not be affected by Brownian motion. They will be in contact with the bubble only if their hydrodynamically determined trajectories come within one particle radius (r_p) of the bubble. This region is called the collision regime. By considering the collision regime in which the Brownian diffusion is negligible [70], the collection efficiency of a bubble can be expressed as:

$$\eta = \eta_1 \times \eta_2, \qquad (5.26)$$

Dissolved Air Flotation (DAF) for Wastewater Treatment

where

η_1 = collision efficiency, that is, the fraction of particles in the bubble's path that collided with the bubble

η_2 = attachment efficiency, that is, the fraction of particles colliding with the bubble that stick to it

Equation 5.26 indicates that η_2 will depend mainly on the chemical nature of the particle surface, the bubble surface, and the thin film of liquid draining from between them. Reay and Ratcliff also reported on the predicted collision efficiency and illustrated it with a graph; $\eta_1 = 1.25 (r_p/R_b)^{1.9}$ for $(\rho_p/\rho_f) = 1$, $\eta_1 = 3.6 (r_p/R_b)^{2.05}$ for $(\rho_p/\rho_f) = 2.5$, and η_1 is roughly proportional to $(r_p/R_b)^2$ over the density range used. The symbols used in the aforementioned expressions are as follows [69]:

r_p = particle radius (cm)
R_b = bubble radius (cm)
ρ_p = particle density (gm/mL)
ρ_f = fluid density (gm/mL)

Since η_1 is proportional to R_b^2, the average number of particles picked up by a bubble (by assuming η_1 is constant) should be roughly independent of bubble size and the flotation rate should be proportional to bubble frequency (i.e., the amount of bubbles rather than bubble diameter over the entire range of particle sizes). This prediction is applicable to bubbles of diameter up to 0.1 mm [71]. However, when latex particles (3–9 μm) having almost the same density as water and larger zeta potential (+10.6 mV) were used in the experiments, they could not get close to the bubble surface [71]. This means the bubble–particle collision model is not appropriate for latex particles.

Flint and Howarth, in their review on the collision efficiency of small particles with spherical air bubbles, reported that the collision of a particle with a bubble would depend on the balance of viscous, inertial, and gravitational forces acting on the bubble [72]. Besides that, the form of streamlines around the bubble also play an important role in whether or not collision takes place. Flint and Howarth formulated an equation of motion of a small spherical particle similar to a spherical bubble rising in an infinite pool of liquid in the *j*th direction as follows [72]:

$$m_p \frac{\partial v_j}{\partial t} = G_j + C_d (u_j - v_j), \qquad (5.27)$$

where
G_j = body force acting on the particle; for raindrop collision, $G_j = 0$. In flotation, there is clearly a component of relative acceleration due to gravity because the bubble and particle are of distinctly different densities.
C_D = dimensional drag coefficient for the particle, depending on the shape of the particle and the Reynolds number past it. For a spherical particle, the drag will be the same in all directions.
V_j = particle velocity.
u_j = velocity the fluid would have at the position of the particle if no particle were there. For fine particles in flotation, it is assumed that the flow around the particle has an insignificant effect compared with the flow pattern due to the bubble; u_j then depends on the shape of the bubble and the Reynolds number around it.
$4t$ = time

By considering the relative two-dimensional motion of a spherical bubble and particle where the bubble is held stationary at the origin of the coordinate system by a liquid flow equal to the bubble

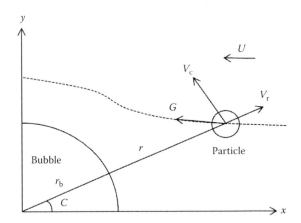

FIGURE 5.10 Geometry of bubble–particle system. (From Adlan, M.N., A study of dissolved air flotation tank design variables and separation zone performance. PhD thesis, University of Newcastle Upon Tyne, 1998.)

rise velocity in the negative direction (Figure 5.10), Flint and Howarth suggested the equation of motion for the particle as follows [72]:

$$\frac{4}{3}\pi r_p^3 \rho_p \frac{\partial v_y}{\partial t} = 6\pi \mu_f r_p \left(u_y - v_y \right) \tag{5.28}$$

$$\frac{4}{3}\pi r_p^3 \rho_p \frac{\partial v_x}{\partial t} = \frac{4}{3}\pi r_p^3 \left(\rho_p - \rho_f \right) g - 6\pi r_p \left(u_x - v_x \right). \tag{5.29}$$

Reducing these equations to their dimensionless form and introducing the variable v, u, and t, and parameters K and G:

$$v_x^* = \frac{v_x}{u} \quad v_y^* = \frac{v_y}{u}$$

$$u_x^* = \frac{u_x}{u} \quad u_y^* = \frac{u_y}{u}$$

$$t^* = \frac{tu}{r_b}$$

and,

$$K = \frac{2\rho_p r_E^2 u}{9\mu_f} r_b$$

$$G = 2\left(\rho_p - \rho_f\right) r_p^2 \frac{g}{9\mu_f} u$$

i.e.,

$$K \frac{\partial v_y^*}{\partial t^*} = u_y^* - v_y^*$$

$$K \frac{\partial v_x^*}{\partial t^*} = -G - u_x^* + v_x^*$$

where
- r_p = particle radius (cm)
- ρ_p = particle density (gm/mL)
- ρ_f = fluid density (gm/mL)
- v = component of particle velocity (cm/s)
- t = time (s)
- μ_f = fluid viscosity (kg/m/s)
- u = component of velocity field due to bubble (cm/s)
- x, y = Cartesian position coordinates (cm)
- v = dimensionless component of particle velocity (cm/s)
- t = dimensionless time (s)
- K = particle inertia parameter
- G = dimensionless settling velocity of particle (cm/s)

According to Flint and Howarth, calculation for K down to 0.001 shows that the collision efficiency remains substantially constant for $0.001 < K < 0.1$, meaning that collision efficiency is virtually independent of K and of whether Stokes or potential flow is assumed [72]. They suggested that for a fine particle characterized by $K < 0.1$, inertial effects of the particle may be neglected, and single bubble collision efficiency η can be calculated from:

$$\eta = \frac{G}{(1+G)}. \tag{5.30}$$

However, Flint and Howarth indicated that in the flotation tank, the collision efficiency may be several times as great as those predicted from single bubble calculations [72]. This may be due to at least three reasons:

1. The presence of hindering effects of the neighboring bubbles that reduced the rising velocity of the bubble. For fine bubbles, this could lead to an increase in collision efficiency.
2. Difference in the shape of liquid streamlines around the bubble. The greater the number of bubbles, the closer the assemblage and straighter the streamlines. This results in the increase of collisions between particles and bubbles.
3. The motion of particles upstream from the target bubble is influenced by the layers of bubbles ahead and is thus no longer parallel to the direction of bubble motion.

The aforementioned opinion, which was expressed by Flint and Howarth, is found to be in agreement with that of Fukushi et al., as the latter showed that a single collector collision model was not appropriate in the DAF process [45]. Furthermore, King indicated that the calculated collision efficiency based on the works of Reay and Ratcliff, Flint and Howarth, and Sutherland and Woodburn et al. were not in agreement with each other [73,71–72,74–75].

5.4.8 Interception and Diffusion

According to Yao et al., a single particle of filter media is a collector, and if any suspended particle is in contact with the collector, then a process known as interception occurs [75]. The contact efficiency of a single media particle or collector is the ratio of the rate at which the particles strike the collector to the rate at which particles flow toward the collector, which can be expressed as follows:

$$\eta = \frac{\text{Rate at which particles strike the collector}}{u_o c_o \left(\frac{\pi d^2}{4} \right)}, \tag{5.31}$$

where

*u*ₒ = water velocity (m/s)
*c*ₒ = suspended particle concentration upstream from the collector where the flow is undisturbed by the presence of the grain
d = grain diameter (cm)

In the case of flotation, the single collector efficiency (η) may be defined as follows [46,77]:

$$\eta = \frac{\text{Particle} - \text{buble collision rate}}{\text{Particle} - \text{bubble approach}}. \quad (5.32)$$

Reay and Ratcliff indicated that submicron particles will reach the bubbles mainly by Brownian diffusion. In the diffusion regime, collection efficiency will decrease with increasing particle radius, r_p [69]. Flotation of these submicron particles could be improved if they were agglomerated into flocs of suitable size in the collision regime. Theoretical calculations were made on particles with diameter < 0.2 μm and bubbles size of 75 μm. At normal temperatures and pressures, particles < 1 μm in diameter suspended in gases or water will exhibit a Brownian motion that is sufficiently intense to produce collision with a surface immersed in the fluid [78]. Yao et al., in describing basic transport mechanisms in water filtration, explained that when a particle in suspension is subjected to random bombardment by molecules of the suspending medium, then a Brownian movement of the particle known as diffusion takes place [76]. Numerical and analytical determinations of single-collector efficiency were discussed by Yao et al. based on the works of previous investigators, and the following equations were established:

$$\eta_D = 4.04\,\text{Pe}^{-2/3} = 0.9\left(\frac{kT}{\mu d_p d v_o}\right)^{2/3} \quad (5.33)$$

$$\eta_I = \tfrac{3}{2}\left(\frac{d_p}{d}\right)^2 \quad (5.34)$$

$$\eta_g = \frac{(\rho_p - \rho)}{18\mu v_o} g d_p^2, \quad (5.35)$$

where η_D, η_I, and η_G are the theoretical values for single-collector efficiency when the sole transport mechanisms are diffusion, interception, and gravity settling, respectively. Pe is the Peclet number (i.e., Pe = $2R_b U_b/D_f$, where R_b is bubble radius, U_b is bubble rising velocity, and D_f is particle diffusivity in cm²/s), k is Boltzmann's constant, T is the absolute temperature, d_p is the diameter of suspended particle, d is the diameter of the collector or the bubble, which is equal to d_p, v_o is the approach velocity of fluid, and ρ is the density of fluid, which is equal to ρ_f.

Then the expression for total single-collector efficiency of a media grain can be written as follows [76,79]:

$$\eta_T = \eta_D + \eta_I + \eta_G. \quad (5.36)$$

Edzwald and Walsh used the same theoretical approach as in filtration to develop a conceptual model for flotation [10,76,79]. Thus, the following equations are introduced:

$$\eta_D = 0.9\left(\frac{k_b T}{\mu d_p d_b U_b}\right)^{2/3} \quad (5.37)$$

Dissolved Air Flotation (DAF) for Wastewater Treatment

$$\eta_I = \tfrac{3}{2}\left(\frac{d_p}{d_b}\right)^2 \tag{5.38}$$

$$\eta_G = (\rho_p - \rho_f)\frac{g d_p^2}{18\mu U_b}. \tag{5.39}$$

Comparing the equations used in filtration to the aforementioned equations, the approach velocity of fluid and Boltzmann's constant have been changed to U_b (bubble rise velocity) and k_b (Boltzmann's constant for bubble), respectively. This is done to suit the mechanisms involved in flotation. Ward, in his review on capture mechanisms, introduced a new form of equations for η_D and η_G by substituting the bubble rise velocity U from Equation 5.4 into Equations 5.37 and 5.39. Thus, we have the following result [56]:

$$\eta_D = 6.18\left(\frac{kT}{\rho_f g d_p}\right)^{2/3}\left(\frac{1}{d_b}\right)^2 \tag{5.40}$$

$$\eta_G = \left(\frac{d_p}{d_b}\right)^2\frac{(\rho_p - \rho_f)}{\rho_f}. \tag{5.41}$$

Results on single-collector efficiency by Edzwald and Walsh show a minimum efficiency occurs at a particle size of around 1 μm [10]. For removal efficiency, Edzwald et al. used the same principle as in Equation 5.26, changing only the symbols of the expression as follows:

$$R = \alpha_E \eta_T (100\%), \tag{5.42}$$

where α_E is the attachment efficiency. If the total number of bubbles (N_b) is considered, then Edzwald et al. suggests the following equation for particle removal:

$$\frac{dN_p}{dt} = -(\alpha_{pb}\eta_T) A_b U_b N_b N_p, \tag{5.43}$$

where A_b is the projected area of the bubble and N_p is the particle number concentration. By having a bubble volume concentration of $\Phi_b = \pi d_b^3 N_b / 6$, and substituting into Equation 5.43, we have the following equation:

$$\frac{dN_p}{dt} = -(\alpha_{pb}\eta_T)\frac{\pi d_b^2}{4} U_b N_p \frac{6\Phi_b}{d_b^3} = -\left(\frac{3}{2}\right)\left(\frac{\alpha_{pb}\eta_T U_b \Phi_b N_p}{d_b}\right). \tag{5.44}$$

The particle number concentration removal in terms of flotation tank depth can be rewritten as:

$$\frac{dN_b}{dH} = -\frac{3}{2}\left(\frac{\alpha_{pb}\eta_T \Phi_b N_p}{d_b}\right). \tag{5.45}$$

Edzwald et al. also produced a summarized table of their model parameters for DAF facilities. This is as shown in Table 5.4. However, this model has not been tested or verified [10].

TABLE 5.4
Model Parameters for DAF Facilities

Parameter	Affected by	Comments
	Pre Treatment	
α_{pb} (particle-bubble attachment efficiency)	Particle-bubble charge interaction and hydrophilic nature of particles	Improve α_{pb} by chemical pretreatment, coagulation, and pH conditions
N_p (particle number concentration)	Coagulation addition and flocculation time	Coagulant may add particles, flocculation may reduce N_p and increase d_p.
η_T (single collector efficiency)	Diffusion and interception	Minimum for d_p of 1 μm
	Flotation Tank	
d_p (bubble diameter)	Saturator pressure	Small bubbles produce large interfacial areas and surface forces between bubbles and particles. Small bubbles, improve η_T
Φ_b (bubble volume concentration)	Saturator pressure and recycle ratio	Large Φ_b ensures collision opportunities and lowering of floc density

Source: Edzwald, J.K., Walsh, J.P., Dissolved air flotation: Laboratory and pilot plant investigation. AWWA Research Foundation and AWWA, 1992.

Ward, in his article on DAF, made an improvement on Equation 5.45 by integrating it over the tank depth H from $N = N_o$ at the surface $H = 0$ to $N = n$ at the tank base $H = H$. Thus, the overall particle removal equation becomes [56]:

$$N = N_o e^{-\left(\frac{3\alpha_{pb}\eta_T \Phi_b H}{2d_b}\right)}. \qquad (5.46)$$

Then, the overall efficiency is given by:

$$\eta = 1 - \frac{N}{N_o}. \qquad (5.47)$$

5.4.9 TANK DESIGN

The usual design procedure for any flotation unit can be based on Figure 5.11. All the suspended solids in the flotation chamber should have a sufficient rise velocity to travel the effective depth D within the specified detention time T. This means the rise rate V_T must be at least equal to the effective depth D divided by the detention time T, or equal to the flow divided by the surface area:

$$V_T = \frac{D}{T} = \frac{Q}{A_S}, \qquad (5.48)$$

where
  V_T = vertical rise rate of suspended solids (m/s)
  D = effective depth of flotation chamber (m)
  T = detention time (s)
  Q = influent flow rate, (m³/s)
  A_S = surface area of flotation chamber (m²)

Dissolved Air Flotation (DAF) for Wastewater Treatment

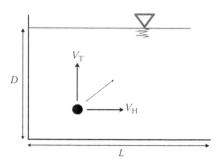

FIGURE 5.11 Basic design concept of flotation unit. (From Adlan, M.N., A study of dissolved air flotation tank design variables and separation zone performance. PhD thesis, University of Newcastle Upon Tyne, 1998.)

The particles to be removed must also have a horizontal velocity

$$V_H = \frac{Q}{A_C}, \quad (5.49)$$

where
  V_H = horizontal velocity (m/s)
  A_C = cross-sectional area of flotation chamber (m²)

If the flotation chamber is in a rectangular shape, then the following equations can be established:

$$W = \frac{A_C}{D} \quad (5.50)$$

$$L = \frac{A_S}{W} = \frac{A_S}{(A_C/D)} = V_H D \frac{A_S}{Q}, \quad (5.51)$$

where
  W is the width of the flotation chamber (m),
  L is the effective length of the flotation chamber (m),
  Q is the influent flow rate (m³/s),

and the value of D/W is usually between 0.3 and 0.5.

The size of the flotation tank can be reduced if the separation rate is increased [80]. Katz and Wullschleger showed that a particle with a bubble attached to it would increase in its rising rate with an increase in the particle size. This finding is similar to that reported by Packham and Richards [38]. However, other factors such as pressure, recycle ratio, temperature, pH, zeta potential of the particles, number and size of bubbles produced, types of nozzles, flocculation process, flow condition, and configuration of the tank are believed to have a significant effect on the separation process [11,45,81].

Longhurst and Graham reported that the surface overflow rate (SOR) or rise rate is the fundamental criterion for tank design [15]. It is defined as the flow rate divided by the surface area of the flotation tank. In practice, the surface area is based on the interfacial area between clarified water and sludge, and not on the total area of the flotation tank [15]. The characteristics of water and bubble size will determine the air/floc aggregate rise velocity. For normal design purposes, rise velocities between 3 and 8 m/h have been used [68]. For laminar flow, the maximum size of bubbles is 130 μm; for bubbles measuring <130 μm, Stokes' law applies [11], and Equation 5.2 can be used to calculate the rise rate. The maximum bubble size for laminar flow can be calculated using Equation 5.3 by assuming limited laminar flow, $Re = 1$, and using the relationship between bubble size and rise rate of air bubble, which has been established in graphical form [11]. A survey carried by Longhurst and Graham showed that the average normal operating SOR is below 6–9 m/h with a maximum rate of up to 11 m/h [15].

Bratby and Marais, in their investigation on the application of DAF in activated sludge, were of the same opinion as Longhurst and Graham regarding the design of flotation units [15,82]. Instead of SOR, Bratby and Marais used the term down flow rate, which is defined as the total flow into the unit divided by the plan area at the outlet. It is the value of limiting down flow rate (V_L), where the bubble–particle agglomerates are carried down with the effluent that controls the design of the tank. Data published by Edzwald on the design and operation parameters of DAF showed that there were still considerable variations in retention time, hydraulic loading, and recycle ratio between different treatment works in different parts of the world [83]. These are shown in Table 5.5.

In terms of shape, Zabel and Melbourne indicated that a rectangular shape has gained greater acceptance due to advantages such as simple design, easy introduction of flocculated water, easy

TABLE 5.5
Summary of DAF Design and Operation Parameters

Parameter	South Africa	Finland	The Netherlands	UK	UK (Edzwald)	Scandinavia
Flocculation						
Intensity						
Time (min)	4–15	20–127	8–16	20–29	18–20	28–44
Flotation						
Reaction zone						
Time (min)	1–4		0.9–2.1			
Hyd. load. (m/h)	40–100		50–100			
Separation zone						
Hyd. load. (m/h)	5–11	2.5–8	9–26			
Total flotation area						
Hyd. load. (m/h)			10–20	5–12	8.4–10	6.7–7
Time (min.)					11–18	
Recycle (%)	6–10	5.6–42	6.5–15	6–10	5–10	10
Unpacked Sat.						
Pressure (kPa)	400–600				400–550	460–550
Hyd. load. (m/h)	20–60					
Time (s)	20–60					
Packed Sat.						
Pressure (kPa)	300–600				400–500	
Hyd. load. (m/h)	50–80					
Packing depth (m)	0.8–1.2					
Saturators[c]						
Pressure (kPa)		300–750	400–800	310–830	480–550	

Source: Edzwald, J.K., *Water Sci. Technol.*, 31, 1–23, 1995.
[c] Unspecified with respect to unpacked or packed saturator, [1,2,15,83].

float removal, small area, and flexibility of scale-up [21]. In addition to that, floc breakup is minimized, hydraulic efficiency is maximized, and engineering and construction is simplified [15]. A tank with SOR in the range 6–12 m/h would have a depth of 1.2–1.6 m and a residence time of 5–15 min [84]. Results from a survey (questionnaire) carried out by Longhurst and Graham in Great Britain showed that, in practice, tank depths range 1–3.2 m with a mean value of 2.4 m, while tank shapes vary from "squarish" to "long and thin"; however, there is a continuing debate in this area [15]. Gregory and Zabel indicated that tank depth of about 1.5 m with an overflow rate of 8–12 m/h (depending on the type of water) is normally used [11]. An effective flotation unit could be between 0.48 and 2.74 m deep [6]. The angle for the inlet baffle is approximately 60° to the horizontal, which ensures minimum disturbance to the bubble–floc agglomerate [18]. However, Longhurst and Graham reported that, in theory, the baffle angle can range from 45° to 90° to the horizontal [15]. "Purac" have used a vertical baffle in the production of the "Flofilter" tank to avoid an eddying current during clarification and hydraulic congestion during filter backwash. But for the conventional filter, they preferred to use an inclined baffle.

The depth of water below the water surface is found to vary across treatment plants, and greater depths than those recommended by the Water Research Centre of 0.3–0.4 m are normally used [15]. In South Africa, the depth varies from 1.5 to 3.5 m, and in the United Kingdom it varies from 1.0 to 3.2 m [85]. This means there is still no agreement about the extent to which depth affects the optimization of design criteria. Regarding the width of the tank, it was observed that widths between 2.4 and 9.4 m are found in practice. However, Gregory and Zabel reported that tank widths are less significant to hydraulic flow and are sometimes restricted by the sludge-removing device [11]. A study carried out by Heinanen on the use of DAF for potable water treatment in Finland showed that the design parameters for the process are still far from ideal and this has resulted in high construction costs [14]. He indicated that the situation could be avoided if research institutes had played an important part in the design work.

Discussions with Noone [82] indicated that there is a need to investigate the optimum shape of the tank. This means further investigations would be useful to justify the arrangements of the nozzles, the distance between the inlet and the baffle, the baffle angle, and the depth of water surface from the baffle. Longhurst and Graham indicated that if the length of the tank runs only up to point A (Figure 5.12), the tank may be too short, and the floc will not achieve its optimum flotation that occurs at point B [15,87]. At point C, the tank is too long, which will cause the floc to settle down.

Noone indicated that even in Severn Trent Water, there is a range of tank sizes with varying rectangular shapes and different arrangements of air nozzles, baffles, depths, and operational

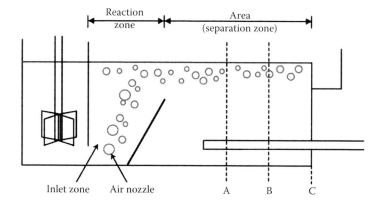

FIGURE 5.12 Arrangements in flotation tank. (From Adlan, M.N., A study of dissolved air flotation tank design variables and separation zone performance. PhD thesis, University of Newcastle Upon Tyne, 1998; Murshed, M.F., Removal of turbidity, suspended solid and aluminum using DAF pilot plant. MSc thesis, Universiti Sains Malaysia, 2007.)

procedures, with no evidence to prove their effectiveness [82]. Thus, it is worth investigating these parameters so that a better fundamental understanding can be developed regarding the optimization of tank dimensions and the flotation process in practice. This could result in the saving of power, chemicals, and operation times and in the development of standard design procedures.

Based on the comments by Noone, Adlan had carried out investigations at several water treatment plants belonging to Severn Trent Water [88]. Acoustic Doppler Velocimeter with three-dimensional down-looking probe was used in the investigation as it is ideal for measuring velocity close to the bottom of the boundary layer. Turbidity data were monitored at the same points where velocity data were taken in the separation zone of DAF tanks. Comparisons were made, and results showed that the design of the rectangular DAF tank could be improved by reducing the length of the separation zone without reduction in turbidity.

5.5 ADVANTAGES OF DAF APPLICATION IN WASTEWATER TREATMENT

The flotation technique has distinctive advantages over the conventional gravity settling technique for the removal of low-density particles that have a tendency to float. Flotation techniques are classified based on the methods of producing bubbles. The advantages of DAF are as follows [23,89–91,106–108]:

- Efficient at removing particles and turbidity, resulting in more economical filter designs. It allows short detention time of about 5–10 min in flocculation tanks.
- Higher hydraulic loading rates can be used than in most settling processes.
- More efficient than sedimentation in removing low-density floc formed from coagulation of Total Organic Compound (TOC).
- It allows lower coagulant dosages resulting in smaller chemical storage and lesser sludge.
- Smaller footprints with stacked flotation over filtration arrangement.
- Improved algae removal and cold water performance.
- Less sensitive to flow variations.
- Process flexibility through air loading.

5.6 APPLICATION OF DAF PROCESS IN WASTEWATER TREATMENT

DAF process is a physical process which has been used widely in wastewater treatment. DAF is generally used as a combination process with coagulation. The industries that implement DAF for their wastewater treatment process are paper mill, chemical-mechanical polishing (CMP) wastewater, meat industry, personnel care product, and seafood industry [5,92–95]. Oily wastewater or oil refineries wastewater also use DAF for their wastewater treatment [90,97–101].

Zoubolis and Avranas applied coagulation and DAF process in treating simulated oily wastewater (using octane) with an initial concentration of 500 mg/L [96]. The coagulants used in this study were anionic polyacrylamide, cationic (K-1384), and $FeCl_3$. Sodium oleate (NaOl) was used as a collector. It is a common anionic flotation collector used to enhance the floatability of coagulated solids and emulsion droplets (flotation collector). The DAF system was operated as a batch study using DAF jar-test with a pressure of 4–5 atm. and recycle ratio of 30%. Results indicated that the use of polyacrylamide and cationic (K-1384) did not offer a good solution. However, using $FeCl_3$ to demulsify and increase the droplets size improved the droplet–bubble adhesion and the overall DAF performance. The best removal rate for oil removal was obtained at 100 mg/L Fe^{3+} with 50 mg/L NaOl at pH 6 and recycle ratio of 30%. The pH was optimum at 6 because at this pH, with the addition of Fe^{3+}, the zeta potential of the hydrocarbon particle is nearly zero, and the electrostatic barrier is greatly reduced, which thus improves the droplet–particle attachment.

Al-Shamrani et al. carried out another study using DAF and coagulation in treating oily wastewater [25]. The study was carried out in a batch process. The sample was a solvent-refined petroleum distillate manufactured from crude petroleum oil. Alum was used as a coagulant agent at 100 mg/L with pH 8. The DAF system was equipped with a packed saturator column with 90% efficiency, and the recycle ratio was set at 10%. Two situations were observed in this study. First, at low concentration of alum with high working pressure for the saturator (80 psi) and high recycle ratio (20%), only 72% of oil was removed. However, increasing the dosage up to 100 mg/L at pH 8 with the same working pressure (80 psi) but at lower recycle ratio (8%), almost all the oil was removed. This finding shows that the increase in alum concentration destroyed the protective action created by the emulsifying agent (compound in oily wastewater). Simultaneously, the repulsive effects of the electrical double layers (zeta potential) were reduced as the concentration of alum increased. This results in the formation of larger droplets or particles through coalescence.

In 2003, Zouboulis et al. studied the removal of humic acid from synthetic landfill leachates as a possible post-treatment stage after biological treatment using coagulation and DAF [100]. This experiment used a column flotation. Samples were prepared with humic acid concentration from 50 to 300 mg/L, with 30% humic acid accounted for in the COD value. It means that after the biological treatment of the landfill leachate, the COD value ranges from 500 to 1000 mg/L, in which humic acids account for around 30%. This compound is a very reluctant organic acid. The coagulation process carried out prior to DAF used ethanol as frother and N-acetyl-N,N,N-trimethylammonium-bromide (CTAB) as a collector. The frother function is used to control the efficiency of bubble generation and to combine effectively with the collector. In the DAF system, the bubbles are produced by injecting air from the bottom of the column through a microporous frit at a flow rate of 200 cm^3/min. According to Zouboulis et al., there are three main steps in the flotation mechanism:

1. Surface modification of the specific "species" (concerned compounds or contaminants) in the water or wastewater using chemicals.
2. Contact and adherence of hydrophobic species with bubbles.
3. Separation of particles.

In this study, 70 mg/L of collector was used with 5 min flotation time resulting in 95% humic acid removal. As a result, this study proved that the flotation process can be used as a post-treatment in the landfill leachate. Besides, they discovered the parameters that control this process: (1) type and dosage of frother, (2) dosage of collector, (3) pH solution, (4) ionic strength (concentration of other salt), and (5) flotation time. Studies conducted by other researchers dealt with all these parameters, except the ionic strength. In the study conducted by Zouboulis et al., two different salts were studied, the Na_2SO_4 and NaCl [100]. Increasing the concentration of these salts above 0.01 M decreases humic acids removal. This proves that the ions (SO_4^{2-} and Cl^-) compete or interact with the cationic collectors. As a result, ions reduce the chances of humic acids to float. However, comparing SO_4^{2-} and Cl^-, the Cl^- effects are not significant and they decrease the overall percentage removal of humic acids to 10% only. This may be due to the charges carried by SO_4^{2-}, which is double, thus reacting more with the collector compared to Cl^- with one charge.

Another important point in this work is the zeta potential measurement at the optimum dosage. Results indicate that reduction in zeta potential (from negative value to zero) increases in percentage the removal of humic acids. However, beyond the optimum points, the results of zeta potential show the opposite charge (positive). This proves that to obtain a good agglomeration, the zeta potential of the concerned compound with collector or coagulant agent should be low or near zero.

Another study by Hami et al. showed the ability of coagulation and DAF in treating wastewater in refineries [98]. The sample was collected from the effluent of the American Petroleum Institute (API) separator. The coagulation process was carried out as a pre-treatment using a mixed coagulant containing alum, polyelectrolyte, and powered activated carbon (PAC). The suitable pH was in the range 6.5–8.0. The DAF process was carried out in a laboratory as a pilot scale. Three different pressures (2, 3, and 5 atm.) were chosen for this study, and the flow rate depended on the flow characteristics of pumps. For the DAF and recycle unit, the flow rate was 1.0–5.0 L/h and 1.0–2.0 L/h, respectively. In the first phase of the study, the DAF process as a single treatment did not offer good COD and BOD removals. Only 19%–64% and 27%–70% of COD and BOD removals were, respectively, obtained using DAF alone. However, increase in percentage removal was achieved by a combination of coagulation and DAF. The COD and BOD removals were 72%–92.5% and 76%–94%, based on the initial concentration of 198 and 95 mg/L, respectively. Alum was believed to reduce the zeta potential of the particle in the influent, which thus provoked the agglomeration process, while the polyelectrolyte worked as a flocculant agent, and the PAC provided a site for the particles to absorb, and subsequently the floc was removed by flotation. Another factor that was addressed in this study was the flow rate through the nozzle in the flotation tank. Hami et al. found that the increase in flow rate causes a decrease in percentage removal because when high flow rate was applied, the resident was reduced (the bubbles do not have enough time to float all the particles). Owing to this, the amount of effluent with lower contaminant concentration was reduced, which resulted in low recycle ratio. Subsequently, it reduced the dispersed air bubbles due to lower gas holdup.

Tsai et al. utilized nano-bubble flotation to treat CMP wastewater [92]. The experimental works were carried out using laboratory and pilot scale experiments. The coagulants, also known as activators, used in this study were alum, ferric chloride, and PACL. The flocculant (collector) types utilized after the coagulation process were CTAB and NaOl. In the DAF system, a special equipment named NBG (nano-bubble generator) was used to produce the nano-sized bubble. The recycled water was injected into the flotation cell through NBG at pressure around 7.7 atm. The bubble size was around 30–5000 nm. Based on an experiment using the laboratory scale DAF unit, PACL performs better than alum. However, a combination of $FeCl_3$ with NaOl is superior to other combinations. Conversely, a combination of PACL–NaOl was used in the NBFT (nano-bubble flotation technology) as a pilot scale due to some advantages of PACL in terms of cost and chemical characteristics. Experimental results using the NBFT process show that chemical cost was reduced four times that of the conventional coagulation–sedimentation process. The addition of the collector shows that it is able to absorb at the air–liquid interface and enhance the resistance of the bubble to burst. The collectors provide an electrical potential to the bubble–particle and enhance the flotation process. Another advantage of adding a collector (NaOl in this study) is that it is able to reduce the bubble size, thereby increasing the flotation process. Moreover, the hydraulic retention time (HT), bubble size, bubble quantity, and saturator pressure considerably influenced the NBFT process. Based on this study, the optimum condition for NBFT process in treating the CMP wastewater was 50–60 mg/L of PACL and 5–10 mg/L of NaOl with 1 h HT, and 10%–20% recycle ratio. The pH was not adjusted, and the raw pH value was 9.4. More than 95% of turbidity, total solids, and total silica were removed.

de Nardi et al. focused on treating slaughterhouse wastewater using DAF process as laboratory and pilot plant scale experiments [93]. Coagulation process was carried out prior to DAF using PACL and anionic polyacrylamide. In the pilot plant scale experiment, the coagulation process used 24 mg Al^{3+}/L PACL with 1.5 mg/L anionic polymer, followed by DAF process using 100% influent pressurization at 300 kPa. The lab-scale experimental process was conducted with and without coagulation. The concentration of the coagulant and flocculant was the same as that in the pilot scale experimental process. However, the pressure was higher compared with the pilot scale, which was 450 kPa. Another difference between pilot and laboratory scale processes in this study was the recycle ratio. In the laboratory scale experimental work, 40% recycle ratio was applied. Results indicated that using the pilot-scale full pressurization, the removal for Suspended solids (SS), oil and grease (O&G) was unsatisfactory at 28%–58% and 41%–57%, respectively. However, in a lab-scale setup with recycle ratio, the removal is higher than the pilot plant scale experiment. This could be due to the recycle ratio and pressure. As mentioned earlier, these two factors significantly affect the DAF efficiency in terms of the bubble size and bubble concentration. Comparison of DAF with and without coagulation in the lab-scale experimental setup showed that the combination of coagulation and DAF gave higher removal than DAF as a single process. Around 54%–60%

and 46%–74% removals of SS and O&G, respectively, were obtained when using DAF as a single treatment, while a combination treatment (coagulation/DAF) gave 74% and 99% removal in their respective order. This proves that chemical treatment is very important for improving DAF performance. Recycle ratio is an important factor for boosting the removal efficiency. Recycle ratio is also efficient in controlling and varying the amount of air supplied.

de Sena et al. utilized coagulation/DAF as pre-treatment to treat the wastewater from the meat industry [94]. Ferric chloride with 80 mg/L Fe^{3+} was used during the coagulation process, and the DAF process was operated at 4.0 bars, with 5 min saturation time and 20% recycle ratio. Application of coagulation/DAF in this study gave 60.5% and 78% removal of BOD_5 and COD from the initial concentration of 1500 and 2900 mg/L, respectively. This research work proves that this chemical or physical process can be used as a pre-treatment process.

Another application of DAF was carried out in treating personnel care product wastewater collected from Unilever Masheq Company, Egypt [95]. Comparison was made between coagulation or precipitation and coagulation/DAF. Three types of coagulants were tested. The coagulants concentration was 600, 800, and 700 mg/L, while the optimum pH was 8.23, 9.1, and 6.9 for ferric chloride, ferric sulphate, and alum, respectively. The DAF process was operated at 4.0 bars. The coagulation/precipitation process gave COD removal of around 75.8%, 77.5%, and 76.7% using ferric chloride, ferric sulphate, and alum, respectively, and the BOD_5 removals were 78%, 78.7%, and 74%, respectively. The coagulation or DAF process gives different outputs, such as differences in BOD removal. Using the three types of coagulants is not significant. However, the COD removals were 71.5%, 67.7%, and 77.5% for ferric chloride, ferric sulphate, and alum, respectively. This experimental study shows that the use of coagulation or DAF offers lower investment and running cost compared with coagulation or precipitation, around 27.3% and 23.7%, respectively. The application of DAF is summarized in Table 5.6.

5.7 APPLICATION OF DAF PROCESS IN LANDFILL LEACHATE TREATMENT

To date, the DAF is used more in industrial wastewater treatment for paper mill, meat processing, and oil-based industries like POME, soap and oil refinery industry. The performance of DAF in numerous research works has been summarized in Table 5.6. In 2000, Zouboulis et al. studied humic acid removal from simulated leachate using DAF process [96]. However, based on the literature review, no study has been carried out for actual landfill leachate treatment using DAF process. This indicates that there is a gap in the knowledge of DAF capability in leachate treatment. Based on the advantages offered by DAF, such as higher hydraulic loading, allowing lower coagulant dosages, being less sensitive to flow variations, and many more (described in Section 5.5), DAF process can be a good alternative for landfill leachate treatment. Recently, study on semi-aerobic landfill leachate treatment using DAF was carried out by Palaniandy et al. and Adlan et al. [101,102].

The study carried out by Palaniandy et al. suggests a way to apply this method as a large-scale application in landfill areas [101]. This research work investigates the application of DAF in leachate treatment with and without alum coagulation. Based on the study, a coagulation process must be introduced to facilitate the destabilization of colloidal particles or emulsions. This is because DAF process in leachate treatment without coagulation shows that the variations in percent removal of turbidity, color, and COD were considerably low, which implies that the main pollutant in the leachate was in soluble organic and inorganic matter such as humic acid, fulvic acid, iron, sodium, potassium, sulfate, and chloride, [102,103]. However, with the addition of coagulant (alum), the percentage removal of the studied parameters increased to 70%, 79%, and 42% for color, COD, and turbidity, respectively. Based on this study, DAF process is able to treat landfill leachate. Further research has been carried out by Adlan et al. in this field by optimizing coagulation and DAF process in semi-aerobic landfill leachate using response surface methodology (RSM). In this research work, ferric chloride ($FeCl_3$) was chosen to induce coagulation [102]. The results show 50%, 75%, 93%, and 41% reduction in turbidity, COD, color, and NH_3–N, respectively. Comparing this research work

TABLE 5.6
Summary of Previous Studies on DAF Process

Parameter Concern	Operating Parameter	Findings	Authors
Oil removal	• 100 mg/L Fe^{3+} • 50 mg/L NaOl • pH 6 • Saturator pressure 4–5 atm. • Recycle ratio 30%	Results indicate that the use of polyelectrolyte did not offer favorable results. However, using the $FeCl_3$ to demulsify and increase the size of droplets and improve the droplet–bubble adhesion, the overall DAF performance increased.	Zouboulis and Avranas [96]
Oil removal	• 100 mg/L alum • pH 8 • Saturator pressure (80 psi) • Recycle ratio (20%) • Packed saturator column with 90% efficiency	Increase in alum concentration destroyed the protection action created by the emulsifying agent (compound in oily wastewater). Simultaneously, the repulsive effects of the electrical double layers (zeta potential) were reduced as the concentration of alum increased.	Al-Shamrani et al. [25]
Humic acid removal	• Ethanol as frother • 70 mg/L CTAB as a collector • Bubbles were produced by injecting the air from the bottom of the column through a microporous frit, at a flow rate of 200 cm³/min. • 5 min flotation time	Removal of humic acid was 95%. This study proves that the flotation process can be used as a post-treatment in landfill leachate. Increasing the concentration of salts (SO_4^{2-} and Cl^-) above 0.01 M resulted in the decrease in humic acids removal. Comparison between SO_4^{2-} and Cl^-, the Cl^- effects was not significant, and it decreased the overall percentage removal of humic acids, which was only 10%. To obtain a good agglomeration, the zeta potential of the concerned compound with collector or coagulant agent should be low or near zero.	Zouboulis et al. [100]
COD and BOD from refineries wastewater	• Mixed coagulant contained alum, polyelectrolyte, and powered activated carbon (PAC). • Suitable pH is in the range of 6.5 and 8.0. • The flow rate depended on the flow characteristics of pumps; for DAF and recycle unit, the flow rate was 1.0–5.0 L/h and 1.0–2.0 L/h, respectively.	19%–64% and 27%–70% removal of COD and BOD, respectively, using DAF alone. Combination of coagulation and DAF, COD, and BOD removals were 72%–92.5% and 76%–94%, respectively, from the initial concentration of 198 and 95 mg/L, respectively. The alum was believed to reduce the zeta potential of the particle in the influent; thus, it provokes an agglomeration process while the polyelectrolyte works as a flocculant agent; the PAC provided a place for the particles to adsorb, and subsequently the floc was removed by flotation.	Hami et al. [98]

(Continued)

TABLE 5.6 (Continued)
Summary of Previous Studies on DAF Process

Parameter Concern	Operating Parameter	Findings	Authors
Turbidity, total solids, and total silica removal from CMP wastewater	• 50–60 mg/L of PACL • 5–10 mg/L of NaOl 1 h HT • 10%–20% recycle ratio • pH was not adjusted, and the raw pH value was 9.4 • NBG (nano bubble generator) was used to produce the nano-sized bubble. • Pressure around 7.7 atm. • Bubble size was around 30–5000 nm	This study indicated 40% greater removal than the conventional coagulation–sedimentation process. More than 95% turbidity, total solids, and total silica removal efficiencies were obtained.	Tsai et al. [92]
O&G SS from slaughterhouse wastewater	Pilot scale; • 24 mg Al^{3+}/L PACL 1.5 mg/L anionic polymer • 100% influent pressurization at 300 kPa • Lab scale (without coagulation); • Pressure at 450 kPa 40% recycle ratio • Lab scale (with coagulation); • 24 mg Al^{3+}/L PACL • 1.5 mg/L anionic polymer • Pressure at 450 kPa • 40% recycle ratio	Using the pilot-scale full pressurization, the removal for SS and O&G was unsatisfactory at 28%–58% and 41%–57%, respectively. Using the lab-scale DAF process (without coagulation), the removal was 54%–60% and 46%–74% for SS and O&G, respectively. Using the lab-scale DAF process (with coagulation), the removal was 74% and 99% for SS and O&G, respectively.	de Nardi et al. [93]
BOD_5 and COD from meat industry wastewater	• Ferric chloride with 80 mg/L Fe^{3+} • Pressure at 4.0 bars • 5 min saturation time • 20% recycle ratio	60.5% and 78% removal for BOD_5 and COD from initial concentration of 1500 and 2900 mg/L, respectively.	de Sena et al. [94]
BOD_5 and COD from personnel care product wastewater	Coagulation/precipitation; • 600, 800, and 700 mg/L, and optimum pH was 8.23%, 9.1%, and 6.9% for ferric chloride, ferric sulphate, and alum, respectively Coagulation/DAF; • 600, 800, and 700 mg/L, and optimum pH was 8.23%, 9.1%, and 6.9% for ferric chloride, ferric sulphate, and alum, respectively • Pressure at 4.0 bars	COD removal; 75.8%, 77.5%, and 76.7% for ferric chloride, ferric sulphate, and alum, respectively. BOD_5 removal; 78%, 78.7%, and 74% for ferric chloride, ferric sulphate, and alum, respectively. COD removal; 71.5%, 67.7%, and 77.5% for ferric chloride, ferric sulphate, and alum, respectively. BOD removal was not significant between these three types of coagulants. The use of coagulation/DAF offers lower investment and running cost compared with coagulation/precipitation, around 27.3% and 23.7%, respectively.	El-Gohary et al. [95]

with previous studies [102,104], coagulation/sedimentation offers a good removal in landfill leachate treatment. However, in terms of chemical usage, the coagulation/sedimentation process needs double that required by the coagulation/DAF process. The proper determination of type and dosage of chemicals will not only improve the process but also influence the running cost as proved by El-Gohary et al. [95]. Based on the study carried out by El-Gohary et al. [95], the initial investment and the running cost for coagulation and sedimentation are higher by 27.3% and 23.7%, respectively, compared with coagulation/DAF. Besides, the land area required for coagulation/DAF is less by 30% compared with coagulation/sedimentation. As a result, coagulation/DAF is more economical compared with coagulation/sedimentation in treating wastewater [95].

LIST OF NOMENCLATURE

A	Area (m^2)
D	Drag force (kN)
μ	Dynamic viscosity (kg/m/s)
U	Terminal velocity (m/s)
A	Radius of sphere (m)
Σ	Density of sphere (kg/m^3)
P	Density of liquid (kg/m^3)
p_g	Density of gas (kg/m^3)
C_D	Drag coefficient
Re	Reynolds number
ν	Kinematic viscosity (m^2/s)
R	Ratio of total removal of solid concentration after flotation to the inflow solid concentration = $1 - C_0/C_i$
C_0	Effluent suspended solids
C_i	Influent suspended solids
V_r	Rising velocity of a single particle/air bubble (m/s)
Q	Flow applied to the flotation unit (m^3/s)
Ah	Horizontal area of the unit (m^2)
V	Volume of bubble (m^3)
$\Delta\rho$	Difference of density between liquid and gas (kg/m^3)
S	Area of bubble (m^2)
PB	Vapour pressure (Pa)
x_B	Mol. fraction of the solute
K_B	Constant
C_S	Concentration of air in saturated liquid (mg/L)
P	Absolute pressure (Pa)
K	Henry's law constant (M/atm)
F	Efficiency factor (kPa/mg/L)
a_p	Concentration of air precipitated on reducing the pressure to atmospheric (mg/L)
P	Saturator pressure in atmospheres
C_r	Concentration of air released in the tank (mg/L)
C_a	Concentration of air that remains in solution at atmospheric pressure (mg/L)
R_r	Recycle ratio
K	Influent saturation factor

ϕ_b	Bubble volume concentration
p_{sat}	Saturated density of air
p_w	Density of water (gm/cm³)
P_A	Dissolved pressure (dyne/cm²)
P_0	Atmospheric pressure (dyne/cm²)
R	Gas constant (erg/K·mol)
T	Absolute temperature (K)
M_w	Molecular weight of water (g/g·mol)
K_H	Henry's law constant (dyne/cm²)
V_b	Volume of air occupied by a single spherical bubble (cm³)
h	Depth from the liquid surface to the bubble (cm)
σ	Surface tension (dyne/cm²)
d_a	Volumetric mean diameter of bubble (cm)
G	Volumetric flow rate of air generated under decrease in pressure (cm³/s)
Q	Volumetric flow rate (cm³/s)
η_1	Collision efficiency
η_2	Attachment efficiency
r_p	Particle radius (cm)
R_b	Bubble radius (cm)
p_p	Particle density (gm/mL)
G_j	Body force acting on the particle
V_j	Particle velocity (cm/s)
u_j	Velocity the fluid would have at the position of the particle if no particle were there (cm/s)
t	Time (s)
V	Component of particle velocity (cm/s)
μ_f	Fluid viscosity (kg/m/s)
x, y	Cartesian position coordinates (cm)
u	Component of velocity field due to bubble (cm/s)
v	Dimensionless component of particle velocity (cm/s)
K	Particle inertia parameter
G	Dimensionless settling velocity of particle (cm/s)
η	Single bubble collision efficiency
u_0	Water velocity (cm/s)
c_0	Suspended particle concentration upstream from the collector where the flow is undisturbed by the presence of the grain
d	Grain diameter (cm)
η_D, η_I, η_G	Theoretical values for single collector efficiency
Pe	Peclet number
R_b	Bubble radius (cm)
U_b	Bubble rising velocity (cm/s)
D_f	Particle diffusivity (cm²/s)
d_p	Diameter of suspended particle (cm)
d	Diameter of collector or bubble (cm)
v_0	Approach velocity of fluid (cm/s)
k_b	Boltzmann's constant for bubble

α_{pb}	Attachment efficiency
N_b	Total number concentration of bubbles
A_b	Projected area of bubble (m^2)
N_p	Particle number concentration
D	Effective depth of flotation chamber (m)
V_T	Vertical rise rate of suspended solids (m/s)
T	Detention time (s)
A_s	Surface area of flotation chamber (m^2)
V_H	Horizontal velocity (m/s)
A_c	Cross-sectional area of flotation chamber (m^2)
W	Width of the flotation chamber (m)
L	Effective length of the flotation chamber (m)

REFERENCES

1. Rodrigues, R.T. and Rubio, J. (2007). DAF-dissolved air flotation: Potential applications in the mining and mineral processing industry. *International Journal of Mineral Processing.* Vol. 82, pp. 1–13.
2. Merrill, C.W. and Pennington, J.W. (1962). The magnitude and significance of flotation in the mineral industries of the United States. In *Froth Flotation 50th Anniversary Volume* (Edited by Fuerstenau, D.W.), Chapter 4. The American Institute of Mining, Metallurgical, and Petroleum Engineers, Inc., New York.
3. Travers, S.M. and Lovett, D.A. (1985). Pressure flotation of abattoir wastewaters using carbon dioxide. *Water Research.* Vol. 19, pp. 1479–1482.
4. Krofta, M. and Wang, L.K. (1982). Potable water treatment by dissolved air flotation. *Journal American Water Works Association.* Vol. 74, pp. 305–310.
5. Krofta, M., Wang, L.K., and Pollman, C.D. (1988). Treatment of seafood processing wastewater by dissolved air flotation, carbon adsorption and free chlorination. *Proceedings of the 43rd Purdue Industrial Waste Conference*, Purdue University, West Lafayette, IN, pp. 535–550.
6. Wang, L.K. and Wang, M.H.S. (1989). Bubble dynamics and air dispersion mechanisms of air flotation process systems. Part A: Materials balances. *Proceedings of the 44th Purdue Industrial Waste Conference*, Purdue University, West Lafayette, IN, pp. 493–504.
7. Van Puffelen, J., Buijs, P.J., Nunn, P.A.N.M., and Hijnen, W.A.M. (1995). Dissolved air flotation in potable water treatment: The Dutch experience. *Water Science Technology.* Vol. 31, pp. 149–157.
8. Childs, A.R., Burnfield, I., and Rees, A.J. (1977). Operational experience with the 2300 m^3/day pilot plant of the Essex Water Company. *Papers and Proceedings of the WRC Conference: Flotation for Water and Waste Treatment, Medmenham.*
9. Nickols, D., Moerschell, G.C., and Broder, M.V. (1995). The first DAF water treatment plant in the United States. *Water Science Technology.* Vol. 31, pp. 239–246.
10. Edzwald, J.K. and Walsh, J.P. (1992). Dissolved air flotation: Laboratory and pilot plant investigation. AWWA Research Foundation and AWWA, Denver, CO.
11. Gregory, R. and Zabel, T.F. (1990). Sedimentation and flotation. In *Water Quality and Treatment: A Handbook of Community Water Supply* (Edited by Pontius, F.W.), 4th edition. McGraw Hill, New York, pp. 367–453.
12. Kitchener, J.A. (1984). The froth flotation process: Past and present and future-in brief. In *The Scientific Basis of Flotation* (Edited by Ives, K.J.). NATO ASI Series. Martinus Nijhoff Publishers, The Hague, the Netherlands, pp. 3–52.
13. Lundgren, H. (1976). Theory and practice of dissolved air flotation. *Journal of Filtration and Separation.* Vol. 13, pp. 24–28.
14. Heinanen, J. (1988). Use of dissolved air flotation in potable water treatment in Finland. *Aqua Fennica.* Vol. 18, pp. 113–123.
15. Longhurst, S.J. and Graham, J.D. (1987). Dissolved air flotation for water treatment: A survey of operational units in Great Britain. *Public Health Engineering.* Vol. 14, pp. 71–76.
16. Zabel, T.F. (1978). Flotation. Twelfth International Water Supply Congress, Kyoto, Japan.

17. Packham, R.F. and Richards, W.N. (1975). Water clarification by flotation 3: Treatment of Thames water in a pilot-scale flotation plant. WRC Technical Report TR2, Medmenham, UK.
18. Zabel, T.F. (1985). The advantage of dissolved-air flotation for water treatment. *Journal American Water Works Association.* Vol. 77, pp. 42–46.
19. Gregory, R., Zabel, T.F., and Edzwald, J.K. (1999). Sedimentation and flotation. In *Water Quality and Treatment: A Handbook of Community Water Supplies* (Edited by Letterman, R.D.). McGraw-Hill, New York, pp. 7.1–7.87.
20. Ho, C.C. and Chan, C.Y. (1986). The application of lead dioxide-coated titanium anode in the electroflotation of palm oil mill effluent. *Water Research.* Vol. 20, pp. 1523–1529.
21. Zabel, T.F. and Melbourne, J.D. (1980). *Flotation. Development in Water Treatment-1.* Applied Science Publishers, London.
22. Barnes, D., Bliss, P.J., Gould, B.W., and Vallenrine, H.R. (1981). *Water and Wastewater Engineering Systems*, Chapter 8. Pitman Books Ltd., London.
23. Zabel, T.F. and Hyde, R.A. (1977). Factors influencing dissolved air flotation as applied to water clarification. *Papers and Proceedings of WRC Conference: Flotation for Water and Waste Treatment, Marlow.*
24. Zabel, T. (1984). Flotation in water treatment. In *The Scientific Basis of Flotation* (Edited by Ives, K.J.), NATO ASI Series. Martinus Nijhoff Publishers, The Hague, the Netherlands.
25. Al-Shamrani, A.A., James, A., and Xiao, H. (2002). Destabilisation of oil–water emulsions and separation by dissolved air flotation. *Water Research.* Vol. 36, pp. 1503–1512.
26. Edzwald, J.K. (2010). Dissolved air flotation and me. *Water Research.* Vol. 44, pp. 2077–2106.
27. Haarhoff, J. and Steinbach, S. (1996). A model for the prediction of the air composition in pressure saturators. *Water Research.* Vol. 30, pp. 3074–3082.
28. Vesilind, P.A. (2003). *Wastewater Treatment Plant Design.* Water Environment Federation, New York.
29. HDR Engineering, Inc. (2001). *Handbook of Public Water Systems.* John Wiley & Sons, New York.
31. Harper, J.F. (1972). The motion of bubbles and drops through liquids. In *Advances in Applied Mechanics* (Edited by Chia, S.Y.). Academic Press, New York, Vol. 12, pp. 59–129.
32. Li, W.H. and Lam, S.H. (1964). *Principles of Fluid Mechanics.* Addison-Wesley Publ. Co., Inc., Reading, MA.
33. Gaudin, A.M. (1939). *Principles of Mineral Dressing*, Chapter VIII. McGraw-Hill, London.
34. Batchelor, G.K. (1988). *An Introduction to Fluid Dynamics.* University Press, Cambridge, MA.
35. Shaw, D.J. (1991). *Introduction to Colloid and Surface Chemistry*, Chapter 1, 4th edition. Butterworth-Heinemann Ltd, Oxford.
36. Clift, R., Grace, J.R., and Weber, M.E. (1978). *Bubbles, Drops, and Particles.* Academic Press, New York.
37. Chorlton, F. (1967). *Textbook of Fluid Dynamics*, Chapter 8. D. Van Nostrand Co. Ltd., London.
38. Packham, R.F. and Richards, W.N. (1972). *Water Clarification by Flotation 1: A Survey of the Literature.* WRA Technical Paper TP87, Medmenham, UK.
39. Moore, W.D. (1965). The Velocity Rise of Distorted Gas Bubbles in a Liquid of Small Viscosity. *Journal of Fluid Mechanics* 23(04):749–766.
40. Jameson, G.J. (1984). Experimental techniques in flotation. In *The Scientific Basis of Flotation* (Edited by Ives, K.J.), NATO ASI Series. Martinus Nijhoff Publisher, The Hague, the Netherlands, pp. 53–78.
41. Shannon, W.T. and Buisson, D.H. (1980). Dissolved air flotation in hot water. *Water Research.* Vol. 14, pp. 759–765.
42. Fair, G.M., Geyer, J.C., and Okun, D.A. (1968). *Water and Wastewater Treatment Engineering*, Vol. 2, Chapter 26. John Wiley & Sons.
43. Vrablik, E.R. (1959). Fundamental principles of dissolved-air flotation of industrial wastes. *Proceedings of the 14th Industrial Waste Conference*, Purdue University, West Lafayette, IN, pp. 743–779.
44. Krofta, M. and Wang, L.K. (1989). Bubble dynamics and air dispersion mechanisms of air flotation process systems, Part B: Air dispersion. *Proceeding of the 44th Purdue Industrial Waste Conference*, Purdue University, West Lafayette, IN, pp. 505–511.
45. Fukushi, K., Tambo, N., and Matsui, Y. (1995). A kinetic model for dissolved air flotation in water and wastewater treatment. *Water Science Technology.* Vol. 31, pp. 37–47.
46. Edzwald, J.K., Malley, J.P, and Yu, C. (1990). A conceptual model for dissolved air flotation in water treatment. *Water Supply.* Vol. 9, pp. 141–150.
47. Fukushi, K., Tambo, N., and Kiyotsuka, M. (1985). An experimental evaluation of kinetic process of dissolved air flotation. *Journal Japan Water Works Association.* Vol. 607, pp. 32–41.
48. Levich, V.G. (1962). *Physicochemical Hydrodynamics.* Prentice-Hall, Inc., Englewood Cliffs, NJ.

49. Howe, R.H.L. (1958). A mathematical interpretation of flotation for solid-liquid separation. In *Biological Treatment of Sewage and Industrial Wastes*, Vol. 2 (Edited by McCabe, B.S. and Eckenfelder, W.W.). Reinhold Publishing Corp., New York, pp. 241–250.
50. Karamanev, G.K. (1994). Rise of bubbles in quiescent liquids. *Journal American Institute Chemical Engineer.* Vol. 40, pp. 1418–1421.
51. Gerrard, W. (1980). *Gas Solubilities Widespread Applications*, Chapter 1. Pergamon Press, Oxford, UK.
52. Lovett, D.A. and Travers, S.M. (1986). Dissolved air flotation for abattoir wastewater. *Water Research.* Vol. 20, pp. 421–426.
53. Backhurst, J.R., Harker, J.H., and Porter, J.E. (1974). *Problems in Mass Transfer.* Edward Arnold Ltd, London.
54. Atkins, P.W. (1994). *Physical Chemistry*, Chapter 7. Oxford University Press, Oxford, UK.
55. Takahashi, T., Miyahara, T., and Mochizuki, H. (1979). Fundamental study of bubble formation in dissolved air pressure flotation. *Journal Chemical Engineering Japan.* Vol. 12, pp. 275–280.
56. Ward, A.S. (1992). Dissolved air flotation for water and wastewater treatment. *Process Safety and Environmental Protection.* Vol. 70, pp. 214–218.
57. Klassen, V.I. and Mokrousov, V.A. (1963). *An Introduction to the Theory of Flotation.* Butterworths, London.
58. Edwards, W.M. (1984). Mass transfer and gas absorption. In *Perry's Chemical Engineers' Handbook* (Edited by Green, D.W.), 6th Edition. McGraw-Hill, New York.
59. Eckenfelder, W.W. Jr., Rooney, T.F., Burger, T.B., and Gruspier, J.T. (1958). Studies on dissolved air flotation on biological sludges. In *Biological Treatment of Sewage and Industrial Wastes*, Vol. 2 (Edited by McCabe, B.S. and Eckenfelder, W.W.). Reinhold Publishing Corp., New York, pp. 241–250.
60. Bratby, J. and Marais, G.V.R. (1974). Dissolved air flotation. *Filtration & Separation.* Vol. 11, pp. 614–624.
61. Rykaart, E.M. and Haarhoff, J. (1995). Behaviour of air injection nozzles in dissolved air flotation. *Water Science Technology.* Vol. 31, pp. 25–35.
62. Gulas, V., Benefield, R.L., Lindsey, R., and Randall, C. (1980). Design considerations for dissolved air flotation. *Water & Wastewater Sewage Works.* Vol. 127, pp. 30–31, 42.
63. Ramirez, E.R. (1980). Comparative physicochemical study of industrial wastewater treatment by electrolytic, dispersed and dissolved air flotation technologies. *Proceedings of the 34th Industrial Waste Conference*, Purdue University, West Lafayette, IN, pp. 699–709.
64. Jones, A.D. and Hall, A.C. (1981). Removal of metal ions from aqueous solutions by dissolved air flotation. *Filtration & Separation*, September/October, pp. 386–390.
65. Bratby, J. and Marais, G.V.R. (1975). Saturator performance in dissolved-air (pressure) flotation. *Water Research.* Vol. 9, pp. 929–936.
66. Cassell, E.A., Kaufman, K.M., and Matijevic, E. (1975). The effects of bubble size on microflotation. *Water Research.* Vol. 9, pp. 1017–1024.
67. Collin, G.L. and Jameson, G.J. (1976). Experiment on the flotation of fine particles: The influence of particle size and charge. *Chemical Engineering Science.* Vol. 31, pp. 985–99.
68. Rovel, J.M. (1977). Experiences with dissolved air flotation for industrial effluent treatment. *Papers and Proceeding of the WRC Conference: Flotation for Water and Waste Treatment,* Bucks.
69. Reay, D. and Ratcliff, G.A. (1973). Removal of fine particles from water by dispersed air flotation. *The Canadian Journal of Chemical Engineering.* Vol. 51, pp. 178–185.
70. Gochin, R.J. (1990). Flotation, In *Solid-Liquid Separation* (Edited by Svarovsky, L.), 3rd Edition. Butterworths, London, pp. 591–613.
71. Reay, D. and Ratcliff, G.A. (1975). Experimental testing of the hydrodynamic collision model of fine particle flotation. *The Canadian Journal of Chemical Engineering.* Vol. 53, pp. 481–486.
72. Flint, L.R. and Howarth, W.J. (1971). The collision efficiency of small particles with spherical air bubbles. *Chemical Engineering Science.* Vol. 26, pp. 1155–1168.
73. King, R.P. (1982). Flotation of fine particles. In *Principles of Flotation* (Edited by King, R.P.), South African Institute of Mining and Metallurgy, Johannesburg, South Africa, pp. 215–225.
74. Sutherland, K.L. (1948). Kinetics of the flotation process. *Journal Physical Colloid Chemistry.* Vol. 52, pp. 394–425.
75. Woodburn, E.T., King, R.P., and Colborn, R.P. (1971). The effect of particle size distribution on the performance of a phosphate flotation process. *Metallurgical and Material Transactions.* Vol. 2, pp. 3163–3174.
76. Yao, K.M., Habibian, M.T., and O'Melia, C.R. (1971). Water and waste water filtration: Concept and applications. *Environmental Science and Technology.* Vol. 5, pp. 1105–1112.

77. Malley, J.P. Jr. and Edzwald, J.K. (1991). Concepts for dissolved air flotation treatment of drinking waters. *Aqua*. Vol. 40, pp. 7–17.
78. Friedlander, S.K. (1967). Particle diffusion in low speed flows. *Journal of Colloid Interface Science*. Vol. 23, pp. 157–164.
79. O'Melia, C.R. (1985). Particles, pretreatment and performance in water filtration. *Journal of Environmental Engineering*. Vol. 111, pp. 874–890.
80. Katz, W.J. and Wullschleger, R. (1957). Studies of some variables which affect chemical flocculation when used with dissolved air flotation. *Proceedings of 12th Purdue Industrial Waste Conference*, Purdue University, West Lafayette, IN, pp. 466–479.
81. Eckenfelder, W.W. Jr. and O'Connor, D.J. (1961). *Biological Waste Treatment*, Chapter 5. Pergamon Press, London.
82. Noone, G. (1995). Personal Communication.
83. Bratby, J. and Marais, G.V.R. (1975). Dissolved air (pressure) flotation: An evaluation of inter-relationships between process variables and their optimisation for design. *Water S.A.* Vol. 1, pp. 57–69.
84. Edzwald, J.K. (1995). Principles and applications of dissolved air flotation. *Water Science and Technology*. Vol. 31, pp. 1–23.
85. Hyde, R.A. (1975). Water clarification by flotation-4, design and experimental studies on a dissolved air flotation plant treating 8.2 m^3/h of River Thames Water. WRC Technical Report TR13, Medmenham, UK.
86. Haarhoff, J. and Vuuren, L.R.J. (1995). Design parameters for dissolved air flotation in South Africa. *Water Science and Technology*. Vol. 31, pp. 203–212.
87. Murshed, M.F. (2007). Removal of turbidity, suspended solid and aluminum using DAF pilot plant. MSc thesis, Universiti Sains Malaysia.
88. Adlan, M.N. (1998). A study of dissolved air flotation tank design variables and separation zone performance. PhD thesis, University of Newcastle Upon Tyne.
89. Ponasse, M., Dupre, V., Aurelle, Y., and Secq, A. (1998). Bubble formation by water release in nozzle II. Influence of various parameters on bubble size. *Water Research*. Vol. 32, pp. 2498–2506.
90. Al-Shamrani, A.A., James, A., and Xioh, H. (2002). Separation of oil from water by dissolved air flotation. *Colloids and Surfaces A: Physicochemical and Engineering Aspects*. Vol. 209, pp. 15–26.
91. Edzwald, J.K. (2006). Dissolved air flotation in drinking water treatment. Chapter 6 in *Interface Science in Drinking Water Treatment*, Vol 10, eds. G. Newcombe and D. Dixon. Elsevier, New York.
92. Tsai, J.C., Kumar, M., Chen, S.Y., and Lin, J.G. (2007). Nano-bubble flotation technology with coagulation process for the cost-effective treatment of chemical mechanical polishing wastewater. *Separation and Purification Technology*. Vol. 58, pp. 61–67.
93. de Nardi, I.R., Fuzi, T.P., and Del Nery, V. (2008). Performance evaluation and operating strategies of dissolved-air flotation system treating poultry slaughterhouse wastewater. *Resources, Conservation and Recycling*. Vol. 52, pp. 533–544.
94. de Sena, R.F., Moreira, R.F.P.M., and Jose, H.J. (2008). Comparison of coagulants and coagulation aids for treatment of meat processing wastewater by column flotation. *Bioresource Technology*. Vol. 99, pp. 8221–8225.
95. El-Gohary, F., Tawfik, A., and Mahmoud, U. (2010). Comparative study between chemical coagulation/precipitation (C/P) versus coagulation/dissolved air flotation (C/DAF) for pre-treatment of personal care products (PCPs) wastewater. *Desalination*. Vol. 252, pp. 106–112.
96. Zouboulis, A.I. and Avranas, A. (2000). Treatment of oil-in-water emulsions by coagulation and dissolved-air flotation. *Colloids and Surfaces A: Physicochemical and Engineering Aspects*. Vol. 172, pp. 153–161.
97. Bensadok, K., Belkacem, M., and Nezzal, G. (2007). Treatment of cutting oil/water emulsion by coupling coagulation and dissolved air flotation. *Desalination*. Vol. 206, pp. 440–448.
98. Hami, M.L., Al-Hash, M.A., and Al-Doori, M.M. (2007). Effect of activated carbon on BOD and COD removal in a dissolved air flotation unit treating refinery wastewater. *Desalination*. Vol. 216, pp. 116–122.
99. Poha, P.E., Onga, W.Y.J., Laub, E.V., and Chonga, M.N. (2014). Investigation on micro-bubble flotation and coagulation for the treatment of anaerobically treated palm oil mill effluent (POME). *Journal of Environmental Chemical Engineering*. Vol. 2, pp. 1174–1181.
100. Zouboulis, A.I., Jun, W., and Katsoyiannis, I.A. (2003). Removal of humic acids by flotation. *Colloids and Surfaces A: Physicochemical and Engineering Aspects*. Vol. 231, pp. 181–193.
101. Palaniandy, P., Adlan, M.N., Aziz, H.A., and Murshed, M.F. (2010). Application of dissolved air flotation (DAF) in semi-aerobic leachate treatment. *Chemical Engineering Journal*. Vol. 157, pp. 316–322.

102. Adlan, M.N., Palaniandy, P., and Aziz, H.A. (2011). Optimization of coagulation and dissolved air flotation (DAF) treatment of semi-aerobic landfill leachate using response surface methodology (RSM). *Desalination*. Vol. 277, pp. 74–82.
103. Aziz, H.A., Alias, S., and Adlan, M. (2007). Colour removal from landfill leachate by coagulation and flocculation processes. *Bioresource Technology*. Vol. 98, pp. 218–220.
104. Lo, I.M.C. (1996). Characteristics and treatment of leachates from domestic landfills. *Environment International*. Vol. 22, pp. 433–442.
105. Ghafari, S., Aziz, H.A., and Bashir, M.J.K. (2010). The use of poly-aluminum chloride and alum for the treatment of partially stabilized leachate: A comparative study. *Desalination*. Vol. 257, pp. 110–116.
106. Bondelind, M., Sasic, S., Bergdahl, L. (2013). A model to estimate the size of aggregates formed in a dissolved air flotation. *Applied Mathematical Modelling*. Vol. 37, pp. 3036–3047.
107. Lakghomi, B., Lawryshyn, Y., and Hofmann, R. (2015). A model of particle removal in a dissolved air flotation tank: Importance of stratified flow and bubble size. *Water Research*. Vol. 68, pp. 262–272.
108. Zhanga, Q., Liu, S., Yang, C., Chen, F., Lu, S. (2014). Bioreactor consisting of pressurized aeration and dissolved air flotation for domestic wastewater treatment. *Separation and Purification Technology*. Vol. 138, pp. 186–190.

6 Restaurant Waste Treatment and Management

Jerry R. Taricska
Hole Montes, Inc.

Jaclyn M. Taricska
Public Utility Company

Yung-Tse Hung
Cleveland State University

Lawrence K. Wang
Lenox Institute of Water Technology, Krofta Engineering Corporation, and Zorex Corporation

CONTENTS

6.1	Introduction	184
6.2	Food Waste Recovery	184
	6.2.1 Source Reduction	184
	6.2.2 Donation of Food	186
	6.2.3 Food Scraps for Animal Feed	188
	6.2.4 Industrial Uses for Food Scraps	188
	6.2.4.1 Anaerobic Digestion Process	189
	6.2.4.2 Biomass Gasification Process	194
	6.2.4.3 Biohydrogen Production	197
	6.2.4.4 Composting Process	198
	6.2.5 Industrial Uses for Waste FOG	199
	6.2.6 Restaurant Waste Treatment and Management on Board Cruise Ships	200
6.3	Summary of Source Reduction Methods and Treatment Processes	202
6.4	Sources of Further Information and Advice	203
List of Nomenclature		203
References		204

Abstract

Restaurant waste is composed of solid and liquid waste. The solid waste stream includes food waste, plastics, paper and paperboard, while the liquid waste stream includes fats, oils, and grease (FOG). In the United States, in 2012, 36 million tons of food waste was generated and only 1.7 million tons were recycled. Approximately 20% of the prepared food in the United States ends up as waste. Of the total waste stream from restaurants, over 50% is food waste. This chapter discusses how restaurant waste can be reduced or recycled. Solid restaurant waste can be reduced through food recovery techniques such as food source recovery, food donation, food scraps for animal food, and food scraps for industrial uses. Food waste from restaurant can be used to produce methane from anaerobic digestion (two-stage and single high-rate digesters including egg-shape digesters), synthetic gas (syngas) from biomass gasification (fixed bed, fluidized bed, and plasma processes), and hydrogen by fermentation (biohydrogen production). Another industrial use of food waste is the production of compost through windrows or forced air composting processes. Restaurant and hotels generate over 3 billion gallons of waste cooking oil per year. Ideally, the waste oil can be collected and converted to biodiesel fuel by transesterification process. Some restaurant wastes from cruise ships can also be recycled for reuse. The current restaurant waste treatment and management practice on board modern cruise ships is introduced in this chapter.

6.1 INTRODUCTION

Generally, waste from restaurants includes solid waste and liquid waste. Solid waste is disposed into municipal solid waste facilities and the liquid stream is disposed into publicly owned wastewater treatment plants. Municipal solid waste facilities include landfills and incinerators. Publicly owned wastewater treatment plants typically include biological, chemical, and physical treatment processes. The solid waste stream includes food waste, plastic, paper, and paperboard. The liquid waste stream includes fats, oils, and grease (FOG) and also discharges from restrooms and kitchens. The treatment of solid waste from restaurants can be accomplished by waste reduction techniques and then disposal, whereas the liquid waste stream treatment can be accomplished by separation, recycling, treatment, and then disposal.

The total municipal solid waste generated in the United States in 2012 was approximately 251 million tons. Food waste from restaurants includes uneaten food and food preparation scraps. The total food waste generated in the United States in 2012 was slightly more than 36 million tons, of which approximately 1.7 million tons (i.e., <5%) was recycled. Therefore, 34.3 million tons were disposed of in landfills or incinerators. A total of 100.4 tons of plastic, paper, and paperboard waste was generated, with approximately 47.2 tons or 47.2% of it being recycled. Food waste is one of the largest components of the waste stream being disposed in landfills and incinerators. Since only <5% is being recycled, it provides the greatest potential for recycling [1].

The generation and disposal of restaurant waste have both economical and environmental effects. The reduction and the recycling of waste from restaurants can reduce disposal cost and minimize potential cost increases in disposal cost. The reduction of restaurant waste that is sent to landfills and incinerators will extend the life of landfills and save energy at the incinerator. Additionally, less food waste disposed at the landfill will reduce the amount of methane gas produced from anaerobic digestion of the organic waste that occurs in the landfill. Diverting the food waste from landfills to food recovery provides many benefits [2]. Waste reduction can be accomplished by utilizing a food recovery program. The food recovery program recommended by the U.S. Environmental Protection Agency (USEPA) and U.S. Department of Agriculture (USDA) includes the following hierarchy activities: source reduction, feed hungry people, feed animals, industrial uses, and composting and landfill/incineration, with landfilling and incinerating as the last option [3–5].

6.2 FOOD WASTE RECOVERY

6.2.1 SOURCE REDUCTION

The National Restaurant Association has estimated that approximately 20% of the prepared food in the United States ends up as waste [6]. Of the total waste stream from restaurants, over 50% is food waste [7]. A problem with restaurant management is that they monitor the amount of food products they order, receive, and sell, but few monitor the amount of food products they purchase that ends up as waste. Management cannot manage what they do not know. Therefore, a waste audit should be conducted prior to initiating a reduction program, and it should continue after the reduction program is initiated. The audit will identify what and how much is being wasted and what is being composted or donated. With this information, management will be able to determine what and how waste can be reduced.

To aid in this audit, a waste log can be used to record the time, food type, reason for loss, and quantity of loss (number of portions or number of quarts or pounds). Additionally, the time, date, special event, weather conditions, and who is logging in the information shall be recorded in the log. The USEPA has developed a log sheet that can be used as such or as a guide for developing a spreadsheet specific to a restaurant [8]. The log shall be used for preconsumer and postconsumer foods. Preconsumer food shall be tracked daily, while postconsumer foods should be measured at least once per month. Other measures that will aid in this audit are organizing the kitchen to improve waste collection and providing various types of containers and bags to improve waste separation and measurement devices such as scales with preweighted containers and volumetric containers [8–10].

To encourage and promote employees to record and measure the amount of food wasted, management must provide areas in the kitchen for measuring and disposing of wasted food. The preconsumer food waste area should provide a means for weighing the amount of food waste. The food waste is placed in a preweighed container; it is weighed on the scale and then emptied into a separate food waste disposal container. Additionally, to verify the individual weight recording, the weight of the disposal container and waste could be used as a check. A separate area can be provided for the postconsumer food. Both areas should be provided with separate color-coded containers for organic waste, recyclables, and inorganic waste. The containers should be color coded to aid workers in identifying the proper container for each type of waste. The following colors could be used: green for organic wastes, blue for recyclable wastes, black for inorganic wastes [9]. With waste logs and designated waste areas for preconsumer and postconsumer foods, restaurant management will be able to quantify the amount of food wasted and identify which foods have the potential for source reduction.

Preconsumer kitchen waste has been estimated to range from 4% to 10% of the food purchased. This waste is composed of food trimming, overproducted food, expired foods, contaminated food, and other foods that do not make it to the consumer but are disposed as waste. The restaurant management should examine and modify the following to reduce the preconsumer kitchen waste:

- Quantities and timing for ordering and purchasing of food.
- Production and handling practices.
- Menu items.
- Secondary uses for excess food (leftovers) [3].

The waste logs and sale records can aid in modifying quantities of food to be purchased. Food purchasing shall be based on historical demands and not on a typical weekly or monthly use. Historical records will help management identify seasonal, holiday, and special event demand variations.

The cost for production and handling of bulk food products versus precut or prepared food products shall be examined and compared. Waste disposal cost shall also be included in this comparison. For example, the cost per pound for bulk lettuce shall include the preparation cost (time spent during the preparation and cleanup) and the disposal cost for the trimmed lettuce compared with the cost of prepared lettuce (cost per lb). This comparison shall also be done for bulk meats versus cost to cut and size meat portions and whole bulk eggs versus eggs shelled in bulk. Additionally, produce shall be checked carefully at delivery for rotten or damaged products. Any substandard product shall not be accepted and will be returned in the delivery truck. Produce shall be rotated in storage so that older products are used first. This rotation of stock shall be done at every delivery of food, which helps to reduce or minimize the amount of spoilage or waste. Several improvements to the handling of preconsumer food can reduce waste; they include the following:

- Store raw perishable foods, vegetables, fruits, and meats in airtight, see through containers.
- Stalking vegetables (celery, lettuce, carrots, and broccoli) can be reconstituted by trimming off the very bottom part of the stalk and immersing them in warm water for 15 min.
- Meat product storage in the freezer shall be vacuum sealed or wrapped tightly to prevent freezer burn [2,3].

Menu items shall be examined to see if any item is producing excessive preparation waste. Additionally, the menu shall be modified to include secondary use of excess or leftover foods, which could be used for soups, stews, omelets, sauces, stuffing, or entrees that use these foods. By using these foods in this way, a portion of them will not end up in the waste stream [2,3,10]. Instituting these reduction methods has been shown to reduce the preconsumer food waste up to 47% by weight [10]. Therefore, waste reduction provides opportunities for restaurants to lower operational cost by reducing both lost raw product and waste disposal.

Additional waste reduction can be accomplished by examining the food audit for postconsumer food waste. This waste includes plate waste and return of ordered items. Plate waste can be reduced by considering portion control and menu modifications [3,11]. Restaurant management must determine if what is left on the plate is due to portion size being too large, quality being poor, and/or the customers' lack of desire for that particular food. Management can reduce portions, improve quality, eliminate items, or have items that are presented to customers as an option and not automatically provided with the meals.

6.2.2 Donation of Food

In 1996, U.S. Public Law No. 101-610, 104 Statue 3183 (codified 42 USC 12671–12673), Section 401, of the U.S. National and Community Service Act was amended by the Emersion Good Samaritan Food Donation Act to Section 402. It provides a law model for states, the District of Columbia, the Commonwealth of Puerto Rico, and the territories and possessions of the United States; it has no enforcement powers. This Act encourages the donation of apparently wholesome food or grocery products to nonprofit organizations for distribution to needy individuals. Several important definitions are provided by the Act to encourage restaurants to donate food as follows [12,13]:

- Apparently fit grocery product: The term "apparently fit grocery product" means a grocery product that meets all quality and labeling standards imposed by federal, state, and local laws and regulations even though the product may not be readily marketable due to appearance, age, freshness, grade, size, surplus, or other conditions.
- Apparently wholesome food: The term "apparently wholesome food" means food that meets all quality and labeling standards imposed by federal, state, and local laws and regulations even though the food may not be readily marketable due to appearance, age, freshness, grade, size, surplus, or other conditions.
- Food: The term "food" means any raw, cooked, processed, or prepared edible substance, ice, beverage, or ingredient used or intended for use in whole or in part for human consumption.

The aforementioned terms define food that is still good for human consumption and that restaurants could donate instead of throwing them into the waste stream. The Act then defines the act of giving, the giver of the food, and the receiver of the food [12,13]:

- Donate: The term "donate" means to give without requiring anything of monetary value from the recipient, except that the term shall include giving by a nonprofit organization to another nonprofit organization, notwithstanding that the donor organization has charged a nominal fee from the donee organization, if the ultimate recipient or user does not require anything of monetary value.
- Person: The term "person" means an individual, corporation, partnership, organization, association, or governmental entity, including a retail grocer, wholesaler, hotel, motel, manufacturer, restaurant, caterer, farmer, and nonprofit food distributor or hospital. In the case of a corporation, partnership, organization, association, or governmental entity, the term includes an officer, director, partner, deacon, trustee, council member, or other elected or appointed individual responsible for the governance of the entity.

- Nonprofit organization: The term "nonprofit organization" means an incorporated or unincorporated entity that:
    a. is operating for religious, charitable, or educational purposes; and
    b. does not provide net earnings to, or operate in any other manner that inures to the benefit of, any officer, employee, or shareholder of the entity.

There are people who help to collect and distribute the food, and there are places where these activities occur. They are defined as follows [12,13]:

- Gleaner: The term "gleaner" means a person who harvests for free distribution to the needy, or for donation to a nonprofit organization for ultimate distribution to the needy, an agricultural crop that has been donated by the owner.
- Collection or gleaning of donations: A person who allows the collection or gleaning of donations on property owned or property occupant, or paid or unpaid representatives of a nonprofit organization, for ultimate distribution to needy individuals shall not be subject to civil or criminal liability that arises due to the injury or death of the gleaner or representative, except that this paragraph shall not apply to an injury or death that results from an act or omission of the person constituting gross negligence or intentional misconduct.

To protect organizations receiving the food, the model law then defines wrongdoing [12,13].

- Gross negligence: The term "gross negligence" means voluntary and conscious conduct by a person with knowledge (at the time of the conduct) that the conduct is likely to be harmful to the health or well-being of another person.
- Intentional misconduct: The term "intentional misconduct" means conduct by a person with knowledge (at the time of the conduct) that the conduct is harmful to the health or well-being of another person.

To protect organizations and people giving the food and the people and organizations distributing the food, the following limits their liability and responsibility [12,13].

- Liability for damages from donated food and grocery products: A person or gleaner shall not be subject to civil or criminal liability arising from the nature, age, packaging, or condition of apparently wholesome food or an apparently fit grocery product that the person or gleaner donates in good faith to a nonprofit organization for ultimate distribution to needy individuals, except that this paragraph shall not apply to an injury to, or death of, an ultimate user or recipient of the food or grocery product that results from an act or omission of the donor constituting gross negligence or intentional misconduct.
- Partial compliance: If some or all of the donated food and grocery products do not meet all the quality and labeling standards imposed by federal, state, and local laws and regulations, the person or gleaner who donates the food and grocery products shall not be subject to civil or criminal liability in accordance with this section if the nonprofit organization that receives the donated food or grocery products
    a. is informed by the donor of the distressed or defective condition of the donated food or grocery products; agrees to recondition the donated food or grocery products to comply with all the quality and labeling standards prior to distribution; and
    b. is knowledgeable of the standards to properly recondition the donated food or grocery product.

The "Emerson Good Samaritan Food Donation Act" provides the foundation for development of state laws to protect all parties (donors, distributors, needy, etc.) involved in the food donation process. Restaurant owners and food rescue programs (food banks, shelters, soup kitchens, youth centers, senior centers, etc.) must work together in order to meet the restaurants need to reduce food waste and the cost of disposal. The food rescue programs need to provide foods and reduce the cost for purchasing foods from wholesalers and retailers. Restaurants who are donating food to recovery programs must ensure food is within the expiration date and is kept at safe temperatures (below 40 F and above 140 F). Additionally, the restaurants must check for signs of spoilage [14]. The recovery programs must encourage restaurants to donate foods by making the process simple and easy. This can be accomplished by having free pickup of the foods and by publicizing donors and acknowledging their support [14].

In addition, for restaurants, reducing the disposal cost of food waste can be achieved by donating foods; the donated foods in the United States can generate tax benefits. Inventory donated to charities by the 1976 Tax Reform Act (Section 2135) becomes an allowable income tax deduction and allows the donor to determine the "fair market value" of their donation not to exceed two times the cost [15].

6.2.3 Food Scraps for Animal Feed

Pinter, a farmer in New Jersey, utilizes food scraps from Rutgers University to feed his hogs and cattle [18,19]. Additionally, poultry, sheep, and goats can also be fed with food scraps [16–20]. Restaurant food scraps benefit both farmers and restaurant owners.

Food scraps benefit farmers by lowering their grain bills and providing a low-cost, high-quality food source for their livestock. The food scraps contributed for feed benefit restaurant owners by lowering their waste disposal cost. It has been reported that a large food scrap program for animal feed can reduce disposal costs from 30% to 50% [18,19]. By removing food scraps from the restaurant waste stream, the waste stream will not attract many vectors, such as flies and rodents, at restaurant waste storage areas. Additionally, reducing the amount of waste that can putrefy at the restaurant waste storage areas will reduce the amount of undesirable odors released in these areas [20]. These reductions can make the restaurant more attractive to its patrons.

Food scraps have to meet the quality standards in order to be used as feed for animals. Food scraps for animals must be low in salt. Coffee grounds must not be added to the food scraps. Additionally, the food scraps must be separate from other materials (napkins, containers, and other nonfood items) and kept covered and refrigerated (or stored in a cool place) prior to the pickup.

The U.S. Federal Swine Health Protection Act (PL 96 468) requires food scraps, which contain meats, animal material, or food that has come in contact with meat or animal products, to be boiled before it is used to feed hogs. Other food scraps are not regulated by this Act, but may be regulated by local and state laws. This Act also requires the facilities preparing the food scraps for hog feed to be registered with either the USDA or the chief agricultural or animal health official in the state in which the facility is located [17,18].

Along with the quality, handling, storage, and permitting requirements of the food scraps for the animal feed program, the program must be coordinated with a local farmer to meet their need. Without farmers who have a need to supplement their animal feed with food scraps, there is no program. Farmers must be informed on how food scraps can lower their grain bills for feed and provide a high-quality food source rich in nutrients for their livestock [20]. Farmers and restaurant owners must work together to make a successful food scrap program for animal feed.

6.2.4 Industrial Uses for Food Scraps

Food scraps can be used in an anaerobic digestion process and gasification process to produce methane and synthetic gas [21–35]. Biohydrogen can also be produced from food scraps [36]. Additionally, food scraps can be converted to compost. The conversions of the food scraps to usable gases or to soil-like materials provide an alternative to landfilling food scraps.

6.2.4.1 Anaerobic Digestion Process

To produce methane gas from an anaerobic process requires the organic material (food scraps) to go through a four-phase microbial process, which is illustrated in Figure 6.1 [21,22]. Numerous species of bacteria are involved in these processes. In the hydrolysis process, food scraps interact with water and bacteria to convert to soluble organic molecules. Once food scraps go through the hydrolysis process, there will be a variety of soluble organic molecules (sugar, amino acids, and fatty acids) available as food sources that will cause various microbial (bacterial) processes to occur simultaneously. These processes are fermentation, acetogenesis, and methanogenesis. Fermentation and acetogenesis bacteria convert soluble organic molecules and volatile fatty acids to hydrogen (H_2), carbon dioxide (CO_2), and acetic acid (CH_3COOH). The organisms responsible for the acetogenesis phase are strictly obligate anaerobes; that is, these microorganisms cannot exist and propagate with free oxygen [26]. Therefore, it is important that influent waste be discharged into the digester in the absence of free oxygen. Methane (CH_4) and carbon dioxide (CO_2) are produced when methanogens (methane-producing bacteria) convert acetic acid, hydrogen, and carbon dioxide to methane [21,22,26].

Two basic types of reactor configuration used to produce methane are the single-stage digester with no mixing and the two-stage digester with mixing only in the first stage. The single-stage or conventional digester does not require mixing. Digestion and gas production and withdrawal, supernatant and withdrawal, and stabilization and withdrawal of stabilized concentrated sludge are accomplished in a single enclosed circular tank with a sloped bottom. The contents in the digester are layered from top to bottom as follows: gas storage and collection, scum layer, supernatant, active digestion, stabilized and concentrated solids layers. In a two-stage digester, commonly known as a high-rate system, primary stabilization of the food scraps is accomplished in the first-stage digester, using mixing. In the second-stage digester (without mixing), the following is accomplished: supernatant separation, sludge concentration, gas production, and gas storage [25,28]. Figures 6.2 and 6.3 illustrate a two-stage and a single-stage high-rate anaerobic digester system, respectively.

Digesters are typically constructed from reinforced concrete. The digester is a low, vertical, cylindrical tank with a conical bottom. The diameter of the tank can vary from approximately 6 to 34 m (20–111.5 ft) in increments to accommodate standard covers for digesters. The sidewall depth ranges from approximately 6 to 12 m (20–39.4 ft). The slope of the conical bottom is at minimum, 1 vertical to 4 horizontal (approximately 14°) for gravity sludge removal and can be reduced to

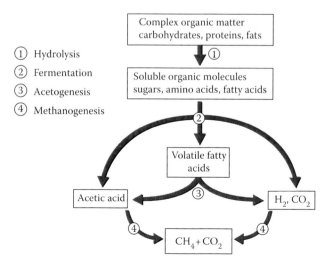

FIGURE 6.1 Conversion of food scraps to methane gas by anaerobic process. (From USEPA, Organics: Anaerobic digestion science, U.S. Environmental Protection-Pacific Southwest, Region 9, U.S. Environmental Protection Agency, Oakland, CA, 2015.)

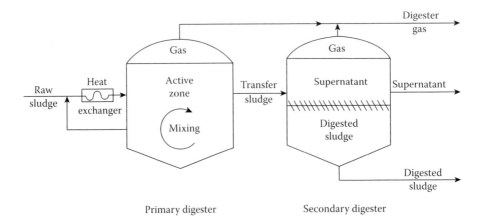

FIGURE 6.2 Two-stage high-rate anaerobic digester system. (From USEPA, Process design manual for sludge treatment and disposal, EPA625/1-79-011, U.S. Environmental Protection Agency, MERL, Cincinnati, OH, 1979.)

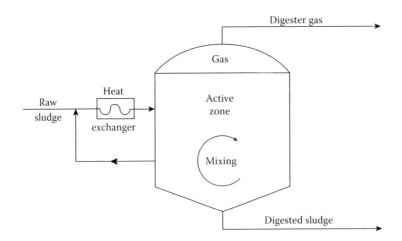

FIGURE 6.3 Single-stage, high-rate anaerobic digester system. (From USEPA, Process design manual for sludge treatment and disposal, EPA625/1-79-011, U.S. Environmental Protection Agency, MERL, Cincinnati, OH, 1979.)

1 vertical to 12 horizontal (approximately 4.9°) when solids are removed with pumps. To improve removal of sludge and grit, scum buildup, and the amount of debris accumulated, egg-shaped digesters (ESDs) have been developed and put into operation in Europe since the 1950s and in the United States since the 1970s [25,28].

The ESD provides a small bottom area with steep side slopes. The conical bottom of the ESD has a slope ranging from 37° to 45°, which concentrates the settled sludge and grit to a smaller area than in a conventional digester, providing a central location to effectively remove sludge and grit. Another benefit of the steep side slopes of the ESD is that the top of the digester has a small surface area, which limits the tendency for scum buildup and debris accumulation at the top. Gas storage would be minimal in an ESD due to the small surface area at the top. Therefore, a separate gas tank must be provided [25,28]. Figure 6.4 shows a 99 ft tall ESD with 37° sloped bottom at a maximum diameter of 66 ft.

There are many factors that must be considered when designing a digester. They include percentage of total solids (TS) and volatile solids (VS) of the waste stream along with the flow rate, time, and operational temperature. TS can range from 50% to 90%, with 60% TS as typical [25]. Digester

Restaurant Waste Treatment and Management

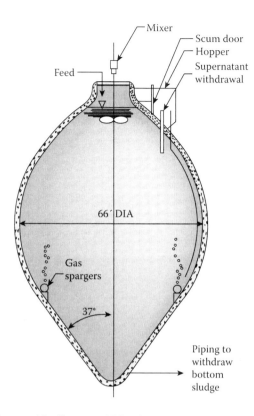

FIGURE 6.4 Egg-shaped anaerobic digester—99 ft tall. (From USEPA, Process design manual for sludge treatment and disposal, EPA625/1-79-011, U.S. Environmental Protection Agency, MERL, Cincinnati, OH, 1979.)

size is based on VS, organic loading, time, and volume. The GLUMBRA Standards [29] recommend the volume of the digester to be determined as follows:

- Completely mixed digester: Loading rate up to 1.28 kg of VS/m³/day (80 lb/1000 ft³/day).
- Mixed digester: Loading rate up to 0.64 kg of VS/m³/day (40 lb/1000 ft³/day).

Two operational parameters for digesters are temperature and solid retention time (SRT). Digesters are designed to operate at mesophilic temperatures ranging from 30°C to 37.7°C (86–100 F) or at thermophilic temperatures ranging from 49°C to 57°C (120–135 F) [25,31]. The equipment types that have been used to increase and maintain the temperature of the sludge are external heat exchangers, jacket draft tube mixers, and internal pipe coils. The most common type of equipment used is the external heat exchanger, because the tubes are readily accessible for maintenance and cleaning. For a mixed and insulated digester, the estimates for heat dissipated or loss from the digester structure range from 1260 W/1000 m₃ (1200 Btu/h/1000 ft³) in warm areas to 4190 W/1000 m₃ (4000 Btu/h/1000 ft³) in cold areas. The actual heat loss for digesters can be determined from the heat transfer coefficients of the construction materials of the digesters. The amount of added heat needed to raise the temperature of the sludge to a desired process temperature is determined as follows:

$$H = wc(T_2 - T_1), \qquad (6.1)$$

where H is the amount of heat (J or Btu) needed to raise the temperature from T_1 to T_2, w is weight of raw sludge being heated (kg or lb), c is the mean specific heat of the raw sludge (4200 J/kg/°C or

1 Btu/lb/F), T_2 is the process temperature (mesophilic or thermophilic digester) of the sludge in the digester (°C or F), and T_1 is the temperature of the raw waste entering the digester (°C or F) [25,31]. The total heat required for the digester is the sum of the heat required to raise the sludge temperature to the process temperature and the heat added to make up for the heat dissipated or loss from the digester structure.

The SRT or mean cell residence time (MCRT) is also used as an operational parameter for digesters. The complete mix processes shown in Figures 6.1 and 6.2 do not return solids; therefore, as a result, the SRT is equal to detention time (θ). The SRT can be determined as follows:

$$\text{SRT} = \theta = \frac{V}{Q}, \tag{6.2}$$

where SRT is the solid retention time (day), V is volume of the digester (m³ or ft³), and Q is the flow rate into the digester (m³/day or ft³/day). It is important to maintain a consistent SRT, which means the flow is constant to and from the digester. The flow from the digester must be equal to the flow to the digester. A food scrap digester will require preparation of the food scraps to a consistent slurry and a storage tank where the slurry can be fed at a consistent flow rate to the digester. The anaerobic sludge and liquor removed must equal the flow into the digester [25].

The anaerobic digester process will produce methane gas (CH_4), carbon dioxide (CO_2), and several impurities. When the digester is properly operated, the gas produced will be composed of approximately 65% CH_4 and 35% CO_2. During digester upset, over or under acidifying, the percentage of CO_2 will increase and CH_4 will decrease. The gas produced from the digester will have an approximate fuel value of 5850 kg-cal/m³ (656 Btu/ft³). The amount of methane gas generated can be estimated from the following equation:

$$C = 0.35(eF - 1.42A), \tag{6.3}$$

where C is the amount of methane generated from the digester (m³/day), 0.35 is the conversion factor for the amount of methane generated for organic loading consumed at 35°C, e is the efficiency of organic utilization, F is the amount of organic matter consumed during the digester process as the 5 day biochemical oxygen demand (BOD_5, kg/day), and A is the VS to the digester (kg/day) [24]. Anaerobic digesters typically produce 0.75–1.12 m3 of gas/kg of VS destruction (12–18 ft³ gas/lb VS destruction). The gas has an approximate heating value of 22,400 kJ/m³ (600 Btu/ft³), with 60%–65% methane content [28,31].

The gas produced from the digestion and gasification process contains impurities that include solids, water, nonmethane fuel components (butane, propane, carbon monoxide, and hydrogen), sulfur and sulfur components (H_2S), and siloxanes (decamethylcyclopentasiloxane, $C_{10}Si_5H_{30}O_5$ and other polydimethylsiloxanes and cyclomethicone). The type and amount of treatment required for gas depends on the amount of gas produced and the end use of the gas. For small digesters, where the gas is only used for maintaining the temperature in the process (by using a boiler with a heat exchanger), gas treatment is minimal. The gas treatment process would include removal of water and solids and flaring of excess gas. As the digesters increase in size with an increase in loading (food scraps), the gas production will increase. The excess gas, the gas not used to heat the digester, can be used for generating electricity. As a result, further cleaning of the gas will be required to remove the remaining impurities to protect the generating equipment from damage and to protect the life of the equipment. The additional cleansing of the gas is accomplished by adding adsorption and absorption beds and filters to the gas cleaning system. For a gas to be used as a fuel for generating equipment, the final gas product must contain the highest percentage of methane gas at the lowest level of impurities. The cost for generating gas must be economically feasible. The cost for the final gas product, amortized capital cost for the digester, generating equipment or other engines, gas treatment equipment, and the operation and maintenance costs must be less than the cost to purchase fuel or other energy sources [26,30].

The San Antonio Water System's Biosolids Facilities have the methane gases (biogas) that are produced from anaerobic digesters processed and transferred into a commercial gas pipeline. The digesters process 140 t/day of biosolids and produce 1 Mft^3/day of biogas. The biogas contains approximately 65% CH_4, 35% CO_2, and trace amounts of other gases (siloxanes and H_2S). The biogas contains 10 ppm of H_2S and is saturated with water. It has an energy value ranging from 550 to 600 Btu/ft^3. The biogas is processed through a pressure swing adsorption (PSA) process. The PSA targets the removal of CO_2, H_2S, and siloxanes from the biogas along with H_2O in the biogas under a high pressure of approximately 100 psig. After processing, the biogas contains 2% CO_2, 4 ppm of H_2S, and 7 lb H_2O/Mft^3. The processed biogas has an energy value of approximately 990 Btu/ft^3 and pressure of 90 psig. It is further pressurized to approximately 700 psig. This pressure is required to enter into the commercial gas pipeline [32].

The USEPA and East Bay Municipal Utilities District (EBMUD) of California conducted a bench-scale study of an anaerobic digester with food waste [23,24]. The study compared the results from an anaerobic digester with food waste with an anaerobic digester with municipal wastewater solids, which is shown in Table 6.1.

Food scraps were collected and processed through the EBMUD digester preparation system to produce feed slurry and remove debris. The conditioned food scrap slurry was then sent to the digester for digestion (see Figure 6.5). A combined thickened waste activated sludge and screened primary sludge were used in the other study for comparison.

The comparison shows that food waste had an 86.3%–89.9% VS range and a 0.55–1.09 lb/ft^3-day Chemical Oxygen Demand (COD) loading range, while the municipal waste had a 77% VS and a 0.06–0.3 lb/ft^3-day COD loading range. This higher organic loading (VS and COD) is expected since the food waste is not processed through a wastewater treatment (activated sludge) process, which would result in a lower organic loading from bio-oxidation. The VS destruction for food waste is high because food scrap waste provides a highly soluble organic food (soluble organic molecules as sugar, amino acids, and fatty acids) available as a food source for the microorganisms, and as a result, hydrolysis, fermentation, acetogenesis, and methanogenesis processes will be more efficient. The results of the study also showed that production of methane (ft^3/lb TS) is higher for the food waste digestion at 15 day MCRT compared with the municipal wastewater solids digestion at 15 day MCRT. When the food digester MCRT was lowered from 15 to 10 days, the gas production

TABLE 6.1
Comparison of Food Scrap and Municipal Wastewater Sludge Bench-Scale Studies

Parameter	Unit	Food Scrap Waste Mesophilic Digester 15-day MCRT	Food Scrap Waste Mesophilic Digester 10-day MCRT	Municipal Wastewater Mesophilic Sludge Digester 15-day MCRT
VS	%	86.3	89.9	77
VS loading	lb/ft^3-day	0.28	0.53	0.2
COD loading	lb/ft^3-day	0.55	1.09	0.06–0.3
VSD	%	73.8	76.4	63
Comparison of gas production[a]	%	133	95	100

Source: Gray, D., et al., Anaerobic digestion of food waste, EPA-R9-WST-06-004, Pacific Southwest, Region 9, U.S. Environmental Protection Agency, Oakland, CA, 2008.

[a] Comparison = $\dfrac{\text{Food waste digester gas production}}{\text{Wastewater sludge gas production}} \times 100$.

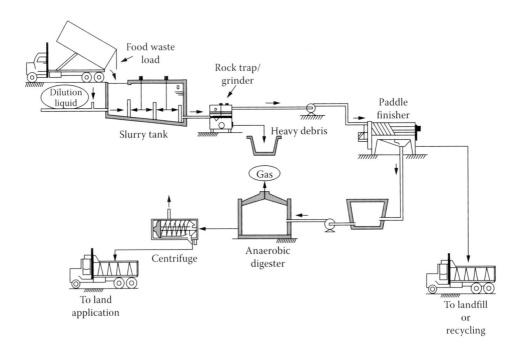

FIGURE 6.5 East Bay Municipal Utilities District (EBMUD) food scrap digester system. (From USEPA and EBMUD, Turning food waste into energy at the East Bay Municipal Utility District: Investigating the anaerobic digestion process to recycle post-consumer food waste, EPA-R9-WST-06-004, Pacific Southwest, Region 9, U.S. Environmental Protection Agency, Prepared by East Municipal Utility District, Oakland, CA, 2008.)

decreased. This indicates that there was not sufficient time for the microorganisms to convert all the available food to methane. Along with the MCRT or SRT, there are several important parameters that must be monitored and controlled to efficiently operate anaerobic digesters. These include temperature, pH, alkalinity, organic loading, VS loading, volatile solids reduction, mixing of the digester, and gas production [23,24].

Food processors can utilize raw food waste to produce energy. The Furmano Foods' production plant in Northumberland, PA, uses food waste from its facility that processes millions of tomatoes, beans, salads, vegetables, and peppers annually to produce biogas by anaerobic digestion. The biogas produced from the digester is dried and scrubbed and then fed to a 250-kW generator to produce electric power for the facility; the excess energy is sold to the power grid [35].

6.2.4.2 Biomass Gasification Process

The conversion of organic (carbonaceous) material by incomplete oxidation (under oxygen-starved conditions that prevent the complete conversion of the organic material to carbon dioxide and water) to a gas is generally referred to as the gasification process. This gas is generally referred to as synthetic gas or syngas, in which the major components of the gas are hydrogen (H_2) and carbon monoxide (CO) and the minor components are carbon dioxide (CO_2), water (H_2O), methane (CH_4), and nitrogen (N_2).

There are several steps involved in the biomass gasification process. The process can be described as a two-step process that includes pyrolysis and gasification. The steps are illustrated in Figure 6.6. In Step 1, the pyrolysis process, the biomass is decomposed by heat, and 75%–90% of the volatile materials are converted to gaseous and liquid forms of hydrocarbons and char. Char is a high carbon material that is nonvolatile. In Step 2, the gasification process converts the hydrocarbons and char to syngas.

The chemical reactions for the gasification process comprise exothermic and endothermic reactions. Some of the reactions are described below [31,32].

Restaurant Waste Treatment and Management

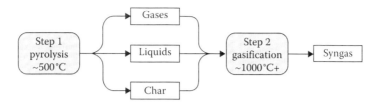

FIGURE 6.6 Steps for gasification. (From Ciferno, J. and Marano, J., *Benchmarking Biomass Gasification Technologies for Fuels, Chemicals and Hydrogen Production*, National Energy Technology Laboratory, U.S. Department of Energy, Pittsburgh, PA, 2002.)

Exothermic reactions (releases energy)

$$\text{Combustion} \rightarrow \text{Biomass volatile/char} + O_2 \rightarrow CO_2 \qquad (6.4)$$

$$\text{Partial oxidation} \rightarrow \text{Biomass volatile/char} + O_2 \rightarrow CO \qquad (6.5)$$

$$\text{Methanation} \rightarrow \text{Biomass volatile/char} + 2H_2 \rightarrow CH_4 \qquad (6.6)$$

$$\text{Water-gas shift} \rightarrow CO + H_2O \rightarrow CO_2 + H_2 \qquad (6.7)$$

$$\text{CO methanation} \rightarrow CO + 3H_2 \rightarrow CH_4 + H_2O \qquad (6.8)$$

Endothermic reactions (absorbs energy)

$$\text{Steam-carbon} \rightarrow \text{Biomass volatile/char} + H_2O \rightarrow CO + H_2 \qquad (6.9)$$

$$\text{Boudouard reaction} \rightarrow \text{Biomass volatile/char} + CO_2 \rightarrow 2CO \qquad (6.10)$$

The endothermic reactions can use direct or indirect heat. In the directly heated gasification system, both the pyrolysis and gasification processes are carried out in a single reactor, while in the indirectly heated gasification system, the combustion process is separated from the remaining reaction.

The pyrolysis process converts the biomass at high temperatures (approximately 500°C) in an oxygen-starved environment, to produce a useful gas that can be burned directly to heat boilers or other equipment. This process releases the volatile components of the biomass in a series of complex reactions [26,27,33,34]. Biomass such as food scraps have been reported to have a VS content ranging from 86.3% to 89.9% [24]. The gasification process converts the gaseous and liquid hydrocarbons and char at higher temperatures (1000°C or higher) to syngas. At the downstream end of the gasification system, a water–gas shift reaction with a catalyst is applied to the gasification system to increase the hydrogen to carbon monoxide ratio (H_2/CO) of the syngas. This increase in the H_2/CO ratio (0.5–0.7) improves the gas so that it can be used as a synthetic fuel substitute for gasoline and diesel fuel [34]. In larger gasification facilities (using coal), leftover tars and char are further gasified and converted to carbon monoxide using steam and/or partial combustion. Gasification of biomass provides benefit over direct-fired systems as follows [27]:

- Syngas from gasification can be used to heat boilers, turbines, engines, and fuel cells.
- Syngas from gasification contains fewer impurities and produces fewer emissions.
- Syngas can be produced from a variety of biomass ranging from woody residues to agricultural residues to dedicated crops to food wastes.
- High overall efficiency results when the syngas is used in a gas turbine to generate power with a system to recover the heat from the exhaust system of the turbine.

The most commonly used gasification processes are the fixed-bed and the fluidized-bed gasifiers. The fixed-bed gasifier is the simpler and less costly option compared with the fluidized bed gasifier, but it produces syngas with a lower heat value than the fluidized bed gasifier [26,33,34]. Biomass can produce syngas with various heating values as follows [33]:

- Sawdust (7.8% moisture): 19.3 MJ/kg (8315 Btu/lb).
- Switchgrass (8.4% moisture): 15.4 MJ/kg (6634 Btu/lb).
- Dry sewage sludge (4.7% moisture): 8 MJ/kg (3446 Btu/lb).

The heat value for food scraps would be slightly higher than the dry sewage sludge, depending on the carbon content of the food scraps. Food scraps containing significant amounts of vegetables like lettuce have low carbon content, and therefore a lower heat value, than food scraps containing significant amounts of vegetables like corn, which has high carbon content and therefore a higher heat value.

A new process being considered in the gasification industry is the plasma process for gasification. Plasma is a gas that is very highly ionized and conducts electrical current, which is similar to lightning and gas at the surface of the sun. Plasma is made by passing an electrical charge through air, which is composed of oxygen, nitrogen, argon, and other gases. As a result of the electrical charge, an electric arc is formed. This arc dissociates the gas into electrons and ions, causing the temperature to exceed 6000°C [33,34]. Figures 6.7 and 6.8 illustrate the plasma torch and plasma torch gasifier, respectively.

The gasifier is lined with a refractory material (nonmetallic material that maintains its strength at high temperature) so that the high temperature generated by the plasma torch will not destroy the gasifier structure. A water jacket is also installed above the torch area of the gasifier to cool the syngas, but maintain its temperature above 1000°C so that tar formation is prevented. The advantages of the plasma torch gasifier is that the high operating temperature (>6000°C) breaks down components in the food scraps into their elemental constituents and increases kinetics of the reactions to produce the syngas [33,34]. It also has been reported that the plasma torch gasifier operates at a high efficiency, with only 2%–5% of the total energy input into the gasifier being used by the plasma torch [33,34].

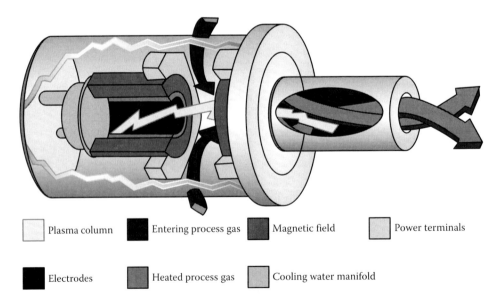

FIGURE 6.7 Plasma torch. (From Westinghouse Plasma; NETL, *Plasma Gasification*, National Energy Technology Laboratory, U.S. Department of Energy, Pittsburgh, PA, 2016.)

Restaurant Waste Treatment and Management

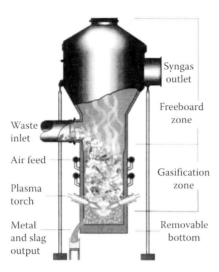

FIGURE 6.8 Plasma torch gasifier. (From Westinghouse Plasma; NETL, *Plasma Gasification*, National Energy Technology Laboratory, U.S. Department of Energy, Pittsburgh, PA, 2016.)

Columbia University proposed a plasma torch gasifier processing system for food scraps collected from restaurants in New York City [27]. Participating restaurants will be asked to separate the food scraps into separate containers to be picked up regularly and transported to the gasification facility. The design of the proposed system is projected to process 10 tons/day of food scrap waste from 400 restaurants. The equipment included in the plasma torch gasifier processing system are as follows:

- Shredder machine
- Pulverizing machine and fluidization water system
- Freeze-drying system (100% removal of water)
- Grinder
- Plasma torch gasifier
- Steam mixing chamber (water–gas shift reaction to increase H_2/CO)
- Gas turbine

For this system to work, various conveyors and piping are required to transport material from one piece of equipment to another. The syngas produced is projected to be used to fuel a gas turbine to generate electricity. This plasma torch gasifier processing system is projected to operate at 45%–50% efficiency [27].

6.2.4.3 Biohydrogen Production

The fermentation process for the production of hydrogen was examined by Wongthanate et al. [36]. The bench-scale study compared production of hydrogen when restaurant wastes were pretreated with one of following methods: addition of a methanogenic inhibitor by using 1M sodium 2-bromoethanesulfonate (Besa:$C_2H_4BrO_3SNa$, 0.2/L), (2) sterilized in an autoclave at 121°C for 20 min, (3) sonicated in an ultrasonic bath for 20 min, and (4) acidified by adding $HClO_4$ to low the pH to 3. The restaurant waste collected from a cafeteria was grinded in a blender and sieved through a 5-mm screen. The screened restaurant waste was then mixed with distilled water at a volume ratio of waste to distilled water of 3:1. The restaurant food waste had a TSS of 75,700 mg/L, VSS of 48,600, COD of 770 mg/L, BOD of 126,000 mg/L, and pH of 4.83.

The batch reactors for each pretreated restaurant waste were run for 168 hours. The cumulative hydrogen productions were 0.57 mL H_2/g COD for methanogenic inhibited pretreated waste,

3.48 mL H$_2$/g COD for sterilized pretreated waste, 2.09 mL H$_2$/g COD for sonicated pretreated waste, and 2.96 mL H$_2$/g COD for acidified pretreated waste. As can be seen from the results, the best pretreatment method for the production of hydrogen from restaurant waste is sterilization, followed by acidification [36].

6.2.4.4 Composting Process

Composting is a biological process in which green organic material and brown organic material are mixed at optimal moisture, oxygen flow, and temperature to produce a product called compost. The green material contains a high amount of nitrogen and the brown material contains a high amount of carbon [37]. The amount of nitrogen and carbon contained in the material is generally determined by the C:N ratio. Materials like grass clippings and food waste have C:N ratios of 17 and 15, respectively. These ratios indicate that these material have a higher nitrogen content than corn stalks, which have a C:N ratio ranging from 60–120. Table 6.2 lists carbon to nitrogen ratios of selected materials.

The composting operation maintains a C:N ratio ranging from 30:1 to 45:1 [37]. A mixture of food waste and straw can be blended to achieve a C:N ratio as follows:

3 parts of food waste by weight	3 × (15:1) = 45:3
1 part of straw by weight	1 × (80:1) = 80:1
Total	125:4 or 31.25:1

The suggested moisture content range for a compost process is 50%–60%. The moisture level of the mixture materials for a compost process can be determined from the following equation.

$$M_T = \frac{(W_1 M_1) + (W_2 M_2) + (W_n M_n) + \cdots}{(W_1 + W_2 + W_n)}, \qquad (6.11)$$

where M_T is the percentage of moisture of the compost material, W_1 is weight of material 1 (kg or lb), M_1 is the percentage of moisture in material 1, W_2 is weight of material 2 (kg or lb), M_2 is the percentage of moisture in material 2, W_n is weight of material n (kg or lb), and M_n is the percentage of moisture in material n [37].

The final operational parameters to be considered are oxygen flow and temperature of the composting process. Oxygen (air) flow to the compost can be accomplished by frequently turning over the compost pile (or row) or by using blowers to force air into or pull air through the pile (or row) of compost. The former is called the windrows composting process, and the latter is called the forced aeration composting process. Air flow to the composting process must be managed to prevent drying out the material below the optimal moisture levels [37,38].

Temperature of the core of the compost pile (or row) shall be maintained in the range 55°C–77°C (131–160° F). This temperature range, within the core of the compost pile (or row), indicates that the organic material is being consumed by the microbial population, which releases energy as heat and pathogens are being killed. Without proper air circulation or mixing, hot spots can develop in the composting mixtures that will dry out the compost and reduce microbial activity [37,38].

Both windrows and the forced aeration composting processes are exposed to the climate if shelters are not provided. The weather can provide too much or too little moisture along with too much heat or too little heat. As a result, manufacturers have developed an in-vessel aerated composting process to overcome the effects of weather and to have better control over the composting process [37,38]. Restaurant waste made up of food can create problems by attracting vectors (varmints and scavengers), emitting odors, and discharging leachates [38–43]. Since the in-vessel composting process is enclosed and the operator of the process can control mixing, air flow, and moisture, the problems with vectors, odor, and leachates can be minimized or controlled. Additional composting processes and restaurant waste management technologies can be found from the literature [44–47].

TABLE 6.2
Carbon to Nitrogen Ratios for Composting Material

Material	C:N	Material	C:N
Apple filter cake	13	Mussel waste	2.2
Apple pomace	48	Newspaper	400–850
Apple-processing sludge	7	Olive husks	30–35
Aquatic plants	15–35	Paper (from domestic refuse)	125–180
Blood meal	3	Paper fiber sludge	250
Broiler manure	14	Paper mill sludge	55
Cardboard (corrugate)	560	Paper pulp	90
Castor pomace	8	Paunch manure	20–30
Cocoa shells	22	Pig manure	14
Coffee grounds	20	Potato tops	28
Compost	15–20	Potato (culled)	18
Corn silage	35–45	Potato-processing sludge	28
Corn stalks	60–73	Poultry carcasses	5
Corn waste	60–120	Rice hulls	110–130
Cottonseed meal	7	Sawdust	442
Cow manure	10–30	Sawmill waste	170
Crab/lobster wastes	4.9	Seaweed	5–27
Cranberry filter cake	31	Sheep manure	16
Cranberry plants	61	Shrimp wastes	3.5
Fish wastes	3.6	Shrub trimmings	53
Fish-breading crumbs	28	Slaughterhouse waste	2–4
Fish-processing sludge	5.2	Softwood bark	100–1000
Food wastes	14–16	Softwood chips and shavings	1300
Fruit wastes	40	Soil	12
Grass clippings	17	Soybean meal	4–6
Hardwood bark	100–400	Straw, general	80
Hardwood chips and shavings	450–800	Straw, oat	60
Hay, general	15–32	Straw, wheat	127
Hay, legume	15–19	Telephone books	772
Hay, nonlegume	32	Tomato-processing waste	11
Hoof and horn meal	3	Tree trimmings	16
Horse manure	30	Turkey litter	16
Laying hen manure	6	Vegetable wastes	11–19
Leaves	40–80	Water hyacinths (fresh)	20–30

Source: USEPA, Composting, waste-resource conservation-common wastes & materials-organic material, U.S. Environmental Protection Agency, Washington, DC, 2012.

Note: Paunch manures are the contents (rumens) in the first stomach chamber in cattle, goats, and sheep.

6.2.5 Industrial Uses for Waste FOG

Another waste generated from restaurants is FOG. This waste is disposed of in sanitary sewers and landfills. Restaurants and hotels in the United States generate over 3 billion gallons of waste cooking oil per year [38]. FOGs are being collected and converted by manufacturers to biodiesel fuel [40–42]. In the United States, PacBio was one of the first viable biodiesel fuel plants; it was established in Maui, Hawaii, in 1996. This facility located at the Central Maui Landfill collected used cooking oils and converted it to biodiesel fuel [40]. Cooking oils have high viscosity and low volatility, which cause problems when used in diesel engines, resulting in engine deposits, injector coking (overheating of injector nozzle causing clogging of the injector tip with carbon deposits), and piston ring sticking. These problems can be reduced or eliminated by the transesterification process. This process removes glycerin and decreases the viscosity, but maintains the octane number (measures the ignitibility of the diesel fuel) and the heating value of oils [42].

Greases and oils are lipids and are chemically classified as triglycerides. Generally, oils are liquids at room temperature and fats are solids at room temperature. Spent cooking oils are classified as yellow grease or brown grease. Yellow grease is spent cooking oil that has been rendered, in which solids are filtered out and the water is removed by heating the grease. The free fatty acid (FFA) content in the oil must be less than 15% to be classified as yellow grease. If the grease FFA content is greater than 15%, then the oil is classified as brown grease. The FFA content of grease plays an important role in the transesterification process. This process produces biodiesel (methyl esters) and removes the glycerin from the triglyceride when reacted with an alcohol in the presence of a catalyst. This reaction is shown in the following equation [42]:

$$\begin{array}{c}CH_2-O-\overset{O}{\underset{\parallel}{C}}-R_1\\ |\\ CH-O-\overset{O}{\underset{\parallel}{C}}-R_2\\ |\\ CH_2-O-\overset{O}{\underset{\parallel}{C}}-R_3\end{array} + 3CH_3OH \xrightarrow{\text{Catalyst}} \begin{array}{c}CH_3-O-\overset{O}{\underset{\parallel}{C}}-R_1\\ \\ CH_3-O-\overset{O}{\underset{\parallel}{C}}-R_2\\ \\ CH_3-O-\overset{O}{\underset{\parallel}{C}}-R_3\end{array} + \begin{array}{c}CH_2-OH\\ |\\ CH-OH\\ |\\ CH_2-OH\end{array}$$

Triglyceride + Methanol ⟶ Methyl esters + Glycerin

Restaurant waste oil was reported to comprise 41.8 % FFA and 24% MIU (moisture, insoluble, and unsaponifiables). In addition to the FFA and MIU levels in the waste oil, it has been reported to have a moisture level above 2%. Because of FFA and MIU levels in the waste oils, the transesterification process requires an acid catalyst to convert the triglyceride to esters. It also has been shown that a moisture level at 2% can reduce the ester conversion from the transesterification process to between 50% and 60%. At moisture levels less than 0.5%, the ester conversion can be around 90% [42]. Restaurant waste oil can have various levels of FFA, MIU, moisture, and other contaminants that will affect the transesterification process depending on the usage of the FOG and the handling and storage prior to receiving the waste oil at the biodiesel processing facility.

6.2.6 Restaurant Waste Treatment and Management on Board Cruise Ships

There are over 15 major cruise lines and over 300 cruise ships traveling around the world, often in pristine ocean or sea waters, such as the beautiful Mediterranean Sea, Gulf of Mexico (Caribbean), Pacific Ocean, Atlantic Ocean, and South China Sea. Each cruise ship is a floating city with about 3000 passengers and crew on board. In the past, the deep blue seas where the cruise ships navigated were often polluted by ships traveling back and forth on the same routes, discharging partially treated or untreated restaurant wastes, room wastes, etc. Now, most of the cruise ship managers employ practices and procedures that are substantially more protective of the aquatic environment than are required by the global cruise industry regulations [48,49]. Specifically, the cruise ship managers follow the environmental stewardship practices recommended by the Cruise Lines International Association (CLIA), Fort Lauderdale, FL: (1) equipping cruise ships with advanced wastewater treatment systems that produce water that is cleaner than most wastewater treatment facilities in U.S. cities; (2) adopting numerous energy-efficient measures, including switching to low-energy LED lights, developing smoother hull coatings to consume less fuel, using recycled hot water to heat passenger cabins and special window tinting that keeps passageways cooler and utilizes less air conditioning (halogen and incandescent light bulbs have been giving way to LED lights, which last 25 times longer, use 80% less energy, and generate 50% less heat); and (3) implementing rigorous recycling programs as many CLIA member lines have comprehensive programs and crew members who are specially trained and responsible for sorting, processing, storing, recycling, and the final disposal of garbage. These programs recycle special paper, glass, plastics, aluminum, scrap metal, fluorescent lamps, batteries, toner cartridges and cooking oil, and even special wastes such as chemicals used in photo processing. It is estimated that CLIA members recycle up to 80,000 tons of garbage (including the processed restaurant food wastes) each year [48–53].

Dining on a cruise ship is an integral part of the cruise travel experience and enjoyment. Figure 6.9 shows an overall view of one of many restaurants of the *Brilliance of the Seas*, which is one of the cruise ships of Royal Carribean International, FL. At dining hours, the same restaurant is quickly filled with over 1000 people as shown in Figure 6.10. There are a lot of leftovers that must be properly processed and managed at the end of a cruising day. Restaurant food waste amounts to about 2 kg per day per passenger, and is usually collected in up to 25 decentralized feeding stations and transported by vacuum or water through a pipe system to a centralized waste treatment system [51,52].

The restaurant's food wastes are mixed with some water and go through a series of oil–water, and water–solid separation process systems. The FOG fraction of the restaurant food waste amounts to (1) about 2–3 m^3/day of sludge oil; and (2) about 100–200 L/day of grease from grease traps such as cooking oil from frying pans, etc. Both FOG from the oil–water separators and the solids from the water–solid separators are put into a pretreatment unit for heating and dewatering the oily solids before they are burned in an on-board high-temperature incinerator. The incinerator ashes are collected and transported to a land facility for proper disposal when the cruise ship reaches a designated port. [51–53]

The water fraction of the restaurant food waste from the oil–water separator and the solid–water separator is mixed with all other gray water and black water from the ship rooms, hospital, laundries, etc. forming a combined wastewater. This combined wastewater in a modern cruise ship is then transported

FIGURE 6.9 An overview of a typical cruise ship restaurant serving over 2500 passengers.

FIGURE 6.10 Restaurant service and operation on board the cruise ship *Brilliance of the Seas*, Royal Caribbean International, FL.

to an advanced wastewater treatment (AWT) system in which the bioreactors treat the wastewater to meet the effluent discharge standards. The AWT effluent, depending on the area of operation, can be discharged directly overboard if the cruise ship is at least 12 miles away from the shore [52].

A cruise ship's waste treatment system can be operated in either sorting mode or nonsorting mode, depending on the operational profile of the cruise ship. Sorting mode means everything that can be recycled will be sorted out and prepared by crushing, compacting, or bagging for onshore landing, and only the remaining material will be given into the incinerator. Nonsorting mode means everything (excluding hazardous or combustible materials) will be given through the incinerator when adequate reception facilities are not available on the foreseeable route. The combined waste is then burned; some valuable residues can still be taken out of the ash and given into a recycling process. The readers are referred to the literature [51] for more information regarding a Deerberg MPWMS system for advanced waste management on board cruise ships.

6.3 SUMMARY OF SOURCE REDUCTION METHODS AND TREATMENT PROCESSES

Restaurant wastes, composed of solid and liquid waste, are disposed of either by landfilling or by wastewater treatment processes, respectively. Food wastes are the largest component of the solid restaurant waste. Reduction of food waste provides economical and environmental benefits and can be achieved by the implementation of food recovery programs such as source reduction, donation of food to feed people, providing food scraps for animal feed, industrial uses, composting, landfilling, and incineration.

To achieve source reduction, a waste audit logging preconsumer and postconsumer food wastes should be utilized to identify causes of food wastes. By reducing waste at the source, restaurants can reduce operational costs by reducing both raw product and food disposal. Unused or extra food can be donated to food rescue programs for distribution to needy individuals. Restaurants must ensure that the donated food is safe for consumption. Providing food scraps to feed animals benefits both restaurant owners and farmers by reducing wastes from the restaurants and giving farmers a low-cost option of providing a high-quality food source to their livestock. Communication and coordination between owners and farmers are key in making this program work.

As an alternative to landfill disposal, industrial uses of food waste can be achieved through the conversion of food waste to usable gases, compost, and biodiesel fuel. Usable gas includes the production of methane gas through anaerobic digestion, the production of synthetic gas by the gasification process of biomass, and hydrogen from fermentation (biohydrogen).

Anaerobic digestion process produces methane gas through a four-phase microbial process including hydrolysis, fermentation, acetogenesis, and methanogenesis. Design configurations for anaerobic digesters are typically a vertical reinforced concrete cylinder with conical (sloped) tank bottoms and can vary between a single-stage digester with no mixing to a two-stage digester with mixing. ESD was developed to improve the removal of sludge and grit, reduce scum buildup, and reduce the amount of debris accumulated. Design is based on the percentage of TS, the VS or organic loading, solids retention time (SRT), and temperature (mesophilic or thermophilic) of the digester.

The biomass gasification process produces synthetic gas (syngas) by incomplete oxidation of organic (carbonaceous) material in a two-step process involving pyrolysis and gasification. The fixed-bed and fluidized-bed gasifiers are the most commonly used gasification processes with a downstream water–gas shift reaction with a catalyst to increase the H_2/CO ratio and to improve the gas for use as a synthetic fuel substitute for gasoline and diesel fuel. A new process being considered in the gasification industry is the plasma process for gasification. Plasma torch gasifier uses electrical current to highly ionizing air to create very high temperatures that exceed 6000°C. This gasification process has been reported to result in high efficiency, with only 2%–5% of the total energy input into the gasifier being used by the plasma torch. A bench-scale study showed that biohydrogen production is improved by pretreating the food waste with sterilization.

Composting is a biological process in which green organic material and brown organic material are mixed at optimal moisture, oxygen flow, and temperature to produce a soil-like product called compost. Oxygen (air) flow to the compost can be accomplished by the windrows composting process or the forced aeration composting process. Manufacturers have developed an in-vessel aerated composting process to overcome the effects of weather (moisture, air flow, temperature) to allow operators to have better control over the composting process.

6.4 SOURCES OF FURTHER INFORMATION AND ADVICE

The USEPA [43] provides a food waste management cost calculator to be used in conjunction with the waste audit and waste log. This calculator is a Microsoft Excel spreadsheet that provides cost for alternatives to food waste disposal, including source reduction, donation, composting, and recycling of yellow grease. Metcalf & Eddy, Inc. [31] and GLUMBRA [29] provide design examples and design standards for various types of anaerobic digestion processes. Additional resources [54–56] also provide technical information for the treatment of wastewater that can be generated from restaurants.

LIST OF NOMENCLATURE

A	Volatile solids (kg/day)
c	Mean specific heat of raw sludge 4200 (J/kg/°C or Btu/lb/F)
C	Amount of methane generated (m³/day)
C:N	Carbon–nitrogen ratio (No units)
e	Efficiency of organic utilization (No units)

F	Amount of organic material (BOD_5) consumed during digestion (kg/day)
H	Amount of heat to raise the temperature of sludge (J or Btu)
M_1	Percentage of moisture of the compost material 1 (%)
M_2	Percentage of moisture of the compost material 2 (%)
M_n	Percentage of moisture of the compost material n (%)
M_T	Percentage of moisture of the compost material (%)
Q	Flow rate (m³/day or ft³/day)
SRT	Solid retention time (day)
T_1	Temperature of raw sludge entering into the digester (°C or F)
T_2	Process temperature of the sludge (°C or F)
V	Volume (m³ or ft³)
W_1	Weight of material 1 in compost (kg or lb)
W_2	Weight of material 2 in compost (kg or lb)
W_n	Weight of material n in compost (kg or lb)
θ	Detention time (day)

REFERENCES

1. USEPA, Municipal solid waste generation, recycling, and disposal in the United States: Fact and figures for 2012, EPA 530-F-14001, U.S. Environmental Protection Agency, Solid Waste and Emergency Response, Washington, DC (2014).
2. IWMB, Restaurant guide to waste reduction and recycling, The Integrated Waste Management Board, City and County of San Francisco, San Francisco, CA (1992).
3. USEPA, Food Waste Reduction, Waste-resource conservation-common wastes & materials-organic material, U.S. Environmental Protection Agency, Washington, DC (2011).
4. USEPA and USDA, Waste not, want not: Feeding the hungry and reducing solid waste through food recovery, EPA 530-R-99-040, U.S. Environmental Protection Agency Office of Solid Waste and U.S. Department of Agriculture, Washington, DC (2011).
5. NCDENR, A fact sheet for restaurant waste reduction, North Carolina Department of Environment and Natural Resources, Division of Pollution Prevention Environmental Assistance, Raleigh, NC (1999).
6. MDEQ, Restaurant pollution prevention and waste reduction, Fact Sheet, Michigan Department of Environmental Quality Environmental Science and Services Division, Lansing, MI (2008).
7. ILG, Restaurant waste management and recycling, Institute for Local Government, County of San Diego, San Diego, CA (2015).
8. USEPA, Waste logbook—Facility, U.S. Environmental Protection Agency Office of Solid Waste, Washington, DC (2011).
9. Valeiras, P.H., Prado, P.V., Chung, S.Y., and Zheng, Z.U., *Food Waste to Energy*, Presentation, Columbia Service-Learning Program, Columbia University, New York (2005).
10. Oregon DEQ and City of Hillisboro, Food waste prevention case study: Intel Corporation's Cafés, Oregon Department of Environmental Quality and City of Hillisboro, Hillisboro, OR (2010).
11. MWCOG, *Recycling Guidebook for the Hospitality and Restaurant Industry*, Metropolitan Washington Council of Governments, Department of Environment Programs, Washington, DC (2000).
12. USGPO, Public Law 104-210, United States Government Printing Office, Washington, DC (1996).
13. USDA, Appendix C: Text of Emerson Good Samaritan Food Donation Act, Public Law 104-210, U.S. Department of Agriculture, Washington, DC (1996).
14. USEPA, Donating surplus food to the needy, Solid Waste and Emergency Response (5306W), EPA530-F-038, U.S. Environmental Protection Agency, Washington, DC (1996).
15. USEPA, Food donation: Feed people-not landfill, Waste-resource conservation-common wastes & materials-organic material, U.S. Environmental Protection Agency, Washington, DC (2012).
16. USEPA, Feed animals, Waste-resource conservation-common wastes & materials-organic material, U.S. Environmental Protection Agency, Washington, DC (2012).
17. USEPA, Managing food scraps as animal feed, EPA530-F-96-037, Solid waste and emergency response, U.S. Environmental Protection Agency, Washington, DC (1996).

18. USEPA, Food scraps go to the animals, Barthold recycling and roll-off services, EPA530-F-06-035, U.S. Environmental Protection Agency, Washington, DC (2006).
19. USEPA, Feeding animals—The business solution to food scraps, EPA530-F-09-022, U.S. Environmental Protection Agency, Washington, DC (2009).
20. ONRCD, Food scrap reduction project, Ottauquechee Natural Resources Conservation District, White River Junction, VT (2012).
21. USEPA, Industrial uses, Waste-Resource Conservation, U.S. Environmental Protection Agency, Washington, DC (2015).
22. USEPA, Organics: Anaerobic digestion science, U.S. Environmental Protection-Pacific Southwest, Region 9, U.S. Environmental Protection Agency, Oakland, CA (2015).
23. Gray, D., Suto, P., and Peck, C., Anaerobic digestion of food waste, EPA-R9-WST-06-004, Pacific Southwest, Region 9, U.S. Environmental Protection Agency, Oakland, CA (2008).
24. USEPA and EBMUD, Turning food waste into energy at the East Bay Municipal Utility District: Investigating the anaerobic digestion process to recycle post-consumer food waste, EPA-R9-WST-06-004, Pacific Southwest, Region 9, U.S. Environmental Protection Agency, Prepared by East Municipal Utility District, Oakland, CA (2008).
25. Taricska, J.R., Huang, J.Y.C., Chen, J.P., Hung, Y.T. and Zou, S.W., Anaerobic digestion. In: *Handbook of Environmental Engineering, Volume 8: Biological Treatment Processes*, Wang, L.K., Pereira, N.C., and Hung, Y.T., (eds), Humana Press, Totowa, NJ (2009).
26. USEPA and CHPP, Biomass combined heat and power catalog of technologies, U.S. Environmental Protection Agency and Combined Heat and Power Partnership, Washington, DC (2007).
27. CSLP, A system for converting food waste into energy, final report, Columbia Service-Learning Program, Primary Facilitator Joseph Weingartner, Project Manager Christian Aucoin, Columbia University, New York (2007).
28. USEPA, Process design manual for sludge treatment and disposal, EPA625/1-79-011, U.S. Environmental Protection Agency, MERL, Cincinnati, OH (1979).
29. GLUMBRA, Recommended standards for sewage works, Great Lakes Upper Mississippi River Board of State Sanitary Engineers, Health Education Service, Albany, NY (2004).
30. DeVos, J., Total BIOGAS. Quality management, Presentation Applied Filter Technology, January 18, 2011.
31. Metcalf & Eddy, Inc., *Wastewater Engineering: Treatment and Reuse*, (Fourth edition), McGraw-Hill, New York (2003).
32. Drzymala, I.E., Titerle, D., and Slack, W., Sustainability in biosolids treatment: San Antonio water system's commercial biogas solution. ASCE Webinar, Presentation October 19, 2010 (2014).
33. NETL, *Plasma Gasification*, National Energy Technology Laboratory, U.S. Department of Energy, Pittsburgh, PA (2016).
34. Ciferno, J. and Marano, J., *Benchmarking Biomass Gasification Technologies for Fuels, Chemicals and Hydrogen Production*, Prepared by Jared Ciferno and John Marano, National Energy Technology Laboratory, U.S. Department of Energy, Pittsburgh, PA (2002).
35. Grant, S., Treatment tackling treatment technology, *Water & Waste Digest Magazine*, April 28, 2016.
36. Wongthanate, J., Chinnacotpong, K., and Khumpong, M., Impact of pH, temperature and pretreatment method on biohydrogen production from organic wastes by sewage microflora, *International Journal of Energy Environmental Engineering*, Vol. 5, article 76 (2014).
37. USEPA, Composting, waste-resource conservation-common wastes & materials-organic material, U.S. Environmental Protection Agency, Washington, DC (2012).
38. USCC, Best management practices (BMPs) for incorporating food residual into existing yard waste composting operations, The U.S. Composting Council, Ronkonkoma, New York (2009).
39. USEPA, Industrial uses, waste-resource conservation-common wastes & materials-organic material, U.S. Environmental Protection Agency, Washington, DC (2011).
40. USEPA, Food to Fuel: Pacific Biodiesel, Inc., EPA530-F-06-037, Waste-resource conservation-common wastes & materials-organic material, U.S. Environmental Protection Agency, Washington, DC (2006).
41. USEPA, Learn about biodiesel, U.S. Environmental Protection-Pacific Southwest, Region 9, San Francisco, CA (2004).
42. Canakci, M., The potential of restaurant waste lipids as biodiesel feedstocks, *Bioresource Technology*, Vol. 98, Issue 1, 183–190 (2007).
43. USEPA, Food waste management cost calculator, Version 1.0, U.S. Environmental Protection Agency, Office of Solid Waste, Washington, DC (2009).

44. Wang, L.K., Hung, Y.T., and Li, K.H. Vermicomposting process. In: *Biosolids Treatment Processes*, Wang, L.K., Shammas, N.K., and Hung, Y.T., (eds), Humana Press, Totowa, NJ, 689–704 (2007).
45. Shammas, N.K., and Wang, LK. Biosolids composting. In: *Biosolids Treatment Processes*, Wang, L.K., Shammas, N.K., and Hung, Y.T., (eds), Humana Press, Totowa, NJ, 645–688 (2007).
46. San Francisco, Restaurant guide waste reduction and recycling, San Francisco City and County, CA. www.calrecycle.ca.gov/publications/Documents/BizWaste%5C44198016.pdf. 1992. Updated 2016.
47. DNR-WI, Recycling and waste reduction in restaurant industry, State of Wisconsin, Department of Natural Resources, Madison, WI. http://dnr.wi.gov/files/pdf/pubs/wa/wa1536.pdf, April (2016).
48. GIWA, Discharges from cruise ships cause problems for the Caribbean Islands. Global International Waters Assessment (GIWA), UNEP, Kalmar, Sweden (2006).
49. Salisbury, S., Cruise ships produce huge amounts of waste and sewage, *Palm Beach Post*, December 15, 2012. www.palmbeachpost.com.
50. Food Cycle Science, Inc., Commercial on-site organic waste reduction and conversion—Cruise ships. Food Cycle Science, December 2014. https://nofoodwaste.com.
51. Editor, Green Ships+Green Ports=Sustainable Shipping. *Green Port Magazine*, Venice, Italy. September 23, 2011.
52. Wang, L.K., and Wang, M.H.S., Personal communications with the restaurant manager of cruise ship, *Brilliance of the Seas*, Royal Caribbean International, Miami, FL, May–June, 2016.
53. Wang, L.K. and Wang, M.H.S. Personal communications with the restaurant manager of cruise ship, Norwegian Spirit, Norwegian Cruise Line, Port Canaveral, FL. January 23–30, 2016.
54. Wang, L.K., Tay, J.H., Tay, S.T.L., and Hung, Y.T. (eds), Environmental Bioengineering, Humana Press, Totowa, NJ, 867 pp. (2010).
55. Wang, L.K., Hung, Y.T., Lo, H.H., Yapijakis, C. and Alberto Larz Ribas (translator) (eds), Tratamiento de los Residuos de la Industria del Procesado de Alimentos (Portuguese), CRC Press, Boca Raton, FL, 398 pp. (2008).
56. Shammas, N.K. and Wang, L.K. (authors). Luiz Claudio de Queiroz Faria and Marco Aurelio Chaves Ferro (editors and translators). Absteciment de Água e Remocão de Residuos. (Portuguese–Spanish), Wiley, Hoboken, NJ, USA and gen LTC. www. Grupogen.com.br. 751 pp. (2013).

7 Treatment of Textile Industry Waste

Siew-Teng Ong, and Sie-Tiong Ha
Universiti Tunku Abdul Rahman

Siew-Ling Lee
Universiti Teknologi Malaysia

Pei-Sin Keng
International Medical University

Yung-Tse Hung
Cleveland State University

Lawrence K. Wang
Lenox Institute of Water Technology, Krofta Engineering Corporation, and Zorex Corporation

CONTENTS

7.1	Textile Industry	240
	7.1.1 Overview of Textile Industry	240
7.2	Process Descriptions of Textile Industry	242
	7.2.1 Yarn Formation	243
	7.2.1.1 Natural Fibers	243
	7.2.1.2 Man-Made Fibers	244
	7.2.2 Fabric Formation	244
	7.2.3 Fabric Preparation	245
	7.2.4 Dyeing	245
	7.2.4.1 Types of Dyes	246
	7.2.5 Printing	246
	7.2.6 Finishing	246
	7.2.7 Fabrication	248
7.3	Subcategory Descriptions of Textile Industry	248
	7.3.1 Wool Scouring Subcategory	248
	7.3.2 Wool Finishing Subcategory	248
	7.3.3 Low Water Use Processing Subcategory	249
	7.3.4 Woven Fabric Finishing Subcategory	249
	7.3.5 Knit Fabric Finishing Subcategory	250
	7.3.6 Carpet Finishing Subcategory	250
	7.3.7 Stock and Yarn Finishing Subcategory	250
	7.3.8 Nonwoven Manufacturing Subcategory	251
	7.3.9 Felted Fabric Processing Subcategory	251

7.4	Wastewater Characterization of Textile Industry	251
	7.4.1 Wastewater from Fabric Formation	251
	7.4.2 Wastewater from Fabric Preparation	253
	7.4.3 Wastewater from Dyeing Operation	253
	7.4.4 Wastewater from Printing Operation	254
	7.4.5 Wastewater from Finishing Operation	255
7.5	Pollutant Removability of Textile Industry Wastewater	255
	7.5.1 Biological Treatment	257
	7.5.2 Chemical Coagulation	258
	7.5.3 Adsorption	260
	7.5.4 Oxidation Methods	263
	7.5.4.1 Hydrogen Peroxide (H_2O_2)	264
	7.5.4.2 Fenton's Reagent (H_2O_2–Fe(II) Salts)	265
	7.5.4.3 Ozonation	267
	7.5.4.4 Chlorine	269
	7.5.5 Combined Biological/Physical–Chemical Treatment	269
	7.5.6 Dyes Removal by Low-Cost Adsorbents	269
	7.5.6.1 Agricultural Waste	280
	7.5.6.2 Industrial By-Products	281
	7.5.6.3 Natural Materials	287
	7.5.7 Cost for Treatment of Textile Wastewaters	295
7.6	Pollution Prevention Opportunities in Textile Industry	298
	7.6.1 Quality Control for Raw Materials	298
	7.6.2 Chemical Substitution	303
	7.6.3 Process Modification	303
	7.6.4 Process Water Reuse and Recycle	304
	7.6.5 Equipment Modification	305
	7.6.6 Good Operating Practices	305
7.7	Textile Point Source Discharge Effluent Limitations, Performance Standard, and Pretreatment Standards	306
	7.7.1 U.S. Environmental Regulations for The Nine Textile Subcategories	306
7.8	Conclusion	307
List of Acronyms		308
References		308

Abstract

This chapter presents a brief overview of the textile industries, discusses the textile processes, and focuses on the wastewater characterization and treatment processes of this industry. The main treatment technologies and emerging techniques using various waste materials from agriculture and industry or naturally occurring biosorbents are highlighted. The discussion on pollution prevention serves as background to later sections.

7.1 TEXTILE INDUSTRY

7.1.1 OVERVIEW OF TEXTILE INDUSTRY

The textile industry is one of the oldest industries in the United States, dating back to the beginning of the American Industrial Revolution in the 1790s. In fact, the primary textile manufacturers that operate in the United States, such as textile knitting, weaving, nonwoven operations, textile dyeing/printing, and all related processes, have a long history in the timeline of civilization [1]. In the early days, this industry was primarily a family-run business. Flax and wool were the major fibers used; however, cotton, grown primarily on southern plantations, became increasingly important [2].

The twentieth century has seen the development of the first man-made fibers (rayon was first produced in 1910). Although natural fibers (wool, cotton, silk, and linen) are still used extensively today, they are more expensive and are often mixed with man-made fibers such as polyester, which is the most widely used synthetic fiber. In addition, segments of the textile industry have become highly automated and computerized [2].

The manufacturing of textiles is categorized by the Office of Management and Budget (OMB) under Standard Industrial Classification (SIC) code 22. The SIC system was established by OMB to track the flow of goods and services in the economy by assigning a numeric code to them. Table 7.1 lists SIC codes and their associated products.

In 1994, textile production was a USD 70 billion industry that employed nearly 670,000 textile workers and over 900,000 apparel workers. With more than 4000 companies (many of which are privately owned) and over 5000 plants making products with many different end-uses, ranging from apparel to air bags to space suits, this industry supplies the largest nondurable consumer product market in the United States. The textile industry also has recorded 2.2 million direct employees (12% of the U.S. workforce), with 700,000 of them employed in the primary textiles segment [4].

TABLE 7.1
Standard Industrial Classifications (SIC 22) within the Textile Industry

3-Digit SIC Code	4-Digit SIC Code
SIC 221—Broad-woven fabric mills: cotton	SIC 2211—Broad-woven fabric mills: cotton
SIC 222—Broad-woven fabric mills: man-made fiber and silk	SIC 2221—Broad-woven fabric mills: man-made fiber and silk
SIC 223—Broad-woven fabric mills: wool (including dyeing and finishing)	SIC 2231—Broad-woven fabric mills: wool (including dyeing and finishing)
SIC 224—Narrow fabric mills: cotton, wool, silk, and man-made fiber	SIC 2241—Narrow fabric mills: cotton, wool, silk, and man-made fiber
SIC 225—Knitting mills	SIC 2251—Women's full-length and knee-length hosiery, except socks
	SIC 2252—Hosiery, not elsewhere classified
	SIC 2253—Knitted outwear mills
	SIC 2254—Knitted underwear and nightwear mills
	SIC 2257—Weft knit fabric mills
	SIC 2258—Lace and warp knit fabric mills
	SIC 2259—Knitting mills, not elsewhere classified
SIC 226—Dyeing and finishing textiles, except wool fabrics and knitted goods	SIC 2261—Finishers of broad-woven fabrics of cotton
	SIC 2262—Finishers of broad-woven fabrics of man-made fiber and silk
	SIC 2269—Finishers of textiles, not elsewhere classified
SIC 227—Carpets and rugs	SIC 2273—Carpets and rugs
SIC 228—Yarn and thread mills	SIC 2281—Yarn spinning mills
	SIC 2282—Yarn texturizing, throwing, twisting, and winding mills
	SIC 2284—Thread mills
SIC 229—Miscellaneous textile goods	SIC 2295—Coated fabrics, not rubberized
	SIC 2296—Tire cord and fabrics
	SIC 2298—Cordage and twine
	SIC 2299—Textile goods, not elsewhere classified

Source: Executive Office of the President of the United States (1987). *Standard Industrial Classification Manual*, Office of Management and Budget, Washington, DC.

TABLE 7.2
Summary of Establishment Sizes within the Textile Industry

Industry SIC Code	Percentage of Establishments[a] with 0–19 Employees	Percentage of Establishments with 20–49 Employees	Percentage of Establishments with 50–99 Employees	Percentage of Establishments with 100 or More Employees
SIC 221	*64*	4	4	28
SIC 222	40	8	6	26
SIC 223	*45*	22	9	23
SIC 224	*49*	14	14	22
SIC 225	*44*	21	14	21
SIC 226	32	22	15	31
SIC 227	*53*	12	9	26
SIC 228	24	11	13	52
SIC 229	*58*	18	11	12

Source: United States Department of Commerce, Census of manufactures, industry series, weaving and floor covering mills, industries 2211, 2221, 2231, 2241, and 2273, Bureau of the Census, Washington, DC, 1995; United States Department of Commerce, Census of manufactures, industry series, knitting mills, industries 2251, 2252, 2253, 2254, 2257, 2258, and 2259, Bureau of the Census, Washington, DC, 1995; United States Department of Commerce, Census of manufactures, industry series, dyeing and finishing textiles, except wool fabrics and knit goods, industries 2261, 2262, and 2269, Bureau of the Census, Washington, DC, 1995; United States Department of Commerce, Census of manufactures, industry series, yarn and thread mills, industries 2281, 2282, and 2284, Bureau of the Census, Washington, DC, 1995; United States Department of Commerce, Census of manufactures, industry series, miscellaneous textile goods, industries 2295, 2296, 2297, 2298, and 2299, Bureau of the Census, Washington, DC, 1995.

Note: The values in italic highlight the large percentage of facilities that have fewer than 20 employees.

[a] An establishment is a physical location where manufacturing takes place. Manufacturing is defined as the mechanical or chemical transformation of substances or materials into new products.

Table 7.2 summarizes establishment sizes within the textile industry. A larger percentage of establishments have fewer than 20 employees, and these do not include yarn and thread mills (SIC 228). Establishments where man-made fiber and silk broad-woven fabric are manufactured (SIC 222) have 100 employees or more per establishment. A summary of these statistics is shown in Table 7.2.

The maquiladora textile and apparel industry is growing stronger every year, with plants now expanding further south into Mexico, beyond the United States–Mexico border. In 1996, about 345 maquiladoras, or about 16% of all maquiladora industries, were involved in some facet of the textile and apparel industry, including the fabrication of apparel, carpeting, automobile accessories, and household goods [10].

In many Asian countries, the textile industry is one of the main revenue-generating sectors. For instance, textile and clothes have been the major export items in Thailand since the mid-1980s. It is estimated that in 1995–1996, there were about 5000 textile mills in Thailand, of which 80% were in the small- and medium-scale industry category. The share of the textile sector to the overall gross domestic product (GDP) increased from about 148 billion Baht in 1991 to about 240 billion Baht in 1996 [11].

7.2 PROCESS DESCRIPTIONS OF TEXTILE INDUSTRY

Owing to the diversity of products and application of the range of textiles, the type of processing used is highly variable and depends on site-specific manufacturing practices, the type of fiber used, and the final physical and chemical properties that are desired. The basic textile manufacturing begins with the production or harvesting of raw fiber. After the raw natural or manufactured fibers have been shipped from the chemical plant or the farm, the textile mill facilities perform any one

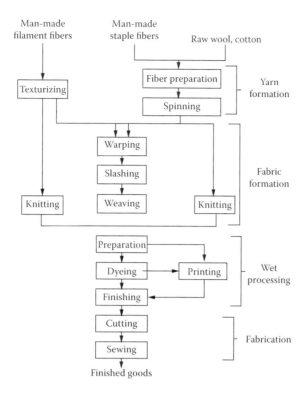

FIGURE 7.1 Typical textile processing flowchart. (From U.S. EPA, EPA office of compliance sector notebook project: Profile of the textile industry, EPA/310-R-97-009. U.S. Environmental Protection Agency, Washington, DC, September, 136 p., 1997.)

of the following operations: (1) yarn formation, (2) fabric formation, (3) fabric preparation, (4) dyeing, (5) printing, (6) final finishing, and (7) product fabrication. These stages are highlighted in the process flowchart shown in Figure 7.1.

7.2.1 Yarn Formation

Textile fibers used in yarn formation can be obtained (1) from natural sources such as wool and cotton, (2) from regenerative cellulosic materials such as rayon and acetate, or (3) from synthetic materials such as polyester and nylon. Although most textile fibers are processed using spinning operations, the processes leading to spinning vary depending on the source of fibers.

7.2.1.1 Natural Fibers

Natural fibers, which predominantly include animal and plant fibers such as cotton and wool, must go through a series of preparation steps before they can be spun into yarn. Although equipment used for cotton is designed somewhat differently from that used for wool, the machinery operates in essentially the same fashion. The following are the main steps used for processing wool and cotton:

1. *Cleaning*: To remove natural fiber contaminants such as natural waxes and oils, metals, agricultural residues, and lubricant residues caused from harvesting and processing.
2. *Opening/blending*: To sort, clean, and blend the fibers. Sorting and cleaning are performed in machines known as openers to remove particles of dirt, twigs, and leaves. Blending the fibers from different bales improves the consistency of the fibers.
3. *Carding*: To separate and remove shorter fibers, which would weaken the yarn. In addition, carding also aligns the fibers into thin and parallel sheets to prepare them for spinning.

4. *Combing*: Similar to carding except that the brushes and needles used are finer and more closely spaced. Worsted wool and combed cotton yarns are finer (smaller) than yarn that has not been combed because of the higher degree of fiber alignment and further removal of short fibers.
5. *Drawing*: To increase fiber alignment. Several slivers are combined into a continuous, ropelike strand and fed to a machine known as a drawing frame, which contains several sets of rollers. As the slivers pass through the drawing frame, they are further drawn out and lengthened from their original size. During drawing, slivers from different types of fibers (e.g., cotton and polyester) also may be combined to form blends.
6. *Drafting*: To stretch the yarn further using a frame. This process also imparts a slight twist as it removes the yarn and winds it onto a rotating spindle. The yarn, now termed a roving in ring spinning operations, is made up of a loose assemblage of fibers drawn into a single strand. Following drafting, the rovings may be blended with other fibers before being processed into woven, knitted, or nonwoven textiles.
7. *Spinning*: To spin the fibers together into either spun yarns or filament yarns. Spun yarns are composed of overlapping staple length fibers that are bound together by twisting, whereas filament yarns are made from continuous fine strands of man-made fiber. The most common spinning system in use today is ring spinning.

7.2.1.2 Man-Made Fibers

Man-made fibers include cellulosic fibers such as rayon and acetate (created by reacting chemicals with wood pulp), and synthetic fibers such as polyester and nylon (synthesized from organic chemicals). Unlike cotton and wool, the yarn formation of man-made fibers does not involve extensive cleaning and combing procedures as they are synthesized from organic chemicals. Man-made fibers, both synthetic and cellulosic, are manufactured using spinning processes, where fibers are formed by forcing a liquid through a small opening beyond which the extruded liquid solidifies to form a continuous filament.

Following spinning, the man-made fibers are drawn, or stretched, to align the polymer molecules and strengthen the filament. Man-made filaments often require additional drawing and are processed in an integrated drawing/twisting machine. Methods for making spun yarn from man-made fibers are similar to those used for natural fibers. Man-made filaments are typically texturized using mechanical or chemical treatments, to curl or crimp straight rodlike filament fibers and simulate the appearance, structure, and feel of natural fibers.

7.2.2 Fabric Formation

Textile fabrics are formed mainly by weaving and knitting processes. Weaving mills classified as broad-woven mills consume the largest portion of textile fiber and produce the raw textile material from which most textile products are made. Knitting, the second most frequently used method of fabric construction, is used largely in the hosiery and sock markets and in the manufacture of underwear, lingerie, and outerwear such as knit sport shirts. The popularity of knitting has increased due to the increased versatility of techniques, the adaptability of man-made fibers, and the growth in consumer demand for wrinkle-resistant, stretchable, snug-fitting fabrics.

Before the actual weaving operation, the yarn must be prepared by one or more of the following operations: winding, spooling, warping, and/or sizing/slashing. These operations are incorporated to transfer the yarn from the type of package that resulted from the spinning or texturizing operation onto a type that is suitable for fabric manufacturing. Slashing, or applying size to the warp yarn, will consolidate the proper number of yarns onto the warp beam for the fabric construction, form a coating that protects the yarn against snagging or abrasion during weaving, and dry the yarn in preparation for the weaving operation. As for the preparation of knitting, the yarn is lubricated to increase the speed and ease of the knitting operation. This step may be accomplished in the knitting mill or during yarn formation (spinning or texturizing).

In the United States, starch is the most common primary sizing agent used, with a consumption of 130 million pounds/year. As for fully synthetic products, polyvinyl alcohol (PVA) has been identified as the leading synthetic size, with an estimated consumption of 70 million pounds/year. PVA is gaining in popularity because it can be recycled. It is used with polyester/cotton yarns and pure cotton yarns either in a pure form or in blends with natural and other synthetic sizes. Other synthetic sizes include acrylic and acrylic copolymer components. Semisynthetic products such as carboxymethyl cellulose (CMC), modified starches, and starch esters are also commonly used sizes. Oils, waxes, and other additives are often used in combination with sizing agents to increase the softness and pliability of the yarns.

7.2.3 Fabric Preparation

In fabric preparation/pretreatment, natural impurities or processing chemicals that may interfere with dyeing, printing, and finishing are removed. This process cleans the fabric and increases its absorbency and whiteness, which ensures better dye uniformity and color fastness. Typical preliminary treatments include singeing, desizing, scouring, and bleaching. However, some of the mentioned processes may be omitted, and the order of the processes may also vary, depending on the requirement of the end product.

1. *Singeing*: A dry and continuous process used on woven goods to remove fibers protruding from yarns or fabrics. This is usually achieved by passing the fibers over a flame or heated copper plates.
2. *Desizing*: A process to remove sizing agents applied to warp yarns to protect against chafing or breakage during weaving. For natural fibers such as cotton, which are often sized with water-insoluble starches, enzymes are used to break these starches into water-soluble sugars. These are then removed by washing prior to scouring. As for man-made fibers, these are generally sized with water-soluble sizes that are easily removed by a hot-water wash or in the scouring process.
3. *Scouring*: A process that involves the usage of a hot alkaline solution (however, in some cases, solvent solutions may also be used) to remove impurities and handling contaminants that are present in the fiber. These impurities and contaminants may include lubricants, dirt, other natural materials, water-soluble sizes, antistatic agents, and fugitive tints.
4. *Bleaching*: A chemical process to decolorize colored impurities from fibers, yarns, or cloth that are not removed during scouring and prepare the cloth for further finishing processes, such as dyeing or printing. About 95% of bleaching operations use hydrogen peroxide and about 5% use calcium hypochlorite [4].
5. *Mercerizing*: A continuous chemical process used for cotton and cotton/polyester goods to increase dyeability, luster, and appearance. This process causes the flat, twisted ribbonlike cotton fiber to swell into a round shape and to contract in length. In comparison with the original fiber, the fiber after this process becomes more lustrous, has better fiber strength, and possesses higher affinity for dyes.

7.2.4 Dyeing

Dyes are applied at various stages of production to add color to textiles (either by the finishing division of vertically integrated textiles companies or by specialty dyehouse). Dyeing operations can be performed using batch or continuous processes. Both techniques involve dye application, dye fixation with heat and/or auxiliary chemicals, and washing. However, for long runs of a particular fabric color, continuous dyeing tends to be more efficient than batch processes, which are generally the only economical method for small runs with many color changes.

The batch dyeing process consists of a certain amount of textile substrate, usually 100–1000 kg, loaded into a dyeing machine and brought to equilibrium, or near equilibrium, with a solution containing the dye. Because of high affinity for the fibers, the dye molecules leave the dye solution and enter the fibers over a period of time, depending on the type of dye and fabric used. The auxiliary chemicals used and the controlled operational conditions (e.g., dyebath's temperature) will accelerate and optimize the process of dye fixation. After this, the colored textile substrate is washed to remove unfixed dyes and chemicals.

In continuous dyeing processes, textiles are fed continuously into a dye range at speeds usually between 50 and 250 m/min. This technique accounts for about 60% of the total yardage of product dyed in the industry [2]. To formulate cost-effective operating conditions, the dyer may have to process 10,000 m of textiles or more per color, although specialty ranges are now being designed to run as little as 2000 m economically. Most woven fabric is colored by pad dyeing, a method in which the fabric is continuously dyed using pad/squeeze type equipment, whereby the dye is transferred to the fabric by passing the fabric across rollers that are partially submerged in the dye solution.

7.2.4.1 Types of Dyes

Dye molecules are made of two major components, namely, chromophore and auxochrome. The chromophores are responsible for producing the color, while the auxochromes render the molecule soluble in water and give enhanced affinity for the fibers. Dyes exhibit considerable structural diversity and can be classified both by their chemical structure (e.g., azo, phthalocyanine, and anthraquinone dyes) and by their application to the fiber type (e.g., reactive and disperse dyes). The former approach is adopted by practicing dye chemists, whereas the latter is used predominantly by the dye users and dye technologists. However, often, both terminologies are used, for example, an azo disperse dye for polyester and a phthalocyanine reactive dye for cotton.

The advantage of using chemical classification is that it readily identifies dyes as belonging to a particular group and having characteristic properties, for example, azo dyes (strong, good all-round properties, cost-effective) and anthraquinone dyes (weak, expensive). As for usage classification (the principal type of classification adopted is the Colour Index, C.I.), there is no need to consider in detail dyes' chemical structures, which are very complex. Dyes may also be classified on the basis of their solubility: soluble dyes include acid, mordant, metal complex, and direct, basic, and reactive dyes; insoluble dyes include azoic, sulfur, vat, and disperse dyes. The classification of dyes according to their applications, principal substrates, and the representative chemical types are summarized in Table 7.3.

The most important types of dyes belong to the category that is used for dyeing cotton and polyester because they are the main textile fibers used in the industry. For cotton, the most commonly used dyes are reactive and direct dyes. Reactive dyes react with fiber molecules to form chemical bonds, whereas direct dyes color the fabric directly in one operation and without the aid of an affixing agent. Disperse dyes, which are predominantly used on polyester, are substantially water-insoluble nonionic dyes. They are dispersed in water, where the dyes are dissolved into fibers. Vat dyes, such as indigo, are also commonly used for cotton and other cellulosic fibers.

7.2.5 PRINTING

In printing, color is deposited and fixed on the fabric by using a variety of techniques such as steam, heat, or chemical treatment. Commercial printing methods include pigment printing, wet printing, discharge printing, and carpet printing. Compared with dyes, pigments are more commonly used in printing operations, as they do not require washing and generate little waste. But, as pigments are typically insoluble and have no affinity for the fibers, resin binders are used to attach the pigments to substrates. Solvents are applied as transporting agents to move the pigment and resin mixture to the substrate. Print pastes are also formulated to ensure proper flow properties during application (thixography).

7.2.6 FINISHING

Most fabrics undergo one or more final finishing processes through chemical or mechanical treatments to enhance their properties such as appearance, durability, texture, or performance.

TABLE 7.3
Classification of Dyes Based on Usages, Principal Substrates, and Chemical Types

Dye Class	Principal Substrates	Type of Chemical Compound
Acid	Nylon, wool, silk, paper, leathers	Azo (including premetallized), anthraquinone, triphenylmethane, azine, xanthene, nitro, nitroso
Basic	Paper, polyacrylonitrile, modified nylon, polyester	Cyanine, hemicyanine, diazahemicyanine, diphenylmethane, triarylmethane, azo, azine, xanthene, acridine, oxazine, anthraquinone
Direct	Cotton, rayon, paper, leather, nylon	Azo, phthalocyanine, stilbene, oxazine
Disperse	Polyester, polyamide, acetate, acrylic, plastics	Azo, anthraquinone, styryl, nitro, benzodifuranone
Reactive	Cotton, wool, silk, nylon	Azo, anthraquinone, phthalocyanine, formazan, oxazine
Solvent	Plastics, gasoline, varnishes, lacquers, stains, inks, fats, oils, and waxes	Azo, triphenylmethane, anthraquinone, phthalocyanine
Sulfur	Cotton, rayon	Undefined structures
Vat	Cotton, rayon, wool	Anthraquinone (including polycyclic quinones), indigoids
Oxidation bases	Hair, fur, and cotton	Aniline black, undefined structures
Fluorescent brighteners	Soaps and detergents, all fibers, oils, paints, and plastics	Stilbene, pyrazoles, coumarin, naphthalimides
Food, drug, and cosmetics	Food, drug, cosmetics	Azo, anthraquinone, carotenoid, triarylmethane
Mordant	Wool, leather, anodized aluminum	Azo, anthraquinone

Mechanical finishes can be used to increase the luster and feel of textiles through the following steps:

1. *Heatsetting*: To stabilize and impart textural properties to synthetic fabrics and to fabrics containing high concentrations of synthetics. This process can also give cloth resistance to wrinkling during wear and ease-of-care properties.
2. *Brushing*: To reduce the luster of fabrics by roughening or raising the fiber surface and change the feel or texture of the fabric.
3. *Softening*: To soften and increase the fabric's sheen by reducing the surface roughness between individual fibers.
4. *Shearing*: To remove surface fibers by passing the fabric over a cutting blade.
5. *Compacting*: To reduce residual shrinkage of fabrics after repeated laundering.

Chemical finishes to textiles can impart a variety of properties ranging from decreasing static cling to increasing flame resistance. Among all, the most popular chemical finishes are those that involve the ease-of-care properties, such as the permanent-press, soil-release, and stain-resistant finishes. The chemical treatment is usually followed by drying, curing, and cooling steps. The application of chemical finishes is often done in conjunction with mechanical finishing steps. Some of the common chemical finishes are as follows:

1. *Optical finishes*: To either brighten or deluster the fabric.
2. *Absorbent and soil release finishes*: To alter surface tension and other properties to increase water absorbency or improve soil release.
3. *Softeners and abrasion-resistant finishes*: To improve feel or to increase the textile's ability to resist abrasion and tearing.
4. *Physical stabilization and crease-resistant finishes*: To stabilize cellulosic fibers to laundering and shrinkage by imparting permanent press properties to fabrics.

7.2.7 Fabrication

The final step in the textile industry involves fabricating finished cloth into a wide variety of apparel, household, and industrial products. Simpler products such as bags, sheets, towels, pillowcases, blankets, and draperies can be produced by the textile mills themselves. However, more complex housewares and apparel are made by the cutting trade. Factors such as pattern layout efficiency, level of expertise of cutting/sewing operators, and information flow will have an influential effect on the amount of waste generated in product fabrication operations.

7.3 SUBCATEGORY DESCRIPTIONS OF TEXTILE INDUSTRY

Based on the raw materials used, products manufactured, production processes employed, mill size and age, waste treatability, location, climate, and treatment costs, the textile industry can be classified into the following nine subcategories:

1. Wool scouring
2. Wool finishing
3. Low water use processing
4. Woven fabric finishing
5. Knit fabric finishing
6. Carpet finishing
7. Stock and yarn finishing
8. Nonwoven manufacturing
9. Felted fabric processing

7.3.1 Wool Scouring Subcategory

Wool scouring subcategory refers to facilities where natural impurities are scoured from raw wool and other animal hair fibers. Since the raw wool and animal hair fibers contain grease along with oils such as lanolin, as well as dirt, dead skin, sweat residue, and vegetable matter, these materials must be cleaned thoroughly before they can be converted into textile products. In the textile industry, scouring is always carried out by using detergents and alkalis. Besides, the cleaning process usually takes a long time as wool is thick. Consequently, wastewaters generated contain considerably higher pollutant concentrations than those of other subcategories [12,13].

At integrated mills with both wool scouring and other finishing operations, wool scouring effluent limitations to the wool scouring production are applied to determine discharge allowances. Alternatively, limitations associated with other finishing operations to the production related to each finishing operation can be used to verify the discharge allowances [12,13].

7.3.2 Wool Finishing Subcategory

Facilities where fabric is finished are classified under this subcategory. Wool, other animal hair fibers, or blends containing primarily wool or other animal hair fibers appear as major components of the facilities. Processes such as carbonizing, fulling, bleaching, scouring (except raw wool scouring), dyeing, rinsing, and application of functional finishing chemicals are employed along with wool finishing. Unlike other finishing categories, wool finishing requires a wide variety of chemicals used to process wool fabrics. As a result, high raw waste loads containing toxic pollutants such as phenols and chromium are generated [12,13].

This subcategory also includes mills where stock or yarn consisting primarily wool, other animal hair fibers, or blends containing primarily wool or other animal hair fibers is finished as well as where the carbonizing process is performed. In contrast, those mills where carbonizing is not employed are included in the stock and yarn finishing subcategory.

Discharge allowances of integrated mills where both wool finishing and other textile operations are carried out should be determined by applying wool finishing effluent limitations to the wool finishing production. It can also be determined by applying limitations associated with other operations to the production corresponding to each operation [12,13].

7.3.3 Low Water Use Processing Subcategory

Products and processes that by themselves do not produce large effluent volumes are considered as low water use processing operations. These operations cover the manufacture of greige goods such as yarn, woven fabric ,and knit fabric; laminating or coating fabrics; texturizing yarn; tufting and backing carpet; tire cord; and fabric dipping. The washing and cleaning of equipment are usually the principal source of effluents from such processes [12,13].

There are two subdivisions under the low water use processing subcategory, namely:

1. General processing that covers all low water use processes, excluding water jet weaving.
2. Water jet weaving that includes the manufacture of woven greige goods using the water jet weaving process. In fact, water jet weaving is not technically a low water use process. Even though the wastewater discharge rate is considerably higher for water jet weaving than that for other low water use processes, the low strength of the wastewater results in low pollutant mass discharge rate.

7.3.4 Woven Fabric Finishing Subcategory

Woven fabric finishing subcategory refers to facilities where primarily woven fabric is finished. This subcategory can be divided into two groups: the first group involves impurities removal, cleaning, and modification of the cloth using the processing operations of desizing, scouring, bleaching, and mercerizing; the second group includes dyeing, printing, resin treatment, and application of functional finishing chemicals. Apparently, these processes generate various effluents containing dyes, chemicals, and associated additives used in woven fabric finishing. This has led to a complicated problem since the presence of these pollutants poses unacceptable chronic and acute health risks to human as well as to aquatic life. There are 600 mills estimated to fall into this subcategory. Among these mills, only 20% treat their own waste. Meanwhile, 75% discharge to municipal systems and 5% have no waste treatment [12,13].

This subcategory also covers integrated mills where primarily woven fabric is finished along with greige manufacturing or other finishing operations, including denim finishing and yarn dyeing. However, woven fabrics composed primarily of wool, other animal hair fibers, or blends containing primarily wool or other animal hair fibers are included in the wool finishing subcategory.

In contrast, weaving is also carried out at many finishing facilities. As compared to finishing wastes, the added hydraulic and pollutant loadings from slasher equipment used for cleanup are not significant. To develop new source performance standards (NSPS), three subdivisions have been identified to account for higher raw waste loads associated with more complex operations and desizing:

1. Simple manufacturing operations that include facilities involving desizing, dyeing, or other fiber preparation processes.
2. Complex manufacturing operations that refer to facilities where the simple unit processes such as desizing, dyeing, and fiber preparation are performed as well as other manufacturing operations including printing, waterproofing, or application of stain resistance or other functional fabric finishes.

3. Desizing that refers to facilities where more than 50% of total production is desized. Other processes such as fiber preparation, scouring, mercerizing, functional finishing, bleaching, dyeing, and printing are employed at these facilities.

The discharge allowances of integrated mills where both woven fabric finishing and other textile operations are executed, ought to be determined by applying woven fabric finishing effluent limitations to the woven fabric finishing production and by applying limitations associated with other operations to the production corresponding to each operation [12,13].

7.3.5 KNIT FABRIC FINISHING SUBCATEGORY

The knitting industry is recognized to involve as a large number of plants and specialized product segments, where the major segments include knit fabric goods, hosiery, outerwear, and underwear. Since both sizing/desizing and mercerizing operations are not required for knits, the knit fabric finishing subcategory is different from the woven fabric finishing subcategory. The raw waste loads generated are also dissimilar for all parameters as compared with woven fabric finishing, as these fabrics contain lubricants and anti-static agents [12,13].

In developing NSPS, three subdivisions have been recognized to explain higher raw waste loads associated with more complex operations and to account for hosiery production. They are as follows:

1. Simple manufacturing operations that include facilities involving fiber preparation and dyeing.
2. Complex manufacturing operations that include facilities involving simple unit processes such as fiber preparation and dyeing as well as manufacturing operations including printing, waterproofing, or application of stain resistance or other functional fabric finishes.
3. Hosiery products that cover facilities where any type of hosiery is dyed or finished. Hosiery finishing mills are generally much smaller in terms of wet production as compared with other knit fabric finishing facilities. Besides, batch processing is more frequently employed and always performs only one major wet processing operation (dyeing) in the hosiery finishing mills. Therefore, lower raw waste loadings result from hosiery production [12,13].

7.3.6 CARPET FINISHING SUBCATEGORY

Carpet finishing subcategory covers facilities where textile-based floor covering products, of which carpet is the primary element, are finished. Usually, processing operations such as scouring, carbonizing, fulling, bleaching, dyeing, printing, and application of functional finishing chemicals are performed [12,13].

This subcategory also includes integrated mills where primarily carpet is finished along with tufting or backing operations or other finishing operations (such as yarn dyeing). In contrast, mills involving only carpet tufting and/or backing are included in the general processing subdivision under the low water use processing subcategory [12,13].

7.3.7 STOCK AND YARN FINISHING SUBCATEGORY

Stock and yarn finishing subcategory includes facilities where stock, yarn or cotton, and/or synthetic fiber thread are finished. Processing operations including scouring, bleaching, mercerizing, dyeing, or application of functional finishing chemicals are performed. Therefore, this subcategory is distinguished from the woven fabric finishing subcategory as no desizing operation is required [12,13].

This subcategory also includes facilities where stock or yarn consisting principally of wool, other animal hair fibers (or blends containing primarily wool or other animal hair fibers) is finished if carbonizing is not employed.

7.3.8 NONWOVEN MANUFACTURING SUBCATEGORY

Facilities where nonwoven textile products of wool, cotton, or synthetics, singly or as blends, are manufactured by mechanical, thermal, and/or adhesive bonding procedures are included in the nonwoven manufacturing subcategory. However, nonwoven products manufactured by fulling and felting processes are included in the felted fabric processing subcategory, which is discussed in the next section.

This subcategory also covers a variety of processing methods and products. Some typical processing operations such as carding, web formation, wetting, bonding (e.g., padding or dipping with latex acrylic or polyvinyl acetate resins), and application of functional finishing chemicals are performed. The processing is dry as mechanical and thermal bonding techniques are involved. Besides, low water use for adhesive bonding may be employed, and it greatly influences process-related waste characteristics resulting from the cleanup of bonding mix tanks and application equipment. In some cases, pigments for coloring the goods are added to the bonding materials. Therefore, this subcategory produces wastewaters that are similar to those discharged from mills in the carpet finishing subcategory [12,13].

7.3.9 FELTED FABRIC PROCESSING SUBCATEGORY

This subcategory includes facilities where primarily nonwoven products are manufactured by employing fulling and felting operations as a means of achieving fiber bonding. Typically, felt is a nonwoven cloth that is produced from wool, rayon, and blends of wool, rayon, and polyester by matting, condensing, and pressing woollen fibers. Therefore, the felting operation is accomplished by subjecting the web or mat to moisture, chemicals (detergents), and mechanical action. In this subcategory, wastewaters are always generated during rinsing steps, which are required to prevent rancidity and spoilage of the fibers. The produced wastewaters are similar to those discharged from mills in the wool finishing subcategory [12,13].

7.4 WASTEWATER CHARACTERIZATION OF TEXTILE INDUSTRY

The textile industry is not only a water-, chemical-, and energy-dependent industry but also its consumption of these three sources is highly intensive. Within the industry, the majority usage of water, chemicals, and energy is intended for wet processing. This processing step involves treatment with chemical baths. As a result, washing, rinsing, and drying are quite common between key treatment steps. Thus, the huge amount of wastewater generated with a wide and diverse range of contaminants must be treated prior to disposal. Table 7.4 lists the release of typical air, water, and solid pollutants associated with various textile manufacturing processes.

Table 7.5 shows characteristics and contents of wastewaters from textile manufacturing processes.

7.4.1 WASTEWATER FROM FABRIC FORMATION

The size solution serves only a transitory need for weaving, resulting in the use and disposal of huge quantities of warp size by the textile manufacturing industry each year. In the United States, the consumption is estimated at about 200 million pounds/year [4]. Starch size can contribute up to 50% of the total BOD loading from processing of woven fabrics. A major problem associated with starch size is that it cannot be reused or recycled because the starch degrades to various sugars during desizing.

TABLE 7.4
Summary of Potential Releases Emitted during Textiles Manufacturing

Process	Air Emissions	Wastewater	Residual Wastes
Fiber preparation	Little or no air emissions generated	Little or no wastewater generated	Fiber waste, packaging waste, and hard waste
Yarn spinning	Little or no air emissions generated	Little or no wastewater generated	Packaging wastes, sized, yarn, fiber waste, cleaning and processing waste
Slashing/sizing	VOCs	BOD, COD, metals, cleaning waste, size	Fiber lint, yarn waste, packaging waste, unused starch-based sizes
Weaving	Little or no air emissions generated	Little or no wastewater generated	Packaging waste, yarn and fabric scraps, off spec fabric, used oil
Knitting	Little or no air emissions generated	Little or no wastewater generated	Packaging waste, yarn and fabric scraps, off spec fabric
Tufting	Little or no air emissions generated	Little or no wastewater generated	Packaging waste, yarn and fabric scraps, off spec fabric
Desizing	VOCs from glycol ethers	BOD from water-soluble sizes, synthetic size, lubricants, biocides, antistatic compounds	Packaging waste, fiber lint, yarn waste, cleaning materials such as wipes, rags, and filters, cleaning and maintaining wastes containing solvents
Scouring	VOCs from glycol ethers and scouring solvents	Disinfectants and insecticide residues, NAOH, detergents, fats, oils, pectin, wax, knitting lubricants, spin finishes, spent solvents	Little or no residual waste generated
Bleaching	Little or no air emissions generated	Hydrogen peroxide, sodium silicate, or organic stabilizer, high pH	Little or no residual waste generated
Singeing	Small amounts of exhaust gases from the burners	Little or no wastewater generated	Little or no residual waste generated
Mercerizing	Little or no air emissions generated	High pH, NaOH	Little or no residual waste generated
Heatsetting	Volatilization of spin finishing agents applied during synthetic fiber manufacture	Little or no wastewater generated	Little or no residual waste generated
Dyeing	VOCs	Metals, salt surfactants, toxics, organic processing assistants, cationic materials, color, BOD, COD, sulfide, acidity/alkalinity, spent solvents	Little or no residual waste generated
Printing	Solvents, acetic acid from drying and curing oven emissions, combustion gases, particulate matter	Suspended solids, urea, solvents, color, metals, heat, BOD, foam	Little or no residual waste generated
Finishing	VOCs, contaminants in purchased chemicals, formaldehyde vapors, combustion gases, particulate matter	BOD, COD, suspended solids, toxics, spent solvents	Fabric scraps and trimming, packaging waste
Product fabrication	Little or no air emissions generated	Little or no wastewater generated	Fabric scraps

Source: U.S. EPA, EPA office of compliance sector notebook project: Profile of the textile industry, EPA/310-R-97-009, U.S. Environmental Protection Agency, Washington, DC, September, 136 p., 1997.

TABLE 7.5
Contents and Characteristics of Wastewaters According to Textile Process

Parameters	\multicolumn{7}{c}{Process Categories}
	1	2	3	4	5	6	7
BOD_5/COD	0.2	0.29	0.35	0.54	0.35	0.3	0.31
BOD_5 (mg/L)	6000	300	350	650	350	300	250
TSS (mg/L)	8000	130	200	300	300	120	75
COD (mg/L)	30000	1040	1000	1200	1000	1000	800
Oil and grease (mg/L)	5500	—	—	14	53	—	—
Total chrome (mg/L)	0.05	4	0.014	0.04	0.05	0.42	0.27
Phenol (mg/L)	1.5	0.5	—	0.04	0.24	0.13	0.12
Sulfide (mg/L)	0.2	0.1	8.0	3.0	0.2	0.14	0.09
Color (ADMI)	2000	1000	—	325	400	600	600
pH	8.0	7.0	10	10	8.0	8.0	11
Temp. (°C)	28	62	21	37	39	20	38
Water usage (L/kg)	36	33	13	113	150	69	150

Source: U.S. EPA, Textile processing industry, EPA-625/778-002, U.S. Environmental Protection Agency, Washington, DC, October, 502 p., 1978.

Note: Categories description: 1, raw wool scouring; 2, yarn and fabric manufacturing; 3, wool finishing; 4, woven fabric finishing; 5, knitted fabric finishing; 6, carpet manufacturing; 7, stock and yarn dyeing and finishing.

7.4.2 Wastewater from Fabric Preparation

The process of removing size chemicals from textiles (desizing) is one of the industry's largest sources of wastewater pollutants. In this process, large quantities of size used in the weaving processes (more than 90%) are typically disposed of in the effluent stream. The remaining (10%) are recycled. Desizing processes often contribute up to 50% of the BOD load in wastewater [2]. A typical BOD load from preparation processes is shown in Table 7.6. Other pollution issues related to this operation are high pH, spent solvents, lubricants, and oils.

7.4.3 Wastewater from Dyeing Operation

A considerable amount of colored wastewater is generated from dyeing operations; the spent dyebath has been identified as the main source for this. The average wastewater generation from a dyeing facility is estimated at between one and two million gallons per day. Disperse dyeing generates about 12 to 17 gallons of wastewater per pound of product. Similar processes for reactive and direct dyeing generate even more wastewater, about 15 to 20 gallons per pound of product [2].

Wastewater from dyeing operations typically contains by-products, residual dye, and auxiliary chemicals. In addition, the presence of additional pollutants such as salts, metals, cleaning solvents, and oxalic acid has also been detected. Of the 700,000 tons of dyes produced annually worldwide, it is estimated that about 10%–15% of the dye is disposed of in effluents from dyeing operations. Salts are used mostly to achieve good dyebath exhaustion of anionic dyes, such as direct and fiber reactive dyes on cotton. Typical cotton batch dyeing operations use quantities of salt that range from 20% to 80% of the weight of goods dyed, and this leads to high salt concentration in the wastewater.

Metals found in textile mill effluents may be due to fiber, incoming water, dyes, plumbing, and chemical impurities. Often, metals are simply impurities generated during dye manufacture, though, in some cases, they form an integral part of the dye molecule. Tables 7.7 and 7.8 present the source of metals in textile effluents and metal concentrations in dyeing wastewater, respectively.

TABLE 7.6
Typical BOD Loads from Preparation Process

Process	Pounds of BOD Per 1000 Pounds of Production
Singeing	0
Desizing	
Starch	67
Starch, mixed size	20
PVA or CMC	0
Scouring	40–50
Bleaching	
Peroxide	3–4
Hypochlorite	8
Mercerizing	15
Heatsetting	0

Source: U.S. EPA, Best management practices for pollution prevention in the textile industry, EPA/625/R96-004, U.S. Environmental Protection Agency, Washington, DC, September, 320 p., 1996.

PVA, polyvinyl alcohol; CMC, carboxymethyl cellulose.

TABLE 7.7
Typical Sources of Metals in Effluent

Metals	Typical Sources
Arsenic	Fibers, incoming water, fugitive emissions from treated timber
Cadmium	Impurity in salt
Chrome	Dyes, laboratory
Cobalt	Dyes
Copper	Dyes, incoming water, fiber
Lead	Dyes, plumbing, shop
Manganese	Permanganate strip (for repairing mildew)
Mercury	Dye/commodity chemical impurity
Nickel	Dyes
Silver	Photo operations
Tin	Finishing chemicals, plumbing
Titanium	Fiber
Zinc	Dyes, impurity in commodity chemicals, incoming water, plumbing

Source: U.S. EPA, Best management practices for pollution prevention in the textile industry, EPA/625/R96-004, U.S. Environmental Protection Agency, Washington, DC, September, 320 p., 1996.

7.4.4 WASTEWATER FROM PRINTING OPERATION

Typical waste generated during various printing techniques comprises color residues, urea, solvents, metals, and substantial amounts of fats, oil, and grease. Metals are found in the wastewater effluents as a result of using metal-based reducing agents in discharge printing processes for polyester fabric.

TABLE 7.8
Concentration of Metals in Dyeing Wastewaters

		Metals (Concentration in mg/L)					
Dye Class	Type of Fiber	Cadmium	Chromium	Copper	Lead	Mercury	Zinc
Acid	Rayon	0.02	0.27	0.05	0.25	0.6	1.41
	Polyamide	0.02	0.08	1.43	0.21	0.38	1.39
	Wool	0.04	0.11	0.07	0.22	0.48	3.43
Acid premetallized	Polyamide	0.02	0.85	0.48	0.12	1.23	1.78
Basic	Wool	7.5	0.21	0.05	0.1	1.53	3.1
	Acrylic	0.03	0.03	0.09	0.12	0.39	1.06
	Polyester	0.05	0.05	0.05	0.26	0.43	0.46
Direct	Cotton	0.16	0.07	12.05	0.42	1.39	0.87
	Viscose	0.18	2.71	8.52	1.95	0.5	1.32
Direct after copperable	Cotton	0.21	0.07	11.61	0.6	0.79	1.02
Reactive	Cotton	0.2	0.12	0.23	0.54	0.62	0.65
Azoic	Cotton	0.02	0.05	0.06	0.16	1.12	2.02
Sulfur	Cotton	0.01	0.08	0.08	0.28	1.15	0.54
Vat	Cotton	0.05	0.07	0.37	0.42	2.2	0.83
Developed	Cotton	0.02	0.04	3.93	0.15	0.5	0.66
Disperse	Polyamide	0.02	0.03	0.04	0.08	0.27	1.06
	Polyester	0.05	0.1	0.16	0.18	0.99	1.53
	Triacetate	0.02	0.14	0.08	0.15	0.58	1.0
Direct + disperse + acid premetallized	Blend of cotton + acetate + rayon + acid dyeable rayon	0.05	0.26	0.05	0.1	0.31	1.04
	Cotton + polyester	0.05	0.04	1.83	0.2	0.79	0.46
Disperse + acid premetallized	Wool + polyester	0.02	1.03	0.4	0.27	0.5	1.54

Source: Horning, R.H., Textile dyeing wastewaters: Characterization and treatment, PB-285 115, U.S. Department of Commerce, National Technical Information Service, Washington, DC, 1978.

7.4.5 Wastewater from Finishing Operation

The presence of spent solvents (e.g., methyl ethyl ketone, toluene, and xylene) in this category of wastewater primarily arises from the solvent coating operation. The use of natural polymers such as starches, modified starches, alginates, and gums as builders can be relatively environmentally safe; however, they have a high BOD.

7.5 POLLUTANT REMOVABILITY OF TEXTILE INDUSTRY WASTEWATER

Effluents from textile industries contain suspended solids, large amount of dissolved solids, unreacted dyestuff, and other chemicals that are used in different stages of dyeing, fixing, washing, and other processing. Table 7.9 lists the concentrations of pollutants in untreated wastewater according to the nine textile subcategories.

The effluents from textile industries tend to contain dyes in sufficient quantities for them to be considered as objectionable type of pollutants, for both toxicological and esthetical reasons. Close study of the dyes has revealed that carcinogenicity is linked to specific types of dye intermediates or metabolites, such as benzidines. Whaley examined 1460 dyes and found that 55% of the known dyes were predicted to be hazardous [17]. Figure 7.2 shows a general overview of the predicted hazardous nature of dyes, and Table 7.10 presents a specific breakdown of dyes by color class.

TABLE 7.9
Untreated Wastewater Concentrations of Conventional and Nonconventional Pollutants

Subcategory	BOD$_5$ (mg/L)	COD (mg/L)	TSS (mg/L)	O & G (mg/L)	Sulfide (µg/L)	Total Phenols (µg/L)	Color (APHA Units)
1. Wool scouring	1830	6900	2740	580	#	#	#
2. Wool finishing	150	650	50	#	#	50	#
3. Low water use processing							
a. General processing	380	1060	220	#	#	#	#
b. Water jet weaving	120	180	25	#	#	#	#
4. Woven fabric finishing							
a. Simple processing	300	900	60	65	55	49	1000
b. Complex processing	350	1170	80	45	100	180	500
c. Desizing	405	1240	160	70	130	146	#
5. Knit fabric finishing							
a. Simple processing	205	765	60	95	55	108	390
b. Complex processing	260	835	50	50	155	107	760
c. Hosiery products	325	1300	80	100	560	62	450
6. Carpet finishing	440	1190	65	20	175	130	490
7. Stock and yarn finishing	190	685	40	25	245	172	570
8. Nonwoven manufacturing	175	2360	80	#	#	#	#
9. Felted fabric processing	205	555	115	30	#	575	#

Source: U.S. EPA, Development document for effluent limitations guidelines and standards for the textile mills point source category (final), EPA-440182022, U.S. Environmental Protection Agency, Washington, DC, June, 546 p., 1982.

Note: Historical data—Median values, #: Insufficient data to report value.

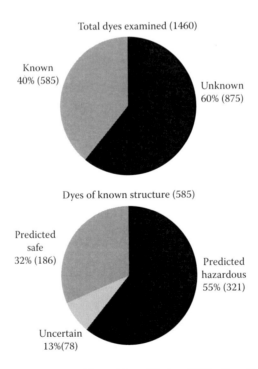

FIGURE 7.2 Predicted hazardous nature of dyes. (From Whaley, W.M., *Dyes Based on Safer Intermediates*, Fiber and Polymer Science Seminar Series, North Carolina State University, Raleigh, NC, 1984.)

TABLE 7.10
Predicted Hazardous Nature of Dyes for Year 1984

Colors	Total Known[a]	Safe (of Known)	Uncertain (of Known)	Hazardous (of Known)	Percent Hazardous (of Known)	Unknown
Yellows	87	39	16	32	37	190
Oranges	54	15	6	33	61	98
Reds	140	47	19	74	53	214
Violets	49	18	7	24	49	40
Blues	98	41	13	44	45	179
Greens	27	10	0	17	63	20
Browns	24	2	1	21	88	94
Blacks	35	9	8	18	51	36
Azoics	71	5	8	58	82	4
Total	585	186	78	321	55	875

Source: Whaley, W.M., *Dyes Based on Safer Intermediates*, Fiber and Polymer Science Seminar Series, North Carolina State University, Raleigh, NC, 1984.

Note: All dyes except vats, sulfurs, foods, and brighteners.

[a] Sixty percent of structures were undisclosed and thus could not be evaluated.

As far as textile wastewater treatment is concerned, color removal and reduction of BOD/COD are the main problems to be addressed in the primary treatment to make the effluent suitable for subsequent treatment.

7.5.1 BIOLOGICAL TREATMENT

Biological treatment involves the degradation or breakdown of organic matter by fungal, microbial culture, microbial biomass (living or dead), or biosorbents. This is one of the most common and widespread techniques used in dye wastewater treatment compared with other physical and chemical processes. Biological treatment is an attractive alternative because it is relatively inexpensive, and under suitable conditions, much of the soluble organic matter can be completely mineralized.

For aerobic digestion, bacteria and fungi are the two microorganism groups that have been intensively studied for their ability to decolorize dye wastewater. This process takes place under a good aeration environment, where the organisms use the organic matter and dissolved oxygen that are present in the wastewater to live and grow. Large amounts of biodegradable organic matter can be removed, leaving a liquid either substantially free from organic matter or containing resistant material that the organisms cannot break down or tolerate.

Anaerobic treatment has also demonstrated its potential in degrading a wide variety of synthetic dyes [18–20]. Anaerobic process occurs in two stages: (1) acidogenic bacteria convert organics (e.g., carbohydrates, fats, or proteins) into metabolites of low molecular weight such as alcohols and short-chain fatty acids; (2) organic acids are then fermented to methane and carbon dioxide. To have a satisfactory decolorization process when anaerobic digestion is being considered as a process option, a few aspects need to be factored in, which include solid retention time, toxicity, long-term performance, and waste sludge withdrawal.

Biological treatability of 20 colored wastes from textile carpet dyeing operations was investigated in laboratory-scale studies. Wastewaters no. 1–5 were industrial wastewaters from the Durham and Wastewaters no. 6–20 were municipal wastewaters from Chapel Hill. Figure 7.3 shows the experimental design of the biological treatment of the 20 dyeing wastewaters. Details of the dyeing wastewaters are listed in Table 7.11.

Tables 7.12 through 7.31 show efficiency of COD, BOD, and color removals using biological treatment from dyeing wastewaters no. 1–20.

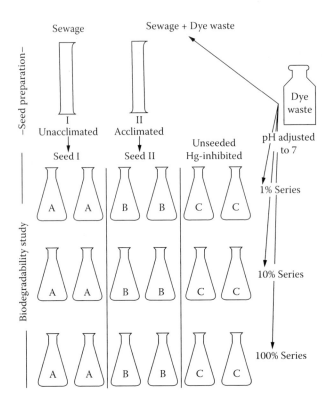

FIGURE 7.3 Experimental design for biological study of 20 dyeing wastewaters. (From Rai, H.S., et al., *Crit. Rev. Environ. Sci. Technol.*, 25, 219–238, 2005.)

Table 7.32 summarizes the effect of dyeing wastewaters on nitrification.

Besides biological methods, treatment processes such as coagulation or precipitation, carbon adsorption, and oxidation may also be used to treat wastewater from dyebaths.

7.5.2 Chemical Coagulation

Chemical coagulation is considered as an economically feasible method for dye removal. Prior to this treatment, a presedimentation step is required to remove high concentrations of settleable solids. The coagulation process is a rapid reaction that involves destabilization of colloidal particles. The particles are essentially coated with a chemically sticky layer that allows them to flocculate (agglomerate) and subsequently be removed either by sedimentation, flotation, or filtration. Typical coagulants are alum (aluminum sulfate) and iron salts (e.g., iron sulfate). The addition of ferrous sulfate and ferric chloride provides an excellent removal of direct dyes from wastewater, but not in the case of acid dyes. The optimum coagulant concentration to be used depends on the static charge of the dye in solution.

The coagulation process was performed on 20 samples from industrial and municipal wastewaters (Table 7.11), and the coagulants used were lime, aluminum salts, and ferric iron salts. The results of the chemical coagulation, adsorption, and ozonation treatment based on these 20 samples are presented in Table 7.33.

One of the limitations in this treatment is the difficulty in handling the sludge produced. The disposal of huge amounts of sludge not only creates an environmental problem, but also indirectly increases the cost of treatment. Table 7.34 lists the characteristics of sludge resulting from the coagulation process with specific doses of alum and iron sulfate coagulants.

TABLE 7.11
List of 20 Dyeing Wastewater Samples from Industrial and Municipal Wastewaters of Durham and Chapel Hill

Wastewater Sample	Dyeing System[a]
Wastewater no. 1	Vat dyes on cotton
Wastewater no. 2	1:2 metal complex dyes on polyamide
Wastewater no. 3	Disperse dyes on polyester
Wastewater no. 4	After-copperable direct dyes on cotton
Wastewater no. 5	Reactive dyes on cotton
Wastewater no. 6	Disperse dyes on polyamide carpet
Wastewater no. 7	Acid chrome dyes on wool
Wastewater no. 8	Basic dyes on polyacrylic
Wastewater no. 9	Disperse dyes on polyester carpet
Wastewater no. 10	Acid dyes on polyamide
Wastewater no. 11	Direct dyes on rayon
Wastewater no. 12	Direct developed dyes on rayon
Wastewater no. 13	Disperse, acid, and basic dyes on polyamide carpet
Wastewater no. 14	Disperse dyes on polyester
Wastewater no. 15	Sulfur dyes on cotton
Wastewater no. 16	Reactive dyes on cotton
Wastewater no. 17	Vat and disperse dyes on 50/50 cotton–polyester blend
Wastewater no. 18	Basic dyes on polyester
Wastewater no. 19	Disperse, acid, and basic dyes on polyamide carpet
Wastewater no. 20	Naphthol on cotton

Source: U.S. EPA, Textile dyeing wastewaters: Characterization and treatment, EPA-600/2-78-098, U.S. Environmental Protection Agency, Washington, DC, May, 310 p., 1978.

[a] Refer U.S. EPA [21] for complete dyeing procedures for wastewater no. 1–20.

TABLE 7.12
Biological Treatment on Dyeing Wastewater from Vat Dyes on Cotton (Wastewater No. 1)

	TOC mg/L Initial	Final	% Removal	BOD$_5$ mg/L Initial	Final	% Removal	Color, ADMI Units Initial	Final	% Removal
Set									
A-1	23	0	100	35	1	97	—	—	—
B-1	27	0	100	37	1	97	—	—	—
C-1	26	28	+8	—	—	—	—	—	—
A-10	50	9	82	50	1	98	270	86	68
B-10	45	7	84	50	<1	>98	252	38	85
C-10	48	18	62	—	—	—	186	238	+28
A-100	251	98	61	246	4	98	1886	1199	17
B-100	249	73	71	222	1	100	1080	937	13
C-100	266	179	33	—	—	—	1458	1528	+5

Source: U.S. EPA, Textile dyeing wastewaters: Characterization and treatment, EPA-600/2-78-098, U.S. Environmental Protection Agency, Washington, DC, May, 310 p., 1978.
Notes: Set A—dilutions of wastewater seeded with activated sludge which organism not acclimated to the specific dyeing wastewater. Diluted to 1, 10, 100% and labeled as A-1, A-10, A-100; Set B—dilutions seeded with microorganisms acclimated to the wastewater being tested; Set C—dilutions unseeded, nut with Hg(II) added to prevent biological activity.

TABLE 7.13
Biological Treatment on Dyeing Wastewater from 1:2 Metal Complex Dye on Polyamide (Wastewater No. 2)

	TOC			BOD$_5$			Color, ADMI Units		
	mg/L			mg/L					
Set	Initial	Final	% Removal	Initial	Final	% Removal	Initial	Final	% Removal
A-1	23	0	100	20	3	85	64	22	66
B-1	23	0	100	20	4	80	48	20	58
C-1	23	24	4	—	—	—	26	8	69
A-10	65	16	77	39	4	90	187	40	79
B-10	71	13	82	41	4	91	91	40	44
C-10	65	74	+14	—	—	—	22	43	
A-100	492	123	67	530	5	99	300	297	1
B-100	452	123	73	560	4	99	242	358	+48
C-100	472	522	+10	—	—	—	261	281	+8

Source: U.S. EPA, Textile dyeing wastewaters: Characterization and treatment, EPA-600/2-78-098, U.S. Environmental Protection Agency, Washington, DC, May, 310 p., 1978.

TABLE 7.14
Biological Treatment on Dyeing Wastewater from Disperse Dyes on Polyester (Wastewater No. 3)

	TOC			BOD$_5$			Color, ADMI Units		
	mg/L			mg/L					
Set	Initial	Final	% Removal	Initial	Final	% Removal	Initial	Final	% Removal
A-1	22	0	100	44	1	98	40	14	65
B-1	26	2	92	44	1	98	55	12	78
C-1	26	31	+19	—	—	—	37	22	41
A-10	49	16	67	76	1	99	94	52	45
B-10	49	16	67	72	1	99	103	40	61
C-10	45	65	+44	—	—	—	62	89	+44
A-100	276	163	41	288	<3	99	348	217	38
B-100	289	183	37	294	3	99	255	246	4
C-100	285	264	7	—	—	—	311	377	+21

Source: U.S. EPA, Textile dyeing wastewaters: Characterization and treatment, EPA-600/2-78-098, U.S. Environmental Protection Agency, Washington, DC, May, 310 p., 1978.

7.5.3 Adsorption

Adsorption is a process that involves the interphase accumulation or concentration of substances at a surface or interface because of unbalanced surfaces forces. The material being adsorbed is termed adsorbate and the adsorbing phase is the adsorbent. Adsorption can be classified into two types, physisorption and chemisorption. It mainly depends on the type of bonding involved. If the attraction between the solid surface and the adsorbed molecules is physical in nature (e.g., van der Waals forces), the adsorption is referred to as physical adsorption (or physisorption) and is reversible. There is a rapid formation of an equilibrium interfacial concentration, followed by slow diffusion into the adsorbent. The overall rate of adsorption is controlled by the rate of diffusion of the solute molecules within the

TABLE 7.15
Biological Treatment on Dyeing Wastewater from After-Copperable Direct Dye on Cotton (Wastewater No. 4)

	TOC			BOD$_5$			Color, ADMI Units		
	mg/L			mg/L					
Set	Initial	Final	% Removal	Initial	Final	% Removal	Initial	Final	% Removal
A-1	24	0	100	29	2	93	52	14	73
B-1	31	1	97	41	0	100	52	30	42
C-1	29	20	31	—	—	—	57	32	47
A-10	43	6	86	44	4	91	403	336	17
B-10	40	6	85	41	4	90	420	306	27
C-10	44	30	32	—	—	—	403	372	8
A-100	143	27	81	132	5	96	577	623	+3
B-100	139	27	81	108	6	94	564	716	+27
C-100	134	145	+8	—	—	—	554	—	—

Source: U.S. EPA, Textile dyeing wastewaters: Characterization and treatment, EPA-600/2-78-098, U.S. Environmental Protection Agency, Washington, DC, May, 310 p., 1978.

TABLE 7.16
Biological Treatment on Dyeing Wastewater from Reactive Dye on Cotton (Wastewater No. 5)

	TOC			BOD$_5$			Color, ADMI Units		
	mg/L			mg/L					
Set	Initial	Final	% Removal	Initial	Final	% Removal	Initial	Final	% Removal
A-1	25	2	91	—	—	—	73	54	26
B-1	31	8	74	—	—	—	121	72	40
C-1	26	26	0	—	—	—	79	57	28
A-10	39	18	54	—	—	—	541	530	2
B-10	49	18	63	—	Inhibitory	—	562		
C-10	—			—	—	—	550	549	0
A-100	184	117	36	—	—	—	4084	3991	2
B-100	111	85	23	—	Inhibitory	—			
C-100	199	175	22	—	—	—	4123	4162	+1

Source: U.S. EPA, Textile dyeing wastewaters: Characterization and treatment, EPA-600/2-78-098, U.S. Environmental Protection Agency, Washington, DC, May, 310 p., 1978.

capillary pores of the adsorbent. In contrast, if the attraction force is due to a stronger bonding such as formation of chemical bonds, the adsorption process is termed chemical adsorption (or chemisorption). This kind of chemical interaction is much stronger, and consequently makes it difficult for chemisorbed species to be removed from the solid surface. An adsorption equilibrium is established when the concentration of contaminants remaining in solution is in a dynamic balance with that at the surface.

Adsorption techniques employing solid adsorbents are widely used for the removal of organic compounds such as volatile organic compounds and dyes. Among all, activated carbon is the most

TABLE 7.17
Biological Treatment on Dyeing Wastewater from Disperse Dyes on Polyamide Carpet (Wastewater No. 6)

	TOC mg/L			BOD$_5$ mg/L			Color, ADMI Units		
Set	Initial	Final	% Removal	Initial	Final	% Removal	Initial	Final	% Removal
A-1	24	4	83	36	4	>97	—	—	—
B-1	27	4	85	37	2	95	—	—	—
C-1	25	24	4	—	—	—	—	—	—
A-10	36	7	80	47	<1	>98	<1	Too low for	
B-10	38	9	76	50	<1	>98	8	analysis	
C-10	37	37	0	—	—	—	—	—	—
A-100	192	72	62	129	<1	>99	86	110	+28
B-100	191	72	62	126	<1	>99	86	139	+62
C-100	195	151	22	—	—	—	85	139	+64

Source: U.S. EPA, Textile dyeing wastewaters: Characterization and treatment, EPA-600/2-78-098, U.S. Environmental Protection Agency, Washington, DC, May, 310 p., 1978.

TABLE 7.18
Biological Treatment on Dyeing Wastewater from Acid Chrome Dye on Wool (Wastewater No. 7)

	TOC mg/L			BOD$_5$ mg/L			Color, ADMI Units		
Set	Initial	Final	% Removal	Initial	Final	% Removal	Initial	Final	% Removal
A-1	30	2	93	24	<1	96	—	—	—
B-1	29	0	100	23	<1	96	—	—	—
C-1	28	24	14	—	—	—	—	—	—
A-10	47	6	87	52	<1	>98	117	147	+26
B-10	46	6	87	44	<1	>98	105	167	+59
C-10	47	42	11	—	—	—	93	148	+59
A-100	210	48	77	216	<1	>99	—	Not measured, see	—
B-100	245	34	86	216	<1	>99	—	10% dilution	—
C-100	200	204	+2	—	—	—	—		—

Source: U.S. EPA, Textile dyeing wastewaters: Characterization and treatment, EPA-600/2-78-098, U.S. Environmental Protection Agency, Washington, DC, May, 310 p., 1978.

widely used adsorbent for wastewater treatment. In the preparation of activated carbon, organic materials such as coal, nut shells, or wood are first pyrolyzed to a carbonaceous residue. The most important property of activated carbon, the porous structure, is formed during the activation step in which residual hydrocarbons on the carbon are oxidized selectively by steam or air. The total number of pores, their shape, and size are decisive factors for the adsorbent's usage, adsorption capacity, and even the dynamic adsorption rate of the material. Commercially available activated carbons usually have surface areas ranging from 500 to 1500 m^2/g. For convenience in practice, the adsorptive capacity of activated carbon is usually characterized by measuring the amount of a colored body or dye that is adsorbed from a liquid solution. Table 7.35 presents the adsorption efficiencies of activated carbon and bentonite for various dyes.

TABLE 7.19
Biological Treatment of Dyeing Wastewater from Basic Dyes on Polyacrylic (Wastewater No. 8)

Set	TOC Initial (mg/L)	TOC Final	% Removal	BOD$_5$ Initial (mg/L)	BOD$_5$ Final	% Removal	Color Initial (ADMI)	Color Final	% Removal
A-1	27	1	96	38	2	95	—	—	—
B-1	26	9	65	26	2	92	—	—	—
C-1	28	22	21	—	—	—	—	—	—
A-10	41	10	73	55	3	94	—	399	77
B-10	44	14	68	60	6	90	1720	599	65
C-10	44	40	9	—	—	—	—	506	70
A-100	255	22	91	198	153	23	11,260	12,790	—
B-100	270	117	57	222	76	66	—	12,750	None
C-100	245	192	22	—	—	—	—	15,930	—

Source: U.S. EPA, Textile dyeing wastewaters: Characterization and treatment, EPA-600/2-78-098, U.S. Environmental Protection Agency, Washington, DC, May, 310 p., 1978.

TABLE 7.20
Biological Treatment on Dyeing Wastewater from Disperse Dyes on Polyester Carpet (Wastewater No. 9)

Set	TOC Initial (mg/L)	TOC Final	% Removal	BOD$_5$ Initial (mg/L)	BOD$_5$ Final	% Removal	Color Initial (ADMI)	Color Final	% Removal
A-1	31	0	100	35	<1	97	—	—	—
B-1	31	0	100	—	—	—	—	—	—
C-1	29	26	10	—	—	—	—	—	—
A-10	49	9	82	50	<1	>98	49	Too low for analysis	
B-10	53	9	79	53	<1	>98	19		
C-10	50	50	0	—	—	—	34	—	—
A-100	280	115	59	234	<2	>99	333	262	21
B-100	255	113	56	174	<2	>99	326	260	20
C-100	250	168	33	—	—	—	334	352	+6

Source: U.S. EPA, Textile dyeing wastewaters: Characterization and treatment, EPA-600/2-78-098, U.S. Environmental Protection Agency, Washington, DC, May, 310 p., 1978.

The adsorption results of 20 dyeing wastewater samples are summarized in Table 7.33. Table 7.36 presents the effect of different types of powdered activated carbon on treatment of dyeing wastewater no. 8 (basic dye on polyacrylic) at pH 4.5.

Figure 7.4 demonstrates the adsorption of color from reactive and basic dyeing wastewaters in accordance with the Langmuir adsorption model.

7.5.4 Oxidation Methods

Oxidation is a method by which wastewater is decolorized using oxidizing agents. It is commonly used because of its simplicity of application, short reaction time, and low quantities requirement. Generally, two forms, namely, chemical oxidation and ultra violet (UV)-assisted oxidation using

TABLE 7.21
Biological Treatment on Dyeing Wastewater from Acid Dye on Polyamide (Wastewater No. 10)

	TOC			BOD$_5$			Color, ADMI Units		
	mg/L			mg/L					
Set	Initial	Final	%Removal	Initial	Final	%Removal	Initial	Final	%Removal
A-1	32	0	100	50	—	>98	—	—	—
B-1	34	2	94	41	<1	98	—	—	—
C-1	32	29	9	—	1	—	—	—	—
A-10	59	26	56	69	<1	>98	326	331	+2
B-10	64	29	55	63	<1	>98	350	355	+1
C-10	62	60	3	—	—	—	320	315	2
A-100	280	211	25	270	<3	>99	3300	3110	6
B-100	330	190	42	216	<3	>99	3529	2990	15
C-100	305	305	0	—	—	—	2970	2925	2

Source: U.S. EPA, Textile dyeing wastewaters: Characterization and treatment, EPA-600/2-78-098, U.S. Environmental Protection Agency, Washington, DC, May, 310 p., 1978.

TABLE 7.22
Biological Treatment on Dyeing Wastewater from Direct Dye on Rayon (Wastewater No. 11)

	TOC			BOD$_5$			Color, ADMI Units		
	mg/L			mg/L					
Set	Initial	Final	% Removal	Initial	Final	% Removal	Initial	Final	% Removal
A-1	32	2	94	44	2	95	239	36	85
B-1	30	1	97	48	<1	98	303	71	77
C-1	39	26	33	—	—	—	221	78	65
A-10	40	11	72	42	<1	>98	1491	1003	33
B-10	43	14	67	46	<1	>98	1508	1591	+6
C-10	53	41	23	—	—	—	1395	1585	+14
A-100	170	153	10	51	<3	>94	12,750	11,350	11
B-100	180	154	14	54	<3	>94	12,107	13,636	+13
C-100	200	185	8	—	—	—	12,675	13,938	+10

Source: U.S. EPA, Textile dyeing wastewaters: Characterization and treatment, EPA-600/2-78-098, U.S. Environmental Protection Agency, Washington, DC, May, 310 p., 1978.

hydrogen peroxide (H_2O_2), Fenton's reagent, ozone, or potassium permanganate ($KMnO_4$), are used for treating effluents, especially those obtained from primary treatment (sedimentation). The oxidation process can degrade the dyes partially (generally to lower molecular weight species such as aldehydes, carboxylates, sulfates, and nitrogen) or completely (reduce the complex dye molecules to carbon dioxide and water). It is worth noting that pH and catalysts play an important role in the oxidation process.

7.5.4.1 Hydrogen Peroxide (H_2O_2)

The main oxidizing agent is usually hydrogen peroxide (H_2O_2), which is known for its strong oxidizing properties. This agent needs to be activated by some means, for example, UV light. The factors influencing H_2O_2/UV treatment are H_2O_2 concentration, the intensity of UV irradiation, pH, dye

TABLE 7.23
Biological Treatment on Dyeing Wastewater from Direct Developed Dye on Rayon (Wastewater No. 12)

Set	TOC Initial (mg/L)	TOC Final	% Removal	BOD$_5$ Initial (mg/L)	BOD$_5$ Final	% Removal	Color Initial (ADMI)	Color Final	% Removal
A-1	23	0	100	27	1	96	84	26	70
B-1	23	0	100	34	1	97	71	26	63
C-1	20	22	+10	—	—	—	80	44	45
A-10	25	2	92	35	1	97	306	182	40
B-10	24	3	88	32	1	97	294	209	29
C-10	25	22	12	—	—	—	295	163	45
A-100	66	45	32	42	1	98	3190	2455	23
B-100	70	43	38	41	1	98	2860	2770	3
C-100	68	—	—	—	—	—	—	—	3

Source: U.S. EPA, Textile dyeing wastewaters: Characterization and treatment, EPA-600/2-78-098, U.S. Environmental Protection Agency, Washington, DC, May, 310 p., 1978.

TABLE 7.24
Biological Treatment on Dyeing Wastewater from Disperse, Acid, and Basic Dyes on Polyamide Carpet (Wastewater No. 13)

Set	TOC Initial (mg/L)	TOC Final	% Removal	BOD$_5$ Initial (mg/L)	BOD$_5$ Final	% Removal	Color Initial (ADMI)	Color Final	% Removal
A-1	24	2	92	22	<1	>95	22	22	0
B-1	19	2	89	23	<1	96	45	26	42
C-1	19	22	+15	—	—	—	16	20	+25
A-10	31	4	87	35	3	91	107	36	66
B-10	30	4	87	34	3	91	140	40	71
C-10	26	22	15	—	—	—	97	39	60
A-100	115	53	54	90	11	88	637	210	67
B-100	119	53	55	102	14	86	642	257	60
C-100	121	77	36	—	—	—	653	410	37

Source: U.S. EPA, Textile dyeing wastewaters: Characterization and treatment, EPA-600/2-78-098, U.S. Environmental Protection Agency, Washington, DC, May, 310 p., 1978.

structure, and dyebath composition. Effective decolorization normally occurs at pH ~7, at higher UV radiation intensity, and with an optimal H_2O_2 concentration that is different for different dye classes. Acid dyes are the easiest to decompose, and with an increasing number of azo groups, the decoloration effectiveness decreases [23]. Yellow and green reactive dyes need longer decoloration times, while other reactive dyes as well as direct, metal-complex and disperse dyes are decolorized quickly [24]. In the group of blue dyes examined, only blue dyes were not vat decolorized, yet their structure changed with the process in such a way that they could be easily filtrated. For pigments, H_2O_2/UV treatment is not suitable, because of the formation of a filmlike coating that is difficult to remove.

7.5.4.2 Fenton's Reagent (H_2O_2–Fe(II) Salts)

Fenton's reagent, a solution of H_2O_2 and an iron catalyst, is a suitable chemical means for treating wastewaters that are resistant to biological treatment or are poisonous to live biomass [25].

TABLE 7.25
Biological Treatment on Dyeing Wastewater from Disperse Dyes on Polyester (Wastewater No. 14)

Set	TOC mg/L Initial	TOC Final	% Removal	BOD$_5$ mg/L Initial	Final	% Removal	Color, ADMI Units Initial	Final	% Removal
A-1	16	2	88	17	1	94	—	—	—
B-1	18	8	56	18	1	94	—	—	—
C-1	17	13	24	—	—	—	—	—	—
A-10	50	25	50	34	<1	>97	160	150	6
B-10	50	29	42	35	<1	>97	177	166	6
C-10	49	42	9	—	—	—	169	200	+18
A-100	320	232	28	180	4	98	1141	1162	+2
B-100	355	223	37	165	22	87	647	1316	—
C-100	445	268	40	—	—	87	1072	1165	+9

Source: U.S. EPA, Textile dyeing wastewaters: Characterization and treatment, EPA-600/2-78-098, U.S. Environmental Protection Agency, Washington, DC, May, 310 p., 1978.

TABLE 7.26
Biological Treatment on Dyeing Wastewater from Sulfur Dye on Cotton (Wastewater No. 15)

Set	TOC mg/L Initial	TOC Final	% Removal	BOD$_5$ mg/L Initial	Final	% Removal	Color, ADMI Units Initial	Final	% Removal
A-1	32	4	88	23	3	87	—	—	—
B-1	32	2	94	25	3	88	—	—	—
C-1	32	32	0	—	—	—	—	—	—
A-10	68	5	93	114	<3	>97	164	33	80
B-10	68	3	96	106	<3	>97	142	36	75
C-10	—	67	0	—	—	—	93	42	55
A-100	510	52	90	810	<1	100	1435	692	52
B-100	460	54	88	870	<1	100	1240	882	29
C-100	—	98	—	—	—	—	1345	1090	19

Source: U.S. EPA, Textile dyeing wastewaters: Characterization and treatment, EPA-600/2-78-098, U.S. Environmental Protection Agency, Washington, DC, May, 310 p., 1978.

Generally, it is effective in decolorization of both soluble and insoluble dyes, for instance, acid, reactive, direct, metal-complex dyes [26], though some dyes like vat and disperse dyes were found to be resistant to it, for example, palanil blue 3RT [27]. Dyes like remazol brilliant blue B, sirrus supra blue BBR, indanthrene blue GCD, irgalan blue FGL, and helizarin blue BGT have been reported to be significantly decolorized [28]. Besides, this process also offers advantages such as reduction in chemical oxygen demand (COD) (except with reactive dyes), total organic carbon (TOC), and toxicity. The applicability of the process is not affected by the presence of high suspended solid concentration. However, one of the major drawbacks is the sludge generation through the flocculation of the reagent and the dye molecules. The generated sludge has conventionally been incinerated to

TABLE 7.27
Biological Treatment on Dyeing Wastewater from Reactive Dyes on Cotton (Wastewater No. 16)

	TOC			BOD$_5$			Color, ADMI Units		
	mg/L			mg/L					
Set	Initial	Final	% Removal	Initial	Final	% Removal	Initial	Final	% Removal
A-1	20	0	100	20	<1	>95	—	—	—
B-1	25	1	96	25	<1	>96	—	—	—
C-1	23	24	+4	—	—	—	—	—	—
A-10	36	4	89	36	<1	>97	256	192	25
B-10	35	4	88	35	<1	>97	282	195	31
C-10	39	55	+41	—	—	—	191	172	10
A-100	118	24	80	118	1	99	719	690	4
B-100	120	30	75	120	1	>99	730	712	2
C-100	131	185	+41	—	—	—	727	724	<1

Source: U.S. EPA, Textile dyeing wastewaters: Characterization and treatment, EPA-600/2-78-098, U.S. Environmental Protection Agency, Washington, DC, May, 310 p., 1978.

TABLE 7.28
Biological Treatment on Dyeing Wastewater from Vat and Disperse Dyes on 50/50 Cotton-Polyester Blend (Wastewater No. 17)

	TOC			BOD$_5$			Color, ADMI Units		
	mg/L			mg/L					
Set	Initial	Final	% Removal	Initial	Final	% Removal	Initial	Final	% Removal
A-1	30	6	80	47	9	81	—	—	—
B-1	27	6	78	40	10	75	—	—	—
C-1	29	32	+10	—	—	—	—	—	—
A-10	63	14	78	>47	10	>78	92	62	33
B-10	58	14	76	>51	7	>86	66	56	15
C-10	67	59	12	—	—	—	52	66	+27
A-100	289	85	70	>128	32	>75	371	302	18
B-100	363	83	77	>128	38	>70	367	312	15
C-100	327	372	+14	—	—	—	364	367	0

Source: U.S. EPA, Textile dyeing wastewaters: Characterization and treatment, EPA-600/2-78-098, U.S. Environmental Protection Agency, Washington, DC, May, 310 p., 1978.

produce power, but the performance is dependent on the final floc formation and its settling quality. Some dyes do not coagulate at all (e.g., cationic dye), whereas acid, direct, vat, mordant, and reactive dyes usually form floc that is of poor quality and does not settle well.

7.5.4.3 Ozonation

The use of ozone in wastewater treatment was pioneered in the early 1970s. Decolorization by ozone occurs in a relatively short time, and the dosage applied to the dye-containing effluent is dependent on the total color and residual COD to be removed with no residue or sludge formation [29] and no toxic metabolites [30]. Often, ozonation leaves the effluent with no color. As far as the ecological

TABLE 7.29
Biological Treatment on Dyeing Wastewater from Basic Dyes on Polyester (Wastewater No. 18)

Set	TOC mg/L Initial	Final	% Removal	BOD$_5$ mg/L Initial	Final	% Removal	Color, ADMI Units Initial	Final	% Removal
A-1	34	2	94	42	4	98	—	—	—
B-1	32	5	84	47	<1	98	—	—	—
C-1	34	34	0	—	—	—	—	—	—
A-10	117	8	93	174	<1	>99	116	97	16
B-10	127	12	90	204	<1	>99	155	146	6
C-10	136	142	+7	—	—	—	—	125	—
A-100	1150	346	70	1530	258	83	1172	650	45
B-100	1170	275	76	1620	216	87	1304	670	50
C-100	1020	1040	+2	—	—	—	1059	1050	0

Source: U.S. EPA, Textile dyeing wastewaters: Characterization and treatment, EPA-600/2-78-098, U.S. Environmental Protection Agency, Washington, DC, May, 310 p., 1978.

TABLE 7.30
Biological Treatment on Dyeing Wastewater from Disperse, Acid and Basic Dyes on Polyamide Carpet (Wastewater No. 19)

Set	TOC mg/L Initial	Final	% Removal	BOD$_5$ mg/L Initial	Final	% Removal	Color, ADMI Units Initial	Final	% Removal
A-1	19	0	100	25	1	96	—	—	—
B-1	19	0	100	19	<1	95	—	—	—
C-1	20	12	40	—	—	—	—	—	—
A-10	27	0	100	30	<1	>97	5	Too low for analysis	
B-10	27	1	96	29	<1	>97	12		
C-10	26	22	15	—	—	—	14		
A-100	150	34	77	129	2	98	28	Too low for analysis	
B-100	146	34	77	144	2	99	24		
C-100	155	92	41	—	—	—	27		

Source: U.S. EPA, Textile dyeing wastewaters: Characterization and treatment, EPA-600/2-78-098, U.S. Environmental Protection Agency, Washington, DC, May, 310 p., 1978.

parameters are concerned (values of COD and BOD), there are several opinions: some believe that the parameters remain unaffected; others believe that they decrease or increase. Experiences with TOC reduction are uniform, for example, ozone treatment does not influence it. Some disadvantages of ozonation include its short half-life (decomposes in about 20 min) and high cost. The time can be further shortened if dyes are present. Ozone stability is affected by the presence of salts, pH, and temperature. Under alkaline conditions, ozone decomposition is accelerated; thus, careful monitoring of the effluent pH is required [25]. Results of the ozonation of 20 dyeing wastewaters are summarized in Table 7.33. Figure 7.5 shows the schematic diagram of the apparatus for ozonation study.

Table 7.37 lists the concentration of textile pollutants before and after ozonation process.

Example of the decolorization ability of ozone for C.I. Disperse Yellow 42, C.I. Basic Yellow 11, and C.I. Acid Red 151 are shown in Figures 7.6 and 7.7.

TABLE 7.31
Biological Treatment on Dyeing Wastewater from Napthol on Cotton on Polyamide Carpet (Wastewater No. 20)

	TOC			BOD$_5$			Color, ADMI Units		
	mg/L			mg/L					
Set	Initial	Final	% Removal	Initial	Final	% Removal	Initial	Final	% Removal
A-1	47	0	100	61	<1	>98	—	—	—
B-1	43	2	95	34	<1	>97	—	—	—
C-1	36	30	17	—	—	—	—	—	—
A-10	52	6	88	59	<1	>98	340	212	38
B-10	47	8	83	44	<1	>98	297	229	23
C-10	46	44	4	—	—	—	349	280	20
A-100	174	58	67	204	3	99	2850	509	82
B-100	176	55	69	216	<1	100	2805	497	82
C-100	173	158	9	—	—	—	2619	537	80

Source: U.S. EPA, Textile dyeing wastewaters: Characterization and treatment, EPA-600/2-78-098, U.S. Environmental Protection Agency, Washington, DC, May, 310 p., 1978.

Table 7.38 is a summary of the decolorization of dyeing wastewaters no. 1–20 by ozonation according to the efficiency of treatment (very treatable, poorly treatable, moderately treatable).

7.5.4.4 Chlorine

Besides being widely used as disinfectant in water treatment, chlorine compounds (e.g., calcium hypochlorite and sodium hypochlorite) are also extensively being applied for reduction of color like in pulp and textile bleaching. Dyes with amino or substituted amino groups on a naphthalene ring decolorize more easily than other dyes due to the electrophilic attack by Cl$^-$. Decolorization of reactive dyes require longer reaction times, while metal-complex dye solutions remain partially colored even after an extended period of treatment. Reaction can be enhanced through control of pH and by using catalysts. For example, in the decomposition of metal-complex dyes, the liberated metals such as iron, copper, nickel, and chromium have a constructive effect in the decolorization process. However, for environmental reasons, chlorine-based decolorization processes should be restricted as they produce organochlorine compounds including toxic trihalomethane. Figure 7.8 shows the flow diagram of chlorination wastewater treatment system.

7.5.5 COMBINED BIOLOGICAL/PHYSICAL–CHEMICAL TREATMENT

Biological treatment can reduce the BOD and TOC of dyeing wastewaters, but it is rather ineffective in decolorizing the wastes. The opposite trend is observed in physical–chemical treatment. Thus, the limitation in both biological and physical–chemical treatment can probably be overcome by coupling these two processes. The results of combined treatments for 3 dyeing wastewaters, no. 17, 18 and 20, are shown in Table 7.39.

The coupling of biological degradation with physical–chemical decolorization produces an effluent with the removal of more than 85% of BOD, 70% of TOC, and a residual color of less than 100 (with the exception of wastewater no. 18, which could have been decolorized further by a higher dosage of carbon).

7.5.6 DYES REMOVAL BY LOW-COST ADSORBENTS

Though activated carbon has proven its efficiency and capability in removing various types of pollutants, associated problems such as high capital investment, regeneration process, and reduction in

TABLE 7.32
Effect of Dyeing Wastewaters Nos 1–20 on Nitrification

		10% Strength				100% Strength			
		TKN, mg/L		NO$_2$–N + NO$_3$–N, mg/L		TKN, mg/L		NO$_2$–N + NO$_3$–N, mg/L	
Dye #	Set	Initial	Final	Initial	Final	Initial	Final	Initial	Final
1	A	4.25	6.0	>0.3	2.1	23.0	26.0	0.5	0.5
	B		7.75		0.6		20.5	<0.3	<0.3
	C	3.6	7.5	<0.3	<0.5	30.0	—	0.7	0.9
2	A	24.0	15.0	0.6	12.2	203	—	1.3	—
	B	24.5	15.2	0.7	12.4	206	—	1.4	—
	C	25.5	26.5	<0.5	<0.3	231	—	1.1	1.1
3	A	7.0	7.75	>0.3	1.3	12.0	19.5	1.2	<0.3
	B	7.5	8.25	<0.3	<0.3	13.5	20.5	0.9	<0.3
	C		7.25	<0.3	<0.3	14.0	23.7	0.9	<0.3
4	A	4.5	5.5	<0.3	<0.3	9.5	11.0	0.6	<0.3
	B	4.5	5.75	<0.3	<0.3	10.0	12.0	0.6	<0.3
	C	4.0	3.0	<0.3	<0.3	10.0	10.5	0.5	0.6
5	A	6.5	<3.0	<0.3	6.0	14.2	16.2	1.0	0.7
	B	5.0	6.8	<0.3	0.8	14.5	16.0	0.9	0.6
	C	7.3	3.2	<0.3	<0.3	13.2	13.0	0.9	0.6
6	A	5.0	2.7	<0.3	6.4	18.0	6.0	0.5	10.6
	B	6.0	2.2	<0.3	6.2	19.0	6.0	0.4	10.8
	C	5.0	4.5	<0.3	<0.5	18.0	17.5	0.4	0.8
7	A	5.5	2.5	0.3	7.2	18.0	13.5	0.7	1.1
	B	5.5	7.8	0.2	1.0	19.5	15.0	0.6	0.8
	C	5.5	4.8	0.2	<0.3	17.5	10.2	0.6	0.8
8	A	4.0	5.5	<0.3	<0.5	17.5	11.0	0.3	0.9
	B	4.5	5.2	<0.3	<0.5	15.5	8.0	<0.3	0.8
	C	5.0	3.5	<0.3	<0.5	14.5	15.7	<0.3	1.6
9	A	6.0	2.2	0.3	6.0	13.5	6.8	0.8	2.4
	B	7.5	2.2	0.3	6.2	13.5	6.2	0.7	3.1
	C	7.0	3.0	<0.3	<0.4	12.5	5.2	0.8	<0.3
10	A	5.0	7.5	<0.4	0.3	12.5	15.0	1.1	<0.4
	B	5.0	9.5	0.4	<0.7	—	13.8	0.95	<0.4
	C	4.5	4.5	<0.4	<0.3	11.5	12.5	0.7	<0.4
11	A	4.5	6.0	0.5	6.9	15.5	15.0	2.0	8.5
	B	7.5	5.0	0.5	5.8	17.0	14.5	2.0	8.2
	C	7.0	5.8	0.2	<0.7	15.0	17.3	2.7	2.8
12	A	5.5	2.5	—	6.8	—	—	—	—
	B	—	—	—	—	—	—	—	—
	C	—	—	—	—	—	—	—	—
13	A	3.0	2.2	<0.5	4.9	5.5	7.0	<0.95	<0.4
	B	3.5	2.2	<0.5	5.0	7.5	8.5	<0.95	2.0
	C	3.0	3.0	<0.5	<0.3	7.0	6.8	<0.95	0.75
14	A	4.5	3.2	<0.3	1.0	13.5	23.8	0.4	0.5
	B	3.0	4.0	<0.3	<0.5	13.5	28.0	0.6	<0.5
	C	3.0	2.2	<0.3	<0.4	13.5	23.5	<0.5	<0.6
15	A	5.5	3.0	<0.4	7.9	12.0	20.2	<0.4	0.2
	B	5.0	4.5	<0.4	7.4	12.5	22.5	<0.4	1.3
	C	5.0	4.2	<0.4	<0.4	12.5	14.0	<0.4	2.3
16	A	19.5	8.5	0.95	15.0	127.5	41.0	3.6	60.5
	B	21.5	8.5	1.1	15.5	122.5	40.0	4.0	63.5
	C	20.5	21.5	<0.6	1.4	122.5	120.5	3.5	3.7
17	A	6.0	2.0	<0.3	6.7	20.0	15.2	1.11	0.35
	B	7.5	2.0	<0.3	6.9	22.0	19.2	1.19	0.27
	C	5.0	3.7	<0.2	<0.1	17.0	19.0	1.01	1.02
18	A	5.0	<2.0	<0.4	3.0	17.5	15.5	<0.4	<0.1
	B	5.0	2.0	<0.4	3.5	16.0	16.0	<0.4	<0.1
	C	5.5	4.0	<0.4	<0.1	15.5	17.0	<0.4	0.2
19	A	2.5	2.0	0.3	3.2	10.0	5.0	0.2	0.3
	B	2.	2.0	0.1	3.6	9.5	7.5	0.2	1.1
	C	2.5	2.0	<0.1	<0.3	7.5	6.8	<0.1	<0.3
20	A	7.0	—	1.6	—	24.5	—	13.0	—
	B	7.5	—	1.7	—	31.5	—	13.0	—
	C	6.0	—	1.5	—	25.0	—	13.0	—

Source: U.S. EPA, Textile dyeing wastewaters: Characterization and treatment, EPA-600/2-78-098, U.S. Environmental Protection Agency, Washington, DC, May, 310 p., 1978.

TABLE 7.33
Summary of Physical–Chemical Treatability Studies of Dyeing Wastewaters Nos 1–20

Dyeing Wastewater No.	Dye Class	Substrate	Color	TOC, mg/L	Lime	Alum	PAC	Ozone
1	Vat	Cotton	1910	265	1000 mg/L reduced color to 350, TOC to 210	140 mg/K (pH 6.3) reduced color to 100, TOC to 80	No visible color change up to 4 g/K (pH 11)	—
2	1:2 Metal complex	Polyamide	370	400	1000 mg/L reduced color to 230, no reduction of TOC	160 mg/L (pH 6.3) reduced color to 230, no reduction of TOC	900 mg/L Nuchar D-16 reduced color to 100, TOC to 320 (at pH 6.8)	Efficient decolorization: 130 mg O$_3$/L reduced color to <100, no change in TOC, slight drop in pH
3	Disperse	Polyester	315	300	No change in color or TOC up to 1000 mg/L	60 mg/L (pH 5) reduced color to 45, TOC to 175, pH 5, better than pH 6, 7	Little change in color up to 1200 mg/L of Nuchar D-16 (pH 7.8)	Poor decolorization: 6 g O$_3$/L reduced color to 180, lag period, no change in TOC, small drop in pH
4	After-copperable Direct	Cotton	525 (1280)[a]	135	200 mg/L reduced apparent ADMI color to 120, little change in TOC	16 mg/L (pH 6) reduced apparent ADMI color to 80, little change in TOC	—	Good decolorization: apparent color reduced to 60 by application of 500 mg O$_3$/L, no change in pH or TOC
5	Reactive	Cotton	3890	150	No effect on color or TOC up to 3 g/L	160 mg/L (pH 5) reduced color to 635, TOC to 60, higher doses showed no improvement	2000 mg/L HD-3000 (pH 3.5) reduced color to 100, TOC to 20, pH 3.5, better than pH 7 or 10.7	Good decolorization: color reduced to 100 by 1.0 g/L O$_3$, reduction more rapid at alkaline pH of raw waste (10.7) than at pH 8 or 2, no removal of TOC
6	Disperse	Polyamide carpet	100	130	200 mg/L reduced color to 50, little removal of TOC	30 mg/L (pH 7) reduced color to less than 50, TOC removal small	200 mg/L HD-3000 reduced color to 65, 1000 mg/L reduced TOC to 60	—

(*Continued*)

TABLE 7.33 (Continued)
Summary of Physical–Chemical Treatability Studies of Dyeing Wastewaters Nos 1–20

Dyeing Wastewater No.	Dye Class	Substrate	Color	TOC, mg/L	Lime	Alum	PAC	Ozone
7	Acid/Chrome	Wool	3200	210	No color removal, some TOC removal up to 2000 mg/L	No color removal, some TOC removal up to 160 mg/L at pH 7	4 g/L HD-3000 (pH 4.2) reduced color to 75, acidic pH most effective, small TOC removal	—
8	Basic	Polyacrylic	5600 (12,000)[a]	255	2000 mg/L ineffective for color or TOC reduction	315 mg/L ineffective for color or TOC reduction at pH 7	1000 mg/L Nuchar D-16 (or 2000 mg/L HD-3000) reduced apparent color to less than 100, pH 4.5 optimal, TOC reduced to 150	Good decolorization, but apparent color leveled off at about 200 at 2 g/L O_3, noticeable lag prior to decolorization, reaction more rapid at pH 4.2, than at neutral pH, no change in TOC, but BOD increased after O_3 application
9	Disperse	Polyester carpet	215 (315)[a]	240	1000 mg/L reduced apparent color to 130, TOC to 190	150 mg/L (pH 5) reduced apparent color to 80, TOC to 110, very light floc	2000 mg/L Nuchar D-16 reduced TOC to 75 but no change in color	Decolorization relatively low, apparent lag, acidification to pH 4 more rapid than at pH 7, apparent color reduced to 70 by 4 g/L O_3, no change in TOC but BOD increased with treatment
10	Acid	Polyamide	4000	315	1000 mg/L reduced color to 2150, no reduction of TOC	160 mg/L (pH 7) reduced color to 2150, TOC to 160	1000 mg/L Darco KB or Nuchar D-14 reduced color to 260, TOC to 175, pH 5.1 optimal, cannot get color below 200 even at higher doses	Poor decolorization: color reduced to 400 by 4 g/L O_3, pH dropped dramatically to (3.5) by ozonation, no change in TOC, but BOD increased

(*Continued*)

Treatment of Textile Industry Waste

TABLE 7.33 (Continued)
Summary of Physical–Chemical Treatability Studies of Dyeing Wastewaters Nos 1–20

Dyeing Wastewater No.	Dye Class	Substrate	Color	TOC, mg/L	Lime	Alum	PAC	Ozone
11	Direct	Rayon	12,500	140	3 g/L gave no appreciable color or TOC removal	30 mg/L (pH 5) reduced color to 2000, TOC to 30, color could not be reduced by higher alum doses	2 g/L gave no color or TOC removal at pH 6.6	Poor decolorization: 4.5 g/L O_3 reduced color to 1150, pH dropped during treatment, small TOC reduction, but BOD increased
12	Direct developed	Rayon	2730	55	400 mg/L reduced color to 300, TOC to 30, higher doses gave no improvement	8 mg/L (pH 5) reduced color to 300, TOC to 30, no improvement at higher doses	1000 mg/L (pH 3.1) gave no color reduction	Good color reduction, 3.5 g/L O_3 reduced color to <100, no apparent pH effect, small TOC reduction
13	Disperse, acid, basic	Polyamide carpet	210 (720)[a]	130	No apparent color reduction up to 4 g/L	No apparent color reduction up to 160 mg/L	500 mg/L Darco KB (pH 3) reduced apparent color to 40, TOC to 50, acid pH best	Good decolorization: apparent color reduced to 100 by 1.3 g/L O_3, no effect on pH, no TOC reduction
14	Disperse	Polyester	1245	360	100 mg/L gave no reduction in color or TOC	65 mg/L (pH 5) reduced color to 230, TOC to 130, no improvement at higher doses	1000 mg/L (pH 3) reduced color to only 500, little TOC removal	Poor decolorization: 3.5 g/L O_3 reduced apparent color to 400, defoamer required, small reduction in TOC
15	Sulfur	Cotton	1450	400	500 mg/L reduced color to 100, no effect on TOC, floc did not settle well	8 mg/L reduced color to less than 100, no effect on TOC, floc settled poorly	1000 mg/L (pH 3.9) gave no apparent color reduction	Color reduced to 150 by 6.0 g/L O_3, no apparent pH effect, defoamer required, no change in TOC
16	Reactive	Cotton	1390	230	1000 mg/K gave no apparent color reduction	160 mg/L (pH 7) gave no apparent color reduction	1700 mg/L Hydrodarco B (pH 6) reduced color to less than 50, TOC to 140, pH 6 optimal	Good decolorization: color reduced to less than 100 with 1.2 g/L O_3, reaction more efficient at acidic pH, no change in TOC

(Continued)

TABLE 7.33 (Continued)
Summary of Physical–Chemical Treatability Studies of Dyeing Wastewaters Nos 1–20

Dyeing Wastewater No.	Dye Class	Substrate	Color	TOC, mg/L	Lime	Alum	PAC	Ozone
17	Vat, disperse	Polyester/Cotton	365 (1100)[a]	350	No floc formation or color removal	50 mg/L reduced apparent color to 80, small TOC removal, no adjustment of pH required, treated sample still foamed	1000 mg/L Darco KB at pH 3.3 reduced apparent color to 325	Poor decolorization: 9 g/L O_3 reduced apparent color to 460, defoamer required, slight TOC reduction
18	Basic	Polyester	1300 (2040)[a]	1120	No apparent color reduction up to 1000 mg/L	No apparent color reduction up to 160 mg/L at pH 5	700 mg/L Darco KB (pH 5) reduced apparent color to 100, TOC reduced to 700, pH 6–7 optimal	Apparent color reduced to 300 by 2 g/L O_3 but could not be reduced below 250, defoamer required, acidic pH better than neutral pH, no reduction in TOC
19	Disperse, acid, basic	Polyamide carpet	<50 (190)[a]	160	No apparent color reduction up to 1000 mg/L	8 mg/L (pH 7) reduced color to 50, no change in TOC	No apparent decolorization	Good decolorization to less than 100 by 500 mg/L O_3, TOC reduced slightly
20	Azoic	Cotton	2415	170	800 mg/g had no effect on supernatant color	125 mg/L had effect on supernatant color	800 mg/L of Darco HD-3000 or Nuchar D-16 at pH 6 reduced color of supernatant to 100, TOC removal slight, acid pH best	Efficient initial decolorization to 260 by 1.5 g/L O_3 followed by slower decolorization to less than 100 by 11 g/L O_3, defoamer required, no change in TOC

Source: U.S. EPA, Textile dyeing wastewaters: Characterization and treatment, EPA-600/2-78-098, U.S. Environmental Protection Agency, Washington, DC, May, 310 p., 1978.

[a] Prefiltration step omitted in ADMI color analysis.

TABLE 7.34
Characteristics of Sludge Resulting from Coagulation Process of 3% Dyeing Wastewaters with Selected Doses of Coagulants

		Coagulant Dosage, mg/dm³		
Determination	Unit	CaO 6000	FeSO₄ 500 CaO 8000	Al₂(SO₄)₃ 250 CaO 6000
Volume of Sludge				
2 h	cm³/dm³	15.2	90.9	39.4
6 h	cm³/dm³	15.2	63.6	33.3
Total solids	g/dm³	102.08	52.56	99.17
Volatile total solids	g/dm³	10.07	17.92	13.17
Volatile total solids	%	9.88	15.1	13.3
Sludge volume index	cm³/g	9.6	17.9	9.5
Coefficient of specific resistance to filtration	m/kg	4×10^{11}	2.7×10^{11}	4.3×10^{11}
Time of capillary filtration CST	s	96.9	52.1	74.1
Sludge Thickening				
0.5 h	cm³/dm³	984	—	—
1.0 h	cm³/dm³	960	910	952
2.0 h	cm³/dm³	928	820	892
3.0 h	cm³/dm³	896	710	856
4.0 h	cm³/dm³	872	600	808
5.0 h	cm³/dm³	840	—	768
6.0 h	cm³/dm³	800	530	720
24 h	cm³/dm³	528	390	472
Sludge hydration estimated after 6 h sedimentation	%	88.5	90.7	91.0
Sludge hydration estimated after specific resistance to filtration measurement	%	58.3	63.2	60.9

Source: U.S. EPA, Development of methods and techniques for final treatment of combined municipal and textile wastewater including sludge utilization and disposal, EPA-600/2-79-160, U.S. Environmental Protection Agency, Washington, DC, December, 152 p., 1979.

adsorption capacity hamper its applicability. Research has therefore been intensified to search for low-cost materials to serve as potential alternatives for removal of dyes from wastewater.

Low-cost materials can be defined as those that are generally available free of cost and are abundant in nature. In addition to naturally occurring materials, a great deal of attention has also been focused on utilizing agricultural waste materials or industrial by-products as potential adsorbents for the removal of dyes from wastewaters. In this section, selected materials from agriculture waste, industrial by-products, and biosorbents are discussed in terms of their efficiency for dye removal. Recent reported adsorption capacities of selected adsorbents in Tables 7.40 through 7.49 are shown to provide some idea of adsorbent effectiveness. However, the reported adsorption capacities must be taken as values that can be attained only under specific conditions since adsorption capacities of the adsorbents presented vary, depending on the characteristics of the adsorbent, the experimental conditions, and also the extent of chemical modifications. The reader is encouraged to refer to the original articles for information on experimental conditions. The utilization of these low-cost adsorbents provides not only an economical approach for dye removal, but also other advantages, such as the possibility of attaining a zero-waste situation in the environment.

TABLE 7.35
Adsorption of Dyeing Baths by Powdered Activated Carbon and Bentonite

Dyes	Dye Concentration %	Dye Concentration mg/dm³	Initial COD mg/dm³	Adsorbent	Adsorbent Dosage g/dm³	Optimal Adsorption Time min	pH	Color Removal %	COD Removal %	Anion Detergents Removal %
Metallic-complex dye-polfalen brown GL	10	2500	7120	Activated carbon	20	30	3.4	97.0	76.0	98.0
Sulfur dye—Schwefelneu blau FPL	5	2000	4150	Activated carbon	20	30	10.5	50.0	47.8	94.4
Acid-chrome dye—Acid-chrome blue GRN	5	150	676	Activated carbon	20	60	8.6	99.8	90.6	88.0
Direct dye—Direct brown B	5	1000	1140	Activated carbon	20	60	9.7	96.0	89.6	x
Reactive dye—Helactyn red F Ban	1	3000	394	Activated carbon	20	60	9.1	96.0	64.9	x
Basic dye—Aniline blue RL 50/100	5	1250	2675	Activated carbon	20	30	5.1	90.0	71.1	xx
Basic dye—aniline blue RL 50/100	3	1250	2675	Bentonite	20	30	2.6	100	81.1	xx
Disperse dye—Synthetic blue P-BL	5	3000	4200	Activated carbon	20	30	9.3	70.0	67.7	xx
Mixture of aforementioned dyeing baths	1	330	790	Activated carbon	20	60	9.2	—	75.3	xx

Source: U.S. EPA, Development of methods and techniques for final treatment of combined municipal and textile wastewater including sludge utilization and disposal, EPA-600/2-79-160. U.S. Environmental Protection Agency, Washington, DC, December, 152 p., 1979.

x, nonionic detergents; xx, the dye reacted with chloroform, giving colored compound.

TABLE 7.36
Effect of Different Types of Powdered Activated Carbon on Treatment of Dyeing Wastewater No. 8 (Basic Dye on Polyacrylic) at pH 4.5

Type of Powdered Carbon	Dosage, mg/L	pH	TOC, mg/L	Apparent ADMI Color Value
DARCO HDB	1500	4.5	132	147
DARCO S–51	1500	4.5	142	207
DARCO KB	1500	4.5	108	66
DARCO HD–3000	1500	4.5	140	153
NUCHAR D–14	1500	4.5	134	62
NUCHAR D–16	1500	4.5	125	76

Source: U.S. EPA, Textile dyeing wastewaters: Characterization and treatment, EPA-600/2-78-098, U.S. Environmental Protection Agency, Washington, DC, May, 310 p., 1978.

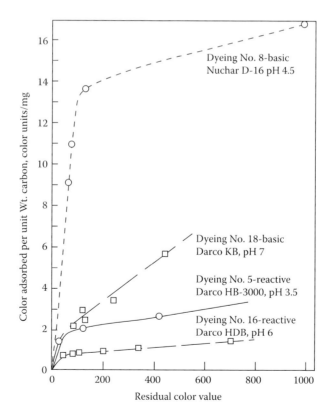

FIGURE 7.4 Adsorption of color from reactive and basic dyeing wastewaters (no. 8) in accordance with Langmuir adsorption model. (From Rai, H.S., et al., *Crit. Rev. Environ. Sci. Technol.*, 25, 219–238, 2005.)

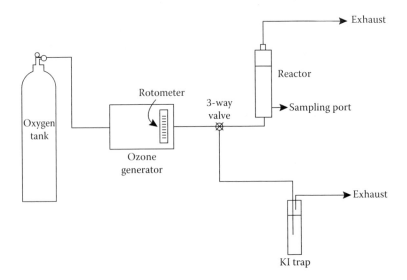

FIGURE 7.5 Schematic diagram of apparatus for ozonation study. (From Rai, H.S., et al., *Crit. Rev. Environ. Sci. Technol.*, 25, 219–238, 2005.)

TABLE 7.37
Concentration of Acid and Base/Neutral Fractions of Textile Pollutants before and after Ozonation Process

		Concentration, μg/L			
		Initial		Final	
Sample	Compound	Dup. #1	Dup. #2	Dup. #1	Dup. #2
A	2-Nitrophenol	40	35	*ND	ND
	Phenol	110	120	ND	ND
	2,4-Dinitrophenol	350	340	ND	ND
	1,4-Dichlorobenzene	3.4	3.1	ND	ND
	Naphthalene	1.6	1.4	ND	ND
B	2-Dichlorophenol	1.3	1.3	ND	ND
	2,4-Dinitrophenol	3200	3300	ND	ND
	1,4-Dichlorobenzene	50	6.2	ND	ND
	Nitrobenzene	51	47	ND	ND
	Di-*n*-butylphthalate	22	24	ND	ND
C	Phenol	1900	a	13	11
	Bis (2-Chloroethyl) ether	2300	—	ND	ND
	N-Nitrosodiphenylamine	29	—	ND	ND
D	1,3-Dichlorobenzene	1.0	—	ND	ND
	Phenol	240	—	15	16
	1,4-Dichlorobenzene	9.4	—	ND	ND
	Naphthalene	16	—	ND	ND
	1,2,3-Trichlorobenzene	8.5	—	ND	ND
	Nitrobenzene	830	—	360	400
E	Phenol	59	57	26	32
	Di-*n*-octylphthalate	1800	1800	360	350

Source: U.S. EPA, Evaluation for Priority Pollutant Removal from Dyestuff Manufacture Wastewaters, EPA/600/S2-84-055, U.S. Environmental Protection Agency, Washington, DC, April, 11p., 1984.
ND, not detected.
[a] Duplicate samples were not run.

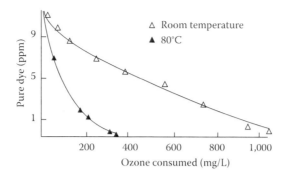

FIGURE 7.6 Ozone decolorization of disperse yellow. (From U.S. EPA, Best management practices for pollution prevention in the textile industry, EPA/625/R96-004. U.S. Environmental Protection Agency, Washington, DC, September, 320 p. 1996.)

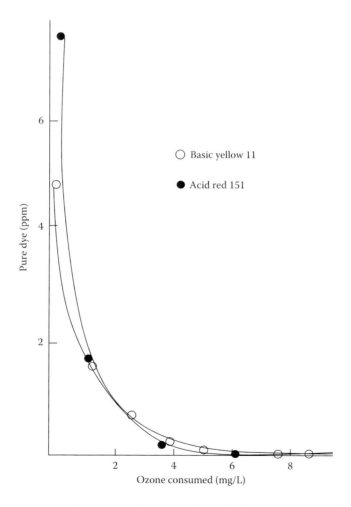

FIGURE 7.7 Ozone decolorization of Basic Yellow 11 and Acid Red 151. (From U.S. EPA, Best management practices for pollution prevention in the textile industry, EPA/625/R96-004. U.S. Environmental Protection Agency, Washington, DC, September, 320 p. 1996.)

TABLE 7.38
Summary of Treatability (Decolorization) of Dyeing Wastewaters Nos 1–20 by Ozonation

Very Treatable by Ozone	Poorly Treatable by Ozone	Moderately Treatable by Ozone
Reactive dyeings—5, 16	Disperse dyeings—3, 9, 14, 17	Basic dyeings—8, 18
1:2 metal-complex dyeing—2	Acid dyeing—10	Sulfur dyeing—15
After-copperable direct dyeing—4	Direct dyeing—11	Azoic dyeing—20
Direct developed dyeing—12		Disperse/acid/basic dyeing—19
Disperse/acid/basic dyeing—13		

Source: U.S. EPA, Textile dyeing wastewaters: Characterization and treatment, EPA-600/2-78-098, U.S. Environmental Protection Agency, Washington, DC, May, 310 p., 1978.

Note: See Table 7.11 for list of dyeing wastewaters nos 1–20; dyeing wastewaters 1, 6, and 7 were not treated by ozone.

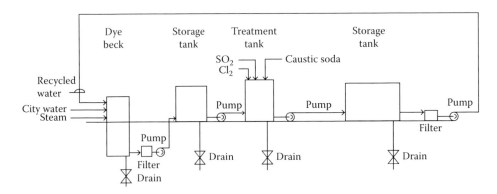

FIGURE 7.8 Flow diagram for chlorination water treatment system. (From U.S. EPA, Best management practices for pollution prevention in the textile industry, EPA/625/R96-004. U.S. Environmental Protection Agency, Washington, DC, September, 320 p., 1996.)

7.5.6.1 Agricultural Waste

Agricultural waste serves as an appealing material for dye removal not only because it is available in large quantities but also due to its physical–chemical properties.

7.5.6.1.1 Sawdust

Sawdust is an abundant by-product of the wood industry, which is used either as cooking fuel or as packing material. It is easily available in the countryside at relatively low or negligible price. It consists of lignin, cellulose, and hemicellulose, with polyphenolic groups playing an important role for binding dyes through different mechanisms. Several interactions occur during the adsorption process, which include complexation, ion exchange due to surface ionization, and formation of hydrogen bonds. The extent of dye removal involving sawdust materials shows that it is pH-dependent. Therefore, it is suggested that the ionic charges on the dyes and the character of the biosorbent play a decisive role on dye adsorption. Table 7.40 summarizes the findings from numerous studies to demonstrate sawdust's potential in removing dyes from wastewater (in both untreated and treated form). Certain chemical treatment introduced extractive sites such as nitrogen-, sulfur- and phosphorous-containing groups, and this is beneficial for dye removal.

TABLE 7.39
Results of Combined Biological/Physical–Chemical Treatment for Dyeing Wastewater Nos 17, 18 and 20

	Initial	After Biological Treatment	Before Physical–Chemical Treatment	After Physical–Chemical Treatment[a]
Dyeing Wastewater No. 17—Vat and Disperse Dyes on Polyester[a] Cotton				
ADMI color	365	313	296	68
Apparent color	1100	—	364	72
BOD_5, mg/L	360	38	—	—
TOC, mg/L	350	83	—	—
Dyeing Wastewater No. 18—Basic Dyes on Polyester[b]				
ADMI color	1300	668	579	87
Apparent color	2040	—	1744	208
BOD_5, mg/L	1470	216	—	—
TOC, mg/L	1120	275	—	—
Dyeing Wastewater No. 20—Azoic Dye on Cotton[c]				
ADMI color	2415	1987	1683	18
BOD_5, mg/L	200	<1	—	—
TOC, mg/L	170	55	—	—

Source: U.S. EPA, Textile dyeing wastewaters: Characterization and treatment, EPA-600/2-78-098, U.S. Environmental Protection Agency, Washington, DC, May, 310 p., 1978.

[a] 55 mg/L Al with no pH adjustment. Final pH 5.9.
[b] 400 mg/L Nuchar D–16 at pH 6.7.
[c] 800 mg/L HD–3000 at pH 6.0.

7.5.6.1.2 Rice Husk

Rice husk, a waste generated from the first stage of rice milling, is a lignocellulosic material that contains approximately 20% silica in combination with a large amount of the structural polymer, lignin. Utilization of this abundant agricultural waste in dye removal is of great significance, as its poor nutritive value and abrasive character make it unattractive to be used as animal feed. The adsorptive properties of rice husk, rice straw, or materials based on its ability to remove colored pollutants from an aqueous environment are shown in Table 7.41.

7.5.6.1.3 Other Agricultural Solid Wastes

Other agricultural solid wastes can also serve as cheap and readily available resources for the removal of dyes, and these include fruit peel, pith, corncob, barley husk, wheat straw, bagasse, and so on. Table 7.42 lists the dye biorsorption capacities of other agricultural wastes.

7.5.6.2 Industrial By-Products

Because of its large quantities and local availability, industrial by-products such as fly ash, metal hydroxide sludge, and red mud can be used as adsorbents for dye removal.

TABLE 7.40
Adsorption Capacities of Sawdust

Adsorbent	Effective Treatment/ Modification	Dye	Adsorption Capacity[a]	References
Sawdust	Carbonization and activation	Bismark brown	1250	[32]
Sawdust	—	Methylene blue	142.36	[33]
Beech sawdust	Calcium chloride treated	Methylene blue	11.4–12.6	[34]
	Zinc chloride treated	Methylene blue	11.7–14.3	
	Magnesium chloride treated	Methylene blue	14.2–17.7	
	Sodium chloride treated	Methylene blue	9.2–10.2	
Beech sawdust	—	Methylene blue	9.78	[35]
	—	Red basic 22	20.2	
	Calcium chloride treated	Methylene blue	13.02–16.05	
	Calcium chloride treated	Red basic 22	23.9–29.1	
Sawdust	Carbonization and steam activation	Acid yellow 36	183.8	[36]
Rattan sawdust	Carbonization	Methylene blue	294.12	[37]
Sawdust	Carbonization	Crystal violet	341	[38]
Sawdust	—	Metanil yellow	3.49–3.57[b]	[39]
	—	Methylene blue	1.17–1.53[b]	
Sawdust	—	Methyl violet	24.6	[40]
Beech sawdust	H_2SO_4 hydrolyzed	Methylene blue	9.78	[41]
	H_2SO_4 hydrolyzed	Red basic 22	20.16	
Beech sawdust	—	Direct brown	526.3	[42]
	—	Direct brown 2	416.7	
	—	Basic blue 86	136.9	
Sawdust	—	Acid blue 25	26.2–60.2	[43]
	—	Methylene blue	27.8–76.9	
Sawdust	—	Acid blue 256	280.3	[44]
	—	Acid yellow 132	398.8	
Sawdust	Sodium hydroxide treated	Brilliant green	52.63–58.48	[45]
Neem sawdust	Acid (HCl) hydrolyzed	Malachite green	3.78–4.35	[46]

Source: U.S. EPA, Textile dyeing wastewaters: Characterization and treatment, EPA-600/2-78-098, U.S. Environmental Protection Agency, Washington, DC, May, 310 p., 1978.

Note: These reported adsorption capacities are values obtained under specific conditions. Readers are encouraged to refer to the original articles for information on experimental conditions.

[a] In the unit of mg/g.
[b] In the unit of mmol/g.

7.5.6.2.1 Metal Hydroxide

Metal hydroxide is a dried waste of the electroplating industry obtained by the precipitation of metal ions in wastewater with calcium hydroxide. It is suggested that metal hydroxide sludge, being a positively charged adsorbent, can remove azo-reactive (anionic) dyes, and the charge of the dyes is an important factor for the adsorption due to the ion exchange mechanism. Table 7.43 lists the dye adsorption capacities of metal hydroxides.

7.5.6.2.2 Fly Ash

Fly ash is a solid and finely divided residue that is produced in large quantities in coal-based thermal power plants. The chemical composition and physical properties of fly ash may vary due to the

TABLE 7.41
Adsorption Capacities of Rice Husk

Adsorbent	Effective Treatment/ Modification	Dye	Adsorption capacity[a]	References
Rice straw	Carbonization	Malachite green	148.74	[47]
Rice husk ash	—	Brilliant green	21.60–26.19	[48]
Rice husk	—	Methylene blue	40.58	[49]
Rice husk	Sodium hydroxide treated carbonization process	Malachite green	56.50–57.14	[50]
	Phosphoric acid treated carbonization process	Malachite green	76.92–92.59	
Rice husk	—	Basic red 2	35.83	[51]
Rice husk	—	Indigo carmine	29.3–65.9	[52]
Rice husk	Treated with acetic acid and hydrogen peroxide	Rhodamine B	0.0533[b]–0.0587[b]	[53]
Rice straw	—	Basic violet 7	1.8–4.1	[54]
	—	Basic blue 3	2.2–4.6	
	—	Direct yellow 50	0.15–0.6	
	—	Acid red 37	0.7–1.6	
	Treated with NaOH and anthraquinone	Basic violet 7	3.1–6.8	
	Treated with NaOH and anthraquinone	Basic blue 3	4.1–7.8	
	Treated with NaOH and anthraquinone	Direct yellow 50	1.0–2.8	
	Treated with NaOH and anthraquinone	Acid red 37	1.8–3.7	
Rice husk	Carbonization, steam digested, and acid treated	Basic blue 9	343.5	[55]
Rice husk	Carbonization and steam activation	Acid yellow 36	86.9	[36]
Rice straw	Oxalic acid and further loaded with Na ion	Basic blue 9	256.4	[56]
	Oxalic acid and further loaded with Na ion	Basic green 4	238.1	
Rice straw	Phosphoric acid and further loaded with Na ion	Basic blue 9	208.33	[57]
	Phosphoric acid and further loaded with Na ion	Basic red 5	188.68	
Rice straw	—	Malachite green	94.34	[58]
	Citric acid and further loaded with Na ion	Malachite green	256.41	
Rice husk	Quaternization with N-(3-chloro-2-hydroxypropyl)-trimethylammonium chloride	Reactive blue 2	130	[59]
Rice husk	—	Basic blue 3 (singly)	13.41	[60]
	—	Reactive orange 16 (singly)	Negligible	
	—	Basic blue (binary)	14.12	
	—	Reactive orange 16 (binary)	Negligible	
	Ethylenediamine	Basic blue 3 (singly)	3.29	
	Ethylenediamine	Reactive orange 16 (singly)	24.88	

(*Continued*)

TABLE 7.41 (Continued)
Adsorption Capacities of Rice Husk

Adsorbent	Effective Treatment/ Modification	Dye	Adsorption capacity[a]	References
	Ethylenediamine	Basic blue (binary)	14.68	
	Ethylenediamine	Reactive orange 16 (binary)	60.24	
Rice husk	Ethylenediamine	Methylene blue (singly)	46.30	[61]
	Ethylenediamine	Methylene blue (binary)	49.51	
Rice husk	—	Neutral red	25.16–32.37	[62]
Rice husk	—	Methylene blue	1347.7	[63]
Rice husk ash	—	Methylene blue	1455.6	[63]
Rice husk	Treated with NaOH	Crystal violet	44.87	[64]
Rice husk	Treated with NaOH	Malachite green	12.16–17.98	[65]
Rice husk ash	—	Methylene blue	18.15	[66]
		Congo red	7.05	
Rice husk	—	Everdirect orange 3GL	27.17–28.41	[67]
	Immobilized with carboxy methyl cellulose (CMC)	Everdirect orange 3GL	20.07–23.36	
	Polyvinyl alcohol (PVA) alginate immobilized	Everdirect orange 3GL	6.81–8.77	
	Treated with HCl	Everdirect orange 3GL	30.01–30.96	
	—	Direct blue 67	44.57–52.63	
	Immobilized with carboxy methyl cellulose (CMC)	Direct blue 67	23.42–35.37	
	Polyvinyl alcohol (PVA) alginate immobilized	Direct blue 67	3.022–3.83	
	Treated with HCl	Direct blue 67	37.001–71.43	
Rice husk	—	Brilliant vital red	10.06	[68]
Rice husk	—	Murexide	15.06	[69]
Rice husk	—	Direct red 31	74.07–129.87	
	—	Direct orange	29.41–66.67	[70]
Rice husk ash	—	Indigo carmine	29.28–65.91	[52]

Note: These reported adsorption capacities are values obtained under specific conditions. Readers are encouraged to refer to the original articles for information on experimental conditions.

[a] In the unit of mg/g.
[b] In the unit of mmol/g.

---

variations in coals from different sources as well as differences in the design of coal-fired boilers. Nevertheless, the primary components that have been recognized are alumina (Al_2O_3), silica (SiO_2), calcium oxide (CaO), and iron oxide (Fe_2O_3). One of the most appealing uses of fly ash in wastewater treatment is its applicability for the removal of dyes. From Table 7.44, it is apparent that this type of material shows different dye adsorption capacities, and this can be attributed to different properties of fly ash, for example, lime content and particle size of fly ash.

7.5.6.2.3 Red Mud

Red mud is a by-product of the aluminum industry obtained from bauxite processing by Bayer process. It is composed of a mixture of solid and metallic oxide-bearing impurities. The red color is caused by the high content of oxidized iron. In addition to iron, the other dominant particles include silica, unleached residual aluminum, and titanium oxide. Table 7.45 shows the dye removal capacity of red mud.

TABLE 7.42
Adsorption Capacities of Other Agricultural Waste

Adsorbent	Dye	Adsorption capacity (mg/g)	References
Sugarcane bagasse	Basic blue 3	23.64	[71]
	Methylene blue	28.25	
	Basic yellow 11	67.11	
Sugarcane bagasse	Basic blue 3	37.59	[72]
	Reactive orange 16	34.48	
Durian peel	Methylene blue	49.50	[73]
Sunflower seed husk	Methylene blue	45.25	[74]
Sunflower seed husk	Reactive black 5	85.71[b] (%)	[75]
Banana peel	Methyl orange	21	
	Basic blue 9	20.8	[76]
	Basic violet 10	20.6	
Orange peel	Methyl orange	20.5	[76]
	Basic blue 9	18.6	
	Basic violet 10	14.3	
Pomelo peel	Basic blue 3	23.87	[77]
Banana pith	Direct red	5.92	[78]
Egyptian bagasse pith	Acid red 114	20	[79]
	Acid blue 25	17.5	
Egyptian bagasse pith	Acid blue 25	14.4	[80]
Treated cotton	Acid blue 25	589	[81]
	Acid yellow 99	448	
Hazelnut shell	Acid blue 25	60.2	[43]
Soy meal hull	Acid blue 92	114.94	[82]
	Acid red 14	109.89	

Note: These reported adsorption capacities are values obtained under specific conditions. Readers are encouraged to refer to the original articles for information on experimental conditions.

[a] In the unit of mg/g.
[b] In the unit of mmol/g.

TABLE 7.43
Adsorption Capacities of Metal Hydroxide

Adsorbent	Dye	Adsorption Capacity (mg/g)	References
Metal hydroxide sludge	Reactive red 2	62.5	[83]
	Reactive red 141	56.18	
	Reactive red 120	48.31	
Fe(III)/Cr(III) hydroxide	Basic blue 9	22.8	[84]
	Direct red 12B	5.0	

Note: These reported adsorption capacities are values obtained under specific conditions. Readers are encouraged to refer to the original articles for information on experimental conditions.

TABLE 7.44
Adsorption Capacities of Fly Ash

Adsorbent	Effective Treatment/Modification	Dye	Adsorption Capacity[a]	References
Bagasse fly ash	—	Orange G	18.796	[85]
	—	Methyl violet	26.248	
Bagasse fly ash	—	Congo red	11.885	[86]
Bagasse fly ash	—	Malachite green	170.33	[87]
Bagasse fly ash	—	Brilliant green	114.76–133.33	[88]
Coal fly ash	—	Acid red 1	92.59–103.09	[89]
	Heat treatment at 600°C	Acid red 1	32.79–52.63	
	NaOH	Acid red 1	12.66–25.12	
Fly ash	—	Methylene blue	5.718	[90]
Fly ash	—	Methylene blue	0.02[b]	[91]
Fly ash	—	Astrazon blue	128.21–152.44	[92]
Bagasse fly ash	—	Auramie O	31.177	[93]
Fly ash	—	Chrysoidine R	1000–2000	[94]
Fly ash	—	Methylene blue	0.006[b]	[95]
	NaOH	Methylene blue	0.008[b]	
	Sonochemically treated with NaOH	Methylene blue	0.016–0.040[b]	
Fly ash	Hydrothermal treated with NaOH	Methylene blue	0.0001[b]	[96]
Coal fly ash	Solid-state fusion method using NaOH	Methylene blue	0.0483[b]	[97]
Fly ash	—	Congo red	0.047[b]	[98]
Fly ash	—	Remazol brilliant blue	179.64	[99]
	—	Remazol red 133	87.41	
	—	Rifacion yellow HED	61.24	
	Heat treatment	Methylene blue	0.014[b]	
Fly ash	HNO_3	Methylene blue	0.025[b]	[100]
Fly ash	Zeolization using NaOH	Methylene blue	0.03[b]	[101]
	Zeolization using NaOH	Crystal violet	0.02[b]	
High lime Soma fly ash	—	Reactive black 5	7.184	[102]
Biomass fly ash	—	Reactive black 5	4.38	[103]
	—	Reactive yellow 176	3.65	
Coal fly ash	Treated with H_2SO_4	Methylene blue	0.0021[b]	[104]
Fly ash	—	Methylene blue	0.0189[b]	[105]
	—	Rhodamine blue	0.0115[b]	
	—	Egacid orange II	0.2364[b]	
	—	Egacid red G	0.1405[b]	
	—	Egacid yellow G	0.0052[b]	
	—	Midlon black VL	0.0033[b]	
Fly ash	Hydrothermal treated using NaOH	Methylene blue	0.00563[b]	[106]
	Hydrothermal treated using NaOH	Rhodamine B	0.00513[b]	
Fly ash	—	Methylene blue	0.012[b]	[107]
	—	Crystal violet	0.008[b]	
	—	Rhodamine B	0.007[b]	
	Microwave	Methylene blue	0.020[b]	
	Microwave	Crystal violet	0.016[b]	
	Microwave	Rhodamine B	0.010[b]	
Municipal solid waste incineration fly ash	—	Methylene blue	10.20	[108]

(Continued)

TABLE 7.44 (*Continued*)
Adsorption Capacities of Fly Ash

Adsorbent	Effective Treatment/Modification	Dye	Adsorption Capacity[a]	References
Fly ash	—	Methylene blue	11.0865	[109]
Coal fly ash	—	Indigo carmine	1.48	[110]
Fly ash	—	Acid black 1	18.939	[111]
		Acid blue 193	22.075	
		Reactive red 23	5.041	
		Reactive blue 171	3.754	

Note: These reported adsorption capacities are values obtained under specific conditions. Readers are encouraged to refer to the original articles for information on experimental conditions.

[a] In the unit of mg/g.
[b] In the unit of mmol/g.

TABLE 7.45
Adsorption Capacities of Red Mud

Adsorbent	Dye	Adsorption Capacity[a]	References
Red mud	Direct red 28	4.05	[112]
Red Mud	Acid violet	1.37	[113]
Activated red mud	Congo red	7.08	[114]
Activated red mud	Acid blue 113	0.172	[115]
$MgCl_2$/Red mud	Reactive blue	27.405	[116]
	Acid red	8.921	
	Direct blue	22.563	
Red mud	Rhodamine B	0.0101–0.0116[b]	
	Fast green	0.0725–0.0935[b]	
	Methylene blue	0.0435–0.0523[b]	[117]
Red mud	Methylene blue	0.0015–0.0078	[118]

Note: These reported adsorption capacities are values obtained under specific conditions. Readers are encouraged to refer to the original articles for information on experimental conditions.

[a] In the unit of mg/g.
[b] In the unit of mmol/g.

7.5.6.3 Natural Materials

Natural materials that are used as low-cost adsorbents for dye removal are those that exist in nature and are used in its natural state or with minor treatment.

7.5.6.3.1 Peat

Peat is a porous and rather complex material containing a wide range of lignin, cellulose, fulvic, and humic acid as its major constituents. Generally, peat can be classified into four groups: moss peat, herbaceous peat, woody peat, and sedimentary peat. It was observed that peat tends to have a high cation exchange capacity and is an effective sorbent for the removal of dyes (Table 7.46). For

TABLE 7.46
Adsorption Capacities of Peat

Adsorbent	Dye	Adsorption Capacity, mg/g	References
Peat	Acid blue 25	12.7	[119]
	Basic blue 69	195	
Peat	Basic violet 14	400	[120]
	Basic green 4	350	
Peat	Basic yellow 21	660	[121]
	Basic blue 3	550	
	Basic red 22	310	
Peat	Basic blue 3	41	[122]
	Basic orange 2	92	
Modified peat-resin particles	Acid orange 2	71.43	[123]
Peat	Reactive black 5	7.003	[124]
Peat	Dark blue CIBA WR	15.92	[125]
	Navy CIBA WB	16.39	
	Yellow CIBA WR 200 %	20.96	
	Red CIBA WB 150 %	28.25	
Peat	Basic blue 3	40.6–41.0	[126]
	Basic orange 2	92.6	
	Basic blue 24	667	
	Basic green 4	526–588	
	Basic violet 4	667–714	
	Acid black 1	25–26	
	Cibacron dark blue WR	15.9	
	Cibacron navy WB	16.4	
	Cibacron yellow WR 200%	21.0	
	Cibacron red WB 150%	28.2	

Note: These reported adsorption capacities are values obtained under specific conditions. Readers are encouraged to refer to the original articles for information on experimental conditions.

the removal of acid and basic dyes, its performance was comparable with that of activated carbon, while for disperse dyes, the performance was much better.

7.5.6.3.2 Chitin and Chitosan

Chitin is a natural biopolymer, which has a chemical structure similar to cellulose, and is generally found in a wide range of natural sources such as exoskeletons of crustaceans, cell wall of fungi, insects, annelids, and molluscs. Chitosan is a poly(aminosaccharide) consisting mainly of poly (1 → 4)-2 amino-2-deoxy-D-glucose unit, produced by deacetylation of chitin. Both chitin and chitosan are mechanically tough polysaccharides that have shown their versatility in the removal of many classes of dyes (Table 7.47).

7.5.6.3.3 Biomass

The term biomass can be interpreted to include a wide range of materials. However, for our discussion, biomass refers to fungal, algae, yeasts, and microbial culture. The use of biomass for decolorization of dyehouse wastewater is increasing because of its availability in large quantities and at low cost. Decolorization using fungal culture can be classified based on their life state: living cells to biodegrade and biosorb dyes, while dead cells to adsorb dyes. Information on the use of biomass to decolorize dyes is presented in Table 7.48.

TABLE 7.47
Adsorption Capacities of Chitosan

Adsorbent	Effective Treatment/Modification	Dye	Adsorption Capacity[a]	References
Chitosan hydrobeads	Conditioned with ammonium sulfate	Eosin Y	80.84	[127]
Chitosan	Grafted with poly(acrylamide)	Remazol yellow Gelb 3RS	1211	[128]
Chitosan	Grafted with poly(acrylic acid)	Basic yellow 37	595	
Chitosan	Chemically modified with acetic acid	Reactive black 5	2.91	[129]
Chitosan	—	Remazol black 13	91.5–130.0	[130]
Chitosan bead	—	Malachite green	82.2–93.6	[131]
Chitosan nanoparticles	Ionic gelation with tripolyphosphate	Eosin Y	3333	[132]
Chitosan cross-linked with glutaraldehyde in the presence of magnetite	Chemically modified with tetraethylenepentamine	Reactive black 5	0.62–0.69[b]	[133]
	Chemically modified with tetraethylenepentamine and glycidyl trimethylammonium chloride	Reactive black 5	0.78–0.90[b]	
Chitosan	Quarternization with glycidyl trimethylammonium chloride	Reactive Orange 16	1060	[134]
Chitosan	Grafted with poly(methylmethacrylate)	Procion yellow MX	250	[135]
	Grafted with poly(methylmethacrylate)	Remazol brilliant violet	357	
	Grafted with poly(methylmethacrylate)	Reactive blue H5G	178	
Chitosan	—	Orange II	0.32–0.33[b]	[136]
	—	Crystal violet	745–1546[b]	
	Monochloroacetic acid	Crystal violet	39900–41400[b]	
Chitosan flakes	—	Reactive red 222	339	[137]
	—	Reactive yellow 145	188	
	—	Reactive blue 222	199	
Chitosan beads	—	Reactive red 222	1653	
	—	Reactive yellow 145	885	
	—	Reactive blue 222	1009	
TiO$_2$–chitosan hybrid materials	—	Methylene blue	0.0016–0.0038	[138]
	—	Benzopurpurin	0.0074–0.016	
Chitosan	—	Acid red 37	128.2	[139]
	—	Acid blue 25	263.2	

(*Continued*)

TABLE 7.47 (Continued)
Adsorption Capacities of Chitosan

Adsorbent	Effective Treatment/Modification	Dye	Adsorption Capacity[a]	References
Chitosan nanoparticles	Cross-linked with ethylene glycol diglycidyl ether	Acid red 37	59.5	
	Cross-linked with ethylene glycol diglycidyl ether	Acid blue 25	142.9	
	Ionic gelation with tripolyphosphate	Acid green 27	2103.6	[140]
Chitosan	Radiation grafted with 2-acrylamido-2-methyl propane sulfonic acid	Methylene blue	74.3	[141]
		Acid red 37	46.1	
Chitosan beads embedded with loofash fibers	Dried form	Reactive red 222	1215	[142]
	Dried form	Acid orange 51	494	
	Dried form	Methylene blue	202	
	Wet form	Reactive red 222	1498	
	Wet form	Acid orange 51	656	
	Wet form	Methylene blue	222	
Chitosan beads	Non cross-linked	Reactive red 189	950	[143]
	Ionic cross-linked with sodium tripolyphosphate and chemical cross-linked with epichlorohydrin	Reactive red 189	1834	
Chitosan beads	Ionic cross-linked with sodium tripolyphosphate and chemical cross-linked with epichlorohydrin	Metanil yellow	1334	[144]
		Reactive blue 15	722	
Chitosan flakes	—	Acid blue 161	471.6	[145]
N,O-carboxymethyl-chitosan/montmorillonite	—	Congo red	74.24	[146]
Chitosan/palm ash composite beads	Cross-linked with epichlorohydrin	Reactive blue 19	416.7–909.1	[147]
Chitosan	—	Congo red	78.90	[148]
	Monochloroacetic acid	Congo red	330.62	
Chitosan	—	Acid orange 10	1.54[b]	[149]
	—	Acid orange 12	2.66[b]	
	—	Acid red 18	1.11[b]	
	—	Acid red 73	1.25[b]	
	Nanoparticles through ionic gelation with tripolyphosphate	Acid orange 10	1.77[b]	

(Continued)

TABLE 7.47 (Continued)
Adsorption Capacities of Chitosan

Adsorbent	Effective Treatment/Modification	Dye	Adsorption Capacity[a]	References
Chitosan hydrogel beads	Nanoparticles through ionic gelation with tripolyphosphate	Acid orange 12	4.33[b]	[150]
	Nanoparticles through ionic gelation with tripolyphosphate	Acid red 18	1.37[b]	[151]
	Nanoparticles through ionic gelation with tripolyphosphate	Acid red 73	2.13[b]	
Chitosan	Impregnated with cetyl trimethyl ammonium bromide	Congo red	433.12	
	—	Acid green 25	645.1	
	—	Acid orange 10	922.9	
	—	Acid orange 12	973.3	
	—	Acid red 18	693.2	
	—	Acid red 73	728.2	
Chitosan	—	Congo red	81.23	[152]
	Chitosan/montmorillonite nanocomposites	Congo red	54.52	
Chitosan	Grafted with 4-formyl-1,3-benzene sodium disulfonate	Basic blue 3	166.5	[153]
Chitosan beads	Non-cross-linked	Reactive red 189	1189	[154]
	Cross-linked with epichlorohydrin	Reactive red 189	1936	
Chitosan flakes	—	Reactive red 222	293–398	[155]
Chitosan beads	—	Reactive red 222	1026–1106	[156]
Chitosan beads	Ionic cross-linked with sodium tripolyphosphate and chemical cross-linked with epichlorohydrin	Acid orange 12	1954	
		Acid orange 7	1940	
		Acid red 14	1940	
		Direct red 81	2383	
		Reactive blue 2	2498	
		Reactive red 2	2422	
		Reactive orange 14	2171	
		Reactive yellow 86	1911	
Chitosan	Modified with tetraethylenepentamine	Eosin Y	292.4	[157]
Chitosan	—	Reactive red	22.48	[158]
	—	Reactive blue	70.08	

(Continued)

TABLE 7.47 (Continued)
Adsorption Capacities of Chitosan

Adsorbent	Effective Treatment/Modification	Dye	Adsorption Capacity[a]	References
Chitosan	25.54 % (w/v) deacetylated chitin	Crystal violet	14.23	[159]
	27.22 % (w/v) deacetylated chitin	Crystal violet	13.09	
	35.06 % (w/v) deacetylated chitin	Crystal violet	31.65	
Chitosan hydrogel/SiO_2	Sol–gel method	Reactive black 5	0.081 ± 0.004[b]	[160]
		Erythrosine B	0.080 ± 0.005[b]	
		Neutral red	0.88 ± 0.02[b]	
		Gentian violet	0.17 ± 0.02[b]	
Chitin hydrogel/SiO_2	Sol–gel method	Reactive black 5	0.0062 ± 0.0004[b]	[160]
		Erythrosine B	0.15 ± 0.01[b]	
		Neutral red	1.06 ± 0.02[b]	
		Gentian violet	0.14 ± 0.01[b]	
Chitosan	Cross-linked with succinyl group	Methylene blue	289.02	[161]
Chitosan hydrogel beads	Formed by sodium dodecyl sulfate gelation	Methylene blue	129.44	[162]
Chitosan microspheres	Modified with poly(methacrylic acid)	Methylene blue	1000	[163]
Hydrophobically modified O-carboxymethyl chitosan	—	Congo red	281.97	[164]
Magnetic chitosan nanoparticles	Ethylenediamine modified	Acid orange 7	3.47[b]	[165]
	Ethylenediamine modified	Acid orange 10	2.25[b]	
Chitosan	Dissolved chitosan	Reactive red 120	81	[166]
	nanodispersed	Reactive red 120	910	
Chitosan	—	Acid blue 9	210	[167]
		Food yellow 3	295	

Note: These reported adsorption capacities are values obtained under specific conditions. Readers are encouraged to refer to the original articles for information on experimental conditions.

[a] In the unit of mg/g.
[b] In the unit of mmol/g.

TABLE 7.48
Adsorption Capacities of Biomass

Adsorbent	Dye	Adsorption capacity[a] (mg/g)	References
Living fungi *Aspergillus niger*	Basic blue 9	1.17	[168]
Living fungi *Rhizopus oryzae*	Rhodamine B	90[b]	[169]
Modified fungi *Aspergillus niger*	Acid blue 29	17.58	[168]
	Disperse red 1	5.59	
Dead fungi *Aspergillus niger*	Basic blue 9	18.54	[168]
	Direct red 28	14.72	
	Acid blue 29	13.82	
Dead fungi *Aspergillus niger*	Synazol	88[b]	[170]
Dead fungi *Aspergillus foetidus*	Reactive black 5	>99[b]	[171]
Dead fungi *Funalia trogii*	Astrazone blue FGRL	48[b]	[172]
	Cibracron red	38[b]	
Dead fungi *P. chrysosporium*	Astrazone blue FGRL	60[b]	[172]
	Cibracron red	51[b]	
Dead fungi *Cunninghamella elegans*	Direct red 80	100[b]	[173]
	Reactive blue 214	99[b]	
	Reactive blue 19	57–63[b]	
Dead fungi *Rhizomucor pusillius*	Direct red 80	100[b]	[173]
	Reactive blue 214	>98[b]	
	Reactive blue 19	84–98[b]	
Dead fungi *Rhizopus arrhizus*	Reactive black 5	588.2	[174]
Dead fungi *Rhizopus arrhizus*	Reactive orange 16	190	[175]
	Reactive red 4	150	
	Reactive blue 19	90	
Dead fungi *Rhizopus stolonifer*	Direct red 80	100[b]	[173]
	Reactive blue 214	98[b]	
	Reactive blue 19	99[b]	
Yeast *Saccharomyces cerevisiae*	Remazol black B	88.5	[176]
	Remazol black	84.6	
	Remazol red RB	48.8	
Yeast *Saccharomyces cerevisiae*	Reactive green	83[b]	[177]
	Reactive blue 38	83[b]	
	Reactive blue 3	91[b]	
Yeast *Candida tropicalis*	Remazol black	182	[178]
Yeast *Candida tropicalis*	Reactive black 5	>9[b]	[179]
Yeast *Candida lipolytica*	Remazol black	250	[178]
Yeast *Candida quilliermendii*	Remazol black	154	[178]
Yeast *Candida utilis*	Remazol black	114	[178]
Yeast *Debaryomyces polymorphus*	Reactive black	>90[b]	[179]
Yeast *Rhodotorula rubra*	Crystal violet	>99[b]	[180]
Bacterial biomass *Streptomyces* BW130	Azo-reactive red 147	29[b]	[181]
	Azo-copper red 171	73[b]	
Bacterial biomass *Streptomyces rimosus*	Methylene blue	86[b]	[182]

(*Continued*)

TABLE 7.48 (Continued)
Adsorption Capacities of Biomass

Adsorbent	Dye	Adsorption capacity[a] (mg/g)	References
Bacterial biomass *Corynebacterium glutamicum*	Reactive black 5	94[b]	[183]
Algal biomass *Chlorella vulgaris*	Reactive red 5	555.6	[184]
	Remazol black B	52.4	
Algal biomass *Enteromorpha prolifera*	Acid red 274	96.4[b]	[185]
Algal biomass *Azolla filiculoides*	Acid red 88	43.6[b]	[186]
	Acid green 3	53.4[b]	
	Acid orange 7	43.8[b]	
Algal bimass *Caulerpa scalpelliformis*	Basic yellow	90[b]	[187]
Biomass *Spirodela polyrrhiza*	Basic blue 9	144.93	[188]

Note: These reported adsorption capacities are values obtained under specific conditions. Readers are encouraged to refer to the original articles for information on experimental conditions.

[a] In the unit of mg/g.
[b] Decolorization percentage (%).

TABLE 7.49
Adsorption Capacities of Clay

Adsorbent	Dye	Adsorption Capacity[a]	References
DTMA–bentonite	Acid blue 193	740.5	[189]
Activated clay/carbons mixture	Acid blue 9	64.7	[190]
Activated clay	Acid blue 9	57.8	[191]
Kaolinite	Metomega chrome orange	0.6506	[192]
Activated bentonite	Sella fast brown	360.5	[193]
Activated clay	Basic blue 69	585	[194]
	Basic red 22	488.4	
Bentonite	Basic blue 9	1667	[195]
Diatomite	Basic blue 9	198	[196]
Diatomite	Basic blue 9	0.42[b]	[197]
Clay	Basic blue 9	6.3	[198]
Bentonite	Basic red 2	274	[199]
Acidic activated clay	Basic red 46	1.968[b]	[200]
	Reactive yellow 181	0.046[b]	
Illite clay	Methyl violet	159.95	[201]
Sepiolite clay	Basic red 46	110	[202]
	Direct blue 85	332	

Note: These reported adsorption capacities are values obtained under specific conditions. Readers are encouraged to refer to the original articles for information on experimental conditions.

[a] In the unit of mg/g.
[b] In the unit of mmol/g.

TABLE 7.50
Advantages and Disadvantages of the Current Methods of Dye Removal from Industrial Effluents

Technology	Advantages	Disadvantages
Oxidation	Relatively simple, rapid, and efficient process	High energy cost, consumption of chemicals, formation of by-products, properly controlled pH and oxidation step are needed
Coagulation and flocculation	Relatively low capital cost and chemicals used are common and easily available	High sludge production, removal is pH-dependent, inefficient for highly soluble dyes
Biological treatment	Economically feasible	Large treatment area needed, slow process, optimal favorable conditions required, not uniformly susceptible to decomposition, less flexibility in design and operation
Membrane filtration	Removes all type of dyes, produces high-quality treated effluents	Expensive, high pressure and energy consumption, expensive membrane and short life span, clogging problems
Adsorption by activated carbon	Efficient adsorbent, high affinity and adsorption capacities	Not effective for disperse, direct and vat dyes, costly regeneration process, loss in adsorption capacity
Low-cost adsorbents	Economically attractive, utilization of natural resources, regeneration is avoidable	Pretreatment or chemical modification required

7.5.6.3.4 Clays

Clay materials possess a layered structure, and they are classified by the differences in their layered structures. There are various types of clays such as bentonite, sepiolite, ball clay, fuller's earth (attapulgite and montmorillonite varieties), and kaolin. The adsorption capabilities result from a net negative charge on the structure of minerals. This negative charge provides clay the competence to adsorb positively charged species. Besides, the high surface area and high porosity are also favorable for the adsorption process. Good removal capability of clay materials to take up dye has been demonstrated (Table 7.49). The adsorption of dyes on clay materials is mainly dominated by ion exchange processes.

Table 7.50 summarizes some of the advantages and disadvantages of the main treatment methods for dye removal.

7.5.7 Cost for Treatment of Textile Wastewaters

Wastewater treatments add extra costs for the textile industry. Cost effectiveness of various treatment and control technologies in seven subcategories within the textile industry has been analyzed. Tables 7.51 through 7.57 illustrate the probable increases in finished product prices for small-, medium-, and some large-size plants in (1) wool scouring, (2) wool finishing, (3) low water use processing, (4) woven fabric finishing, (5) knit fabric finishing, (6) carpet finishing, and (7) stock and yarn finishing subcategories, as they are required to pay for wastewater treatment [12]. These costs were developed to reflect the conventional use of technologies in this industry.

TABLE 7.51
Wastewater Treatment Cost for Wool Scouring (Subcategory 1)

	Alternative B	Alternative C	Alternative D	Alternative E	Alternative F
Production					
1000 kg/day	20.4	6.0	20.4	6.0	6.0
(1000 lb/day)	45.0	13.3	45.0	13.3	13.3
	48.0	30.3	48.0	30.3	30.3
	105.6	66.7	105.6	66.7	66.7
Water consumption					
1000 L/day	257.4	75.7	257.4	75.7	257.4
(1000 gal/day)	68.0	20.0	68.0	20.0	20.0
	719.1	378.9	719.1	378.9	719.1
	190.0	100.1	190.0	100.1	100.1
Capital investment (USD 1,000)	151.0	15.0	107.0	151.0	392.0
	265.0	38.0	156.0	480.0	768.0
Annual cost (USD 1,000)	41.0	4.4	28.0	41.0	190.0
	71.0	11.2	43.0	135.6	398.0
Estimated cost					
¢/kg product	0.8	0.3	0.5	2.7	12.7
(¢/lb product)	0.4	0.1	0.2	1.2	5.7
	0.6	0.3	0.4	1.8	5.3
	0.3	0.1	0.2	0.8	2.4

Source: U.S. EPA, Development document for effluent limitations guidelines and new source performance standards for the textile mills point source category. EPA-440174022A, U.S. Environmental Protection Agency, Washington, DC, June, 260 p., 1974.

Notes: Alternative B = Preliminary and biological treatment; Alternative C = Multi-media filtration; Alternative D = Chemical coagulation/chlorification and multi-media filtration; Alternative E = Activated carbon adsorption; Alternative F = Multiple effect evaporation and incineration.

TABLE 7.52
Wastewater Treatment Cost for Wool Finishing (Subcategory 2)

	Alternative B		Alternative C		Alternative D		Alternative E		Alternative F	
Production										
1000 kg/day	4.3	25.0	8.2	24.7	4.3	15.2	8.5	24.7	8.2	24.7
(1000 lb/day)	9.5	55.0	18.1	54.3	9.5	33.5	18.1	54.3	18.1	54.3
Water consumption										
1000 L/day	495.8	2872.8	946.0	2840.0	495.8	1748.7	943.0	2840.0	943.0	2840.0
(1000 gal/day)	131.0	759.0	250.0	750.0	131.0	462.0	250.0	750.0	250.0	750.0
Capital investment (USD 1,000)	98.0	278.0	60.0	135.0	197.0	349.0	450.0	910.0	1316.0	2991.0
Annual cost (USD 1,000)	30.0	79.0	17.7	39.8	49.0	89.0	132.8	292.5	759.0	2087.0
Estimted cost										
¢/kg product	2.8	1.3	0.9	0.6	4.6	2.3	6.5	4.7	37.0	33.8
(¢/lb product)	1.2	0.6	0.4	0.3	2.0	1.1	2.9	2.1	16.7	15.3

Source: U.S. EPA, Development document for effluent limitations guidelines and new source performance standards for the textile mills point source category, EPA-440174022A, U.S. Environmental Protection Agency, Washington, DC, June, 260 p., 1974.

Notes: Alternative B = Preliminary and biological treatment; Alternative C = Multi-media filtration; Alternative D = Chemical coagulation/chlorification and multi-media filtration; Alternative E = Activated carbon adsorption; Alternative F = Multiple effect evaporation and incineration.

TABLE 7.53
Wastewater Treatment Cost for Low Water Use Processing (Subcategory 3)

	Alternative B	Alternative C	Alternative F
Production			
1000 kg/day	1.5	1.5	1.5
(1000 lb/day)	3.3	3.3	3.3
Water consumption			
1000 L/day	18.9	18.9	18.9
(1000 gal/day)	5.0	5.0	5.0
Capital investment (USD 1000)	10.2	10.0	196.0
Annual cost (USD 1000)	3.9	3.0	95.0
Estimated cost			
¢/kg product	1.0	0.8	25.3
(¢/lb product)	0.4	0.3	9.6

Source: U.S. EPA, Development document for effluent limitations guidelines and new source performance standards for the textile mills point source category, EPA-440174022A, U.S. Environmental Protection Agency, Washington, DC, June, 260 p., 1974.
Notes: Alternative B = Preliminary and biological treatment; Alternative C = Multi-media filtration; Alternative F = Multiple effect evaporation and incineration.

Each table lists the increased cost attributed to biological treatment (alternative B) and the additional cost increases in finished product prices for multi-media filtration (alternative C), chemical coagulation/chlorification followed by multi-media filtration (alternative D), activated carbon adsorption (alternative E), and multiple effect evaporation and incineration (alternative F).

7.6 POLLUTION PREVENTION OPPORTUNITIES IN TEXTILE INDUSTRY

To minimize environmental impacts while at the same time improve efficiency and increase profits, some companies have creatively implemented pollution prevention techniques. The possible actions taken include improving management practices, reducing material inputs, employing substitution of toxic chemicals, and reengineering processes to reuse by-products. In fact, some smaller facilities tend to reduce pollutant releases through aggressive pollution prevention policies to enable themselves get below regulatory thresholds [12,15]. Several pollution prevention opportunities for textile facilities are discussed in this section.

7.6.1 Quality Control for Raw Materials

The implementation of quality control programs can ensure and maintain the quality of the raw materials. This can be achieved through prescreening and conducting testing on the materials. Besides retaining the quality, this policy also offers advantages such as increased product consistency, decreased production of off-quality goods, and less rework [2,15].

TABLE 7.54
Wastewater Treatment Cost for Woven Fabrics (Subcategory 4)

	Alternative B		Alternative C		Alternative D		Alternative E		Alternative F	
Production										
1000 kg/day	4.1	68.1	2.5	12.6	4.1	32.9	2.5	12.6	2.5	12.6
(1000 lb/day)	9.0	150.0	5.6	27.8	9.0	72.5	5.6	27.8	5.6	27.8
Water consumption										
1000 L/day	605.6	10220.0	382.0	1893.0	605.6	4920.0	382.0	1893.0	382.0	1893.0
(1000 gal/day)	160.0	2700.0	101.0	500.0	160.0	1300.0	101.0	500.0	101.0	500.0
Capital investment (USD 1000)	86.0	442.0	38.0	102.0	217.0	570.0	450.0	860.0	768.0	2197.0
Annual cost (USD 1000)	27.0	123.0	11.2	30.1	54.0	152.0	145.8	372.7	398.0	1472.0
Estimated cost										
¢/kg product	2.6	0.7	1.8	1.0	5.3	1.8	23.3	11.8	63.7	48.7
(¢/lb product)	1.2	0.3	0.8	0.4	2.4	0.8	10.4	5.4	28.4	21.2

Source: U.S. EPA, Development document for effluent limitations guidelines and new source performance standards for the textile mills point source category, EPA-440174022A, U.S. Environmental Protection Agency, Washington, DC, June, 260 p., 1974.

Notes: Alternative B = Preliminary and biological treatment; Alternative C = Multi-media filtration; Alternative D = Chemical coagulation/chlorification and multi-media filtration; Alternative E = Activated carbon adsorption; Alternative F = Multiple effect evaporation and incineration.

TABLE 7.55
Wastewater Treatment Cost for Knit Fabrics (Subcategory 5)

	Alternative B		Alternative C		Alternative D		Alternative E		Alternative F	
Production										
1000 kg/day	6.8	54.5	6.8	18.2	6.8	54.5	6.8	18.2	6.8	18.2
(1000 lb/day)	15.0	120.0	15.0	40.0	15.0	120.0	15.0	40.0	15.0	40.0
Water consumption										
1000 L/day	1136.0	9084.0	1136.0	3028.0	1136.0	9084.0	1136.0	3028.0	1136.0	3028.0
(1000 gal/day)	300.0	2400.0	300.0	800.0	300.0	2400.0	300.0	800.0	300.0	800.0
Capital investment (USD 1000)	117.0	397.0	74.0	140.0	286.0	770.0	480.0	910.0	1496.0	3148.0
Annual cost (USD 1000)	35.0	110.0	21.8	41.3	72.0	213.0	135.6	267.5	960.0	2210.0
Estimated cost										
¢/kg product	1.7	0.7	1.1	0.8	3.5	1.3	6.6	4.9	47.1	40.5
(¢/lb product)	0.8	0.3	0.5	0.3	1.6	0.6	3.0	2.2	21.3	18.4

Source: U.S. EPA, Development document for effluent limitations guidelines and new source performance standards for the textile mills point source category, EPA-440174022A, U.S. Environmental Protection Agency, Washington, DC, June, 260 p., 1974.

Notes: Alternative B = Preliminary and biological treatment; Alternative C = Multi-media filtration; Alternative D = chemical coagulation/chlorification and multi-media filtration; Alternative E = Activated carbon adsorption; Alternative F = Multiple effect evaporation and incineration.

TABLE 7.56
Wastewater Treatment Cost for Carpet Mills (Subcategory 6)

	Alternative B	Alternative C	Alternative D	Alternative E	Alternative F
Production					
1000 kg/day	7.0	5.4	7.0	5.4	5.4
(1000 lb/day)	15.5	11.9	15.5	11.9	11.9
	43.2	43.2	43.2	43.2	43.2
	95.2	95.2	95.2	95.2	95.2
Water consumption					
1000 L/day	492.0	378.5	495.0	378.5	378.5
(1000 gal/day)	130.0	100.0	130.0	100.0	100.0
	3028.0	3028.0	3028.0	3028.0	3028.0
	800.0	800.0	800.0	800.0	800.0
Capital investment (USD 1000)	98.0	38.0	197.0	400.0	768.0
	200.0	140.0	452.0	1050.0	3148.0
Annual cost (USD 1000)	30.0	11.2	49.0	116.0	398.0
	57.0	41.3	118.0	404.8	2210.0
Estimated cost					
¢/kg product	1.4	0.7	2.3	7.2	24.6
(¢/lb product)	0.7	0.3	1.1	3.2	11.1
	0.4	0.3	0.9	3.1	17.1
	0.2	0.1	0.4	1.4	7.7

Source: U.S. EPA, Development document for effluent limitations guidelines and new source performance standards for the textile mills point source category, EPA-440174022A, U.S. Environmental Protection Agency, Washington, DC, June, 260 p., 1974.

Notes: Alternative B = Preliminary and biological treatment; Alternative C = Multi-media filtration; Alternative D = Chemical coagulation/chlorification and multi-media filtration; Alternative E = Activated carbon adsorption; Alternative F = Multiple effect evaporation and incineration.

TABLE 7.57
Wastewater Treatment Cost for Stock and Yarn (Subcategory 7)

	Alternative B		Alternative C		Alternative D		Alternative E		Alternative F	
Production										
1000 kg/day	5.0	27.2	4.1	12.4	5.0	27.2	4.1	12.4	4.1	12.4
(1000 lb/day)	11.0	60.0	9.1	27.3	11.0	60.0	9.1	27.3	9.1	27.3
Water consumption										
1000 L/day	916.0	4996.0	752.0	2275.0	916.0	4996.0	752.0	2275.0	752.0	2275.0
(1000 gal/day)	242.0	1320.0	200.2	600.6	242.0	1320.0	200.2	600.6	200.2	600.6
Capital investment (USD 1000)	110.0	293.0	59.0	120.0	256.0	574.0	400.0	730.0	1132.0	2521.0
Annual cost (USD 1000)	33.0	83.0	17.4	35.4	64.0	153.0	116.0	221.4	638.0	11721.0
Estimated cost										
c/kg product	2.2	1.5	1.4	1.0	4.3	1.9	9.4	6.0	51.9	46.3
(c/lb product)	1.0	0.7	0.6	0.4	1.9	0.9	4.2	2.7	23.3	21.0

Source: U.S. EPA, Development document for effluent limitations guidelines and new source performance standards for the textile mills point source category, EPA-440174022A, U.S. Environmental Protection Agency, Washington, DC, June, 260 p., 1974.

Notes: Alternative B = Preliminary and biological treatment; Alternative C = Multi-media filtration: Alternative D = Chemical coagulation/chlorification and multi-media filtration; Alternative E = Activated carbon adsorption; Alternative F = Multiple effect evaporation and incineration.

TABLE 7.58
Typical Water Savings Using Countercurrent Washing

Number of Washing Steps	Water Savings (%)
2	50
3	67
4	75
5	80

Source: U.S. EPA, Best management practices for pollution prevention in the textile industry, EPA/625/R96-004. U.S. Environmental Protection Agency, Washington, DC, September, 320 p., 1996.

TABLE 7.59
Example of Costs and Savings for Dyebath Reuse

Description of Cost/Savings	Value (USD)
Total Costs	
Lab and support equipment	9,000
Machine modifications, tanks, pumps, pipes	15,000–25,000
Annual operating costs	1,000–2,000
Total Savings (Annual)	
Dyes and chemicals	15,000
Water	750
Sewer	750
Energy	4,500

Source: U.S. EPA, Best management practices for pollution prevention in the textile industry, EPA/625/R96-004. U.S. Environmental Protection Agency, Washington, DC, September, 320 p., 1996.

7.6.2 CHEMICAL SUBSTITUTION

Most of the processes in textile manufacturing are chemically demanding processes. Therefore, the substitution of less polluting chemicals for textile processes could be rewarding. This approach reduces not only the hazardous effect of the chemical waste but also the cost of the treatment involved. However, it may vary substantially among mills because of differences in surroundings, process conditions, product, and raw materials [2,15].

7.6.3 PROCESS MODIFICATION

Optimization processes could reduce waste and increase production efficiency, whereas a combination of certain operations such as scouring and bleaching can save sources (e.g., energy and water). The introduction of low bath ratio dyeing can also save energy and reduce chemical usage. The use of pad batch (cold) dyeing for cotton, rayon, and blends offers benefits such as conservation of energy, water, dyes and chemicals, labor, and floor space. This method does not require salt or chemical specialties and appears to be a good way for facilities to reduce waste and save money.

TABLE 7.60
Effluent Limitations of Subcategories 1–7 That Represent the Degree of Effluent Reduction Attainable by the Application of the Best Practicable Control Technology (BPT) Currently Available

Maximum Thirty Day Average Effluent limitations Guidelines[a] for July 1, 1977

Subcategory	BOD$_5$	TSS	COD	Total Chromium	Phenol	Sulfide
Wool scouring[c,d]	5.3	16.1	69.0	0.05	0.05	0.10
Wool finishing[d]	11.2	17.6	81.5	0.07	0.07	0.14
Dry processing[c]	0.7	0.7	1.4	—	—	—
Woven fabric finishing[d]	3.3	8.9	30–60	0.05	0.05	0.10
Knit fabric finishing[d]	2.5	10.9	30–50	0.05	0.05	0.10
Carpet mills	3.9	5.5	35.1–45.1	0.02	0.02	0.04
Stock and yarn dyeing and finishing[d]	3.4	8.7	42.3	0.06	0.06	0.12

Source: U.S. EPA, Development document for effluent limitations guidelines and new source performance standards for the textile mills point source category, EPA-440174022A, U.S. Environmental Protection Agency, Washington, DC, June, 260 p., 1974.

[a] Expressed as $\frac{\text{kg (lb) pollutant}}{\text{kkg (1000 lb) product}}$ except wool scouring as $\frac{\text{kg (lb) pollutant centre}}{\text{kkg (1000 lb) raw grease wool}}$ and carpet mills as $\frac{\text{kg (lb) pollutant centre}}{\text{kkg (1000 lb) primary backed carpet}}$

[b] Oil and grease limitation for wool scouring is $\frac{3.6 \text{ kg (lb)}}{\text{kkg (1000 lb)}}$ raw grease wool

[c] Fecal coliform limit for dry processing is 400 MPN per 100 mL.

[d] For those plants identified as commission finishers, an additional allocation of 100% of the guidelines is to be allowed for the 30 day maximum levels.

Countercurrent washing is a simple, easy to implement, and relatively inexpensive process modification that decreases wastewater from preparation processes. This technique involves the reuse of least contaminated water from the final stage for the next-to-last wash and so on, until the water reaches the first wash stage, where it is then discharged. It is a useful technique that can be retrofitted to any multi-stage continuous washing processes such as dyeing, desizing, scouring, or bleaching. Flow optimization is usually a good pollution prevention activity to run in conjunction with countercurrent washing [2,15]. Table 7.58 lists typical water savings based on the number of times the water is reused.

7.6.4 Process Water Reuse and Recycle

Recovery, recycling, and reuse can be effective tools for minimizing pollutant releases to the environment. Reduction in the cost of raw materials and pollution can be realized by recovering solvents and raw materials used in textile mills. The method is applicable in wet processing in which the consumption of water is very high.

TABLE 7.61
Effluent Limitations of Water Jet Weaving, Nonwoven Manufacturing and Felted Fabric Processing Subcategories That Represent the Degree of Effluent Reduction Attainable by the Application of the Best Practicable Control Technology (BPT) Currently Available

	Conventional Pollutants			
	Maximum for Any 1 Day		Average of Daily Values for 30 Consecutive Days	
Subcategory	BOD_5	TSS	BOD_5	TSS
Low water use processing				
Water jet weaving	8.9	5.5	4.6	2.5
Nonwoven manufacturing	4.4	6.2	2.2	3.1
Felted fabric processing	35.2	55.4	17.6	27.7

	Toxic and Nonconventional Pollutants							
	Maximum for Any 1 Day				Average of Daily Values for 30 Consecutive Days			
Subcategory	COD	Sulfide	Phenols	Total Chromium	COD	Sulfide	Phenols	Total Chromium
Low water use processing								
Water jet weaving	21.3	—	—	—	13.7	—	—	—
Nonwoven manufacturing	40.0	0.046	0.023	0.023	20.0	0.023	0.011	0.011
Felted fabric processing	256.8	0.44	0.22	0.22	128.4	0.22	0.11	0.11

Source: U.S. EPA, Development document for effluent limitations guidelines and standards for the textile mills point source category (final), EPA-440182022, U.S. Environmental Protection Agency, Washington, DC, June, 546 p., 1982.
Notes: pH shall be within the range 6.0–9.0 at all times; Expressed as kg pollutant/kkg of product (lb/1000 lb)

Dyebath reuse involves the process of analyzing, replenishing, and reusing exhausted hot dyebaths to dye further batches of material. This process provides an alternative that is less costly compared with pretreatment plant construction and offers attractive return for investment in the form of dyes, chemicals, and energy. Also, in some cases, it can reduce the effluent volume and pollution concentrations. Under properly controlled conditions, dyebaths can be reused for 15 or more cycles, with an average of 5–25 times [2,15]. Costs and savings of dyebath reuse per dyeing machine are presented in Table 7.59.

7.6.5 Equipment Modification

Modification, retrofitting, or replacing equipment could provide source and waste reduction. Computer-controlled dyeing systems allow one to analyze the process continuously and respond more quickly and accurately than manually controlled systems [2,15].

7.6.6 Good Operating Practices

Incorporation of pollution prevention strategies into company's management policies can help to improve production efficiency and maintain low operating costs. By taking this into consideration, the reduction of waste at ongoing or even at new facilities is possible. Good record keeping, providing training, and establishing incentive programs are some other options to prevent pollution

TABLE 7.62
Effluent Limitations of the Nine Subcategories That Represent the Degree of Effluent Reduction Attainable by the Application of the Best Available Technology (BAT) Currently Available

	Maximum for Any 1 Day				Average of Daily Values for 30 Consecutive Days			
Subcategory	COD	Sulfide	Phenols	Total Chromium	COD	Sulfide	Phenols	Total Chromium
Wool scouring[b]	138.0	0.20	0.10	0.10	69.0	0.10	0.05	0.05
Wool finishing[b]	163.0	0.28	0.14	0.14	81.5	0.14	0.07	0.07
Low Water Use Processing[b]								
General processing	2.8	—	—	—	1.4	—	—	—
Water jet weaving	21.3	—	—	—	13.7	—	—	—
Woven fabric finishing[b]	60.0	0.20	0.10	0.10	30.0	0.10	0.05	0.05
Knit fabric finishing[b]	60.0	0.20	0.10	0.10	30.0	0.10	0.05	0.05
Carpet finishing	70.2	0.08	0.04	0.04	35.1	0.04	0.02	0.02
Stock and yarn finishing	84.6	0.24	0.12	0.12	42.3	0.12	0.06	0.06
Nonwoven manufacturing	40.0	0.046	0.023	0.023	20.0	0.023	0.011	0.011
Felted fabric processing	256.0	0.44	0.22	0.22	128.4	0.22	0.11	0.11

Source: U.S. EPA, Development document for effluent limitations guidelines and standards for the textile mills point source category (final), EPA-440182022, U.S. Environmental Protection Agency, Washington, DC, June, 546 p., 1982.
Notes: Expressed as kg pollutant/kkg of product (lb/1000 lb) except for wool scouring, which is expressed as kg pollutant/kkg of wool processed and wool finishing which is expressed as kg pollutant/kkg of fiber processed; For comission finishers, an additional allocation of 100% of the limitations is allowed.

without changing industrial processes. Proper maintenance of production equipment (e.g., minimizing or avoiding leaks and spills) and identification of unnecessary washing of both fabric and equipment are also important means to save water [2,15].

7.7 TEXTILE POINT SOURCE DISCHARGE EFFLUENT LIMITATIONS, PERFORMANCE STANDARD, AND PRETREATMENT STANDARDS

7.7.1 U.S. ENVIRONMENTAL REGULATIONS FOR THE NINE TEXTILE SUBCATEGORIES

Tables 7.60 through 7.62 show the final (September 1982) effluent limitations of the nine subcategories that represent the degree of effluent reduction attainable by application of either the best practicable control technology (BPT) or best available technology (BAT) currently available [2,15].

Table 7.63 lists the NSPS of the nine subcategories. Any new source must achieve NSPS.

The discharge of pollutants into publicly owned treatment works (POTWs) from any existing and new sources of the nine subcategories must achieve the pretreatment standards listed in General Pretreatment Regulations found at 40 CFR Part 403. In 1979, both revised pretreatment standards for new source (PSNS) and pretreatment standards for existing source (PSES) were proposed [2,15]. Under the proposed pretreatment standards, the controls on total chromium, total copper, and total zinc have been incorporated.

TABLE 7.63
New Source Performance Standards (NSPS) of the Nine Textile Subcategories

Subcategory	Maximum for Any 1 Day		Average of Daily Values for 30 Consecutive Days	
	BOD$_5$	TSS	BOD$_5$	TSS
Wool scouring	3.6	30.3	1.9	13.5
Wool finishing	10.7	32.3	5.5	14.4
Low Water Use Processing				
General processing	1.4	1.4	0.7	0.7
Water jet weaving	8.9	5.5	4.6	2.5
Woven Fabric Finishing				
Simple operations	3.3	8.8	1.7	3.9
Complex operations	3.7	14.4	1.9	6.4
Desizing	5.5	15.6	2.8	6.9
Knit Fabric Finishing				
Simple operations	3.6	13.2	1.9	5.9
Complex operations	4.8	12.2	2.5	5.4
Hosiery products	2.3	8.4	1.2	3.7
Carpet finishing	4.6	8.6	2.4	3.8
Stock and yarn finishing	3.6	9.8	1.9	4.4
Nonwoven manufacturing	2.6	4.9	1.4	2.2
Felted fabric processing	16.9	50.9	8.7	22.7

Source: U.S. EPA, Development document for effluent limitations guidelines and standards for the textile mills point source category (final), EPA-440182022, U.S. Environmental Protection Agency, Washington, DC, June, 546 p., 1982.
Notes: Any new source must achieve the NSPS. Expressed as kg pollutant/kkg of product (lb/1000 lb) except for wool scouring which is expressed as kg pollutant/kkg of wool processed and wool finishing which is expressed as kg pollutant/kkg of fiber processed; For all subcategories, pH within the range 6.0–9.0 at all times.

7.8 CONCLUSION

Readers are referred to the literature [203–205] for additional information of textile industry wastes, treatment technologies and environmental engineering technologies. A short conclusion is presented in this section.

The textile industry has become one of the most important industrial sectors in terms of economic value. This industry is primarily associated with the manufacture of yarn and cloth as well as the production and design of clothing and their distribution. Starting from the production of raw fiber, several processing operations such as yarn formation, fabric formation and preparation, dyeing, printing, final finishing, and product fabrication are employed in the textile mills.

In the textile industry, nine subcategories have been identified according to the raw materials used, products and production process involved, treatability of waste and treatment cost, and so on. The subcategories are wool scouring, wool finishing, low water use processing, woven fabric finishing, knit fabric finishing, carpet finishing, stock and yarn finishing, nonwoven manufacturing, and felted fabric processing. In general, all of these subcategories generate wastewater, rinse water, chemicals, and energy are consumed in large quantities in the industry, especially during wet processing.

In fact, effluents from textile industries consist of large amounts of suspended and dissolved solids, dyestuff, and other chemicals that are used in different stages of processing operations. Therefore, treatment of huge quantities of wastewater with a variety of contaminants and toxic pollutants prior to disposal is necessary to protect environment and human beings. For this purpose, several treatment methods have been developed, including biological treatment, chemical coagulation, adsorption, oxidation, and combined biological/physical–chemical treatment. Besides, the feasibility of applying low cost adsorbents from agriculture waste, industrial by-products, and biomass and natural materials in wastewater treatment has received increasing attention.

It is desirable to minimize environmental impacts without affecting production efficiency and profits. Thus, some pollution prevention techniques such as improving management practices, substituting toxic chemicals, and modifying processes to reuse by-products are widely implemented in the textile industry.

LIST OF ACRONYMS

VOCs	Volatile organic compounds
BOD	Biochemical oxygen demand (mg/L)
COD	Chemical oxygen demand (mg/L)
PVA	Polyvinyl alcohol
CMC	Carboxymethyl cellulose
TSS	Total suspended solids (mg/L)
TOC	Total organic carbon (mg/L)
TKN	Total Kjeldahl nitrogen (mg/L)
PAC	Powdered activated carbon

REFERENCES

1. U.S. EPA (2001). To riches from rags: Profiting from waste reduction. A best practices guide for textile and apparel manufacturers, EPA/region2-p2. U.S. Environmental Protection Agency, Washington, DC, April, 39 p.
2. U.S. EPA (1997). EPA office of compliance sector notebook project: Profile of the textile industry, EPA/310-R-97-009. U.S. Environmental Protection Agency, Washington, DC, September, 136 p.
3. Executive Office of the President of the United States (1987). Standard Industrial Classification Manual, Office of Management and Budget, Washington, DC.
4. U.S. EPA (1998). Preliminary industry characterization: Fabric printing, coating and dyeing, EPA/pic-fabr. U.S. Environmental Protection Agency, Washington, DC, July, 85 p.
5. United States Department of Commerce (1995). Census of manufactures, industry series, weaving and floor covering mills, industries 2211, 2221, 2231, 2241, and 2273, Bureau of the Census, Washington, DC.
6. United States Department of Commerce (1995). Census of manufactures, industry series, knitting mills, industries 2251, 2252, 2253, 2254, 2257, 2258, and 2259, Bureau of the Census, Washington, DC.
7. United States Department of Commerce (1995). Census of manufactures, industry series, dyeing and finishing textiles, except wool fabrics and knit goods, industries 2261, 2262, and 2269, Bureau of the Census, Washington, DC.
8. United States Department of Commerce (1995). Census of manufactures, industry series, yarn and thread mills, industries 2281, 2282, and 2284, Bureau of the Census, Washington, DC.
9. United States Department of Commerce (1995). Census of manufactures, industry series, miscellaneous textile goods, industries 2295, 2296, 2297, 2298, and 2299, Bureau of the Census, Washington, DC.
10. U.S. EPA (1996). Pollution prevention in the textile industry, EPA/textile ind. U.S. Environmental Protection Agency, Washington, DC, September, 136 p.
11. Kumar, S., Visvanathan, C., and Priambodo, A. (1999). *Energy and Environmental Indicators in the Thai Textile Industry*, Sustainable Energy and Environmental Technologies, Asian Institute of Technology, Pathum Thani, Thailand, 524–528.

12. U.S. EPA (1974). Development document for effluent limitations guidelines and new source performance standards for the textile mills point source category, EPA-440174022A. U.S. Environmental Protection Agency, Washington, DC, June, 260 p.
13. U.S. EPA (1982). Development document for effluent limitations guidelines and standards for the textile mills point source category (final), EPA-440182022. U.S. Environmental Protection Agency, Washington, DC, June, 546 p.
14. U.S. EPA (1978). Textile processing industry, EPA-625/778-002. U.S. Environmental Protection Agency, Washington, DC, October, 502 p.
15. U.S. EPA (1996). Best management practices for pollution prevention in the textile industry, EPA/625/R96-004. U.S. Environmental Protection Agency, Washington, DC, September, 320 p.
16. Horning, R.H. (1978). Textile dyeing wastewaters: Characterization and treatment, PB-285 115. U.S. Department of Commerce, National Technical Information Service, Washington, DC.
17. Whaley, W.M. (1984). *Dyes Based on Safer Intermediates*, Fiber and Polymer Science Seminar Series, North Carolina State University, Raleigh, NC.
18. Delee, W., O'Neil, C., Hawkes, F.R., and Pinheiro, H.M. (1998). Anaerobic treatment of textile effluents: A review. *J Chem. Technol. Biotechnol.*, Vol. 73, 323–335.
19. Forgacs, E., Cserhati, T., and Oros, G. (2004). Removal of synthetic dyes from wastewaters: A review. *Environ. Int.*, Vol. 30, 953–971.
20. Rai, H.S., Bhattacharyya, M.S., Singh, J., Bansal, T.K., Vats, P., and Banerjee, U.C. (2005). Removal of dyes from the effluent of textile and dyestuff manufacturing industry: A review of emerging techniques with reference to biological treatment. *Crit. Rev. Environ. Sci. Technol.*, Vol. 25, 219–238.
21. U.S. EPA (1978). Textile dyeing wastewaters: Characterization and treatment, EPA-600/2-78-098. U.S. Environmental Protection Agency, Washington, DC, May, 310 p.
22. U.S. EPA (1979). Development of methods and techniques for final treatment of combined municipal and textile wastewater including sludge utilization and disposal, EPA-600/2-79-160. U.S. Environmental Protection Agency, Washington, DC, December, 152 p.
23. Shu, H.Y. and Huang, C.R. (1995). Ultraviolet enhanced oxidation for color removal of azo dye wastewater. *Am. Dyestuff Rep.*, Vol. 84, 30–34.
24. Pittroff, M. and Gregor, K.H. (1992). Decolorization of textile waste waters by UV-radiation with hydrogen peroxide. *Melliand Engl.*, Vol. 6, 73.
25. Slokar, Y.M. and Le Marechal, A.M. (1997). Methods of decoloration of textile wastewaters. *Dyes Pigments*, Vol. 37, 335–356.
26. Kim, T.H., Park, C., Yang, J.M., and Kim, S. (2004). Comparisons of disperse and reactive dye removals by chemical coagulation and Fenton oxidation. *J. Hazard. Mater.*, Vol. 112, 95–103.
27. Gregor, K.H. (1992). Oxidative decolorization of textile waste water with advanced oxidation process. In: *Chemical Oxidation—Technologies for the Nineties*. Proceedings of the First International Symposium (Eckenfelder, W.W., Bowers, A.R., Roth, J.A., eds), Vanderbilt University, Nashville, TN, 161–193.
28. Gupta, V.K. and Suhas (2009). Application of low-cost adsorbents for dye removal: A review. *J. Environ. Manag.*, Vol. 90, 2313–2342.
29. Ince, N.H. and Gonenc, D.T. (1997). Treatability of a textile azo dye by UV/H_2O_2. *Environ. Technol.*, Vol. 18, 179–185.
30. Gahr, F., Hermanutz, F., and Opperman, W. (1994). Ozonation: An important technique to comply with new German law for textile wastewater treatment. *Water Sci. Technol.*, Vol. 30, 255–263.
31. U.S. EPA (1984). Evaluation for priority pollutant removal from dyestuff manufacture wastewaters, EPA/600/S2-84-055. U.S. Environmental Protection Agency, Washington, DC, April, 11 p.
32. Kumar, K.B.G., Shivakamy, K., Mirande, L.R., and Velan, M. (2006). Preparation of steam activated carbon from rubberwood sawdust (*Hevea brasiliensis*) and its adsorption kinetics. *J. Hazard. Mater.*, Vol. B136, 922–929.
33. Hamdaoui, O. (2006). Dynamic sorption of methylene blue by cedar sawdust and crushed brick in fixed bed columns. *J. Hazard. Mater.*, Vol. 138, 293–303.
34. Batzias, F.A. and Sidiras, D.K. (2007). Dye adsorption by prehydrolysed beech sawdust in batch and fixed-bed systems. *Bioresour. Technol.*, Vol. 98, 1208–1217.
35. Batzias, F.A. and Sidiras, D.K. (2004). Dye adsorption by calcium chloride treated beech sawdust in batch and fixed-bed systems. *J. Hazard. Mater.*, Vol. 114, 167–174.
36. Malik, P.K. (2003). Use of activated carbons prepared from sawdust and rice-husk for adsorption of acid dyes: A case study of Acid Yellow 36. *Dyes Pigments*, Vol. 56, 239–249.

37. Hameed, B.H., Ahmad, A.L., and Latiff, K.N.A. (2007). Adsorption of basic dye (methylene blue) onto activated carbon prepared from rattan sawdust. *Dyes Pigments*, Vol. 75, 143–149.
38. Chakraborty, S., De, S., DasGupta, S., and Basu, J.K. (2005). Adsorption study for the removal of a basic dye: Experimental and modeling. *Chemosphere*, Vol. 58, 1079–1086.
39. Pekkuz, H., Uzun, I., and Guzel, F. (2008). Kinetics and thermodynamics of the adsorption of some dyestuffs from aqueous solution by poplar sawdust. *Bioresour. Technol.*, Vol. 99, 2009–2017.
40. Ofomaja, A.E. and Ho, Y.S. (2008). Effect of temperature and pH on methyl violet biosorption by Mansonia wood sawdust. *Bioresour. Technol.*, Vol. 99, 5411–5417.
41. Batzias, F.A. and Sidiras, D.K. (2007). Simulation of methylene blue adsorption by salts-treated beech sawdust in batch and fixed-bed systems. *J. Hazard. Mater.*, Vol. 149, 8–17.
42. Dulman, V. and Cucu-Man, S.M. (2009). Sorption of some textile dyes by beech wood sawdust. *J. Hazard. Mater.*, Vol. 162, 1457–1464.
43. Ferrero, F. (2007). Dye removal by low cost adsorbents: Hazelnut shells in comparison with wood sawdust. *J. Hazard. Mater.*, Vol. 142, 144–152.
44. Ozacar, M. and Sengil, I.A. (2005). Adsorption of metal complex dyes from aqueous solutions by pine sawdust. *Bioresour. Technol.*, Vol. 96, 791–795.
45. Mane, V.S. and Babu, P.V.V. (2011). Studies on the adsorption of Brilliant Green dye from aqueous solution onto low-cost NaOH treated saw dust. *Desalination*, Vol. 273, 321–329.
46. Khattri, S.D. and Singh, M.K. (2009). Removal of malachite green from dye wastewater using neem sawdust by adsorption. *J. Hazard. Mater.*, Vol. 167, 1089–1094.
47. Hameed, B.H. and El-Khaiary, M.I. (2008). Kinetics and equilibrium studies of malachite green adsorption on rice straw-derived char. *J. Hazard. Mater.*, Vol. 153, 701–708.
48. Mane, V.S., Mall, I.D., and Srivastava, V.C. (2007). Kinetic and equilibrium isotherm studies for the adsorptive removal of Brilliant Green dye from aqueous solution by rice husk ash. *J. Environ. Manag.*, Vol. 84, 390–400.
49. Vadivelan, V. and Vasanth Kumar, K. (2005). Equilibrium, kinetics, mechanism, and process design for the sorption of methylene blue onto rice husk. *J. Colloids Interface Sci.*, Vol. 286, 90–100.
50. Rahman, I.A., Saad, B. Shaidan, S., and Sya Rizal, E.S. (2005). Adsorption characteristics of malachite green on activated carbon derived from rice husks produced by chemical–thermal process. *Bioresour. Technol.*, Vol. 96, 1578–1583.
51. Vasanth Kumar, K. and Sivanesan, S. (2007). Sorption isotherm for safranin onto rice husk: Comparison of linear and non-linear methods. *Dyes Pigments*, Vol. 72, 130–133.
52. Lakshmi, U.R., Srivastava, V.C., Mall, I.D., and Lataye, D.H. (2009). Rice husk ash as an effective adsorbent: Evaluation of adsorptive characteristics for Indigo Carmine dye. *J. Environ. Manag.*, Vol. 90, 710–720.
53. Jain, R., Mathur, M., Sikarwar, S., and Mittal, A. (2007). Removal of the hazardous dye rhodamine B through photocatalytic and adsorption treatments. *J. Environ. Manag.*, Vol. 85, 956–964.
54. Abdel-Aal, S.E., Gad, Y.H., and Dessouki, A.M. (2006). Use of rice straw and radiation-modified maize starch/acrylonitrile in the treatment of wastewater. *J. Hazard. Mater.*, Vol. 129, 204–215.
55. Kannan, N. and Sundaram, M.M. (2001). Kinetics and mechanism of removal of methylene blue by adsorption on various carbons—A comparative study. *Dyes Pigments*, Vol. 51, 25–40.
56. Gong, R.M., Jin, Y.B., Sun, J., and Zhong, K.D. (2008). Preparation and utilization of rice straw bearing carboxyl groups for removal of basic dyes from aqueous solution. *Dyes Pigments*, Vol. 76, 519–524.
57. Gong, R.M., Jin, Y.B., Chen, J., Hu, Y., and Sun, J. (2007). Removal of basic dyes from aqueous solution by sorption on phosphoric acid modified rice straw. *Dyes Pigments*, Vol. 73, 332–337.
58. Gong, R.M., Jin, Y.B, Chen, F.Y., Chen, J., and Liu, Z.L. (2006). Enhanced malachite green removal from aqueous solution by citric acid modified rice straw. *J. Hazard. Mater.*, Vol. 137, 865–870.
59. Low, K.S. and Lee, C. K. 1997. Quartenized rice husk as sorbent for reactive dyes. *Bioresour. Technol.*, Vol. 61, 121–125.
60. Ong, S.T., Lee, C.K., and Zainal, Z. (2007). Removal of basic and reactive dyes using ethylenediamine modified rice hull. *Bioresour. Technol.*, Vol. 98, 2792–2799.
61. Ong, S.T., Lee, W.N., Keng, P.S., Lee, S.L., Hung, Y.T., and Ha, S.T. (2010). Equilibrium studies and kinetics mechanism for the removal of basic and reactive dyes in both single and binary systems using EDTA modified rice husk. *Int. J. Phys. Sci.*, Vol. 5, 582–595.
62. Zow, W., Han, P., Li, Y., Liu, X., He, X., and Han, R. (2009). Equilibrium, kinetic and mechanism study for the adsorption of neutral red onto rice husk. *Desalination Water Treat.*, Vol. 12, 210–218.
63. Sharma, P., Kaur, R., Baskar, C., and Chung, W.J. (2010). Removal of methylene blue from aqueous waste using rice husk and rice husk ash. *Desalination*, Vol. 259, 249–257.

64. Chakraborty, S., Chowdhury, S., and Saha, P.D. (2011). Adsorption of crystal violet from aqueous solution onto NaOH-modified rice husk. *Carbohydr. Polym.*, Vol. 86, 1533–1541

65. Chowdhury, S., Mishra, R., Saha, P., and Kushwaha, P. (2011). Adsorption thermodynamics, kinetics and isosteric heat of adsorption of malachite green onto chemically modified rice husk. *Desalination*, Vol. 265, 159–168.

66. Chowdhury, A.K., Sarkar, A.D., and Bandyopadhyay, A. (2009). Rice husk ash as a low cost adsorbent for the removal of methylene blue and congo red in aqueous phases. *Clean Soil Air Water*, Vol. 37, 581–591

67. Safa, Y. and Bhatti, H.N. (2011). Adsorptive removal of direct dyes by low cost rice husk: Effect of treatments and modifications. *Afr. J. Biotechnol.*, Vol. 10, 3128–3142.

68. Rehman, R., Anwar, J., and Mahmud, T. (2011). Influence of operating conditions on the removal of brilliant vital red dye from aqueous media by biosorption using rice husk. *J. Chem. Soc. Pak.*, Vol. 33, 515–521.

69. Rehman, R., Anwar, J.A., Mahmud, T.A., Salman, M.U., Shafique, U., Uz-zaman, W. (2011). Removal of murexide (dye) from aqueous media using rice husk as an adsorbent. *J. Chem. Soc. Pak.*, Vol. 33, 598–603.

70. Safa, Y. and Bhatti, H.N. (2011). Kinetic and thermodynamic modeling for the removal of Direct Red-31 and Direct Orange-26 dyes from aqueous solutions by rice husk. *Desalination*, Vol. 272, 313–322.

71. Ong, S.T., Khoo, E.C., Hii, S.L., and Ha, S.T. (2010). Utilization of sugarcane bagasse for removal of basic dyes from aqueous environment in single and binary systems. *Desalination Water Treat.*, Vol. 20, 86–95.

72. Wong, S.Y, Tan, Y.P., Abdullah, A.H., and Ong, S.T. (2009). The removal of basic and reactive dyes using quartenized sugar cane bagasse. *J. Phys. Sci.*, Vol. 20, 59–74.

73. Ong, S.T., Keng, P.S., Voon, M.S., and Lee, S.L. (2011). Application of durian peel (*Durio zibethinus Murray*) for the removal of methylene blue from aqueous solution. *Asian J. Chem.*, Vol. 23, 2898–2902.

74. Ong, S.T., Keng, P.S., Lee, S.L., and Hung, Y.T. (2014). Low cost adsorbents for sustainable dye containing-wastewater treatment. *Asian. J. Chem.*, Vol. 26, 1873–1881.

75. Osma, J.F., Saravia, V., Toca-Herrera, J.L., and Couto, S.R. (2007). Sunflower seed shells: A novel and effective low-cost adsorbent for the removal of the diazo dye Reactive Black 5 from aqueous solutions. *J. Hazard. Mater.*, Vol. 147, 900–905.

76. Annadurai, G., Juang, R.S., and Lee, D.J. (2002). Use of cellulose-based wastes for adsorption of dyes from aqueous solutions. *J. Hazard. Mater.*, Vol. B92, 263–274.

77. Ong, S.T. and Liew, S.W. (2014). Immobilization of pomelo peel onto inert supporting material for the removal of Basic Blue 3. *Asian J. Chem.*, Vol. 26, 3808–3814.

78. Namasivayam, C., Prabha, D., and Kumutha, M. (1998). Removal of direct red and acid brilliant blue by adsorption on to banana pith. *Bioresour. Technol.*, Vol. 64, 77–79.

79. Chen, B., Hui, C.W., and McKay, G. (2001). Film-pore diffusion modeling and contact time optimisation for the adsorption of dyestuffs on pith. *Chem. Eng. J.*, Vol. 84, 77–94.

80. Ho, Y.S. and McKay, G. (2003). Sorption of dyes and copper ions onto biosorbents. *Process Biochem.*, Vol. 38, 1047–1061.

81. Bouzaida, I. and Rammah, M.B. (2002). Adsorption of acid dyes on treated cotton in a continuous system. *Mater. Sci. Eng.*, Vol. 21, 151–155.

82. Arami, M., Limaee, N.Y., Mahmoodi, N.M., and Tabrizi, N.S. (2006). Equilibrium and kinetics studies for the adsorption of direct and acid dyes from aqueous solution by soy meal hull. *J. Hazard. Mater.*, Vol. 135, 171–179.

83. Netpradit, S., Thiravetyan, P., and Towprayoon, S. (2003). Application of waste metal hydroxide sludge for adsorption of azo reactive dyes. *Water Res.*, Vol. 37, 763–772.

84. Namasivayam, C. and Sumithra, S. (2005). Removal of direct red 12B and methylene blue from water by adsorption onto Fe (III)/Cr (III) hydroxide, an industrial solid waste. *J. Environ. Manag.*, Vol. 74, 207–215.

85. Mall, I.D., Srivastava, V.C., and Agarwal, N.K. (2006). Removal of Orange-G and Methyl Violet dyes by adsorption onto bagasse fly ash: Kinetic study and equilibrium isotherm analyses. *Dyes Pigments*, Vol. 69, 210–223.

86. Mall, I.D., Srivastava, V.C., Agarwal, N.K., and Mishra, I.M. (2005). Removal of congo red from aqueous solution by bagasse fly ash and activated carbon: Kinetic study and equilibrium isotherm analyses. *Chemosphere*, Vol. 61, 492–501.

87. Mall, I.D., Srivastava, V.C., Agarwal, N.K., and Mishra, I.M. (2005). Adsorptive removal of malachite green dye from aqueous solution by bagasse fly ash and activated carbon-kinetic study and equilibrium isotherm analyses. *Colloids Surf. A*, Vol. 264, 17–28.
88. Mane, V.S., Mall, I.D., and Srivastava, V.C. (2007). Use of bagasse fly ash as an adsorbent for the removal of brilliant green dye from aqueous solution. *Dyes Pigments*, Vol. 73, 269–278.
89. Hsu, T.C. (2008). Adsorption of an acid dye onto coal fly ash. *Fuel*, Vol. 87, 3040–3045.
90. Vasanth Kumar, K., Ramamurthi, V., and Sivanesan, S. (2005). Modeling the mechanism involved during the sorption of methylene blue onto fly ash. *J. Colloids Interface Sci.*, Vol. 284, 14–21.
91. Wang, S.B., Li, L., Wu, H.W., and Zhu, Z.H. (2005). Unburned carbon as a low-cost adsorbent for treatment of methylene blue-containing wastewater. *J. Colloids Interface Sci.*, Vol. 292, 336–343.
92. Karagozoglu, B., Tasdemir, M., Demirbas, E., and Kobya, M. (2007). The adsorption of basic dye (Astrazon Blue FGRL) from aqueous solutions onto sepiolite, fly ash and apricot shell activated carbon: Kinetic and equilibrium studies. *J. Hazard. Mater.*, Vol. 147, 297–306.
93. Mall, I.D., Srivastava, V.C., and Agarwal, N.K. (2007). Adsorptive removal of Auramine-O: Kinetic and equilibrium study. *J. Hazard. Mater.*, Vol. 143, 386–395.
94. Matheswaran, M. and Karunanithis, T. (2007). Adsorption of Chrysoidine R by using fly ash in batch process. *J. Hazard. Mater.*, Vol. 145, 154–161.
95. Wang, S.B. and Zhu, Z.H. (2005). Sonochemical treatment of fly ash for dye removal from wastewater. *J. Hazard. Mater.*, Vol. 126, 91–95.
96. Woolard, C.D., Strong, J., and Erasmus, C.R. (2002). Evaluation of the use of modified coal ash as a potential sorbent for organic waste streams. *Appl. Geochem.*, Vol. 17, 1159–1164.
97. Lin, L., Wang, S.B., and Zhu, Z.H. (2006). Geopolymeric adsorbents from fly ash for dye removal from aqueous solution. *J. Colloids Interface Sci.*, Vol. 300, 52–59.
98. Acemioglu, B. (2004). Adsorption of Congo red from aqueous solution onto calcium-rich fly ash. *J. Colloids Interface Sci.*, Vol. 274, 371–379.
99. Dizge, N., Aydiner, C., Demirbas, E., Kobya, M., and Kara, S. (2008). Adsorption of reactive dyes from aqueous solutions by fly ash: Kinetic and equilibrium studies. *J. Hazard. Mater.*, Vol. 150, 737–746.
100. Wang, S.B., Boyjoo, Y., Choueib, A., and Zhu, Z.H. (2005). Removal of dyes from aqueous solution using fly ash and red mud. *Water Res.*, Vol. 39, 129–138.
101. Wang, S.B., Boyjoo, Y., and Choueib, A. (2005). Zeolitisation of fly ash for sorption of dyes in aqueous solutions. *Stud. Surf. Sci. Catal.*, Vol. 158, 1661–1668.
102. Eren, Z. and Acar, F.N. (2007). Equilibrium and kinetic mechanism for Reactive Black 5 sorption onto high lime Soma fly ash. *J. Hazard. Mater.*, Vol. 143, 226–232.
103. Pengthamkeerati, P., Satapanajaru, T., and Singchan, O. (2008). Sorption of reactive dye from aqueous solution on biomass fly ash. *J. Hazard. Mater.*, Vol. 153, 1149–1156.
104. Lin, J.X., Zhan, S.L., Fang, M.H., Qian, X.Q., and Yang, H. (2008). Adsorption of basic dye from aqueous solution onto fly ash. *J. Environ. Manag.*, Vol. 87, 193–200.
105. Janos, P., Buchtova, H., and Ryznarova, M. (2003). Sorption of dyes from aqueous solutions onto fly ash. *Wat. Res.*, Vol. 37, 4938–4944.
106. Wang, S.B, Soudi, M., Li, L., and Zhu, Z.H. (2006). Coal ash conversion into effective adsorbents for removal of heavy metals and dyes from wastewater. *J. Hazard. Mater.*, Vol. 133, 243–251.
107. Wang, S.B., Boyjoo, Y., and Choueib, A. (2005). A comparative study of dye removal using fly ash treated by different methods. *Chemosphere*, Vol. 60, 1401–1407.
108. Liu, Q., Zhou, Y., Zou, L., Deng, T., Zhang, J., Sun, Y., Ruan, X., Zhu, P., and Qian, G. (2011). Simultaneous wastewater decoloration and fly ash dechlorination during the dye wastewater treatment by municipal solid waste incineration fly ash. *Desalination Water Treat.*, Vol. 32, 179–186.
109. Fan, C. (2011). Adsorption behaviors and characteristics of methylene blue on fly ash in aqueous solutions. *Proceedings of International Symposium on Water Resource and Environmental Protection* (ISWREP), Xi'an, Shaanxi, China, 2213–2216.
110. de Carvalho, T.E.M., Fungaro, D.A., Magdalena, C.P., and Cunico, P. (2011). Adsorption of indigo carmine from aqueous solution using coal fly ash and zeolite from fly ash. *J. Radioanal. Nucl. Chem.*, Vol. 289, 617–626.
111. Sun, D., Zhang, X., Wu, Y., and Liu, X. (2010). Adsorption of anionic dyes from aqueous solution on fly ash. *J. Hazard. Mater.*, Vol. 181, 335–342.
112. Namasivayam, C. and Arasi, D.J.S.E. (1997). Removal of Congo red from wastewater by adsorption onto red mud. *Chemosphere*, Vol. 34, 401–417.
113. Namasivayam, C., Yamuna, R.T., and Arasi, D.J.S.E. (2001). Removal of acid violet from wastewater by adsorption on waste red mud. *Environ. Geol.*, Vol. 41, 269–273.

114. Tor, A. and Cengeloglu, Y. (2006). Removal of congo red from aqueous solution by adsorption onto acid activated red mud. *J. Hazard. Mater.*, Vol. 138, 409–415.
115. Shokohi, R., Jafari, S.J., Siboni, M., Gamar, N., and Saidi, S. (2011). Removal of acid blue 113(AB113) dye from aqueous solution by adsorption onto activated red mud: A kinetic and equilibrium study. *Sci. J. Kurdistan Univ. Med. Sci.*, Vol. 16, 55–65.
116. Wang, Q., Luan, Z., Wei, N., Li, J., and Liu, X. (2009). The color removal of dye wastewater by magnesium chloride/red mud (MRM) from aqueous solution. *J. Hazard. Mater.*, Vol. 170, 690–698.
117. Gupta, V.K., Suhas, Ali, I., and Saini, V.K. (2004). Removal of rhodamine B, fast green and methylene blue from wastewater using red mud, an aluminum industry waste. *Ind. Eng. Chem. Res.*, Vol. 43, 1740–1747.
118. Wang, S., Boyjoo, Y., Choueib, A., and Zhu, Z.H. (2005). Removal of dyes from aqueous solution using fly ash and red mud. *Water Res.*, Vol. 39, 129–138.
119. Ho, Y.S. and McKay, G. (1998). Sorption of dye from aqueous solution by peat. *Chem. Eng. J.*, Vol. 70, 115–124.
120. Sun, Q.Y. and Yang, L.Z. (2003). The adsorption of basic dyes from aqueous solution on modified peat–resin particle. *Wat. Res.*, Vol. 37, 1535–1544.
121. Allen, S.J., McKay, G., and Porter, J.F. (2004). Adsorption isotherm models for basic dye adsorption by peat in single and binary component systems. *J. Colloids Interface Sci.*, Vol. 280, 322–333.
122. Contreras, E.G., Martinez, B.E., Sepúlveda, L.A., and Palma, C.L. (2007). Kinetics of basic dye adsorption onto sphagnum magellanicum peat. *Adsorption Sci. Technol.*, Vol. 25, 637–646.
123. Sun, Q.Y. and Yang, L.Z. (2007). Adsorption of acid orange II from aqueous solution onto modified peat-resin particles. *Huan Jing Ke Xue*, Vol. 28, 1300–1304.
124. Ip, A.W.M., Barford, J.P., and McKay, G. (2009). Reactive Black dye adsorption/desorption onto different adsorbents: Effect of salt, surface chemistry, pore size and surface area. *J. Colloids Interface Sci.*, Vol. 337, 32–38.
125. Sepulveda, L., Troncoso, F., Contreras, E., and Palma, C. (2008). Competitive adsorption of textile dyes using peat: Adsorption equilibrium and kinetic studies in monosolute and bisolute systems. *Environ. Technol.*, Vol. 29, 947–957.
126. Sepúlveda, L., Contreras, E., and Palma, C. (2008). Magellan peat (*Sphagnum magallanicum*) as natural adsorbent of recalcitrant synthetic dyes. *J. Soil Sci. Plant Nutr.*, Vol. 8, 31–43.
127. Chatterjee, S., Chatterjee, S., Chatterjee, B.P., Das, A.R., and Guha, A.K. (2005). Adsorption of a model anionic dye, eosin Y, from aqueous solution by chitosan hydrobeads. *J. Colloids Interface Sci.*, Vol. 288, 30–35.
128. Kyzas, G.Z. and Lazaridis, N.K. (2009). Reactive and basic dyes removal by sorption onto chitosan derivatives. *J. Colloids Interface Sci.*, Vol. 331, 32–39.
129. Ong, S.T. and Seou, C.K. (2014). Removal of reactive black 5 from aqueous solution using chitosan beads: Optimization by Plackett–Burmann design and response surface analysis. *Desalination Water Treat.*, Vol. 52, 7673–7684.
130. Annadurai, A., Ling, L.Y., and Lee, J.F. (2008). Adsorption of reactive dye from an aqueous solution by chitosan: Isotherm, kinetic and thermodynamic analysis. *J. Hazard. Mater.*, Vol. 152, 337–346.
131. Becki, Z., Ozveri, C., Seki, Y., and Yurdakoc, K. (2008). Sorption of malachite green on chitosan bead. *J. Hazard. Mater.*, Vol. 154, 254–261.
132. Du, W.L., Xu, Z.R., Han, X.Y., Xu, Y.L., and Miao, Z.G. (2008). Preparation, characterization and adsorption properties of chitosan nanoparticles for eosin Y as a model anionic dye. *J. Hazard. Mater.*, Vol. 153, 152–156.
133. Elwakeel, K.Z. (2009). Removal of reactive Black 5 from aqueous solutions using magnetic chitosan resins. *J. Hazard. Mater.*, Vol. 167, 383–392.
134. Rosa, S., Laranjeira, M.C.M., Riela, H.G., and Favere, V.T. (2008). Cross-linked quaternary chitosan as an adsorbent for the removal of the reactive dye from aqueous solutions. *J. Hazard. Mater.*, Vol. 155, 253–260.
135. Singh, V., Sharma, A.K., Tripathi, D.N., and Sanghi, R. (2009). Poly(methylmethacrylate) grafted chitosan: An efficient adsorbent for anionic azo dyes. *J. Hazard. Mater.*, Vol. 161, 955–966.
136. Uzun, I. and Guzel, F. (2005). Rate studies on the adsorption of some dyestuffs and *p*-nitrophenol by chitosan and monocarboxymethylated(mcm)-chitosan from aqueous solution. *J. Hazard. Mater.*, Vol. 118, 141–154.
137. Wu, F.C., Tseng, R.L., and Juang, R.S. (2001). Enhanced abilities of highly swollen chitosan beads for color removal and tyrosinase immobilization. *J. Hazard. Mater.*, Vol. 81, 167–177.
138. Zubieta, C.E., Messina, P.V.M., Luengo, C., Dennehy, M., Pieroni, O., and Schulz, P.C. (2008). Reactive dyes remotion by porous TiO_2-chitosan materials. *J. Hazard. Mater.*, Vol. 152, 765–777.

139. Kamari, A., Wan Ngah, W.S., and Liew, L.K. (2009). Chitosan and chemically modified chitosan beads for acid dyes sorption. *J. Environ. Sci.*, Vol. 21, 296–302.
140. Hu, Z.G., Zhang, J., Chan, W.L., and Szeto, Y.S. (2006). The sorption of acid dye onto chitosan nanoparticles. *Polymer*, Vol. 47, 5838–5842.
141. Gad, Y.H. (2008). Preparation and characterization of poly(2-acrylamido-2-methylpropanesulfonic acid)/chitosan hydrogel using γ irradiation and its application in wastewater treatment. *Radiat. Phys. Chem.*, Vol. 77, 1101–1107.
142. Chang, M.Y. and Juang, R.S. (2005). Equilibrium and kinetic studies on the adsorption of surfactant, organic acids and dyes from water onto natural biopolymers. *Colloids Surface A*, Vol. 269, 35–46.
143. Chiou, M.S. and Li, H.Y. (2003). Adsorption behavior of reactive dye in aqueous solution on chemical cross-linked chitosan beads. *Chemosphere*, Vol. 50, 1095–1105.
144. Chiou, M.S. and Chuang, G.S. (2006). Competitive adsorption of dyes Metanil yellow and RB 15 in acid solutions on chemically cross-linked chitosan beads. *Chemosphere*, Vol. 62, 731–740.
145. Aksu, Z., Tatli, A.I., and Tunc, O. (2008). A comparative adsorption/biosorption study of Acid Blue 161: Effect of temperature on equilibrium and kinetic parameters. *Chem. Eng. J.*, Vol. 142, 23–39.
146. Wang, L. and Wang, A.Q. (2008). Adsorption behaviors of Congo red on the N,O-carboxymethyl-chitosan/montmorillonite nanocomposite. *Chem. Eng. J.*, Vol. 143, 43–50.
147. Hasan, M., Ahmad, A.L., and Hameed, B.H. (2008). Adsorption of reactive dye onto cross-linked chitosan/oil palm ash composite beads. *Chem. Eng. J.*, Vol. 136, 164–172.
148. Wang, L. and Wang, A.Q. (2008). Adsorption properties of congo red from aqueous solution onto N,O-carboxymethyl-chitosan. *Bioresour. Technol.*, Vol. 99, 1403–1408.
149. Cheung, W.H., Szeto, Y.S., and McKay, G. (2009). Enhancing the adsorption capacities of acid dyes by chitosan nano particles. *Bioresour. Technol.*, Vol. 100, 1143–1148.
150. Chatterjee, S., Lee, D.S., Lee, M.W., and Woo, S.H. (2009). Enhanced adsorption of congo red from aqueous solutions by chitosan hydrogel beads impregnated with cetyl trimethyl ammonium bromide. *Bioresour. Technol.*, Vol. 100, 2803–2809.
151. Wong, Y.C., Szeto, Y.S., Cheung, W.H. and McKay, G. (2004). Adsorption of acid dyes on chitosan—Equilibrium isotherm analyses. *Process Biochem.*, Vol. 39, 693–702.
152. Wang, L. and Wang, A.Q. (2007). Adsorption characteristics of Congo red onto the chitosan/montmorillonite nanocomposite. *J. Hazard. Mater.*, Vol. 147, 979–985.
153. Crini, G., Gimbert, F., Robert, C., Martel, B., Adam, O., Crini, N.M., Giorgi, F.D., and Badot, P.M. (2008). The removal of Basic blue 3 from aqueous solutions by chitosan-based adsorbent: Batch studies. *J. Hazard. Mater.*, Vol. 153, 96–106.
154. Chiou, M.S. and Li, H.Y. (2002). Equilibrium and kinetic modeling of adsorption of reactive dye on cross-linked chitosan beads. *J. Hazard. Mater.*, Vol. 93, 233–248.
155. Wu, F.C., Tseng, R.L., and Juang, R.S. (2000). Comparative adsorption of metal and dye on flake and bead-types of chitosans prepared from fishery wastes. *J. Hazard. Mater.*, Vol. 73, 63–75.
156. Chiou, M.S., Ho, P.Y., and Li, H.Y. (2004). Adsorption of anionic dyes in acid solutions using chemically cross-linked chitosan beads. *Dyes Pigments*, Vol. 60, 69–84.
157. Huang, X.Y., Mao, X.Y., Bu, H.T., Yu, X.Y., Jiang, G.B., and Zeng, M.H. (2011). Chemical modification of chitosan by tetraethylenepentamine and adsorption study for anionic dye removal. *Carbohydr. Res.*, Vol. 346, 1232–1240.
158. Sreelatha, G., Ageetha, V., Parmar, J., and Padmaja, P. (2011). Equilibrium and kinetic studies on reactive dye adsorption using palm shell powder (an agrowaste) and chitosan. *J. Chem. Eng. Data*, Vol. 56, 35–42.
159. Ling, S.L.Y., Yee, C.Y., and Eng, H.S. (2011). Removal of a cationic dye using deacetylated chitin (chitosan). *J. Appl. Sci.*, Vol. 11, 1445–1448.
160. Copello, G.J., Mebert, A.M., Raineri, M., Pesenti, M.P., and Diaz, L.E. (2011). Removal of dyes from water using chitosan hydrogel/SiO$_2$ and chitin hydrogel/SiO$_2$ hybrid materials obtained by the sol–gel method. *J. Hazard. Mater.*, Vol. 186, 932–939.
161. Huang, X.-Y., Bu, H.-T, Jiang, G.-B., and Zeng, M.H. (2011). Cross-linked succinyl chitosan as an adsorbent for the removal of methylene blue from aqueous solution. *Int. J. Biol. Macromol.*, Vol. 49, 643–651.
162. Chatterjee, S., Chatterjee, T., Lim, S.-R., and Woo, S.H. (2011). Adsorption of a cationic dye, methylene blue, on to chitosan hydrogel beads generated by anionic surfactant gelation. *Environ. Technol.*, Vol. 32, 1503–1514.
163. Xing, Y., Zhang, L., Li, B., Sun, X., and Yu, J. (2011). Adsorption of methylene blue on poly (methacrylic acid) modified chitosan and photocatalytic regeneration of the adsorbent. *Sep. Sci. Technol.*, Vol. 46, 2298–2304.

164. Debrassi, A., Largura, M.C.T., and Rodrigues, C.A. (2011). Adsorption of congo red dye by hydrophobic O-carboxymethyl chitosan derivatives. *Quim. Nova*, Vol. 34, 764–770.
165. Zhou, L.-M., Shang, C., and Liu, Z.-R. (2011). Acid dye adsorption properties of ethylenediamine-modified magnetic Chitosan nanoparticles. *Acta Phys. Chim. Sin.*, Vol. 27, 677–682.
166. Momenzadeh, H., Tehrani-Bagha, A.R., Khosravi, A., Gharanjig, K., and Holmberg, K. (2011). Reactive dye removal from wastewater using a chitosan nanodispersion. *Desalination*, Vol. 271, 225–230.
167. Dotto, G.L. and Pinto, L.A.A. (2011). Adsorption of food dyes onto chitosan: Optimization process and kinetic. *Carbohydr. Polym.*, Vol. 84, 231–238.
168. Fu, Y.Z. and Viraraghavan, T. (2002). Dye biosorption sites in *Aspergillus niger. Bioresour.Technol.*, Vol. 82, 139–145.
169. Das, S.K., Bhowal, J., Das, A.R., and Guha, A.K. (2006). Adsorption behavior of Rhodamine B on *Rhizopus oryzae* biomass. *Langmuir*, Vol. 22, 7265–7272.
170. Khalaf, M.A. (2008). Biosorption of reactive dye from textile wastewater by nonviable biomass of *Aspergillus niger* and Spirogyra sp. *Bioresour. Technol.*, Vol. 99, 6631–6634.
171. Patel, R. and Suresh, S. (2008). Kinetic and equilibrium studies on the biosorption of Reactive Black 5 dye by *Aspergillus foetidus. Bioresour. Technol.*, Vol. 99, 51–58.
172. Asma, D., Kahraman, S., Cing, S., and Yesilada, O. (2006). Adsorptive removal of textile dyes from aqueous solutions by dead biomass. *J. Basic Microbiol.*, Vol. 46, 3–9.
173. Prigione, V., Varese, G.C., Casieri, L., and Marchisio, V.F. (2008). Biosorption of simulated dyed effluents by inactivated fungal biomasses. *Bioresour. Technol.*, Vol. 99, 3559–3567.
174. Aksu, Z. and Tezer, S. (2000). Equilibrium and kinetic modeling of biosorption of Remazol Black B by *Rhizopus arrhizus* in a batch system: Effect of temperature. *Process Biochem.*, Vol. 36, 431–439.
175. O'Mahony, T., Guibal, E., and Tobin, J.M. (2002). Reactive dye biosorption by *Rhizopus arrhizus* biomass. *Enzyme Microbiol. Technol.*, Vol. 31, 456–463.
176. Aksu, Z. (2003). Reactive dye bioaccumulation by *Saccharomyces cerevisiae. Process Biochem.*, Vol. 38, 1437–1444.
177. Kumari, K. and Abraham, E. (2007). Biosorption of anionic textile dyes by nonviable biomass of fungi and yeast. *Bioresour. Technol.*, Vol. 98, 1704–1710.
178. Aksu, Z. and Donmez, G. (2003). A comparative study on the biosorption characteristics of some yeasts for Remazol Blue reactive dye. *Chemosphere*, Vol. 50, 1075–1083.
179. Yang, Q., Yediler, A., Yang, M., and Kettrup, A. (2005). Decolorization of an azo dye, Reactive Black 5 and MnP production by yeast isolate: *Debaryomyces polymorphus. Biochem. Eng. J.*, Vol. 24, 249–253.
180. Kwasniewska, K. (1985). Biodegradation of crystal violet (hexamethyl-p-rosaniline chloride) by oxidation of red yeasts. *Bull. Environ. Contam. Toxicol.*, Vol. 34, 323–330.
181. Zhou, W. and Zimmermann, W. (1993). Decolorization of industrial effluents containing reactive dyes by actinomycetes. *FEMS Microbiol. Lett.*, Vol. 107, 157–162.
182. Nacera, Y. and Aicha, B. (2006). Equilibrium and kinetc modeling of Methylene Blue biosorption by pretreated dead *Streptomyces rimosus*: Effect of temperature. *Chem. Eng. J.*, Vol. 119, 121–125.
183. Vijayaraghavan, K. and Yun, Y.S. (2007a). Utilization of fermentation waste (*Corynebacterium glutamicum*) for biosorption of Reactive Black 5 from aqueous solution. *J. Hazard. Mater.*, Vol. 141, 45–52.
184. Aksu, Z. and Tezer, S. (2005). Biosorption of reactive dyes on the green alga *Chlorella vulgaris. Process Biochem.*, Vol. 40, 1347–1361.
185. Ozer, A., Akkaya, G., and Turabik, M. (2005). Biosorption of Acid Red 274 on *Enteromorpha prolifera* in a batch system. *J. Hazard. Mater.*, Vol. 126, 119–127.
186. Padmesh, T.V.N., Vijayaraghavan, K., Sekaran, G., and Velan, M. (2005). Batch and column studies on biosorption of acid dyes on fresh water macro alga *Azolla filiculoides. J. Hazard. Mater.*, Vol. 125, 121–129.
187. Aravindhan, R., Raghava Roa, J., and Unni Nair, B. (2007). Removal of basic dye from aqueous solution by sorption on green alga *Caulerpa scalpelliformis. J. Hazard. Mater.*, Vol. 142, 68–76.
188. Waranusantigul, P., Pokethitiyook, P., Kruatrachue, M., and Upatham, E.S. (2003). Kinetics of basic dye (methylene blue) biosorption by giant duckweed (*Spirodela polyrrhiza*). *Environ. Pollut.*, Vol. 125, 385–392.
189. Ozcan, A.S., Erdem, B., and Ozcan, A. (2004). Adsorption of Acid Blue 193 from aqueous solutions onto Na-bentonite and DTMA-bentonite. *J. Colloids Interface Sci.*, Vol. 280, 44–54.
190. Ho, Y.S. and Chiang, C.C. (2001). Sorption studies of acid dye by mixed sorbents. *Adsorption*, Vol. 7, 139–147.
191. Ho, Y.S., Chiang, C.C., and Hsu, Y.C. (2001). Sorption kinetics for dye removal from aqueous solution using activated clay. *Sep. Sci. Technol.*, Vol. 36, 2473–2488.

192. Gupta, G.S. and Shukla, S.P. (1996). An inexpensive adsorption technique for the treatment of carpet effluents by low cost materials. *Adsorption Sci. Technol.*, Vol. 13, 15–26.
193. Espantaleon, A.G., Nieto, J.A., Fernandez, M., and Marsal, A. (2003). Use of activated clays in the removal of dyes and surfactants from tannery waste waters. *Appl. Clay Sci.*, Vol. 24, 105–110.
194. El-Guendi, M.S., Ismail, H.M., and Attyia, K.M.E. (1995). Activated clay as an adsorbent for cationic dyestuffs. *Adsorption Sci. Technol.*, Vol. 12, 109–117.
195. Ozacar, M. and Sengil, I.A. (2006). A two stage batch adsorber design for methylene blue removal to minimize contact time. *J. Environ. Manag.*, Vol. 80, 372–379.
196. Al-Ghouti, M.A., Khraisheh, M.A.M., Allen, S.J., and Ahmad, M.N. (2003). The removal of dyes from textile wastewater: A study of the physical characteristics and adsorption mechanisms of diatomaceous earth. *J. Environ. Manag.*, Vol. 69, 229–238.
197. Shawabkeh, R.A. and Tutunji, M.F. (2003). Experimental study and modeling of basic dye sorption by diatomaceous clay. *Appl. Clay Sci.*, Vol. 24, 111–120.
198. Gurses, A., Karaca, S., Dogar, C., Bayrak, R., Acikyildiz, M., and Yalcin, M. (2004). Determination of adsorptive properties of clay/water system: Methylene blue sorption. *J. Colloids Interface Sci.*, Vol. 269, 310–314.
199. Hu, Q.H., Qiao, S.Z., Haghseresht, F., Wilson, M.A., and Lu, G.Q. (2006). Adsorption study for removal of basic red dye using bentonite. *Ind. Eng. Chem. Res.*, Vol. 45, 733–738.
200. Bouatay, F., Dridi-Dhaouadi, S., Drira, N., and Farouk, M.M. (2016). Application of modified clays as an adsorbent for the removal of Basic Red 46 and Reactive Yellow 181 from aqueous solution. *Desalination Water Treat.*, Vol. 57, 13561–13572.
201. Fil, B.A., Korkmaz, M., and Ozmetin, C. (2016). Application of nonlinear regression analysis for methyl violet dye adsorption from solutions onto illite clay. *J. Dispersion Sci. Technol.*, Vol. 37, 991–1001.
202. Santos, S.C.R. and Boaventura, R.A.R. (2016). Adsorption of cationic and anionic dyes on sepiolite clay: Equilibrium and kinetic studies in batch mode. *J. Environ. Chem. Eng.*, Vol. 4, 1473–1483.
203. Wang, L.K., Hung, Y.T., Lo, H.H., and Yapijakis, C. (eds.) *Handbook of Industrial and Hazardous Wastes Treatment.* CRC Press and Marcel Dekker, New York, pp. 379–414 (2004).
204. Shammas, N.K. and Wang, L.K. *Water Engineering: Hydraulics, Distribution, and Treatment.* Wiley, Hoboken, NJ, pp. 806 (2016).
205. Wang, M.H.S. and Wang, L.K. Glossary and conversion factors for water resources engineers. In: *Modern Water Resources Engineering.* Wang, L.K. and Yang, C.T. (eds). Humana Press, Totowa, NJ. 759–851 pp. (2014).

8 BOD Determination, Cleaning Solution Preparation, and Waste Disposal in Laboratories

Mu-Hao Sung Wang and Lawrence K. Wang
Lenox Institute of Water Technology

Eugene De Michele
Water Environment Federation

CONTENTS

8.1	Biochemical Oxygen Demand Determination	286
8.2	BOD Determination Interferences	286
8.3	BOD Testing Apparatus	287
8.4	BOD Testing Reagents	287
8.5	Pretreatment of Wastewater Samples for BOD Testing	288
	8.5.1 Procedure for Neutralization of Samples Containing Caustic Alkalinity or Acidity	288
	8.5.2 Procedure for Handling Toxic Substances	288
	8.5.3 Procedure for Sampling of Water Samples Supersaturated with DO	288
	8.5.4 Procedure for Temperature Adjustment	288
	8.5.5 Procedure for Nitrification Inhibition	288
	8.5.6 Procedure for Dechlorination	289
8.6	Procedure for BOD Determination without Seeding	289
	8.6.1 Sample Preparation and DO Determinations	289
	8.6.2 BOD Calculations When Dilution Water Is Not Seeded	290
	8.6.3 Reliability Observation	291
8.7	Procedures for BOD Determination with Seeding	291
	8.7.1 Sample Preparation and DO Determinations	291
	8.7.2 BOD Calculations When Dilution Water Is Seeded	291
	8.7.3 Reliability Observation	292
8.8	BOD Bottles and Glassware Cleaning	292
	8.8.1 Dichromate Acid Cleaning Solution	292
	8.8.2 Other Alternative Cleaning Solutions	292
	8.8.2.1 Hydrochloric Acid (HCl) Cleaning Solution	292
	8.8.2.2 H_2SO_4 Cleaning Solution	292
	8.8.2.3 Nitric Acid Cleaning Solution	292
	8.8.2.4 Piranha Cleaning Solution	293
	8.8.2.5 Fuming H_2SO_4 Solution	293
	8.8.2.6 Nochromix Cleaning Solution	293
8.9	Laboratory Waste Management and Disposal	293
	8.9.1 BOD Testing Wastewater	293
	8.9.2 Small Quantity Generator Status and Commercial Waste Disposal Option	293

8.9.3 On-Site Waste Treatment by Laboratory Personnel 293
        8.9.3.1 Treatment of BOD Testing Wastewater Containing Chromium 293
        8.9.3.2 Treatment of BOD Testing Wastewater Containing No
                Heavy Metals 294
References 294

Abstract

Under the sponsorship of the Water Environment Federation, simplified laboratory procedures for biological oxygen demand (BOD) determination as outlined in the APHA/AWWA/WPCF Standard Methods for the Examination of Water and Wastewater were developed for use by the waste treatment plant operators, researchers, and engineers. The step-by-step, simplified BOD testing procedures with or without seeding have been documented. All BOD bottles and other glassware must be thoroughly cleaned before and after testing. Some highly toxic, hazardous wastewaters are generated from BOD testing in the laboratory. These wastes must be either pretreated for sewer discharge or stored and delivered to a licensed hazardous waste handler for further processing. This chapter discusses a laboratory manager's decision process for waste management and preparation methods and disposal processes of various cleaning solutions, such as dichromate acid (chromatic acid), hydrochloric acid, sulfuric acid, nitric acid, Piranha solution, fuming sulfuric acid solution and Nochromix solution. Dichromate acid cleaning solution is used when the generated waste contains toxic chromium (VI) and chromium (III). A special process system consisting of oxidation–reduction, neutralization, precipitation, and filtration may be used by a BOD testing laboratory for successful treatment of chromium-containing waste, if the laboratory is a government-approved small quantity generator. Otherwise, the waste must be stored and hauled away for off-site treatment by a commercial hazardous waste handler. When an alternative cleaning solution containing no hazardous heavy metals is used for BOD bottles and glassware cleaning, the resulting BOD testing wastewater can be easily pretreated by an elementary neutralization process prior to sewer discharge, again, if the laboratory is an approved small quantity generator.

8.1 BIOCHEMICAL OXYGEN DEMAND DETERMINATION

Biochemical oxygen demand (BOD) is defined as the amount of oxygen required to decompose a given amount of organic matter [1]. Accordingly, a BOD test determines the amount of organic material in wastewater by measuring the oxygen consumed by microorganisms in biodegrading the organic constituents of the waste. The BOD test consists of measuring the dissolved oxygen (DO) prior to and following a 5-day incubation of the wastewater sample at 20°C to determine the amount of oxygen used biochemically [2–5]. The wastewater sample and its dilutions are made with standard dilution water. After the DO is determined before and after the 5-day incubation period, the BOD is calculated. In case the wastewater sample to be tested has been chlorinated, ozonated, oxidized, heated, intoxicated, acidified, etc., certain pretreatment and microorganisms seeding of the sample will be required [4,5].

8.2 BOD DETERMINATION INTERFERENCES

Caustic alkalinity, mineral acid, free chlorine, and heavy metals are among the factors that may influence the accuracy of the BOD test. All BOD bottles must be extremely clean. The cleaning solutions described in this chapter are recommended.

    The extent of oxidation of nitrogenous compounds during the 5-day incubation period depends on the presence of microorganisms capable of carrying out this oxidation. Such organisms usually are not present in raw wastewater or primary effluent in sufficient numbers to oxidize significant quantities of reduced nitrogen forms in the 5-day BOD test. Currently, many biological wastewater treatment plant (WWTP) effluents contain significant numbers of nitrifying organisms. Because oxidation of nitrogenous compounds can occur in such samples, inhibition of nitrification is recommended for samples of WWTP secondary effluent, for samples seeded with WWTP secondary effluent, and for samples of polluted waters.

8.3 BOD TESTING APPARATUS

1. Burette graduated to 0.1 mL with a 50 mL capacity.
2. Glass-stoppered BOD bottles with collar, 300 mL.
3. Wide-mouth Erlenmeyer flask, 250 mL.
4. Measuring pipette, 10 mL.
5. Large-tipped volumetric pipette
6. Incubator is thermostatically controlled at 20°C; light is not allowed to prevent the possibility of photosynthetic production of DO.
7. Graduated cylinder, 250 mL.
8. Optional DO volumetric flask graduated to deliver 201 mL.

8.4 BOD TESTING REAGENTS

All the reagents necessary for the determination of DO can be found in the *Standard Methods for the Examination of Water and Wastewater*, 22nd edition [2]. In addition to the reagents for DO testing, the following reagents are required for BOD testing.

1. *Distilled water*: Water used in the preparation of the solution must be of highest quality. It must not contain copper or decomposable organic triatter. Ordinary battery distilled water is not good enough.
2. *Phosphate buffer solution*: Dissolve 8.5 g monobasic potassium phosphate (KH_2PO_4), 21.75 g dibasic potassium phosphate (K_2HPO_4), 33.4 nine dibasic sodium phosphate crystals ($Na_2HPO_4 \cdot 7H_2O$), and 1.7 g ammonium chloride (NH_4Cl) in distilled water and make up to 1 liter (L). The pH of this buffer should be 7.2 without further adjustment.
3. *Magnesium sulfate ($MgSO_4$) solution*: Dissolve 22.5 g magnesium sulfate crystals ($MgSO_4 \cdot 7H_2O$) in distilled water and make up to 1 L.
4. *Calcium chloride ($CaCl_2$) solution*: Dissolve 27.5 g anhydrous $CaCl_2$ in distilled water and make up to 1 L.
5. *Ferric chloride ($FeCl_3$) solution*: Dissolve 0.25 g ferric chloride ($FeCl_3 \cdot 6H_2O$) in distilled water and make up to 1 L.
6. *Dilution water*: Add 1 mL each of phosphate buffer (reagent 4.2), magnesium sulfate (reagent 4.3), calcium chloride (reagent 4.4), and ferric chloride solutions (reagent 4.5) for each liter of distilled water. Store at a temperature as close to 20°C as possible. This water should not show a drop in DO of more than 0.2 mg/L on incubation for 5 days. Seed dilution water, if desired, as described in Section 8.7 (Figure 8.1).
7. *Sodium sulfite solution (Na_2SO_3), 0.025 N*: Dissolve 1.575 g Na_2SO_3 in 1 L distilled water. This solution is not stable; prepare daily when dechlorination of wastewater sample is required.
8. *Nitrification inhibitor (CTCMP)*: Use reagent grade 2-chloro-6-(trichloromethyl) pyridine. Commercial products such as N-Serve developed by Dow Chemical Co., or Nitrification Inhibitor 2533 developed by Hach Chemical Co. can also be used instead. See the procedure for nitrification inhibition in Section 8.5.
9. *Sodium hydroxide solution (NaOH), 1 N*: Required only when pH adjustment of wastewater sample is necessary.
10. *Sulfuric acid (H_2SO_4) solution (112504), 1 N*: Required only when pH adjustment of wastewater sample is necessary.

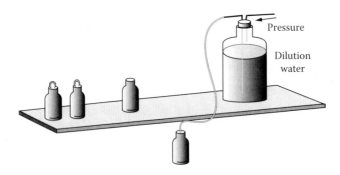

FIGURE 8.1 Addition of dilution water to BCD bottles carefully to avoid air bubbles.

8.5 PRETREATMENT OF WASTEWATER SAMPLES FOR BOD TESTING

One or more of the procedures outlined in this section shall be followed if one of the following conditions exist: (1) sample's pH value is not in the range of 6.5–7.5; (2) sample has been chlorinated or oxidized by ozone or other oxidation agent; (3) samples require nitrification inhibition (such as biologically treated effluents, samples seeded with biologically treated effluents, river waters); (4) industrial wastewater sample contains no microorganisms and requires seeding; (5) wastewater has extremely high or low temperature; (6) toxic substances are present in wastewater; and (7) cold wastewater sample contains supersaturated DO.

8.5.1 Procedure for Neutralization of Samples Containing Caustic Alkalinity or Acidity

Neutralize samples to pH 6.5–7.5 with a solution of 1.0 N H_2SO_4 or 1.0 N NaOH of such strength that the quantity of reagent does not dilute the sample by more than 0.5%. The pH of seeded dilution water should not be affected by the lowest sample dilution [5].

8.5.2 Procedure for Handling Toxic Substances

Certain industrial wastes such as plating wastes contain toxic metals. Such samples often require special study and treatment by experienced water quality laboratories. The sample must be "reseeded" with organisms for BOD determination after toxic metals are removed.

8.5.3 Procedure for Sampling of Water Samples Supersaturated with DO

Samples containing more than 9 mg/L of DO at 20°C may be encountered in cold waters or in water where photosynthesis occurs. To prevent loss of oxygen during incubation of such samples, reduce DO to saturation at 20°C by bringing the sample to about 20°C in a partially filled bottle while agitating by vigorous shaking, or by aerating with compressed air. Seeding of sample is not required.

8.5.4 Procedure for Temperature Adjustment

Bring samples to 20°C ± 1°C before making dilutions. If the samples have been frozen or boiled earlier, after temperature adjustment, the samples must be "reseeded" with organisms.

8.5.5 Procedure for Nitrification Inhibition

If nitrification inhibition is needed, add 10 mg/L of 2-chloro-6 (trichloro methyl) pyridine (CTCMP) of dilution water while adding nutrient and buffer solutions. Note the use of nitrogen inhibition in reporting results. Seeding of sample is not required because CTCMP is added.

8.5.6 Procedure for Dechlorination

Whenever chlorinated wastewater samples are collected for the determination of BOD, sufficient reducing agent must be added to the sample to destroy the chlorine. After dechlorination, the sample must be "reseeded" with organisms [4].

Before reseeding the sample, use the following procedures to check for the presence of chlorine in the composite sample as follows: (1) carefully measure 100 mL of well-mixed sample and take in a 250-mL Erlenmeyer flask; (2) add a few crystals of potassium iodide to the sample and dissolve the crystals; (3) add 1 mL of concentrated H_2SO_4 and mix well; and (4) add five drops of starch.

If no blue color is produced, chlorine is absent and the BOD of the composite may be determined without dechlorination; however, it must be seeded.

If blue color is produced, titrate the 100 mL well-mixed composite sample with 0.025 N sodium sulfite (Na_2SO_3) to the endpoint between the last trace of blue color and a colorless solution. Make the titration very slowly, counting the drops of 0.025 N Na_2SO_3 used and recording the number (n).

To dechlorinate a sample for BOD testing, measure another 100 mL portion of the well-mixed composite sample into a clean 250 mL-Erlenmeyer flask. Add drops of 0.025 N Na_2SO_3 determined (n) necessary for dechlorination in the earlier step and mix well. Use this sample for determination of BOD. If more sample is needed, place a larger sample (measure carefully) into a clean container and add a proportionate number of drops of the 0.025 N Na_2SO_3 for dechlorination.

8.6 PROCEDURE FOR BOD DETERMINATION WITHOUT SEEDING

8.6.1 Sample Preparation and DO Determinations

1. First, the pH value, residual chlorine (or any other known oxidizing agent, toxic substances, etc.), and the sample's case history must be determined; if the wastewater sample's pH value is in the range 6.5–7.5, it has not been chlorinated, ozonated, acidified, heated, etc. and the laboratory procedure outlined in this section shall be followed. (Note: Otherwise, the "Procedures for BOD Determination with Seeding" in Section 8.7 shall be followed). Next, the amount of sample (B) added to the 300-mL BOD bottle must be determined. To make this calculation, one should understand that dilution water at room temperature contains approximately 8 mg/L, of DO. Consequently, if the oxygen demand of the sample to be tested is >8 mg/L it has to be diluted. It is desirable to have at least 1 mg/L of oxygen left unused after 5-day incubation. A DO uptake of at least 2 mg/L after 5 days of incubation produces most reliable results. Table 8.1 is an aid to estimate the amount of sample to be added to a 300-mL BOD bottle. Raw municipal wastewater usually contains about 100–300 mg/L 5-day BOD so that 3–6 mL of sample is generally used. The 5-day BOD of settled wastewater usually ranges from 50 to 200 mg/L, and 6–12 mL of sample is commonly used. For trickling filter effluents, use 15–30 mL of sample. For activated sludge effluents, use 30–150 mL of sample depending on the quality of the effluent. Very strong wastewater or industrial wastes are diluted with one part of wastewater to nine parts of dilution water before adding 3–6 mL of the diluted wastewater sample into a BOD bottle for analysis. In this way, a range of 1000–3000 mg/L 5-day BOD is covered.
2. Fill two 300 mL-DOD bottles about half full with dilution water (Figure 8.1). Then use a large-tipped pipette to dispense the precalculated amount of sample (S) into each of the two 300-mL BOD bottles. Fill each bottle with dilution water and insert stoppers. Exclude all air bubbles.
3. Fill additional two 300-mL BOD bottles with only dilution water and insert stoppers in the same way.
4. Incubate the BOD bottles at 20°C with one bottle containing diluted sample (Step 2) and one bottle containing only dilution water (Step 3). Be sure the area above the stopper contains distilled water. Check daily or use bottle covers to prevent evaporation.

TABLE 8.1
An Aid in Selection of Sample size for BOD$_5$ Determination

	Expected BOD$_5$ Range	
Sample Added to 300-mL bottle (mL)	Min. (mg/L)	Max. (mg/L)
3	210	560
6	105	280
9	70	187
12	53	140
15	42	112
18	35	94
21	30	80
24	26	70
27	24	62
30	21	56
45	14	37
60	11	28
75	8	22
150	4	12

Note: Intial DO assumed to be 7–8 mg/L. Initial DO is the concentration of dissolved oxygen in mg/L of the mixture of the dilution water and the sample immediately after initial mixing.

5. Run a DO determination on the remaining BOD bottle from Step 2 and record the initial DO content as D_1. Run another DO determination on the remaining BOD bottle from Step 3, and record the initial DO content for control as C_1.
6. After 5 days, run DO determination on the two incubated BOD bottles (Step 4). Record the DO content of the incubated diluted sample (Steps 2 and 4) as D_2. Record the DO content of the incubated dilution water for control (Steps 3 and 4) as C_2. Note that the increase or decrease of DO in the bottles with just dilution water should not be used for correction of the diluted sample results. It is only a measure of dilution water quality and there should not be an increase or decrease of more than 0.2 mg/L of DO between C_1 and C_2. Larger changes may be caused by improper testing techniques or contaminated dilution water.

Note: It will be very disappointing if on the 5th day, one finds that there is no DO (0 mL of 0.025 N sodium thiosulfate or PAO solution required in titration, azide modification method for DO test) in the incubated sample and the results are lost. It is better to prepare two diluted samples, one at one-half the concentration of the other, so that the chemist will be sure to get a result. For example, if one estimates that 2% is about right (that is, 6 mL/300 mL, 0.02 when the 5-day BOD is assumed to be 105–280 mg/L), it is *safer* also to prepare a 1% diluted sample (i.e., 3 mL/300 mL = 0.01) so that one can cover a 210–560 mg/L 5-day BOD range. With effluents, if they look bad, it is always safer to prepare dilutions of 5% and 10% rather than only 10%.

8.6.2 BOD Calculations When Dilution Water Is Not Seeded

$$\text{mg/L 5-day BOD} = \frac{100\,(D_1 - D_2)}{P} = 300\,\frac{(D_1 - D_2)}{S}, \tag{8.1}$$

where
- D_1 = DO of diluted sample immediately after sample preparation, mg/L
- D_2 = DO of diluted sample after 5-day incubation at 20°C, mg/L
- S = volume of sample added to the 300-mL BOD bottle, mL
- P = percentage of sample added = $(S/300) \times 100$

8.6.3 Reliability Observation

The 5-day BOD data will be considered to be reliable if minimum residual DO level of 1 mg/L and minimum DO depletion $(C_1 - C_2)$ of 2 mg/L are observed.

8.7 PROCEDURES FOR BOD DETERMINATION WITH SEEDING

8.7.1 Sample Preparation and DO Determinations

Secure about 1 L of an unchlorinated sample of raw wastewater or primary effluent about 24 h before one expects to set up pretreated (such as dechlorinated and pH adjusted) and seeded samples for determination of BOD. Let it stand at room temperature overnight. Pour off the clear portion of the sample and use it for the "seed."

For seeding of the sample, add one milliliter of the aged seed (read the earlier step) to each of the BOD bottles containing pretreated (such as dechlorinated) sample. Also set up samples of the seed in 300-mL BOD bottles for determination of the BOD using 6, 9, and 12 mL seed and calculate the 5-day depletion caused by 1 mL of seed. After "seeding" is complete, use the following procedures for BOD determination:

1. First, determine the sample size (S) using Table 8.1 as a guide.
2. Seed the dilution water (see Reagent 4.6). Record the mL/L of seed in dilution water.
3. Fill two 300-mL BOD bottles with desired volume of wastewater sample (S) and the seeded dilution water; insert stoppers. One bottle is for DO determination. Another bottle is for incubation. The DO of the diluted sample immediately after preparation is recorded as D_1. The DO of the diluted sample after 5-day incubation at 20°C is recorded as D_2. Calculate the "percentage of seed in D_1."
4. Fill an additional two 300-mL BOD bottles with only the seeded dilution water and insert stoppers the same way as for "seed control." One bottle is for immediate DO determination. Another bottle is for incubation. The DO of seed control before incubation is recorded as B_1. The DO of seed control after 5-day incubation at 20°C is recorded as B_2. Calculate the "% seed in B_1."

8.7.2 BOD Calculations When Dilution Water Is Seeded

$$\text{mg/L 5-day BOD} = \frac{100\left[(D_1 - D_2) - (B_1 - B_2)f\right]}{P}, \quad (8.2)$$

$$\text{mg/L 5-day BOD} = \frac{300\left[(D_1 - D_2) - (B_1 - B_2)f\right]}{S}, \quad (8.3)$$

where
- D_1 = DO of diluted sample immediately after sample preparation (mg/L)
- D_2 = DO of diluted sample after 5-day incubation at 20°C (mg/L)
- P = percentage of sample added = $(S/300) \times 100$

S = volume of sample added to the 300-mL BOD bottle (mL)
B_1 = DO of seed control before incubation (mg/L)
B_2 = DO of seed control after incubation (mg/L)
f = ratio of seed in sample to seed in control = (% seed in D_1)/(% seed in B_1)

8.7.3 Reliability Observation

If more than one sample dilution meets the criteria of a residual DO of at least 1 mg/L and a DO depletion of at least 2 mg/L and there is no evidence of toxicity at higher sample concentrations or the existence of any obvious problem, average results are in the acceptable range.

8.8 BOD BOTTLES AND GLASSWARE CLEANING

8.8.1 Dichromate Acid Cleaning Solution

A conventional method of cleaning BOD bottles and other glassware involves soaking glass in a bath containing dichromate cleaning solution (also known as chromatic acid solution) followed by tap water rinses, distilled water rinses, and finally double-distilled water rinses. The dichromate cleaning solution is commercially available [6,7], but can also be prepared easily by a chemist or engineer in the laboratory.

The following is the preparation procedure for dichromate acid cleaning solution:

1. Wear personal protective equipment (PPE; such as nitrile gloves, goggles, and laboratory coat, and set up the required chemicals and equipment under hood in a well-ventilated area.
2. Dissolve 400 g of potassium dichromate in 4-L distilled water in a big glass container and use a magnetic stirring bar only if necessary. Remove the magnetic stirring bar, if used, after potassium dichromate is totally dissolved.
3. Carefully and slowly add 400 mL of concentrated H_2SO_4 to the mixture in Step 2.
4. Keep the solution mixture of Step 3 in a safe place until solution turns dark brown.
5. Label the solution by adding the date and initials of the operator/chemist's name.
6. *Caution*: Potassium dichromate is toxic when inhaled and ingested. H_2SO_4 is corrosive. Wear PPE and use the dichromate acid cleaning solution (or chromatic acid solution) under hood or in a well-ventilated area. Consult with appropriate safety officials at your institution for proper procedures and disposal.

8.8.2 Other Alternative Cleaning Solutions

8.8.2.1 Hydrochloric Acid (HCl) Cleaning Solution

To prepare a 10% solution, wear PPE and slowly add 100 mL of concentrated reagent grade HCl to 900 mL distilled water in a 1-L beaker under the hood with ventilation. Place a beaker cover over the beaker and let it cool before pouring into a storage container. Alternatively, a 15% HCl solution (with a ratio of 150 mL HCl/850 mL water) can also be prepared and used for cleaning. A 0.15% HCl solution is commonly used for rinsing BOD bottles and other glassware.

8.8.2.2 H_2SO_4 Cleaning Solution

To prepare a 10% solution, wear PPE and slowly add 100 mL of concentrated reagent grade H_2SO_4 to 900 mL distilled water in a 1-L beaker under the hood with ventilation. Place a beaker cover over the beaker and let it cool before pouring into a storage container.

8.8.2.3 Nitric Acid Cleaning Solution

To prepare a 20% solution, wear PPE and slowly add 200 mL of concentrated reagent grade nitric acid (HNO_3) to 800 mL distilled water in a 1-L beaker under the hood with ventilation. Place a beaker cover over the beaker and let it cool before pouring into a storage container.

8.8.2.4 Piranha Cleaning Solution

It is a mixture of concentrated H_2SO_4 and 30% hydrogen peroxide (H_2O_2) in the ratio 3:1. The mixture becomes hot when the H_2O_2 is slowly added to the concentrated H_2SO_4. The authors recommend that the laboratory purchase the Piranha solution from a manufacturer because it is dangerous to prepare the solution in the laboratory.

8.8.2.5 Fuming H_2SO_4 Solution

It is concentrated H_2SO_4 containing sulfur trioxide (10%–20%), which makes pyrosulfuric acid. The authors recommend that the laboratory purchase the fuming H_2SO_4 solution from a manufacturer directly.

8.8.2.6 Nochromix Cleaning Solution

Nochromix is a commercial formulation that contains ammonium persulfate. It is a powder that is mixed with water and added to 98% H_2SO_4, forming a clear solution. The cleaning solution can be safely disposed of (after neutralization) via the sanitary sewer, if not contaminated with other heavy metals or toxic substances.

8.9 LABORATORY WASTE MANAGEMENT AND DISPOSAL

8.9.1 BOD Testing Wastewater

The BOD testing wastewater contains no heavy metals if alternative cleaning solutions, such as HCl, H_2SO_4, nitric acid, Piranha solution, fuming H_2SO_4 solution, or Nochromix solution, are used for cleaning the BOD bottles and other glassware.

However, the strong acid and oxidation agents in the wastewater will still label the BOD testing wastes as hazardous wastes. Proper disposal of such laboratory hazardous wastes is required.

8.9.2 Small Quantity Generator Status and Commercial Waste Disposal Option

In most laboratories that conduct BOD testing, small-quantity generators probably qualify as conditionally exempt if they generate no more than 100 kg of hazardous waste in a month. Nevertheless, the laboratory manager should contact the state or local government to see whether or not the laboratory is indeed qualified as a small-quantity generator, so it may treat its own chemical wastes legally [8,9].

If on-site waste treatment in the laboratory is not legal, the laboratory must then store its wastes in lined 55-gallon (208-L) drums for disposal by a commercial hazardous waste handler. The costs of disposal will range from USD 4000 to USD 7000 per drum, depending on the heavy metals content.

8.9.3 On-Site Waste Treatment by Laboratory Personnel

If it is legal for the laboratory to treat its waste prior to sewer discharge, the waste disposal cost can be significantly reduced.

8.9.3.1 Treatment of BOD Testing Wastewater Containing Chromium

If the BOD bottles and other glassware are cleaned by dichromate acid cleaning solution (chromatic acid solution), the wastewater will contain toxic heavy metals, chromium(VI) and chromium(III). The former is much more toxic than the latter. The laboratory manager should first separate the common BOD testing wastewater from the highly toxic and concentrated spent dichromate acid cleaning solution (spent chromatic acid waste).

The common BOD testing wastewater can be pretreated by an elementary neutralization process [10] with industrial grade NaOH solution prior to sewer discharge. The highly toxic spent chromatic acid wastewater may be neutralized by the following: special oxidation–reduction, neutralization, precipitation, and filtration process system [10–13]:

1. Add industrial grade low-cost ferrous chloride or ferrous sulfate to reduce residual chromium(VI) in the chromatic acid waste to chromium(III).
2. Before neutralization of the chromatic acid wastewater (18 M hydrogen ion) by concentrated NaOH solution, the wastewater should be diluted with water by at least sixfold. With the sixfold dilution and adequate mixing, the neutralization process increases wastewater temperature to about 40°C when NaOH is added to acidic wastewater carefully over a 5-min period to avoid boiling. The wastewater must be diluted by a factor of 6 also to avoid crystallization and precipitation of $Na_2SO_4 \cdot 10H_2O$. When $Na_2SO_4 \cdot 10H_2O$ does precipitate, a larger dilution factor is needed.
3. The trivalent chromium ions can easily be precipitated as insoluble chromium hydroxide by the aforementioned chemical neutralization process using NaOH (but not calcium hydroxide, nor calcium oxide) at pH >7. The small amount of precipitated chromium hydroxide sludge can be removed by filtration and collected for further disposal.
4. Chemical precipitation of $Cr(OH)_3$ in sludge is a rapid process. The precipitated and settled wastewater can be filtered within minutes of pH adjustment or chemical addition. The treated wastewater can be discharged to the laboratory sewer (with permit, if required), and the very small amount of $Cr(OH)_3$ in sludge can be delivered to a licensed hazardous waste handler for further processing or recovery [8,14].

8.9.3.2 Treatment of BOD Testing Wastewater Containing No Heavy Metals

If the BOD bottles and other glassware are cleaned by any alternative cleaning solutions listed in Section 8.8, the wastewater will not contain any toxic heavy metals. The laboratory manager still should first separate the common BOD testing wastewater from the highly acidic and concentrated spent cleaning solution.

Again, the common BOD testing wastewater can be pretreated by elementary neutralization process [10] with industrial grade NaOH solution prior to sewer discharge. The highly acidic spent cleaning solution may be neutralized by the following special neutralization process:

1. Before neutralization of the spent acid cleaning solution (i.e., wastewater) by concentrated NaOH solution, the wastewater should be diluted with water at least sixfold. With the sixfold dilution and adequate mixing, neutralization process may increase wastewater temperature to about 40°C when NaOH is added to acidic wastewater carefully over a 5-min period to avoid boiling. The wastewater must be diluted by a factor of 6 also to avoid crystallization and precipitation of $Na_2SO_4 \cdot 10H_2O$. When $Na_2SO_4 \cdot 10H_2O$ does precipitate, a larger dilution factor is needed.
2. The treated, neutralized wastewater can be discharged to the laboratory sewer (with permit, if required).

REFERENCES

1. Wang, MHS and Wang, LK. Environmental water engineering glossary. In: *Advances in Water Resources Engineering*, Yang, CT and Wang, LK (editors). Springer, New York, 471–556 (2015).
2. APHA, AWWA, WEF, Rice, EW, Baird, RB, Eaton, AD, and Clesceri, LS. *Standard Methods for the Examination of Water and Wastewater*, 22nd edition. American Public Health Association, Washington, DC, 1496 p. (2012).
3. Adams, VD. *Water and Wastewater Examination Manual*. Lewis Publishers, Chelsea, MI, 247 p. (1990).
4. Wang, LK, DeMichele, E, and Wang, MHS. Simplified laboratory procedure for BOD determination. Lenox Institute of Water Technology, Lenox, MA. Technical report LIR/06-85/144 (1985).

5. New York State Department of Health, *Laboratory Procedures for Wastewater Treatment Plant Operators*, Albany, NY, 141 p. (ND).
6. University of Wisconsin-Madison. *The Chemical Safety Mechanism—Laboratory Glassware Cleaning*. University of Wisconsin-Madison, Office of Chemical Safety, Madison, WI (2013).
7. Fisher Scientific. *Material Safety Data Sheet—Dichromate Cleaning Solution*. Fisher Scientific Company, LLC, Middletown, VA (2016).
8. Holm, TR. Treatment of spent chemical oxygen demand solutions for safe disposal. University of Illinois, Waste Management and Research Center, Illinois State Water Survey, Urbana-Champaign, IL. Technical Report TR 20, 28 p. (1996).
9. Wang, LK. In-plant management and disposal of industrial hazardous substances. In: *Handbook of Industrial and Hazardous Wastes Treatment*, Wang, LK, Hung, YT, Lo, HH, and Yapijakis, C (editors), 2nd edition. Marcel Dekker and CRC Press, New York, 515–584 (2004).
10. Goel, RK, Flora, JRV, and Chen, JP. Flow equalization and neutralization. In: *Physicochemical Treatment Processes*, Wang, LK, Hung, YT, and Shammas, NK (editors). Humana Press/Springer, Totowa, NJ, 21–44 (2005).
11. Wang, LK and Li, Y. Chemical reduction/oxidation. In: *Advanced Physicochemical Treatment Processes*, Wang, LK, Hung, YT, and Shammas, NK (editors). Humana Press/Springer, Totowa, NJ, 483–519 (2006).
12. Wang, LK, Vaccari, DA, Li, Y, and Shammas, NK. Chemical precipitation. In: *Physicochemical Treatment Processes*, Wang, LK, Hung, YT, and Shammas, NK (editors). Humana Press/Springer, Totowa, NJ, 141–174 (2005).
13. Chen, JP, Chang, SY, Huang, JYC, Baumann, ER, and Hung, YT. Gravity filtration. In: *Physicochemical Treatment Processes*, Wang, LK, Hung, YT, and Shammas, NK (editors). Humana Press/Springer, Totowa, NJ, 501–544 (2005).
14. Wang, LK, Shammas, NK, Aulenbach, DB, and Selke, WA. Treatment of nickel-chromium plating wastes. In: *Handbook of Advanced Industrial and Hazardous Wastes Treatment*, Wang, LK, Shammas, NK, and Hung, YT (editors). CRC Press, Boca Raton, FL, 231–258 (2010).

9 Principles, Procedures, and Heavy Metal Management of Dichromate Reflux Method for COD Determination in Laboratories

Mu-Hao Sung Wang and Lawrence K. Wang
Lenox Institute of Water Technology

Eugene De Michele
Water Environment Federation

CONTENTS

9.1	Introduction	299
9.2	Principles	299
	9.2.1 Chemical Oxidation Equation	299
	9.2.2 Example of Chemical Oxidation	300
	9.2.3 Remaining Dichromate ($Cr_2O_7^{2-}$) Determination and Standard Ferrous Ammonium Sulfate (FAS) Standardization	300
9.3	Interference and Its Chemical Equation	301
9.4	Apparatus	301
9.5	Reagents	301
	9.5.1 H_2SO_4–Ag_2SO_4 (Sulfuric Acid-Silver Sulfate) Solution	301
	9.5.2 Standard $K_2Cr_2O_7$ Solution, 0.25 N, or 0.0417 M	301
	9.5.3 Ferroin Indicator Solution	301
	9.5.4 FAS, 0.25 N, or 0.25 M	302
	9.5.4.1 Standardization Procedure	302
	9.5.5 Mercuric Sulfate ($HgSO_4$)	302
	9.5.6 H_2SO_4	302
	9.5.7 COD Standard Solution, 500 mg/L	303
	9.5.8 Sulfamic Acid	303
	9.5.9 Sodium Hydroxide (NaOH)	303
9.6	Analytical Procedure	303
9.7	Treatment of COD Testing Wastewater	304
	9.7.1 On-Site or Off-Site Treatment of COD Testing Wastewater	304
	9.7.2 Neutralization Process and Decision-Making Process	304
	9.7.3 Combined Removal of Toxic Chromium and Mercury by Sodium Hydroxide Neutralization, Precipitation, Sedimentation, and Filtration	304

9.7.4 Combined Removal of Toxic Silver and Mercury by Sodium Sulfide Precipitation, Sedimentation, and Filtration 305
9.7.5 Effluent Discharge and Sludge Disposal 305
References 305

Abstract

Under sponsorship of Water Environment Federation, the standard procedure for chemical oxygen demand (COD) determination as outlined in the *APHA/AWWA/WPCF Standard Methods for the Examination of Water and Wastewater* was simplified for use by the waste treatment plant operators and researchers. Specifically, the principles, analytical procedures, and waste management processes of dichromate reflux method for rapid COD determination in laboratories are presented and discussed. Proper handling and treatment of spent COD liquid wastes for safe disposal are discussed and emphasized in this chapter. The spent COD testing wastewater contains high concentrations of sulfuric acid, mercuric ions (Hg^{2+}), silver ions (Ag^+), and dichromate salts, which are considered to be hazardous wastes and must be handled properly by the operators, chemists, and engineers in the laboratories. Normally a laboratory is a conditionally exempt small quantity generator if it generates no more than 100 kg of hazardous waste in a calender month. Such facilities may be allowed to treat hazardous wastes on-site. This chapter discusses several on-site treatment processes for possible adoption by small laboratories. It is important to note that even if federal regulations allow on-site treatment of COD testing wastewater, the State or local regulations may prohibit it. Hazardous sulfuric acid in the spent COD waste can be neutralized by sodium hydroxide or equivalent alkaline chemicals, such as potassium hydroxide; sodium hydroxide is cheaper. Before neutralization of COD wastewater (18 M hydrogen ion) by concentrated sodium hydroxide solution, the wastewater should be diluted with water by at least sixfold. With the sixfold dilution and adequate mixing, neutralization process increases wastewater temperature to about 40°C when sodium hydroxide is added to acidic COD wastewater carefully over a 5-min period to avoid boiling. The COD wastewater must be diluted by a factor of 6 also to avoid crystallization and precipitation of Na_2SO_4–$10H_2O$. When Na_2SO_4–$10H_2O$ does precipitate, a larger dilution factor is needed. In case the laboratory hesitates to process the COD testing wastewater by neutralization with sodium hydroxide, the authors recommend that the laboratory manager store all COD testing wastewater in a 55-gallon drum for off-site treatment by a licensed commercial hazardous-waste handler instead. The recent price for commercial disposal of a 55-gallon (208-liter) drum of COD wastewater is US$7,250. If the laboratory manager has confidence to conduct his/her own on-site neutralization process, then the following heavy metal removal processes for removal of chromium, mercury, and silver can be carried out by the laboratory personnel for significant cost saving. The dichromate $Cr_2O_7^{2-}$ in the COD reagent oxidizes most organic substances nearly quantitatively under high temperature and in the presence of concentrated acid. If final detection of COD is done by titration, all chromium should be in the form of Cr^{3+}. The trivalent chromium ions can easily be precipitated as insoluble chromium hydroxide by the above chemical neutralization process using sodium hydroxide (but not calcium hydroxide, nor calcium oxide) at pH >7. The precipitated chromium hydroxide sludge can be removed by filtration. If final COD detection is by colorimetry, there will still be unknown amount of $Cr_2O_7^{2-}$ ions left. The hexavalent chromium in $Cr_2O_7^{2-}$ can be reduced to trivalent chromium Cr^{3+} with any reducing agents (such as industrial grade ferrous sulfate, or steelmaking by-product ferrous chloride) at pH below 3, that is, before the COD wastewater is neutralized with sodium hydroxide. After chromium reduction process is complete, ferrous ions are converted to ferric ions at the same time. Both trivalent chromium ions and trivalent ferric ions can then be easily precipitated by sodium hydroxide as chromium hydroxide and ferric hydroxide, respectively (brown sludge), at pH >7. The chemical reduction and neutralization process can reduce chromium with >99.99% efficiency. The effluent soluble chromium concentration after chemical precipitation, sedimentation, and filtration can be <100 µg/L. The pH range for minimum chromium solubility is 7–11. Neutralizing the sulfuric acid with sodium hydroxide and adjusting the pH to >5 causes the precipitation of mercury oxide (HgO), which removes 96% of the Hg. HgO precipitation may be useful as either pretreatment or the first step of a multistep processing method to treat COD testing wastewater. Neutralizing the sulfuric acid and precipitating most of the mercury may lower the price charged by a hazardous waste service to dispose of the treated wastewater. Straight-chain aliphatic compounds are oxidized more effectively when silver sulfate (Ag_2SO_4) is added as a catalyst in the reagent. Accordingly there are toxic Ag^+ in the COD wastewater. Silver has been removed from the United States Environmental Protection Agency (USEPA) priority-pollutant list, but it is still regulated by many local governments. The Urbana-Champaign Sanitary District (UCSD), IL, for instance, has a silver discharge standard of 0.3 mg/L. It is known that Ag^+ can be precipitated by chloride ions (or any halides), and thus, its catalyzing effect in COD tests can be significantly reduced. For chloride concentrations up to 2000 mg/L, the chloride interference in silver ion's catalyzing effect can be overcome by adding a large excess of Hg^{2+} to complex the chloride. Both Ag^+ and Hg^{2+} can be removed together with sodium sulfide (Na_2S) or equivalent. Regardless of the final COD detection method being used (titration or colorimetry), toxic soluble mercury and silver in the COD wastewater can be effectively removed by sulfide precipitation, sedimentation, and filtration. In the wastewater treatment process system, excess Na_2S is added to precipitate soluble mercury and soluble silver as HgS and Ag_2S, respectively, and then excess zinc ion (Zn^{2+}) is added to precipitate the excess sulfide ion (S^{2-}). Addition of Zn^{2+} reduces concentrations of H_2S at low pH values and concentrations of Hg and Ag at high pH values to very low levels. Addition of Zn^{2+} (such as $ZnCl_2$) after Na_2S allows complete reaction of Hg^{2+} and Ag^+ with HS^- before ZnS precipitation. The final mercury concentration in the filtrate is reduced to <3 µg/L µg/L (which is the mercury discharge standard for the UCSD, IL), representing a 99.999% reduction. The UCSD silver discharge standard of 0.3 mg/L is also met. Chemical precipitation of $Cr(OH)_3$, HgS and Ag_2S sludge is a rapid process. The precipitated and settled wastewater can be filtered within minutes of pH adjustment or chemical addition. The treated wastewater can be discharged to the laboratory sewer (with permit, if required), and the very small amount of $Cr(OH)_3$, HgS, and Ag_2S sludge can be delivered to a licensed hazardous-waste handler for further processing or recovery.

9.1 INTRODUCTION

The chemical oxygen demand (COD) determination measures the oxygen equivalent of that portion of organic matter in a liquid sample that can be oxidized by a strong chemical oxidizing agent. The COD value is important in industrial and domestic waste studies and wastewater treatment plant (WWTP) operations. Generally, biochemical oxygen demand (BOD) is about 50%–70% of the COD values, depending on the type of organic matter in the water. COD values can be determined more quickly than BOD (3 h versus 5 days) and can therefore be used for a much more rapid estimate of WWTP performance. The COD test is most useful for monitoring and control, especially after correlations with constituents such as BOD and total organic carbon have been developed [1–4].

The majority of organic compounds are oxidized completely (95%–100%) in this COD test.

9.2 PRINCIPLES

9.2.1 CHEMICAL OXIDATION EQUATION

The COD is used as a measure of the oxygen equivalent of the organic substance content of a water or wastewater sample that is susceptible to chemical oxidation by a strong oxidizing agent. Many oxidizing agents can be used, but the standard and recommended methods in the United States adopt potassium dichromate ($K_2Cr_2O_7$) to be the oxidant [1–4]. In the COD determination procedures, the organic matters in a wastewater sample are oxidized by a boiling mixture of chromic and sulfuric acid (H_2SO_4). In the presence of oxidizing agent ($K_2Cr_2O_7$) and strong acid (H_2SO_4) under heat, the organic matter in wastewater sample is oxidized to carbon dioxide and water, and at the same time, the Cr(VI) in $K_2Cr_2O_7$ is reduced to Cr(III) ion. The chemical reaction involved in the standard methods for COD determination can be represented by Equation 9.1, assuming $C_aH_bO_c$ is the organic matter in wastewater, as follows:

$$C_aH_bO_c + n\ Cr_2O_7^{2-} + 8n\ H^+ \xrightarrow{\text{under heat}} a\ CO_2 + \left[(b+8n)/2\right] H_2O + 2n\ Cr^{3+}, \quad (9.1)$$

where $n = (2a/3) + (b/6) - (c/3)$.

9.2.2 Example of Chemical Oxidation

Assume $C_{10}H_{19}O_3$ is the empirical formula of organic matter in an industrial wastewater, and write the chemical oxidation equation if the organic matter is oxidized by $K_2Cr_2O_7$ in the presence of H_2SO_4 under heat. Check the material balances of the resulting equation.

Solution

$$a = 10$$
$$b = 19$$
$$c = 3$$
$$n = (2a/3) + (b/6) - (c/3) = (2 \times 10/3) + (19/6) - (3/3) = 8.83.$$

Therefore, the chemical reaction is written as:

$$C_{10}H_{19}O_3 + 8.83\ Cr_2O_7^{2-} + 70.64\ H^+ \xrightarrow{\text{under heat}} 10\ CO_2 + 44.82\ H_2O + 17.66\ Cr^{3+}.$$

Check the material balances of the above-written equation:

C = 10 before reaction and 10 after;
H = (19+70.4) before reaction and (44.82×2) after;
O = (3+7×8.83) before reaction and (20+44.82) after; and
Cr = (2×8.83) before reaction and (20+44.82) after.

Note that everything is correct.

9.2.3 Remaining Dichromate ($Cr_2O_7^{2-}$) Determination and Standard Ferrous Ammonium Sulfate (FAS) Standardization

In any COD determination, an excess of oxidizing agent must be present to ensure that all the organic substances are completely oxidized, within the power of the oxidizing agent. In this case, the remaining oxidizing agent is $Cr_2O_7^{2-}$, which can be determined by a back titration method using a reducing agent (FAS) of known concentration.

FAS concentration must be measured before the back titration of the remaining $Cr_2O_7^{2-}$ because all reducing agents (including FAS) are gradually oxidized by oxygen from the air. Standardization

COD Determination in Laboratories

of FAS is made with 0.25 N solution of $Cr_2O_7^{2-}$. The chemical reaction between FAS (Fe^{2+}) and $Cr_2O_7^{2-}$ may be expressed as follows:

$$6\ Fe^{2+} + Cr_2O_7^{2-} + 14\ H^+ \rightarrow 2\ Fe^{3+} + 2\ Cr^{3+} + 7\ H_2O. \quad (9.2)$$

In the above equation, Cr(VI) in $Cr_2O_7^{2-}$ is reduced to Cr(III) as Cr^{3+} ion, while Fe(II) as Fe^{2+} ion is oxidized to Fe(III) as Fe^{3+} ion at the same time [3].

9.3 INTERFERENCE AND ITS CHEMICAL EQUATION

Generally, COD cannot be measured accurately in wastewater samples containing more than 2000 mg/L of chloride because high chloride concentrations and certain other reducing inorganic ions can cause erroneously high COD results when they are oxidized in the COD test procedure. The following two equations show how chloride ion can be oxidized by $Cr_2O_7^{2-}$ and how this problem can be overcome by adding mercuric sulfate:

$$6\ Cl^- + Cr_2O_7^{2-} + 14\ H^+ \rightarrow 2\ Cr^{3+} + 3\ Cl_2 + 7\ H_2O \quad (9.3)$$

$$HgSO_4 + 2\ Cl^- \rightarrow HgCl_2(solid) + SO_4^{2-}. \quad (9.4)$$

In the presence of excess mercuric ions (Hg^{2+}), the chloride interference can be eliminated. However, the added mercuric sulfate is a toxic heavy metal that will remain in the COD wastewater.

9.4 APPARATUS

1. A reflux apparatus consisting of a 500-mL Erlenmeyer flask with ground-glass neck size 24/40 and a 300 mm jacket Liebig-, West-, or Friedrichs-style condenser with 24/40 ground-glass joint (Figure 9.1).
2. A hot plate with 1.4 W/cm² of heating surface
3. Erlenmeyer flask, 500 mL, with ground-glass neck size 24/40
4. Graduated cylinder, 50 mL
5. Assorted pipettes
6. Burette, graduated to 0.1 mL
7. Glass beads, reagents.

9.5 REAGENTS

9.5.1 H₂SO₄–Ag₂SO₄ (Sulfuric Acid-Silver Sulfate) Solution

Dissolve 22.55 g Ag_2SO_4 in a bottle containing 4.1 kg of concentrated H_2SO_4 or at a ratio of 5.5 g Ag_2SO_4/kg H_2SO_4. It will take Ag_2SO_4 1–2 days to get dissolved. Label bottle to avoid confusion with pure H_2SO_4.

9.5.2 Standard K₂Cr₂O₇ Solution, 0.25 N, or 0.0417 M

Dissolve 12.259 g $K_2Cr_2O_7$, previously dried at 103°C for 2 h in distilled water and make up to 1 L.

9.5.3 Ferroin Indicator Solution

Dissolve 1.485 g of 1,10-phenanthroline monohydrate ($C_{12}H_8N_2$–H_2O) with 0.695 g ferrous sulfate crystals ($FeSO_4$·$7H_2O$) in sufficient distilled water and dilute to 100 mL. This indicator solution may be purchased or prepared.

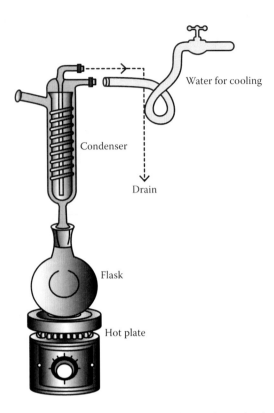

FIGURE 9.1 Reflex apparatus for chemical oxygen demand (COD) determination.

9.5.4 FAS, 0.25 N, or 0.25 M

Dissolve 98 g FAS (FeSO$_4$(NH$_4$)$_2$SO$_4 \cdot$ 6H$_2$O) in distilled water. Add 20 mL concentrated H$_2$SO$_4$, cool, and make up to 1 L. This solution must be standardized against K$_2$Cr$_2$O$_7$ daily.

9.5.4.1 Standardization Procedure

Dilute 10 mL 0.25 N K$_2$Cr$_2$O$_7$ solution to about 100 mL. Add 30 mL pure concentrated H$_2$SO$_4$ and allow it to cool. Titrate against the FAS titrant using two or three drops (0.10–0.1.5 mL) of the ferroin indicator. Use the same volume of ferroin indicator for all titrations. Take the first sharp color change from blue-green to reddish brown as the end point of the titration. Record the volume of FeSO$_4$(NH$_4$)$_2$SO$_4$ used in titration. The following is the equation for standardization:

$$\text{Normality (or molarity) of ferrous ammonium sulfate or FAS (C)}$$
$$= (\text{mL standard 0.25 N potassium dichromate solution})$$
$$\times 0.25 / (\text{mL FeSO}_4 (\text{NH}_4)_2 \text{SO}_4 \text{ used})$$
$$= 10 \times 0.25 / (\text{mL FeSO}_4 (\text{NH}_4)_2 \text{SO}_4 \text{ used}). \tag{9.5}$$

9.5.5 Mercuric Sulfate (HgSO$_4$)

It should be of analytical grade in the form of crystals or powders.

9.5.6 H$_2$SO$_4$

It should be concentrated, 36 N, containing no Ag$_2$SO$_4$.

9.5.7 COD Standard Solution, 500 mg/L

Lightly crush and then dry potassium hydrogen phthalate or KHP (HOOCC$_6$H$_4$COOK) to constant weight at 120°C. Dissolve 425 mg KHP in distilled water, and dilute to 1000 mL. KPH in solid form has a theoretical COD of 1.176 g O$_2$/g, and this solution has a theoretical COD of 500 mg O$_2$/L. The KHP solution is stable when refrigerated for up to 3 months in the absence of visible biological growth.

9.5.8 Sulfamic Acid

It is required only if the interference of nitrites is to be eliminated. Nitrite (NO$_2^-$) exerts a COD of 1.1 mg O$_2$/mg NO$_2^-$. Because concentrations of nitrite in waters rarely exceed 1 or 2 mg/L, the nitrite interference is considered insignificant and is usually ignored. To eliminate a significant interference due to nitrite, add 10 mg sulfamic acid for each milligram of nitrite-N present in the sample volume used; add the same amount of sulfamic acid to the reflux vessel containing the distilled water blank.

9.5.9 Sodium Hydroxide (NaOH)

Inexpensive industrial grade NaOH, in either solid or liquid form, will be sufficient for COD wastewater treatment.

9.6 ANALYTICAL PROCEDURE

1. Add a few clean glass beads to a dry 500-mL Erlenmeyer refluxing flask. Place a 50-mL sample in the flask and add 1.0 g mercuric sulfate and 10–25 mL of 0.25 N K$_2$Cr$_2$O$_7$ solution. (Note: If COD is expected to be high, sample size smaller than 50 mL shall be used, but it shall be diluted to 50 mL in order to accommodate 25 mL K$_2$Cr$_2$O$_7$ solution). Carefully and slowly add 75 mL H$_2$SO$_4$–Ag$_2$SO$_4$ solution, a little at a time, mixing after each addition to dissolve HgSO$_4$. Cool while mixing to avoid possible loss of volatile materials. Make sure that the reflux mixture is mixed thoroughly before heat is applied. Attach the flask to the condenser, turn on cooling water, and reflux the mixture for 2 h. Cover the open end of the condenser with a small beaker to prevent foreign material from entering the refluxing mixture. Cool and then wash down the condenser with about 25 mL distilled water.
2. Disconnect reflux condenser and dilute the mixture to about 140 mL, cool to room temperature, and titrate the excess Cr$_2$O$_7^{2-}$ with FAS, using ferroin indicator. Generally, add two or three drops (0.1–0.15 mL) of the indicator to the sample just before titrating. Take the sharp color change from blue-green to reddish brown as the end point, even though the blue-green color may reappear within minutes. Record the volume of ferrous ammonium sulfate solution used for the sample (B).
3. Reflux in the same manner a blank consisting of 50 mL distilled water instead of the sample, together with the required reagents. Record the volume of FAS solution used for the blank (A).
4. COD wastewater containing toxic heavy metals must be collected together for safe disposal in accordance with Section 9.7.
5. A chemist's skill for COD determination can be evaluated using 500 mg/L COD standard solution.

Calculations

$$\text{mg/L COD} = (A - B) \times C \times \frac{8000}{(\text{mL sample})}, \qquad (9.6)$$

where

A = FAS used for blank, mL,
B = FAS used for sample, mL,
C = normality (or molarity) of FAS.

9.7 TREATMENT OF COD TESTING WASTEWATER

9.7.1 ON-SITE OR OFF-SITE TREATMENT OF COD TESTING WASTEWATER

Proper handling and treatment of spent COD liquid wastes for safe disposal are discussed and emphasized in this chapter. The spent COD testing wastewater contains high concentrations of H_2SO_4, Hg^{2+}, silver ions (Ag^+), and $Cr_2O_7^{2-}$ salts, which are considered to be hazardous wastes and must be handled properly by the operators, chemists, and engineers in the laboratories. Normally, a laboratory is a conditionally exempt small quantity generator if it generates no more than 100 kg of hazardous waste in a calendar month. Such facilities may be allowed to treat hazardous wastes on-site. This chapter discusses several on-site treatment processes for possible adoption by small laboratories. It is important to note that even if federal regulations allow on-site treatment of COD testing wastewater, the state or local regulations may prohibit it [3,5,6].

9.7.2 NEUTRALIZATION PROCESS AND DECISION-MAKING PROCESS

Hazardous H_2SO_4 in the spent COD waste can be neutralized by NaOH or equivalent alkaline chemicals, such as potassium hydroxide; NaOH is cheaper. Before neutralization of COD wastewater (18 M hydrogen ion) by concentrated NaOH solution, the wastewater should be diluted with water by at least sixfold. With the sixfold dilution and adequate mixing, neutralization process increases wastewater temperature to about 40°C when NaOH is added to acidic COD wastewater carefully over a 5-min period to avoid boiling. The COD wastewater must be diluted by a factor of 6 to also avoid crystallization and precipitation of Na_2SO_4–$10H_2O$. When Na_2SO_4–$10H_2O$ does precipitate, a larger dilution factor is needed. In case the laboratory hesitates to process the COD testing wastewater by neutralization with NaOH, the authors recommend that the laboratory manager store all COD testing wastewater in a 55-gal drum for off-site treatment by a licensed commercial hazardous waste handler instead. The recent price for commercial disposal of a 55-gal (208 L) drum of COD wastewater is US$7250. If the laboratory manager has confidence to conduct his/her own on-site neutralization process, then the following heavy metal removal processes for removal of chromium, mercury (Hg), and silver can be carried out by the laboratory personnel for significant cost saving [5–10].

9.7.3 COMBINED REMOVAL OF TOXIC CHROMIUM AND MERCURY BY SODIUM HYDROXIDE NEUTRALIZATION, PRECIPITATION, SEDIMENTATION, AND FILTRATION

The $Cr_2O_7^{2-}$ in the COD reagent oxidizes most organic substances nearly quantitatively under high temperature and in the presence of concentrated acid. If final detection of COD is done by titration, all chromium should be in the form of Cr^{3+}. The trivalent chromium ions can easily be precipitated as insoluble chromium hydroxide by the above chemical neutralization process using NaOH (but not calcium hydroxide, nor calcium oxide) at pH >7. The precipitated chromium hydroxide sludge can be removed by filtration [10].

If final COD detection is by colorimetry, there will still be an unknown amount of $Cr_2O_7^{2-}$ ions left. The hexavalent chromium in $Cr_2O_7^{2-}$ can be reduced to trivalent chromium Cr^{3+} with any reducing agents (such as industrial grade ferrous sulfate or steelmaking by-product ferrous chloride) at pH <3, that is, before the COD wastewater is neutralized with NaOH. After chromium reduction process is complete, ferrous ions are converted to ferric ions at the same time. Both trivalent chromium ions and trivalent ferric ions can then be easily precipitated by NaOH as chromium hydroxide and ferric hydroxide, respectively (brown sludge), at pH >7. The chemical reduction and neutralization process can reduce chromium with >99.99% efficiency. The effluent soluble chromium concentration after chemical precipitation, sedimentation, and filtration can be <100 µg/L. The pH range for minimum chromium solubility is 7–11 [7–10].

Neutralizing the H_2SO_4 with NaOH and adjusting the pH to >5 causes the precipitation of mercury oxide (HgO), which removes 96% of the Hg. HgO precipitation may be useful as either pretreatment or the first step of a multistep processing method to treat COD testing wastewater. Neutralizing the H_2SO_4 and precipitating most of the Hg may lower the price charged by a hazardous waste service to dispose of the treated wastewater [3,5].

9.7.4 Combined Removal of Toxic Silver and Mercury by Sodium Sulfide Precipitation, Sedimentation, and Filtration

Straight-chain aliphatic compounds are oxidized more effectively when Ag_2SO_4 is added as a catalyst in the reagent. Accordingly, there is toxic Ag^+ in the COD wastewater. Silver has been removed from the USEPA priority-pollutant list, but it is still regulated by many local governments. The Urbana-Champaign Sanitary District (UCSD), IL, for instance, has a silver discharge standard of 0.3 mg/L. It is known that Ag^+ can be precipitated by chloride ions (or any halides), and thus, its catalyzing effect in COD tests can be significantly reduced. For chloride concentrations up to 2000 mg/L, the chloride interference in silver ion's catalyzing effect can be overcome by adding a large excess of Hg^{2+} to complex the chloride. Both Ag^+ and Hg^{2+} can be removed together with sodium sulfide (Na_2S) or equivalent [1,5].

Regardless of the final COD detection method being used (titration or colorimetry), toxic soluble Hg and silver in the COD wastewater can be effectively removed by sulfide precipitation, sedimentation, and filtration. In the wastewater treatment process system, excess Na_2S is added to precipitate soluble Hg and soluble silver as HgS and Ag_2S, respectively, and then excess zinc ion (Zn^{2+}) is added to precipitate the excess sulfide ion (S^{2-}). Addition of Zn^{2+} reduces concentrations of H_2S at low pH values and concentrations of Hg and Ag at high pH values to very low levels. Addition of Zn^{2+} (such as $ZnCl_2$) after Na_2S allows complete reaction of Hg^{2+} and Ag^+ with HS^- before ZnS precipitation. The final Hg concentration in the filtrate is reduced to <3 µg/L (which is the Hg discharge standard for the UCSD, IL), representing a 99.999% reduction. The UCSD silver discharge standard of 0.3 mg/L is also met [5].

9.7.5 Effluent Discharge and Sludge Disposal

Chemical precipitation of $Cr(OH)_3$, HgS, and Ag_2S sludge is a rapid process. The precipitated and settled wastewater can be filtered within minutes of pH adjustment or chemical addition. The treated wastewater can be discharged to the laboratory sewer (with permit, if required), and the very small amount of $Cr(OH)_3$, HgS, and Ag_2S sludge can be delivered to a licensed hazardous waste handler for further processing or recovery [3,5,6].

REFERENCES

1. APHA, AWWA, WEF (authors), Rice, EW, Baird, RB, Eaton, AD, and Clesceri, LS (editors). *Standard Methods for the Examination of Water and Wastewater*, 22nd edition, American Public Health Association, Washington, DC, 1496 p. (2012).
2. Adams, VD. *Water and Wastewater Examination Manual*, Lewis Publishers, Chelsea, MI, 247 p. (1990).
3. Wang, LK, DeMichele, E, and Wang, MHS. Simplified laboratory procedure for COD determination using dichromate reflux method. Lenox Institute of Water Technology, Lenox, MA. Technical Report LIR/06-85/143 (1985).

4. New York State Department of Health, Laboratory procedures for wastewater treatment plant operators, Albany, NY, 141 p. (ND).
5. Holm, TR, Treatment of spent chemical oxygen demand solutions for safe disposal, University of Illinois, Waste Management and Research Center, Illinois State Water Survey, Urbana-Champaign, IL. Technical Report TR 20, 28 p. (1996).
6. Wang, LK. In-plant management and disposal of industrial hazardous substances. In: *Handbook of Industrial and Hazardous Wastes Treatment*, 2nd edition, Wang, LK, Hung, YT, Lo, HH, and Yapijakis, C (editors), Marcel Dekker and CRC Press, New York, 515–584 (2004).
7. Wang, LK and Li, Y. Chemical reduction/oxidation. In: *Advanced Physicochemical Treatment Processes*, Wang, LK, Hung, YT, and Shammas, NK (editors), Humana Press/Springer, Totowa, NJ, 483–519 (2006).
8. Goel, RK, Flora, JRV, and Chen, JP. Flow equalization and neutralization. In: *Physicochemical Treatment Processes*, Wang, LK, Hung, YT, and Shammas, NK (editors), Humana Press/Springer, Totowa, NJ, 21–44 (2005).
9. Wang, LK, Vaccari, DA, Li, Y, and Shammas, NK. Chemical precipitation. In: *Physicochemical Treatment Processes*, Wang, LK, Hung, YT, and Shammas, NK (editors), Humana Press/Springer, Totowa, NJ, 141–174 (2005).
10. Chen, JP, Chang, SY, Huang, JYC, Baumann, ER, and Hung, YT. Gravity filtration. In: *Physicochemical Treatment Processes*, Wang, LK, Hung, YT and Shammas, NK (editors), Humana Press/Springer, Totowa, NJ, 501–544 (2005).

10 Treatment and Management of Metal Finishing Industry Wastes

Nazih K. Shammas
Lenox Institute of Water Technology and Krofta Engineering Corporation

Lawrence K. Wang
Lenox Institute of Water Technology, Krofta Engineering Corporation, and Zorex Corporation

CONTENTS

10.1 Industry Description	308
10.1.1 General Description	309
10.1.2 Subcategory Descriptions	312
10.2 Wastewater Characterization	313
10.2.1 Common Metals Subcategory	314
10.2.2 Precious Metals Subcategory	315
10.2.3 Complexed Metals Subcategory	315
10.2.4 Cyanide Subcategory	315
10.2.5 Hexavalent Chromium Subcategory	318
10.2.6 Oils Subcategory	319
10.2.7 Solvent Subcategory	319
10.3 Source Reduction	319
10.3.1 Chemical Substitution	321
10.3.2 Waste Segregation	323
10.3.3 Process Modifications to Reduce Drag-Out Loss	323
10.3.3.1 Wetting Agents	323
10.3.3.2 Longer Drain Times	324
10.3.3.3 Other Drag-Out Reduction Techniques	324
10.3.4 Capture/Concentration Techniques	326
10.3.5 Waste Reduction Costs and Benefits	327
10.4 Pollutant Removability	328
10.4.1 Common Metals	329
10.4.2 Precious Metals	331
10.4.3 Complexed Metal Wastes	331
10.4.4 Hexavalent Chromium	331
10.4.5 Cyanide	332
10.4.6 Oils	332
10.4.7 Solvents	332
10.5 Treatment Technologies	333
10.5.1 Neutralization	333
10.5.2 Cyanide-Containing Wastes	333
10.5.3 Chromium-Containing Wastes	336

10.5.4 Arsenic- and Selenium-Containing Wastes 336
　　　10.5.4.1 Chemical Precipitation and Sedimentation 337
　　　10.5.4.2 Complexation 337
　10.5.5 Other Metals Wastes 338
10.6 Costs 338
　10.6.1 Typical Treatment Options 339
　10.6.2 Costs 339
10.7 CFR for Metal Finishing Effluent Discharge Management 341
　10.7.1 Applicability and Description of the Metal Finishing Point Source Category 341
　10.7.2 Monitoring Requirements of Metal Finishing Effluent Discharges 342
　10.7.3 Effluent Limitations Based on the BPT 342
　10.7.4 Effluent Limitations Based on the BAT 343
　10.7.5 Pretreatment Standards for Existing Sources 343
　10.7.6 New Source Performance Standards 344
　10.7.7 Pretreatment Standards for New Sources 345
10.8 Specialized Definitions 346
References 350

Abstract

The metal finishing industry comprises 44 unit operations involving the machining, fabrication, and finishing of metal products. There are approximately 160,000 manufacturing facilities in the United States, which are classified as being part of the metal finishing industry. The facilities vary in size, from small job shops employing fewer than 10 people to large plants employing thousands of production workers. Unit operations with significant water usage include electroplating, electroless plating, anodizing, conversion coating, etching, cleaning, machining, grinding, tumbling, heat treating, welding, sand blasting, salt bath descaling, paint stripping, painting, electrostatic painting, electroplating, and testing. This chapter covers industry description; wastewater characterization; source reduction; pollutant removability; technologies for treating organic and inorganic wastes, cyanide-containing wastes, chromium-containing wastes, arsenic- and selenium-containing wastes, and other metals wastes; and costs for waste reduction, treatment and disposal. Also introduced in this chapter is the U.S. Code of Federal Regulations (CFR) Title 40, Part 433 (40 CFR part 433) for effluent discharge management of metal finishing point source category. Special coverage includes the following: (a) the applicability and description of the metal finishing point source category; (b) the monitoring requirements of metal finishing effluent discharges; (c) the effluent limitations representing the degree of effluent reduction attainable by applying the best practicable control available; (d) the effluent limitations representing the degree of effluent reduction attainable by applying the best available technology economically achievable; (e) the pretreatment standards for existing sources; (f) the new source performance standards; and (g) the pretreatment standards for new sources.

10.1 INDUSTRY DESCRIPTION

The metal finishing industry is one of many industries subject to regulation under the Resource Conservation and Recovery Act [1,2] and the Hazardous and Solid Waste Amendments (HSWA) [3]. The metal finishing industry has also been subject to extensive regulation under the Clean Water Act [4]. Compliance with these regulations requires highly coordinated regulatory, scientific, and engineering analyses to minimize costs [5].

10.1.1 General Description

The metal finishing industry comprises 44 unit operations involving the machining, fabrication, and finishing of metal products (SIC groups 34–39). There are approximately 160,000 manufacturing facilities in the United States, which are classified as being part of the metal finishing industry [6]. These facilities are engaged in the manufacturing of a variety of products that are constructed primarily by using metals. The operations performed usually begin with a raw stock in the form of rods, bars, sheets, castings, forgings, and so forth, and can progress to sophisticated surface finishing operations. The facilities vary in size, from small job shops employing fewer than 10 people to large plants employing thousands of production workers. Wide variations also exist in the age of the facilities and the number and type of operations performed within facilities. Because of the differences in size and processes, production facilities are custom-tailored to the specific needs of each plant. The possible variations in unit operations within the metal finishing industry are extensive. Some complex products could require the use of nearly all of the 44 possible unit operations, while a simple product might require only a single operation. Each of the 44 individual unit operations is listed with a brief description below [7].

1. *Electroplating* is the production of a thin coating of one metal upon another by electrodeposition.
2. *Electroless plating* is a chemical reduction process that depends upon the catalytic reduction of a metallic ion in an aqueous solution containing a reducing agent and the subsequent deposition of metal without the use of external electric energy.
3. *Anodizing* is an electrolytic oxidation process that converts the surface of the metal to an insoluble oxide.
4. *Chemical conversion coatings* are applied to previously deposited metal or basis material for increased corrosion protection, lubricity, preparation of the surface for additional coatings, or formulation of a special surface appearance. This operation includes chromating, phosphating, metal coloring, and passivating.
5. *Etching and chemical milling* are used to produce specific design configurations and tolerances on parts by controlled dissolution with chemical reagents or etchants.
6. *Cleaning* involves the removal of oil, grease, and dirt from the surface of the basis material using water with or without a detergent or other dispersing material.
7. *Machining* is the general process of removing stock from a workpiece by forcing a cutting tool through the workpiece, removing a chip of basis material. Machining operations such as turning, milling, drilling, boring, tapping, planing, broaching, sawing and cutoff, shaving, threading, reaming, shaping, slotting, hobbing, filing, and chamfering are included in this definition.
8. *Grinding* is the process of removing stock from a workpiece by using a tool consisting of abrasive grains held by a rigid or semirigid binder. The processes included in this unit operation are sanding (or cleaning to remove rough edges or excess material), surface finishing, and separating (as in cutoff or slicing operations).
9. *Polishing* is an abrading operation used to remove or smooth out surface defects (scratches, pits, tool marks, etc.) that adversely affect the appearance or function of a part. The operation usually referred to as buffing is included in the polishing operation.
10. *Barrel finishing* or tumbling is a controlled method of processing parts to remove burrs, scale, flash, and oxides and to improve surface finish.
11. *Burnishing* is the process of finish sizing or smooth finishing a workpiece (previously machined or ground) by displacement, rather than removal, of minute surface irregularities. It is accomplished with a smooth point or line-contact and fixed or rotating tools.
12. *Impact deformation* is the process of applying an impact force to a workpiece such that the workpiece is permanently deformed or shaped. Impact deformation operations include shot peening, forging, high energy forming, heading, and stamping.

13. *Pressure deformation* is the process of applying force (at a slower rate than an impact force) to permanently deform or shape a workpiece. Pressure deformation includes operations such as roiling, drawing, bending, embossing, coining, swaging, sizing, extruding, squeezing, spinning, seaming, staking, piercing, necking, reducing, forming, crimping, coiling, twisting, winding, flaring, or weaving.
14. *Shearing* is the process of severing or cutting a workpiece by forcing a sharp edge or opposed sharp edges into the workpiece, stressing the material to the point of shear failure and separation.
15. *Heat treating* is the modification of the physical properties of a workpiece through the application of controlled heating and cooling cycles. Such operations as tempering, carburizing, cyaniding, nitriding, annealing, normalizing, austenitizing, quenching, austempering, siliconizing, martempering, and malleabilizing are included in this definition.
16. *Thermal cutting* is the process of cutting, slotting, or piercing a workpiece using an oxyacetylene oxygen lance or electric arc cutting tool.
17. *Welding* is the process of joining two or more pieces of material by applying heat, pressure, or both, with or without filler material, to produce a localized union through fusion or recrystallization across the interface. Included in this process are gas welding, resistance welding, arc welding, cold welding, electron beam welding, and laser beam welding.
18. *Brazing is the process* of joining metals by flowing a thin (capillary thickness) layer of nonferrous filler metal into the space between them. Bonding results from the intimate contact produced by the dissolution of a small amount of base metal in the molten filler metal, without fusion of the base metal. The term brazing is used where the temperature exceeds 425°C (800 F).
19. *Soldering* is the process of joining metals by flowing a thin (capillary thickness) layer of nonferrous filler metal into the space between them. Bonding results from the intimate contact produced by the dissolution of a small amount of base metal in the molten filler metal, without fusion of the base metal. The term soldering is used where the temperature range falls <425°C (800 F).
20. *Flame spraying* is the process of applying a metallic coating to a workpiece using finely powdered fragments of wire and suitable fluxes, which are projected together through a cone of flame onto the workpiece.
21. *Sand blasting* is the process of removing stock, including surface films, from a workpiece by the use of abrasive grains pneumatically impinged against the workpiece. The abrasive grains used include sand, metal shot, slag, silica, pumice, or natural materials such as walnut shells.
22. *Abrasive jet machining* is a mechanical process for cutting hard, brittle materials. It is similar to sand blasting, but uses much finer abrasives carried at high velocities [150–910 m/s (500–3000 ft/s)] by a liquid or gas stream. Uses include frosting glass, removing metal oxides, deburring, and drilling and cutting thin sections of metal.
23. *Electrical discharge machining* is a process that can remove metal with good dimensional control from any metal. It cannot be used for machining glass, ceramics, or other nonconducting materials. Electrical discharge machining is also known as spark machining or electronic erosion. The operation was developed primarily for machining carbides, hard nonferrous alloys, and other hard-to-machine materials.
24. *Electrochemical machining* is a process based on the same principles used in electroplating except the workpiece is the anode and the tool is the cathode. Electrolyte is pumped between the electrodes and a potential applied, resulting in rapid removal of metal.
25. *Electron beam machining* is a thermoelectric process in which heat is generated by high-velocity electrons impinging the workpiece, converting the beam into thermal energy. At the point where the energy of the electrons is focused, the beam has sufficient thermal energy to vaporize the material locally. The process is generally carried out in a vacuum. The process results in X-ray emission, which requires that the work area be shielded to absorb radiation.

At present, the process is used for drilling holes as small as 0.05 mm (0.002 in.) in any known material, cutting slots, shaping small parts, and machining sapphire jewel bearings.

26. *Laser beam machining* is the process of using a highly focused, monochromatic collimated beam of light to remove material at the point of impingement on a workpiece. Laser beam machining is a thermoelectric process, and material removal is largely accomplished by evaporation, although some material is removed in the liquid state at high velocity. Since the metal removal rate is very small, this process is used for such jobs as drilling microscopic holes in carbides or diamond wire drawing dies and for removing metal in the balancing of high-speed rotating machinery.

27. *Plasma arc machining* is the process of material removal or shaping of a workpiece by a high-velocity jet of high-temperature ionized gas. A gas (nitrogen, argon, or hydrogen) is passed through an electric arc, causing it to become ionized and raising its temperatures in excess of 16,000°C (30,000 F). The relatively narrow plasma jet melts and displaces the workpiece material in its path.

28. *Ultrasonic machining* is a mechanical process designed to remove material by the use of abrasive grains, which are carried in a liquid between the tool and the work and which bombard the work surface at high velocity. This action gradually chips away minute particles of material in a pattern controlled by the tool shape and contour. Operations that can be performed include drilling, tapping, coining, and the making of openings in all types of dies.

29. *Sintering* is the process of forming a mechanical part from a powdered metal by fusing the particles together under pressure and heat. The temperature is maintained below the melting point of the basis metal.

30. *Laminating* is the process of adhesive bonding of layers of metal, plastic, or wood to form a part.

31. *Hot dip coating* is the process of coating a metallic workpiece with another metal by immersion in a molten bath to provide a protective film. Galvanizing (hot-dip zinc coatings) is the most common hot-dip coating.

32. *Sputtering* is the process of covering a metallic or nonmetallic workpiece with thin films of metal. The surface to be coated is bombarded with positive ions in a gas discharge tube, which is evacuated to a low pressure.

33. *Vapor plating* is the process of decomposition of a metal or compound upon a heated surface by reduction or decomposition of a volatile compound at a temperature below the melting point of either the deposit or the basis material.

34. *Thermal infusion* is the process of applying fused zinc, cadmium, or other metal coating to a ferrous workpiece by imbuing the surface of the workpiece with metal powder or dust in the presence of heat.

35. *Salt bath descaling* is the process of removing surface oxides or scale from a workpiece by immersion in a molten salt bath or a hot salt solution. The workpiece is immersed in the molten salt [temperatures range from 400°C to 540°C (750 to 1000 F)], quenched with water, and then dipped in acid. Oxidizing, reducing, and electrolytic baths are available, and the type needed depends on the oxide to be removed.

36. *Solvent degreasing* is a process of removing oils and grease from the surfaces of a workpiece by the use of organic solvents, such as aliphatic petroleum, aromatics, oxygenated hydrocarbons, halogenated hydrocarbons, and combinations of these classes of solvents. However, ultrasonic vibration is sometimes used with liquid solvent to decrease the required immersion time with complex shapes. Solvent cleaning is often used as a precleaning operation, such as prior to the alkaline cleaning that precedes plating, as a final cleaning of precision parts, or as a surface preparation for some painting operations.

37. *Paint stripping* is the process of removing an organic coating from a workpiece. The stripping of such coatings is usually performed with caustic, acid, solvent, or molten salt.

38. *Painting* is the process of applying an organic coating to a workpiece. This process includes the application of coatings such as paint, varnish, lacquer, shellac, and plastics by methods such as spraying, dipping, brushing, roll coating, lithographing, and wiping. Other processes included under this unit operation are printing, silk screening, and stenciling.
39. *Electrostatic painting* is the application of electrostatically charged paint particles to an oppositely charged workpiece followed by thermal fusing of the paint particles to form a cohesive paint film. Both waterborne and solvent-borne coatings can be sprayed electrostatically.
40. *Electropainting* is the process of coating a workpiece by making it either anodic or cathodic in a bath that is generally an aqueous emulsion of the coating material. The electrodeposition bath contains stabilized resin, dispersed pigment, surfactants, and sometimes organic solvents in water.
41. *Vacuum metalizing* is the process of coating a workpiece with metal by flash heating metal vapor in a high-vacuum chamber containing the workpiece. The vapor condenses on all exposed surfaces.
42. *Assembly* is the fitting together of previously manufactured parts or components into a complete machine, unit of a machine, or structure.
43. *Calibration* is the application of thermal, electrical, or mechanical energy to set or establish reference points for a component or complete assembly.
44. *Testing* is the application of thermal, electrical, or mechanical energy to determine the suitability or functionality of a component or complete assembly.

Table 10.1 presents an industry summary for the metal finishing industry, including the total number of subcategories, number of subcategories studied, and the type and number of dischargers.

10.1.2 Subcategory Descriptions

The primary purpose of subcategorization is to establish groupings within the metal finishing industry such that each subcategory has a uniform set of quantifiable effluent limitations. Several bases were considered in establishing subcategories within the metal finishing industry. These included the following:

1. Raw waste characteristics
2. Manufacturing processes
3. Raw materials

TABLE 10.1
Metal Finishing Industry Summary

Item	Number
Total subcategories	51
Subcategories studied	28
Discharges in industry	98,418
Direct	20,632
Indirect	77,586
Zero discharge	200

Source: USEPA, *Treatability Manual, Volume II. Industrial Descriptions,* report # EPA-600/2-82-001b, U.S. Environmental Protection Agency, Washington, DC, September 1981.

4. Product type or production volume
5. Size and age of facility
6. Number of employees
7. Water usage
8. Individual plant characteristics

After these subcategorization bases were evaluated, raw waste characterization was selected as the basis for subcategorization. The raw waste characterization is divided into two components, inorganic and organic wastes. These components are further subdivided into the specific types of wastes that occur within the components. Inorganics include common metals, precious metals, complexed metals, hexavalent chromium, and cyanide. Organics include oils and solvents.

Table 10.2 lists the unit operations associated with each of the seven industry subcategories (raw waste characteristics). Common metals are found in the raw waste of all 44 unit operations. Precious metals are found in only seven unit operations, complexed metals in three unit operations, hexavalent chromium in seven unit operations, and cyanide in eight unit operations. Within the organics, oils are found in 22 unit operations and solvents in nine unit operations. A unit operation will often be found in more than one subcategory.

10.2 WASTEWATER CHARACTERIZATION

In this section, the water uses in the metal finishing industry are presented, and the waste constituents are identified and quantified.

Water is used for rinsing workpieces, washing away spills, air scrubbing, process fluid replenishment, cooling and lubrication, washing of equipment and workpieces, quenching, spray booths, and assembly and testing. Unit operations with significant water usage include electroplating, electroless plating, anodizing, conversion coating, etching, cleaning, machining, grinding, tumbling, heat treating, welding, sand blasting, salt bath descaling, paint stripping, painting, electrostatic painting, electroplating, and testing. Unit operations with zero discharge are electron beam machining, laser beam machining, plasma arc machining, ultrasonic machining, sintering, sputtering, vapor plating, thermal infusion, vacuum metalizing, and calibration [7].

Table 10.3 displays the ranges of flows in the metal finishing industry. Approximately 81% of the plants have flows of between 1.9 and 57 m^3/h (67–2000 ft^3/h). For those plants with common metals waste streams, the average contribution of these streams to the total wastewater flow within a particular plant is 62.4% (range: 0.007%–100%). All of the plants have a waste stream requiring common metals treatment.

In all, 4.8% of the plants have production processes that generate precious metals wastewater. The average precious metals wastewater flow is 21.5% of total plant flow.

The average contribution of the complexed metal streams to total plant flow is 22.2%. The percentage was computed from data for plants whose complexed metal streams could be segregated from the total stream.

Of the plants, 42.5% have segregated hexavalent chromium waste streams. The average flow contribution of these waste streams to the total wastewater stream is 28.7%. At plants with cyanide wastes, the average contribution of the cyanide-bearing stream to the total wastewater generated is 28.8% (range: 0.1%–100%). Of the plants, 31.2% have segregated cyanide-bearing wastes.

Segregated oily wastewater is defined as oil waste collected from machine sumps and process tanks. The water is segregated from other wastewaters until it has been treated by an oily waste removal system. Of the plants, 12.4% are known to segregate their oily wastes. The average contribution of these wastes to the total plant wastewater flow is 6.6% (range: approximately 0.0%–55.4%).

To characterize the waste streams in each subcategory, raw waste data were collected. Discrete samples of raw wastes were taken for each subcategory, and analyses on the samples were performed. The results of these analyses are presented for each subcategory in Tables 10.4–10.9. In each table,

TABLE 10.2
Subcharacterization of Unit Operations

Industry Subcategory (Raw Waste Characteristics)	Unit Operations	
Common Metals		
All 44 unit operations		
Precious Metals		
Electroplating	Etching	Burnishing
Electroless plating	Cleaning	
Conversion coating	Polishing	
Complexed Metals		
Electroless plating		
Etching		
Cleaning		
Hexavalent Chromium		
Electroplating	Etching	Electrostatic painting
Anodizing	Cleaning	
Conversion coating	Tumbling	
Cyanide		
Eloctroplating	Cleaning	Heat treating
Electroless plating	Tumbling	Electrochemical machining
Conversion coating	Burnishing	
Oils		
Cleaning	Pressure deformation	Solvent degressing
Machining	Shearing	Paint stripping
Grinding	Heat treating	Painting
Polishing	Other abr. jet machining	Assembly
Tumbling	Electrostatic painting	Calibration
Burnishing	Elec. discharge machining	Testing
Impact deformation	Electrochemical machining	
Solvents		
Cleaning	Solvent degreasing	Electrostatic painting
Heat treating	Paint stripping	Electropainting
Electrochemical machining	Painting	Assembly

Source: USEPA, *Treatability Manual, Volume II. Industrial Descriptions,* report # EPA-600/2-82-001b, U.S. Environmental Protection Agency, Washington, DC, September 1981.

data are presented on the number of detections of a pollutant, the number of samples analyzed, the median concentration, the range in concentrations, and the mean concentration of those samples detected. The minimum detection limit for the toxic pollutants in the sampling program is 1 µg/L, and any value below this is listed in the six tables as BDL (below detection limit).

10.2.1 COMMON METALS SUBCATEGORY

Pollutant parameters found in the common metals subcategory raw waste streams from sampled plants are shown in Table 10.4. The major constituents shown are parameters that originate in process solutions (such as from plating or galvanizing) and enter wastewaters by drag-out to rinses. These metals appear in waste streams in widely varying concentrations.

TABLE 10.3
Wastewater Flow Characterization of the Metal Finishing Industry

Flow of Plants (m³/h)	Percentage of Plants Represented by this Flow
<0.38	2.8
0.38–1.9	5.0
1.9–3.8	13
3.8–9.5	17
9.5–19	20.7
19–28	10.7
28–38	10.7
38–57	9.1
57–95	5.0
95–190	3.8
190–380	0.7
>380	1.5

Source: USEPA, *Treatability Manual, Volume II. Industrial Descriptions*, report # EPA-600/2-82-001b, U.S. Environmental Protection Agency, Washington, DC, September 1981.

10.2.2 Precious Metals Subcategory

Table 10.5 shows the concentrations of pollutant parameters found in the precious metals subcategory raw waste streams. The major constituents are silver and gold, which are much more commonly used in metal finishing industry operations than palladium and rhodium. Because of their high cost, precious metals are of special interest to metal finishers.

10.2.3 Complexed Metals Subcategory

The concentrations of metals found in complexed metals subcategory raw waste streams are presented in Table 10.6. Complexed metals may occur in a number of unit operations but come primarily from electroless and immersion plating. The most commonly used metals in these operations are copper, nickel, and tin. Wastewaters containing complexing agents must be segregated and treated independently of other wastes to prevent further complexing of free metals in the other streams.

10.2.4 Cyanide Subcategory

Cyanide has been used extensively in the surface finishing industry for many years; however, it is a hazardous substance that must be handled with caution. The use of cyanide in plating and stripping solutions stems from its ability to weakly complex many metals typically used in plating. Metal deposits produced from cyanide plating solutions are finer grained than those plated from an acidic solution. In addition, cyanide-based plating solutions tend to be more tolerant of impurities than other solutions, offering preferred finishes over a wide range of conditions:

1. Cyanide-based strippers are used to selectively remove plated deposits from the base metal without attacking the substrate.
2. Cyanide-based electrolytic alkaline descalers are used to remove heavy scale from steel.

TABLE 10.4
Concentrations of Pollutants Found in the Common Metals Subcategory Raw Wastewater

Pollutant	Number of Samples	Number of Detections	Range of Detections	Median of Detections	Mean of Detections
			Toxic Pollutants (µg/L)		
Metals and Inorganics					
Antimony	106	22	1–430	6	34
Arsenic	105	31	2–64	10	16
Beryllium	27	23	1–44	5	9
Cadmium	108	60	BDL–19,000	8	1,000
Chromium	105	89	3–35,000	180	16,000
Copper	108	105	3–500,000	180	16,000
Lead	108	73	3–42,000	120	1,400
Mercury	99	32	BDL—400	10	18
Nickel	108	88	4–420,000	200	24,000
Selenium	26	21	1–60	5	9
Thallium	26	21	1–62	3	10
Zinc	108	107	9–330,000	290	19,000
Phthalates					
Bis (2-ethylhexyl) phthalate	93	91	BDL—1,900	6	57
Butyl benzyl phthalate	65	38	BDL—10	BDL	1
Di-n-butyl phthalate	89	79	BDL—10	BDL	BDL
Di-n-octyl phthalate	65	25	BDL—10	BDL	BDL
Diethyl phthalate	83	66	BDL—240	5	31
Dimethyl phthalate	65	7	BDL—10	BDL	2
Nitrogen Compounds					
3, 3-dichlorobenzidene	4	1	BDL		
N-nitroso-di-n-propylamine	4	1	570		
Phenols					
2-Nitrophenol	4	1	24		
Phenol	23	15	BDL—1,000	45	240
Aromatics					
Benzene	6	4	BDL—16	7	8
Ethylbenzene	37	9	BDL—1,200	250	340
Toluene	39	17	2–690	77	140
Polycyclic Aromatic Hydrocarbons					
Fluoranthene	4	1	74		
Isophorone	4	4	13–310	180	170
Napthalene	89	61	BDL—2,000	1	83
Anthracene	82	56	BDL—30	1	2
Fluorene	2	2	BDL—160		80
Phenanthrene	71	55	BDL—30	1	2
Pyrene	4	1	190		
Halogenated Aliphatics					
Carbon tetrachloride	57	37	BDL—1	BDL	BDL
1, 2-Dichloroethane	4	1	3		
1, 1, 1-Trichloroethane	57	43	BDL—550	BDL	18
1, 1, 2-Trichloroethane	57	21	BDL—3	BDL	BDL
Chloroform	65	48	BDL—140	BDL	5
1, 1-Dichloroethylene	58	4	BDL—110	BDL	20
1, 2-Trans-dichloroethylene	5	3	1–5	2	3
1, 2-Dichloropropylene	4	1	2		
Methylene chloride	80	27	BDL—570	BDL	53
Methyl chloride	74	3	BDL—60	3	21
Methyl bromide	4	1	2		
Dichlorobromomethane	5	2	3–8		6
Chlorodibromomethane	4	1	8		
Tetrachloroethylene	59	23	BDL—66	BDL	6
Trichloroethylene	77	49	BDL—480	BDL	22
Posticides and Metabolites					
Dieldrin	4	1	BDL		
Alpha-endosulfan	4	1	9		
Endrin aldehyde	4	1	BDL		
Alpha-BHC	4	1	BDL		

(Continued)

TABLE 10.4 (*Continued*)
Concentrations of Pollutants Found in the Common Metals Subcategory Raw Wastewater

Pollutant	Number of Samples	Number of Detections	Range of Detections	Median of Detections	Mean of Detections
			Toxic Pollutants (µg/L)		
Beta-BHC	4	1	4		
Delta-BHC	4	1	BDL		
			Concentration (mg/L)		
Classical Pollutants					
TSS	107	104	0.56 — 11,000	63	520
Aluminum	8	6	0.03 — 200	0.29	62
Barium	4	3	0.027 — 0.071	0.03	0.043
Calcium	3	3	25 — 76	52	51
Cobalt	4	4	0.009 — 0.023	0.02	0.017
Fluorides	7	3	0.021 — 36	1.1	5.3
Iron	85	76	0.035 — 490	1.9	28
Magnesium	88	87	5.6 — 31	14	16
Manganese	4	4	0.059 — 0.5	0.085	0.22
Molybdenum	7	7	0.031 — 0.3	0.27	0.2
Phosphorous	4	3	0.007 — 77	3	7.9
Sodium	4	3	17 — 310	140	160
Tin	4	4	0.002 — 15	0.86	3.7
Titanium	5	2	0.006 — 0.08	0.03	0.039
Vanadium	7	3	0.01 — 0.22	0.036	0.087
Yttrium	4	3	0.002 — 0.02	0.018	0.013

Source: USEPA, Development document for effluent limitations guidelines and standards for the metal finishing point source category, report # EPA-440/1-80/091, U.S. Environmental Protection Agency, Washington, DC, 1980.

TABLE 10.5
Concentrations of Pollutants Found in the Precious Metals Subcategory Raw Wastewater

Pollutant	Number of Samples	Number of Detections	Range of Detections	Median of Detections	Mean of Detections
			Concentration (mg/L)		
Classical Pollutants					
Silver	15	12	0.033–600	0.38	86
Gold	15	9	0.56–43	0.86	15
Palladium	13	3	0.09–0.12	0.09	0.10
Rhodium	12	1	0.22		

Source: USEPA, Development document for effluent limitations guidelines and standards for the metal finishing point source category, report # EPA-440/1-80/091, U.S. Environmental Protection Agency, Washington, DC, 1980.

3. Cyanide-based dips are often used before plating or after stripping processes to remove metallic smuts on the surface of parts.

Cyanide-based metal finishing solutions usually operate at basic pH levels to avoid decomposition of the complexed cyanide and the formation of highly toxic hydrogen cyanide gas.

The cyanide concentrations found in cyanide subcategory raw waste streams are shown in Table 10.7. The levels of cyanide range from 0.045 to 500 µg/L. Streams with high cyanide

TABLE 10.6
Concentrations of Pollutants Found in the Complexed Metals Subcategory Raw Wastewater

Pollutant	Number of Samples	Number of Detections	Range of Detections	Median of Detections	Mean of Detections
			Concentration (μg/L)		
Toxic Pollutants					
Cadmium	31	9	1–3,600	67	850
Copper	31	28	10–63,000	6,700	11,000
Lead	31	10	2–3,600	420	1,200
Nickel	31	25	26–290,000	3,200	28,000
Zinc	31	31	23–18,000	210	3,000
			Concentration (mg/L)		
Classical Pollutants					
Aluminum	1	1	0.1		
Calcium	1	1	17		
Iron	31	31	0.038–99	0.74	9.9
Magnesium	1	1	2		
Manganese	1	1	0.1		
Phosphorus	31	31	0.023–100	8.2	23
Sodium	1	1	110		
Tin	31	10	0.013–6	0.68	1.6

Source: USEPA, Development document for effluent limitations guidelines and standards for the metal finishing point source category, report # EPA-440/1-80/091, U.S. Environmental Protection Agency, Washington, DC, 1980.

TABLE 10.7
Concentrations of Pollutants Found in the Cyanide Subcategory Raw Wastewater

Pollutant	Number of Samples	Number of Detections	Range of Detections	Median of Detections	Mean of Detections
			Concentration (μg/L)		
Toxic Pollutants					
Cyanide	20	20	45–500,000	45,000	110,000
Cyanide, Amn.	19	18	5–460,000	4,500	86,000

Source: USEPA, Development document for effluent limitations guidelines and standards for the metal finishing point source category, report # EPA-440/1-80/091, U.S. Environmental Protection Agency, Washington, DC, 1980.

concentrations normally originate in electroplating and heat treating processes. Cyanide-bearing waste streams should be segregated and treated before being combined with other raw waste streams.

10.2.5 Hexavalent Chromium Subcategory

Concentrations of hexavalent chromium from metal finishing raw wastes are shown in Table 10.8. Hexavalent chromium enters wastewater as a result of many unit operations and can be very

TABLE 10.8
Concentrations of Pollutants Found in the Hexavalent Chromium Subcategory Raw Wastewater

Pollutant	Number of Samples	Number of Detections	Range of Detections	Median of Detections	Mean of Detections
			Concentration (µg/L)		
Toxic Pollutants					
Chromium, hexavalent	49	41	5–13,000,000	20,000	420,000

Source: USEPA, Development document for effluent limitations guidelines and standards for the metal finishing point source category, report # EPA-440/1-80/091, U.S. Environmental Protection Agency, Washington, DC, 1980.

concentrated. Because of its high toxicity, it requires separate treatment so that it can be efficiently removed from wastewater.

10.2.6 OILS SUBCATEGORY

Pollutant parameters and their concentrations found in the oily waste subcategory streams are shown in Table 10.9. The oily waste subcategory for the metal finishing industry is characterized by both concentrated and dilute oily waste streams that consist of a mixture of free oils, emulsified oils, greases, and other assorted organics. Applicable treatment of oily waste streams is dependent on the concentration levels of the wastes, but oily wastes normally receive specific treatment for oil removal prior to waste treatment for solids removal.

The majority of the pollutants listed in Table 10.9 are priority organics that are used either as solvents or as oil additives to extend the useful life of the oils. Organic priority pollutants, such as solvents, should be segregated and disposed of or reclaimed separately. However, when they are present in wastewater streams, they are most often at the highest concentration in the oily waste stream because organics generally have a higher solubility in hydrocarbons than in water. Oily wastes will normally receive treatment for oil removal before being directed to waste treatment for solids removal.

10.2.7 SOLVENT SUBCATEGORY

The solvent subcategory raw waste streams is generated in the metal finishing industry by the dumping of spent solvents from degreasing equipment (including sumps, water traps, and stills). These solvents predominately comprise compounds classified by the U.S. Environmental Protection Agency (USEPA) as toxic pollutants. Spent solvents should be segregated, hauled for disposal or reclamation, or reclaimed on site. Solvents that are mixed with other wastewaters tend to appear in the common metals or the oily waste streams.

10.3 SOURCE REDUCTION

It is not currently feasible to achieve a zero discharge of chemical pollutants from metal finishing operations. However, substantial reductions in the type and volume of hazardous chemicals wastes from most metal finishing operations are possible [8]. Because end-of-pipe waste detoxification is costly for small- and medium-sized metal finishers, and the cost and liability of residuals disposal have increased for all metal finishers, management and production personnel may be more willing to consider production process modifications to reduce the amount of chemicals lost to waste.

TABLE 10.9
Concentrations of Pollutants Found in the Oils Subcategory Raw Wastewater

Pollutant	Number of Samples	Number of Detections	Range of Detections	Median of Detections	Mean of Detections
			Concentration (µg/L)		
Toxic Pollutants					
Phthalates					
Bis(2-ethylhexyl) phthalate	37	20	2–9,300	73	820
Butyl benzyl phthalate	37	9	1–10,000	130	1,600
Di-n-butyl phthalate	37	15	1–3,100	16	270
Di-n-octyl phthalate	37	2	4–120		62
Diethyl phthalate	37	9	1–1,900	48	420
Dimethyl phthalate	37	34	1–1,300	1	400
Ethers					
Bis(chloromethyl) ether	37	1	9		7
Bis(2-chloroethyl) ether	37	2	4–10		
Bis(2-chloroisopropyl) ether	37	1	4		
Bis(2-chloroethoxy) methane	37	1	3		
Nitrogen Compounds					
1,2-Diphenylhydrazine	37	2	3–12		8
Phenols					
2,4,6-Trichlorophenol	37	3	10–1,000	30	610
Parachloronatecrosal	37	8	4–800,000	2,300	100,000
2-Chlorophenol	37	2	76–620		350
2,4-Dichlorophenol	37	2	10–60		39
2,4-Dimethylphenol	37	6	1–31,000	10	5,200
2-Nitrophenol	37	3	10–320	35	120
4-Nitrophenol	37	1	10		
2,4-Dinitrophenol	37	3	10–10,000	13	3,300
N-Nitrosodiphenylamine	37	5	4–900	750	490
Pentachlorophenol	37	3	10–50,000	3,500	18,000
Phenol	37	3	3–6,600	440	1,700
4,6-Dinitro-o-cresol	37	2	10–5,700		2,200
Aromatics					
Benzene	37	18	1–110	8	12
Chlorobenzene	37	2	11–610		310
Nitrobenzene	37	2	1–10		5
Toluene	37	25	1–37,000	33	1,800
Ethylbenzene	37	16	1–3,500	12	300
Polynuclear Aromatic Hydrocarbons					
Acenaphthene	37	2	57–5,700		2,900
Z-Chloronaphthalene	37	1	130		
Fluoranthene	37	8	1–55,000	110	6,300
Naphthalene	37	10	1–260	100	36
Benzo(a)pyrene	37	1	10		
Chrysene	37	3	1–73	2	25
Acenaphthylene	37	3	77–1,000	140	410
Anthraone	43	7	1–8,000	34	360
Fluorene	37	7	1–760	75	160
Phenanthrene	37	8	2–2,000	80	400
Pyrene	37	5	31–150	75	79
Metaganated Hydrocarbons					
Carbon tetrachloride	37	5	1–10,000	97	8,600
1,2-Dichloroethane	37	6	9–2,100	1,400	1,100
1,1,1-Trichloroethane	37	18	1–1,300,000	260	75,000
1,1-Dichloroethane	37	11	2–1,100	600	460
1,1,2-Trichloroethane	37	4	6–1,300	10	330
1,1,2,2-Tetrachloroethane	37	2	6–570		290
Chloroform	37	19	2–690	10	30
1,1-Dichloroethylene	37	12	2–10,000	200	1,500
1,2-trans-dichloroethylene	43	9	4–1,700	88	510
Methylene chloride	37	29	5–7,600	92	600
Methyl chloride	37	4	1–4,700	9	1,200
Bromoform	37	1	10		
Dichlorobromomethane	37	2	1–10		5
Trichlorofluoromethane	37	2	260–290		200
Chlorodibromomethane	37	3	1–10	2	4
Tetrachloroethylene	37	18	1–110,000	10	8,900
Trichloroethylene	37	11	1–130,000	110	23,000
Pesticides and Metabolites					
Aldrin	37	2	4–11		7
Dieldrin	37	1	3		

(Continued)

TABLE 10.9 (*Continued*)
Concentrations of Pollutants Found in the Oils Subcategory Raw Wastewater

Pollutant	Number of Samples	Number of Detections	Range of Detections	Median of Detections	Mean of Detections
			Concentration (µg/L)		
Pestisides and Metabolites					
Chlordane	37	2	1–13		7
4,4-DDT	37	2	2–10		6
4,4-DDE	37	4	BDL–53	2	14
4,4-DDD	37	3	1–10	4	5
Alpha-endosulfan	37	2	4–28		18
Beta-endosulfan	37	2	BDL–6		3
Endosulfan sulfate	37	4	1–16	11	10
Endrin	37	2	7–10		8
Endrin aldehyde	37	2	10–14		12
Heptachlor	37	1	IDL		
Heptachlor epoxide	37	1	BDL		
Alpha-SHC	37	3	4–10	13	12
Gamma-SHC	37	3	1–9	7	6
Delta-SHC	37	2	4–11		7
Polychlorinated Hydroxis					
Arochlor 1254	37	2	76–1100		590
Arochlor 1248	37	2	160–1100		900
			Concentration (µg/L)		
Classical Pollutants					
Ammonia	37	10	0.46–270	7.9	46
BOD	37	21	10–17,000	1400	3300
COD	37	16	310–1500000	12000	120000
Oil and grease	37	37	66–808660	6100	41000
Phenols, total	37	34	8002–49	0.24	2.5
Total dissolved solids	37	9	250–4900	1600	2000
Total organic carbon	37	37	3–360000	1600	28000
TSS	37	35	35–16,600	600	2700

Source: USEPA, Development document for effluent limitations guidelines and standards for the metal finishing point source category, report # EPA-440/1-80/091, U.S. Environmental Protection Agency, Washington, DC, 1980.

This section provides guidance for reducing waterborne wastes from metal finishing operations to avoid or reduce the need for waste detoxification and the subsequent off-site disposal of detoxification residuals. Waste reduction practices may take the form of the following[5]:

1. Chemical substitution
2. Waste segregation
3. Process modifications to reduce drag-out loss
4. Capture/concentration techniques

10.3.1 Chemical Substitution

The incentive for substituting process chemicals containing nonpolluting materials has been present only in recent years with the advent of pollution control regulations. Chemical manufacturers are gradually introducing such substitutes. By eliminating polluting process materials such as hexavalent chromium- and cyanide-bearing cleaners, and deoxidizers, the treatments required to detoxify these wastes are also eliminated. It is particularly desirable to eliminate processes employing hexavalent chromium and cyanide, since special equipment is needed to detoxify both.

Substituting nonpolluting cleaners for cyanide cleaners can avoid cyanide treatment entirely. For a 7.6 L/min rinse water flow, this means a saving of about USD22,400 in equipment costs and USD12/kg in cyanide treatment chemical costs. In this case, treatment chemical costs are about four times the cost of the raw sodium cyanide cleaner.

There can be disadvantages in using nonpolluting chemicals. Before making a decision, the following questions should be asked of the chemical supplier [5]:

- Are substitutes available and practical?
- Will substitution solve one problem but create another?
- Will tighter chemical controls be required of the bath?
- Will product quality and/or production rate be affected?
- Will the change involve any cost increases or decreases?

Based on a survey of chemical suppliers and electroplaters who use nonpolluting chemicals, some commonly used chemical substitutes are summarized in Table 10.10.

The chemical supplier can also identify any regulated pollutants in the facility's treatment chemicals and offer available substitutes. The federally regulated pollutants are cyanide, chrome, copper, nickel, zinc, lead, cadmium, and silver. Local and/or state authorities may regulate other substances, such as tin, ammonia, and phosphate. The current status of cyanide and noncyanide substitute plating processes is shown in Table 10.11.

TABLE 10.10
Chemical Substitutes

Polluting	Substitute	Comments
Fire dip (NaCN)	Muriatic acid with additives	Slower acting than $+H_2O_2$ traditional fire dip
Heavy copper cyanide plating bath	Copper sulfate	Excellent throwing power with a bright, smooth, rapid finish
		A copper cyanide strike may still be necessary for steel, zinc, or tin-lead base metals
		Requires good preplate cleaning
		Noncyanide process eliminates carbonate buildup in tanks
Chromic acid pickles, deoxidizers, and bright dips	Sulfuric acid and hydrogen peroxide	Nonchrome substitute
		Nonfuming
Chrome-based anti-tarnish	Benzotriazole (0.1%–1.0% solution in methanol) or water-based proprietaries	Nonchrome substitute
		Extremely reactive, requires ventilation
Cyanide cleaner	Trisodium phosphate or ammonia	Noncyanide cleaner
		Good degreasing when hot and in an ultrasonic bath
		Highly basic
		May complex with soluble metals if used as an intermediate rinse between plating baths where metal ion may be dragged into the cleaner and cause wastewater treatment problems
Tin cyanide	Acid tin chloride	Works faster and better

Source: USEPA, Meeting hazardous waste requirements for metal finishers, report # EPA/625/4-87/018, U.S. Environmental Protection Agency, Cincinnati, OH, 1987.

TABLE 10.11
Cyanide and Noncyanide Plating Processes

Metal	Cyanide	Non-cyanide
Brass	Proven	No
Bronze	Proven	No
Cadmium	Proven	Yes
Copper	Proven	Proven
Gold	Proven	Developing
Indium	Proven	Yes
Silver	Proven	Developing
Zinc	Proven	Proven

Source: USEPA, Managing cyanide in metal finishing. Capsule report # EPA 625/R-99/009 U.S. Environmental Protection Agency, Cincinnati, OH, December, 2000.

10.3.2 WASTE SEGREGATION

After eliminating as many pollutants as possible, the next step is segregating polluting streams from nonpolluting streams. Nonpolluting streams can go directly to the sewer, although pH adjustment may be necessary. The segregation process will likely require some physical re-layout and/or repiping of the shop. These potentially nonpolluting rinse streams represent about one-third of all plating process water. Caution must be exercised to make certain that the so-called nonpolluting baths contain no dissolved metal. The cost savings in segregating polluting from nonpolluting streams is realized through wastewater treatment equipment and operating costs. The remaining polluting sources, which require some form of control, involve all dumped spent solutions, including tumble finishing and burnishing washes, cyanide cleaner rinses, plating rinses, rinses after "bright dips," and aggressive cleaning solutions.

10.3.3 PROCESS MODIFICATIONS TO REDUCE DRAG-OUT LOSS

Plating solution that is wasted by being carried over into the rinse water as a workpiece emerges from the plating bath is defined as drag-out, and it is the largest volume source of chemical pollutant in the electroplating shop. Numerous techniques have been developed to control drag-out; the effectiveness of each method varies as a function of the plating process, operator cooperation, racking, barrel design, transfer dwell time, and plated part configuration.

Wetting agents and longer workpiece withdrawal/drainage times are two techniques that significantly control drag-out. These and other techniques are discussed below.

10.3.3.1 Wetting Agents

Wetting agents lower the surface tension of process baths. To remove plating solution dragged out with the plated part, gravity-induced drainage must overcome the adhesive force between the solution and the metal surface. The drainage time required for racked parts is a function of the surface tension of the solution, part configuration, and orientation. Lowering the surface tension reduces the drainage time and also minimizes the edge effect (the bead of liquid adhering to the part edge); thus, there is less drag-out. Plating baths such as nickel and heavy copper cyanide also use wetting agents to maintain grain quality and provide improved coverage. The chemical supplier should be asked if the baths he or she supplies contain wetting agents and, if not, whether wetting agents can be added. In some baths, the use of wetting agents has the potential to reduce drag-out by 50%.

10.3.3.2 Longer Drain Times

With slower withdrawal rates and/or longer drain times, drag-out of process solutions can be reduced by up to 50%. In cases where high-temperature plating solutions are used, slow withdrawal of the rack may also be necessary to prevent evaporative "freezing," which can actually increase drag-out. In the extreme case, too rapid a withdrawal rate causes "sheeting," where huge volumes of drag-out are lost to waste. Figure 10.1 shows the drainage rates for plain and bent-shaped pieces. Drainage for all shapes is almost complete within 15 s after withdrawal, indicating that this is an optimum drain time for most pieces.

One of the best ways to control drag-out loss from rack plating on hand lines is to provide drain bars over the tank from which the rack can be hung to drain for a brief period. Hanging and removing the racks from the drain bars ensures an adequate drain time. Slightly jostling the racks helps shake off adhering solution.

In barrel plating, the barrel should be rotated for a time just above the plating tank to reduce the volume of dragged-out chemical. Holes in the barrels should be as large as possible to improve solution drainage while still containing the pieces. A fog spray directed at the barrel or its contents can also help drag-out drainage. Deionized water is recommended to minimize bath contamination.

The combined application of wetting agents and longer withdrawal/drainage times can significantly reduce the amount of drag-out for many cleaning or plating processes. For example, a typical nickel drag-out can be reduced from 1 to 1/4 L/h by these techniques.

10.3.3.3 Other Drag-Out Reduction Techniques

Rinse elimination. The rinse between a soak cleaner and an electrocleaner may be eliminated if the two baths are compatible.

Low-concentration plating solutions. Low-concentration plating solutions reduce the total mass of chemicals being dragged out. The mass of chemicals removed from a bath is a function of the solution concentration and the volume of solution carried from the bath. Traditionally, the bath concentration is maintained at a midpoint within a range of operating

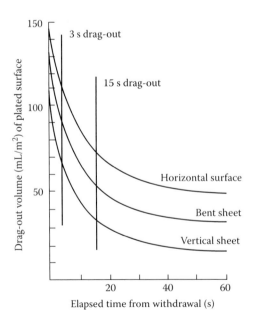

FIGURE 10.1 Typical drag-out drainage rates. (From USEPA, Meeting hazardous waste requirements for metal finishers, report # EPA/625/4-87/018, U.S. Environmental Protection Agency, Cincinnati, OH, 1987.)

conditions. With the high cost of replacement, treatment, and disposal of dragged-out chemicals, the economics of low-concentration baths are favorable.

As an illustration, a typical nickel-plating operation with five nickel tanks has an annual nickel drag-out of about 10,000 L. Assuming the nickel baths are maintained at the midpoint operating concentration, as shown in Table 10.12, the annual costs of chemical replacement, treatment, and disposal are about USD24,000 in terms of 2014 USD. If the bath is converted to the modified operating condition, as shown in Table 10.12, the annual costs of chemical replacement, treatment, and disposal are approximately USD21,700, a saving of over USD2000/year. Generally, any percent decrease in bath chemical concentration results in the same percent reduction in the mass of chemicals lost in the drag-out. The disadvantage of low-concentration baths may be lowered plating efficiencies, which may require higher current densities and closer process control. The reduction in plating chemical replacement, treatment, and disposal costs could be partially offset by the added labor and power costs associated with the use of the lower-concentration baths.

Clean plating baths. Contaminated plating baths, such as carbonate buildup in cyanide baths, can increase drag-out as much as 50% by increasing the viscosity of the bath. Excessive impurities also make the application of recovery technology difficult, if not impossible.

Low-viscosity conducting salts. Bath viscosity indexes are available from chemical suppliers. As the bath viscosity increases, drag-out volume also increases.

High-temperature baths reduce surface tension and viscosity, thus decreasing drag-out volume. Disadvantages to be considered are more rapid solution decomposition, higher energy consumption, and possible dry-on pattern on the workpiece.

No unnecessary components. Additional bath components (chemicals) tend to increase both viscosity and drag-out.

Fog sprays or air knives may be used over the bath to remove drag-out from pieces as they are withdrawn. The spray of deionized water or air removes plating solution from the part and returns as much as 75% of the drag-out back to the plating tank. Fog sprays, located just above the plating bath surface, dilute and drain the adhering drag-out solution, thus reducing the concentration and mass of chemicals lost. Fog sprays are best when tank evaporation rates are sufficient to accommodate the added volume of spray water. Air knives, also located just above the plating bath surface, reduce the volume of drag-out by mechanically scouring the adhering liquid from the workpiece. The drag-out

TABLE 10.12
Standard Nickel Solution Concentration Limits

Chemical	Concentration Range (g/L)	Midpoint Operating Condition (g/L)	Modified Operating Condition (g/L)
		Nickel Sulfate	
$NiSO_4$–$6H_2O$	300–375	338	308
As $NiSO_4$	—	200	182
		Nickel Chloride	
$NiCl_2$–$6H_2O$	60–90	75	64
As $NiCl_2$	—	41	35
		Boric acid	
H_3BO_3	45–49	47	46

Source: USEPA, Meeting hazardous waste requirements for metal finishers, report # EPA/625/4-87/018, U.S. Environmental Protection Agency, Cincinnati, OH, 1987.

concentration remains constant, but the mass of chemicals lost is reduced. Air knives are best when the surface evaporation rates of the bath are too low to allow additional spray water. In some cases, use of supplementary atmospheric evaporators may be justified by economic considerations.

Air knives can be installed for about USD750–800 per bath if an oil-free, compressed air source is available. Fog sprays can also be installed for about USD750–800 per bath if a deionized water source is available. The spray should be actuated only when work is in the spraying position. Properly designed spray nozzles distribute the water evenly over the work, control the volume of water used, and avoid snagging workpieces as they are withdrawn from the tank.

Proper racking. Every piece has at least one racking position in which drag-out will be at a minimum. In general, to minimize drag-out:
- Parts should be racked with major surfaces vertically oriented.
- Parts should not be racked directly over one another.
- Parts should be oriented so that the smallest surface area of the piece leaves the bath surface last.

The optimum orientation will provide faster drainage and less drag-out per piece. However, in some cases, this may reduce the number of pieces on a rack, or the optimum draining configuration may not be the optimum plating configuration. In addition, the user should maintain rack coatings, replace rack contacts when broken, strip racks before plating buildup becomes excessive, and ensure that all holes on racks are covered or filled.

10.3.4 Capture/Concentration Techniques

10.3.4.1 Capture/Concentration with Full Reuse of Drag-Out

The pioneer in simple, low-cost methods of reducing waste in the plating shop was Dr. Joseph B. Kushner. In *Water and Waste Control for the Plating Shop* (1972), he describes a "simple waste recovery system," which captures drag-out in a static tank or tanks for returning to the plating bath. The drag-out tanks are followed by a rinse tank, which flows to the sewer with only trace amounts of polluting salts and is often in compliance with sewer discharge standards. A simplified diagram of this reuse system is shown in Figure 10.2. It is not difficult to automate the direct drag-out recovery process, and commercial units are available.

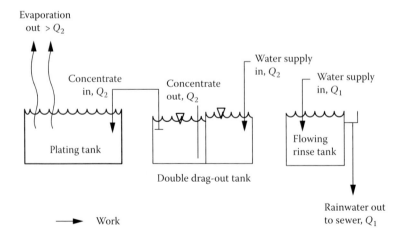

FIGURE 10.2 Kushner method of double drag-out for full reuse. (From USEPA, Meeting hazardous waste requirements for metal finishers, report # EPA/625/4-87/018, U.S. Environmental Protection Agency, Cincinnati, OH, 1987.)

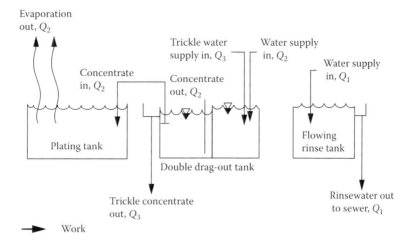

FIGURE 10.3 Modified method of double drag-out for partial reuse. (From USEPA, Meeting hazardous waste requirements for metal finishers, report # EPA/625/4-87/018, U.S. Environmental Protection Agency, Cincinnati, OH, 1987.)

The Kushner concept is easily applicable to hot plating baths where the bath evaporation rate equals or exceeds the pour-back rate, Q_2. The drag-out concentration depends on the bath drag-out rate, the number of drag-out tanks, the rinse water flow rate, Q_2, the plating bath evaporation rate, and drag-out return rate. The number of drag-out tanks must be based on the available space. The higher the number of counterflowed drag-out tanks, the smaller will be the return rate necessary to obtain good rinsing. The Kushner multiple drag-outs are not feasible if there is no room for the required drag-out tanks. If there is little or no evaporation from the bath, supplementary evaporation should be considered. Bath contamination must be minimized by using purified (reverse osmosis, RO) water for Q_2.

10.3.4.2 Capture/Concentration with Partial Reuse of Drag-Out

By adding a trickling water supply and drain, Q_3, to the drag-out tank, the application of Kushner's concept can be extended to other metal finishing processes, which may not be amenable to full reuse but can allow partial reuse. Figure 10.3 depicts the partial reuse scheme. The trickle concentrate can also be batch-treated in a small volume on-site, recycled at a central facility, or mixed with Q_1, for discharge, if the combined metal content is below sewer discharge standards.

10.3.5 Waste Reduction Costs and Benefits

Benefits of waste reduction in the metal finishing shop include the following:

1. Reduced chemical cost
2. Reduced water cost
3. Reduced volume of "hazardous" residuals
4. Reduced pretreatment cost

The benefits of saving valuable chemicals and water and reducing sludge disposal costs can best be illustrated by an example. An electroplating operation discharges 98,400 L/day of wastewater containing 0.91 kg of copper, 1.14 kg of nickel, and 0.91 kg of cyanide. The shop can reduce its generation of cyanide and copper waste by about 50% by eliminating cyanide cleaners and utilizing pour-back of copper cyanide solution; generation of nickel waste can be reduced 90% by pour-back of the nickel solution. Reducing wasted salts also allows a reduced rinse water flow rate, thus saving water and sewer use fees. The chemical costs of treatment are given in Table 10.13, and the annual replacement costs of chemicals are given in Figure 10.4. Calculations of the annual dollar savings are shown in Table 10.14. All costs have been converted into 2014 USD using U.S. ACE Yearly Average Cost Index for Utilities [9].

10.4 POLLUTANT REMOVABILITY

This section reviews the technologies currently available and which are used to remove or recover pollutants from the wastewater generated in the metal finishing industry [5–7,10]. Treatment options are presented for each subcategory within the metal finishing industry. Table 10.15 lists the treatment techniques available for treating wastes from each subcategory.

TABLE 10.13
Chemical Costs of Treatment and Disposal in 2014 USD

Chemical Cost, 2014 USD[a]/kg

Pollutant	Treatment[b]	Disposal[c]
Nickel	3.17	7.77
Copper	3.17	7.77
Cyanide	20.50	NA

Source: USEPA, Meeting hazardous waste requirements for metal finishers, report # EPA/625/4-87/018, U.S. Environmental Protection Agency, Cincinnati, OH, 1987.

[a] Costs were converted from 1979 USD to 2014 USD using U.S. ACE Yearly Average Cost Index for Utilities [9].
[b] Cost of NaOH at USD 1.16/kg and NaOCL at USD 2.73/kg.
[c] Cost of disposal at USD 2.13/kg of sludge (USD 464/drum) at 30% solids content.

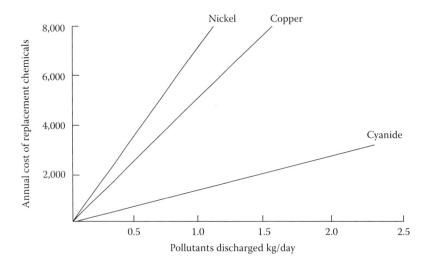

FIGURE 10.4 Annual replacement cost of chemicals in 2007 USD. *Note:* To obtain cost values in 2014 USD, multiply annual cost by 1.16. (From USEPA, Meeting hazardous waste requirements for metal finishers, report # EPA/625/4-87/018, U.S. Environmental Protection Agency, Cincinnati, OH, 1987.)

TABLE 10.14
Illustration of Annual Cost Savings for Waste Reduction

Item	Cost Saving[d] 2014 USD
Process Chemical Savings[a]	
Copper	2,813
Cyanide	563
Nickel	9,000
Treatment Chemical Saving[b]	
Copper	360
Cyanide	2,320
Nickel	812
Reduced Treatment Sludge Disposal[b]	
Copper	882
Cyanide	0
Nickel	1,972
Water and Sewer Use Fee Reduction[c]	5,058
Total annual savings	23,780

Source: USEPA, Meeting hazardous waste requirements for metal finishers, report # EPA/625/4-87/018, U.S. Environmental Protection Agency, Cincinnati, OH, 1987.

[a] From Figure 10.4.
[b] From Table 10.12 and Figure 10.4.
[c] USD 0.89/m^3.
[d] Costs were converted from 1979 USD to 2014 USD using U.S. ACE Yearly average Cost Index for Utilities [9].

10.4.1 COMMON METALS

The treatment methods used to treat wastes within the common metals subcategory fall into two groups:

1. Recovery techniques
2. Solids removal techniques

Recovery techniques are treatment methods used for recovering or regenerating process constituents, which would otherwise be discarded. Included in this group are the following [5–7]:

1. Evaporation
2. Ion exchange
3. Electrolytic recovery
4. Electrodialysis
5. RO

Solids removal techniques are employed to remove metals and other pollutants from process wastewaters to make these waters suitable for reuse or discharge. These methods include the following [5–7]:

1. Hydroxide and sulfide precipitation
2. Sedimentation
3. Diatomaceous earth filtration

TABLE 10.15
Treatment Methods in Current Use or Available for Use in the Metal Finishing Industry

Subcategory/Technology	Number of Plants
Common Metals	
Hydroxide followed by sedimentation	103
Hydroxide followed by sedimentation and filtration	30
Evaporation (metal recovery, bath concentrates, rinse waters)	41
Ion exchange	63
Electrolytic recovery	11
Electrodialysis	3
Reverse osmosis	8
Peet adsorption	0
Insoluble starch xanthate	2
Sulfide precipitation	3
Flotation	29
Membrane flotation	7
Precious Metals	
Evaporation	1
Ion exchange	NR
Electrolyte recovery	NR
Complexed Metals	
High pH precipitation with sedimentation	NR
Membrane filtration	NR
Hexavalent Chromium	
Chemical chrome reduction	343
Electrochemical chromium reduction	2
Electrochemical chromium regeneration	0
Advanced electrodialysis	NR
Evaporation	1
Ion exchange	1
Cyanide	
Oxidation by chlorine	201
Oxidation by ozone	2
Oxidation by ozine with UV radiation	NR
Oxidation by hydrogen peroxide	3
Electrochemical cyanide oxidation	4
Chemical precipitation	3
Reverse osmosis	NR
Evaporation	NR
Oils (Segregated)	
Emulsion breaking	28
Skimming	94
Emulsion breaking and skimming	
Ultrafiltration	20
Reverse osmosis	3
Carbon adsorbtion	10
Coalescing	3
Flotation	29
Centrifugation	5
Integrated adsorption	0
Resin adsorption	0
Ozonation	0
Chemical oxidation	0
Aerobic decomposition	14
Thermal emulsion breaking	0
Solvent Wastes	
Segregation	NR
Contract handling	NR
Sludges	
Gravity thickening	76
Pressure filtration	66
Vacuum filtration	68
Centrifugation	55
Sludge bed drying	77
In Process Control	
Flow reduction	NR

Source: USEPA, *Treatability Manual, Volume II. Industrial Descriptions*, report # EPA-600/2-82-001b, U.S. Environmental Protection Agency, Washington, DC, September 1981.
NR, not reported.

4. Membrane filtration
5. Granular bed filtration
6. Peat adsorption
7. Insoluble starch xanthate treatment
8. Flotation

Three treatment options are used in treating common metals wastes:

Option 1 system consists of hydroxide precipitation [11] followed by sedimentation [12]. This system accomplishes the end-of-pipe metals removal from all common metal-bearing wastewater streams that are present at a facility. The recovery of precious metals, the reduction of hexavalent chromium, the removal of oily wastes, and the destruction of cyanide must be accomplished prior to removal of common metals.

Option 2 system is identical to the Option 1 treatment system, with the addition of filtration devices [13] after the primary solids removal devices. The purpose of these filtration units is to remove suspended solids, such as metal hydroxides, which do not settle out in the clarifiers. The filters also act as a safeguard against pollutant discharge should an upset occur in the sedimentation device. Filtration techniques applicable to Option 2 systems are diatomaceous earth and granular bed filtration [14,15].

Option 3 treatment system for common metals wastes consists of the Option 2 end-of-pipe treatment system plus the addition of in-plant controls for lead and cadmium. In-plant controls include evaporative recovery, ion exchange, and recovery rinses [15].

In addition to these three treatments, there are several alternative treatment technologies applicable to the treatment of common metals wastes. These technologies include electrolytic recovery, electrodialysis, RO, peat adsorption, insoluble starch xanthate treatment, sulfide precipitation, flotation, and membrane filtration [14,15].

10.4.2 Precious Metals

Precious metal wastes can be treated using the same treatment alternatives as those described for treatment of common metals wastes. However, due to the intrinsic value of precious metals, every effort should be made to recover them. The treatment alternatives recommended for precious metal wastes are the recovery techniques—evaporation, ion exchange, and electrolytic recovery.

10.4.3 Complexed Metal Wastes

Complexed metal wastes within the metal finishing industry are a product of electroless plating, immersion plating, etching, and printed circuit board manufacture. The metals in these waste streams are tied up or complexed by particular complexing agents whose function is to prevent metals from coming out of solution. This counteracts the technique employed by most conventional solids removal methods. Therefore, segregated treatment of these wastes is necessary. The treatment method well suited to treating complexed metal wastes is high pH precipitation. An alternative method is membrane filtration [16] that is primarily used in place of sedimentation for solids removal.

10.4.4 Hexavalent Chromium

Hexavalent chromium-bearing wastewaters are produced in the metal finishing industry in chromium electroplating, in chromate conversion coatings, in etching with chromic acid, and in metal finishing operations carried out on chromium as a basis material.

The selected treatment option involves the reduction of hexavalent chromium to trivalent chromium either chemically or electrochemically. The reduced chromium can then be removed using a conventional precipitation-solids removal system. Alternative hexavalent chromium treatment techniques include chromium regeneration, electrodialysis, evaporation, and ion exchange [15].

10.4.5 CYANIDE

Cyanides are introduced as metal salts for plating and conversion coating or are active components in plating and cleaning baths. Cyanide is generally destroyed by oxidation. Chlorine, in either elemental or hypochlorate form, is the primary oxidation agent used in industrial waste treatment to destroy cyanide. Alternative treatment techniques for the destruction of cyanide include oxidation by ozone, ozone with ultraviolet (UV) radiation (oxyphotolysis), hydrogen peroxide, and electrolytic oxidation [17]. Treatment techniques that remove cyanide but do not destroy it include chemical precipitation, RO, and evaporation [15,17].

10.4.6 OILS

Oily wastes and toxic organics that combine with the oils during manufacturing include process coolants and lubricants, wastes from cleaning operations, wastes from painting processes, and machinery lubricants. Oily wastes are generally of three types: free oils, emulsified or water-soluble oils, and greases. Oil removal techniques commonly employed in the metal finishing industry include skimming, coalescing, emulsion breaking, flotation, centrifugation, ultrafiltration, RO, carbon adsorption, and aerobic decomposition [17–19].

Because emulsified oils and processes that emulsify oils are used extensively in the metal finishing industry, the exclusive occurrence of free oils is nearly nonexistent.

Treatment of oily wastes can be carried out most efficiently if oils are segregated from other wastes and treated separately. Segregated oily wastes originate in the manufacturing areas and are collected in holding tanks and sumps. Systems for treating segregated oily wastes consist of separation of oily wastes from the water. If oily wastes are emulsified, techniques such as emulsion breaking or dissolved air flotation [20] with the addition of chemicals are necessary to remove oil. Once the oil–water emulsion is broken, the oily waste is physically separated from the water by decantation or skimming. After the oil–water separation has been carried out, the water is sent to the precipitation/sedimentation unit used for removal of metals. There are three options for oily waste removal:

- *Option 1* system incorporates the emulsion breaking process followed by surface skimming (gravity separation is adequate if only free oils are present).
- *Option 2* system consists of the Option 1 system followed by ultrafiltration.
- *Option 3* treatment system consists of the Option 2 system with the addition of either carbon adsorption or RO.

In addition to these three treatment options, several alternative technologies are applicable to the treatment of oily wastewater. These include coalescing, flotation, centrifugation, integrated adsorption, resin adsorption, ozonation, chemical oxidation, aerobic decomposition, and thermal emulsion breaking [17–19].

10.4.7 SOLVENTS

Spent degreasing solvents should be segregated from other process fluids to maximize the value of the solvents, to preclude contamination of other segregated wastes, and to prevent the discharge of priority pollutants to any wastewaters. This segregation may be accomplished by providing and identifying the necessary storage containers, establishing clear disposal procedures, training personnel in the use of these techniques, and checking periodically to ensure that proper segregation is occurring. Segregated waste solvents are appropriate for on-site solvent recovery or may be contract hauled for disposal or reclamation.

Alkaline cleaning is the most feasible substitute for solvent degreasing. The major advantage of alkaline cleaning over solvent degreasing is the elimination or reduction in the quantity of priority pollutants being discharged. Major disadvantages include high energy consumption, the tendency to dilute oils removed, and to discharge these oils as well as the cleaning additive.

10.5 TREATMENT TECHNOLOGIES

10.5.1 Neutralization

One technique used in a number of facilities that utilize molten salt for metal surface treatment prior to pickling is to take advantage of the alkaline values generated in the molten salt bath while treating other wastes generated in the plant. When the bath is determined to be spent, it is in many instances manifested, hauled off-site, and land disposed. One technique is to take the solidified spent molten salt (molten salt is sold at ambient temperatures) and circulate acidic wastes generated in the facility over the material prior to entry to the waste treatment system. This, in effect, neutralizes the acid wastes and eliminates the requirements of manifesting and land disposal.

10.5.2 Cyanide-Containing Wastes

There are eight methods applicable to the treatment of cyanide wastes for metal finishing [5,21]:

1. Alkaline chlorination
2. Electrolytic decomposition
3. Ozonation
4. UV/Ozonation
5. Hydrogen peroxide
6. Thermal oxidation
7. Acidification and acid hydrolysis
8. Ferrous sulfate precipitation

Alkaline chlorination is the most widely applied method in the metal finishing industry. A schematic for cyanide reduction via alkaline chlorination is provided in Figure 10.5. This technology is generally applicable to wastes containing <1% cyanide, generally present as free cyanide. It is conducted in two stages; the first stage is operated at a pH >10, and the second stage is operated at a pH in the range of 7.5–8. Alkaline chlorination is performed using sodium hypochlorite and chlorine.

Electrolytic decomposition technology was applied to cyanide-containing wastes in the early part of this century. It fell from favor as alkaline chlorination came into use in large-scale facilities. However, as wastes become more concentrated, this technology may find more widespread application in the future. The reason is that it is applicable to wastes containing cyanide in excess of 1%. The basis of this technology is electrolytic decomposition of the cyanide compounds at an elevated temperature (200 F) to yield nitrogen, CO_2, ammonia, and amines (see Figure 10.6).

Ozonation treatment can be used to oxidize cyanide, thereby reducing the concentration of cyanide in wastewater. Ozone, with an electrode potential of +1.24 V in alkaline solutions, is one of the most powerful oxidizing agents known. Cyanide oxidation with ozone is a two-step reaction similar to alkaline chlorination [21]. Cyanide is oxidized to cyanate, with ozone reduced to oxygen per the following equation:

$$CN^- + O_3 \to CNO^- + O_2. \tag{10.1}$$

NaCN + Cl$_2$ → CNCl + NaCl
CNCl + 2 NaOH → NaCNO + H$_2$O
2 NaCNO + 3 Cl$_2$ + 4NaOH → 2 CO$_2$ + N$_2$ + 6 NaCl + 2 H$_2$O

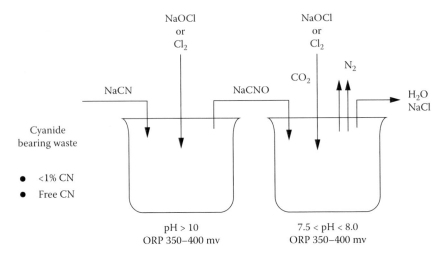

FIGURE 10.5 Cyanide reduction via alkaline chlorination. (From USEPA, Meeting hazardous waste requirements for metal finishers, report # EPA/625/4-87/018, U.S. Environmental Protection Agency, Cincinnati, OH, 1987.)

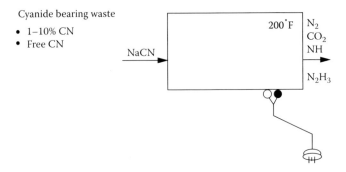

FIGURE 10.6 Cyanide reduction via electrolytic decomposition. (From USEPA, Meeting hazardous waste requirements for metal finishers, report # EPA/625/4-87/018, U.S. Environmental Protection Agency, Cincinnati, OH, 1987.)

Then, cyanate is hydrolyzed, in the presence of excess ozone, to bicarbonate and nitrogen and oxidized per the following reaction:

$$2\,CNO^- + 3\,O_3 + H_2O \rightarrow N_2 + 2\,HCO_3^- + 3\,O_2. \tag{10.2}$$

The reaction time for complete cyanide oxidation is rapid in a reactor system, with 10- to 30-min retention times being typical. The second-stage reaction is much slower than the first-stage reaction. The reaction is typically carried out at the pH range of 10–12 where the reaction rate is relatively constant. Temperature does not influence the reaction rate significantly.

One interesting variation in ozonation technology is augmentation with UV radiation. This is a technology that has been applied on wastes in the coke by-product manufacturing industry.

A significant development has been made that has resulted in significantly less ozone consumption through the use of UV radiation. UV absorption has the following effects:

- Ozone and cyanide are raised to higher energy status
- Free radicals are formed
- More rapid reaction
- Less ozone is required

Cyanide reduction with hydrogen peroxide is effective in reducing cyanide. It has been applied on a less frequent basis within this industry, due to the fact that there are high operating costs associated with hydrogen peroxide generation. The reduction of cyanide with peroxide occurs in two steps and yields CO_2 and ammonia:

$$NaCN + H_2O_2 \rightarrow NaCNO + H_2O \qquad (10.3)$$

$$NaCNO + 2\,H_2O \rightarrow CO_2 + NH_3 + NaOH. \qquad (10.4)$$

Thermal oxidation is another alternative for destroying cyanide. Thermal destruction of cyanide can be accomplished through either high-temperature hydrolysis or combustion. At temperatures between 140°C and 200°C and a pH of 8, cyanide hydrolyzes quite rapidly to produce formate and ammonia [22]. Pressures up to 100 bar are required, but the process can effectively treat waste streams over a wide concentration range and is applicable to both rinse water and concentrated solutions [21].

$$CN^- + 2\,H_2O \rightarrow HCOO^- + NH_3. \qquad (10.5)$$

In the presence of nitrates, formate and ammonia can be destroyed in another reactor at 150°C, according to the following equations:

$$NH_4^+ + NO_2^- \rightarrow N_2 + 2H_2O \qquad (10.6)$$

$$3\,HCOOH + 2\,NO_2^- + 2\,H^+ \rightarrow 3\,CO_2 + 4\,H_2O. \qquad (10.7)$$

Direct acidification of cyanide waste streams was once a relatively common treatment. Cyanide is acidified in a sealed reactor that is vented to the atmosphere through an air emission control system. Cyanide is converted to gaseous hydrogen cyanide, treated, vented, and dispersed.

Acid hydrolysis of cyanates is still commonly used, following a first-stage cyanide oxidation process. At pH 2, the reaction proceeds rapidly, while at pH 7 cyanate may remain stable for weeks [23]. This treatment process requires specially designed reactors to ensure that HOCN is properly vented and controlled. The hydrolysis mechanisms are as follows [21]:

In acid medium:

$$HOCN + H^+ \rightarrow NH_4^+ + CO_2 \text{ (rapid)} \qquad (10.8)$$

$$HOCN + H_2O \rightarrow NH_3 + CO_2 \text{ (slow)}. \qquad (10.9)$$

In strongly alkaline medium:

$$NCO^- + 2\,H_2O \rightarrow NH_2 + HCO_3^- \text{ (very slow)}. \qquad (10.10)$$

Each of the technologies described above is effective in treating wastes containing free cyanides, that is, cyanides present as CN in solution. There are instances in metal finishing facilities where complex cyanides are present in wastes. The most common are complexes of iron, nickel, and zinc. A technology that has been applied to remove complex cyanides from aqueous wastes is ferrous sulfate precipitation. The technology involves a two-stage operation in which ferrous sulfate is first added at a pH of 9 to complex any trace amounts of free cyanide. In the second stage, the complex cyanides are precipitated through the addition of ferrous sulfate or ferric chloride at a pH range of 2–4 [5].

10.5.3 Chromium-Containing Wastes

There are three treatment methods applicable to wastes containing hexavalent chromium. Wastes containing trivalent chromium can be treated using chemical precipitation and sedimentation. The following are the three methods used for treatment of hexavalent chromium:

1. Sulfur dioxide
2. Sodium metabisulfite
3. Ferrous sulfate

Hexavalent chromium reduction through the use of sulfur dioxide and sodium metabisulfite has found the widest application in the metal finishing industry. It is not truly a treatment step, but a conversion process in which the hexavalent chromium is converted to trivalent chromium. The hexavalent chromium is reduced through the addition of the reductant at a pH in the range of 2.5–3, with a retention time of approximately 30–40 min (see Figure 10.7).

Ferrous sulfate has not been as widely applied. However, it is particularly applicable in facilities where ferrous sulfate is produced as part of the process or is readily available. The basis for this technology is that the hexavalent chromium is reduced to trivalent chromium, and the ferrous iron is oxidized to ferric iron.

10.5.4 Arsenic- and Selenium-Containing Wastes

It may be necessary to segregate waste streams containing elevated concentrations of arsenic and selenium, especially those that contain these pollutants at concentrations in excess of 1 mg/L. Arsenic and selenium form anionic acids in solution (most other metals act as cations) and require special preliminary treatment prior to conventional metals treatment. Lime, a source of calcium ions, is effective in reducing arsenic and selenium concentrations when the initial concentration is <1 mg/L. However, preliminary treatment with sodium sulfide at a low pH (i.e., 1–3) may be required

$$SO_2 + H_2O \rightarrow H_2SO_3$$
$$H_2CrO_4 + 3H_2SO_3 \rightarrow Cr_2(SO_4)_3 + 5H_2O$$

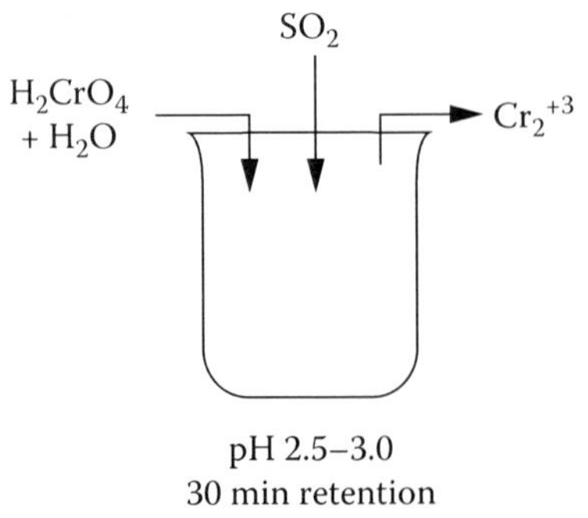

FIGURE 10.7 Hexavalent chromium reduction. (From USEPA, Meeting hazardous waste requirements for metal finishers, report # EPA/625/4-87/018, U.S. Environmental Protection Agency, Cincinnati, OH, 1987.)

for waste streams with concentrations in excess of 1 mg/L [21]. The sulfide reacts with the anionic acids to form insoluble sulfides, which are readily separated by filtration.

10.5.4.1 Chemical Precipitation and Sedimentation

The most important technology in metals treatment is chemical precipitation and sedimentation. It is accomplished through the addition of a chemical reagent to form metal precipitants, which are then removed as solids in a sedimentation step. The options available to a facility as precipitation reagents are lime $Ca(OH)_2$, caustic NaOH, carbonate $CaCO_3$ and Na_2CO_3, sulfide NaHS and FeS, and sodium borohydride $NaBH_4$. The advantages and disadvantages of these reagents are summarized below [21]:

1. Lime
   - Least expensive precipitation reagent
   - Generates highest sludge volume
   - Sludges generally cannot be sold to smelter/refiners
2. Caustic
   - More expensive than lime
   - Generates smaller volume of sludge
   - Sludges can be sold to smelter/refiners
3. Carbonates
   - Applicable for metals whose solubility within a pH range is not sufficient to meet treatment standards

Lime is the least expensive reagent; however, it generates the highest volume of residue. It also generates a residue that cannot be resold to smelters and refiners for reclaiming because of the presence of the calcium ion. Caustic is more expensive than lime; however, it generates a smaller volume of residue. One key advantage of caustic is that the resulting residues can be readily reclaimed. Carbonates are particularly appropriate for metals whose solubility within a pH range is not sufficient to meet a given set of treatment standards. The sulfides offer the benefit of achieving effective treatment at lower concentrations due to lower solubilities of the metal sulfides. Sodium borohydride has application where small volumes of sludge that are suitable for reclamation are desired.

It is appropriate to look at reagent use in the context of the current regulatory framework under HSWA. Historically, lime has been the reagent of choice. It was relatively inexpensive and simple to handle. The phrase "Lime and Settle" refers to the application of lime precipitation and sedimentation technology. In the 1970s, new designs made use of caustic as the precipitation reagent because of the reduction in residue volume realized and the ability for reclamation. In the 1980s, lime and the use of combined reagent techniques were back in use.

One obvious question is why lime was used again as a treatment reagent, given that caustic results in a smaller residue volume and a waste that can undergo reclamation. The answer lies in the three points that result from the implementation of the HSWA hierarchy. As source reduction and material reuse and recovery techniques are applied, facilities will be generating the following:

- More concentrated wastes
- Wastes with a varied array of constituents
- Wastes with a greater degree of complexation

10.5.4.2 Complexation

Complexation is a phenomenon that involves a coordinate bond between a central atom, the metal, and a ligand, the anions. In a coordinate bond, the electron pair is shared between the metal and the ligand. A complex containing one coordinate bond is referred to as a monodentate complex. Multiple coordinate bonds are characteristic of polydentate complexes. Polydentate complexes are also referred to as chelates. An example of a monodentate-forming ligand is ammonia. Examples of chelates are oxylates (bidentates) and ethylene diamine tetraacetic acid (hexadentates).

The return to lime was due to the calcium ion present in lime. The calcium ion that comes into solution through the addition of lime is very effective in competing with the ligand for the metal ion. The sodium ion contributed by caustic is not effective. As such, lime dramatically reduces complexation and is more effective in treating complexed wastes. The term "high lime treatment" has been applied in cases where excess calcium ions are introduced into solution. This is accomplished through the addition of lime to raise the pH to approximately 11.5 or through the addition of calcium chloride (which has a greater solubility than lime).

The use of precipitation reagents in different combinations has been most effective in taking advantage of the attributes of caustic as well as the advantages of lime. As an example, a system may use caustic in the first stage to make a coarse pH adjustment, followed by the addition of lime to make a fine adjustment. This achieves an overall reduction in the sludge volume through the use of the caustic, and more effective metal removal through the use of lime. Sulfide reagents are used in a similar fashion in combination with caustic or lime to facilitate additional metal removal, due to the lower solubility of the metal sulfides. Sulfides are also applicable to wastes containing elevated concentrations (i.e., in excess of 2 mg/L) of selenium and arsenic compounds [21].

10.5.5 Other Metals Wastes

There are three techniques applicable to managing solids generated in metal finishing. These are as follows:

1. Dewatering
2. Stabilization
3. Incineration

There are four dewatering techniques that have been applied in metal processing. The most widely applied techniques are vacuum and belt filtration [24]. They have a higher relative capital cost but generally have a lower relative operating cost. Plate and frame filter presses have experienced less widespread application. Belt filters generally have a lower relative capital cost and higher relative operating costs. The higher operating costs are due to the fact that the units are more labor-intensive. Centrifuges [24] have been applied in specific instances, but are more difficult to operate when a widely varying mix of wastes is treated.

Experience has shown that companies are most successful in applying a dewatering technique that they have successfully designed and operated in similar applications within the company. As an example, many companies operate plate and frame filter presses as a part of metal manufacturing operations. The knowledge gained in metal processing has been successfully transferred to treatment of metal finishing wastes.

There are six stabilization techniques currently available; however, only two of them have found widespread application. These are cementation and stabilization through the addition of lime and fly ash [24,25]. Currently developmental work is being undertaken to make use of bitumen, paraffin, and polymeric materials to reduce the degree to which metals can be taken into solution. Encapsulation with inert materials is also under development.

10.6 COSTS

The investment cost, operating and maintenance costs [26,27], and energy costs for the application of control technologies to the wastewaters of the metal finishing industry have been analyzed. These costs were developed to reflect the conventional use of technologies in this industry. The detailed presentation of the cost methodology and cost data is available in a USEPA publication [6]. The available industry-specific cost information is characterized below.

Treatment and Management of Metal Finishing Industry Wastes

10.6.1 Typical Treatment Options

Many waste treatment, recycle and management options are available [27–39] based on the degree of treatment required by the government [32]. Only several unit operation/unit process configurations have been analyzed for the cost of application to the wastewater of this industry. The components included in these configurations are as follows:

- *Option 1*: Emulsion breaking and oil separation by skimming, cyanide oxidation, chromium reduction, chemical precipitation and sedimentation, and sludge drying beds
- *Option 2*: All of Option 1 plus multimedia filtration
- *Option 3*: All of Option 2 plus ultrafiltration and carbon adsorption for oily waste, zero discharge of any processes using either cadmium or lead by employing evaporative system

The flow diagram for suggested Option 1 is shown in Figure 10.8. The flow diagram for the other options would be similar.

10.6.2 Costs

The cost estimates prepared for the treatment technologies commonly used in this industry are briefly described below. More detail on the factors considered in the cost analysis is available in the source [6].

1. Emulsion breaking and oil separation
   Method: Emulsion is broken by mixing oily waste with alum and a chemical emulsion breaker, followed by gravity oil separation in a tank.
   System component: Small mixing tank, two chemical feed tanks, a mixer, and a large tank equipped with an oil skimmer and a sludge pump. The mixing tank has a retention time of 15 min, and oil skimming tank has a retention time of 2.5 h.
2. Cyanide Oxidation
   Method: Cyanide is destroyed by reaction with sodium hypochlorite under alkaline conditions.

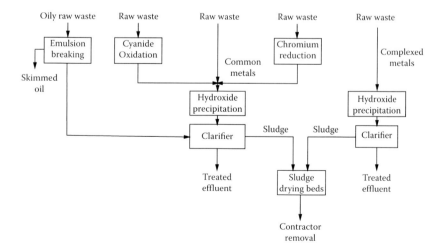

FIGURE 10.8 Metal finishing wastewater treatment flow diagram. (From USEPA, *Treatability Manual, Volume II. Industrial Descriptions*, report # EPA-600/2-82-001b, U.S. Environmental Protection Agency, Washington, DC, September 1981.)

System component: Reaction tanks, reagent storage and feed system, mixers, sensors, and controls. Two identical reaction tanks sized as above-ground cylindrical tank with a retention time of 4 h. Chemical storage consists of covered concrete tanks to store 60 days' supply of sodium hypochlorite and 90 days' supply of sodium hydroxide.

3. Chromium Reduction

    Method: Chemical reduction of hexavalent chromium is carried out by sulfur dioxide under acid conditions for continuous operating system and by sodium bisulfite under acid condition for batch operating system. The reduced trivalent form of chromium is subsequently removed by precipitation as hydroxide.

    System component: Reaction tanks, reagent storage and feed system, mixers, sensors, and controls for continuous chromium reduction. A single above-ground concrete tank with retention time of 45 min is provided. For batch operation, dual above-ground concrete tanks with 4 h retention time are provided.

4. Lime Precipitation and Sedimentation

    Method: Chemical precipitation of dissolved and complexed metals is carried out by reaction with lime and subsequent removal of the precipitated solids by gravity settling in a clarifier. Alum and polyelectrolyte are added for coagulation and flocculation.

    System component: Continuous treatment system includes reagent storage and feed equipment, a mix tank for reagent feed addition, sensors and controls, and clarification basin with associated sludge rakes and pumps. Lime is fed as 30% lime slurry prepared by using hydrated lime. The mix tank is sized for retention time of 45 min, and the clarifier is sized for hydraulic loading of 1360 L/m² and a retention time of 4 h. Batch treatment includes dual reaction-settling tanks sized for 8 h retention time and sludge pumps.

5. Drying Beds

    Method: Sludge is dewatered by gravity drainage and natural evaporation.

    System component: Beds of highly permeable gravel and sand underlain by drain pipes [28].

6. Multimedia Filter

    Method: Polishing treatment after chemical precipitation and sedimentation by filtration through a bed of particles of several distinct size ranges.

    System component: Filter beds, media, backwash mechanism, pumps, and controls. Filter beds are sized for hydraulic loading of 81 L/min/m² (2 gpm/ft²).

7. Ultrafiltration

    Method: Process used for oily waste stream after emulsion breaking–gravity oil separation.

    System component: Filter modules sized on the basis of hydraulic loading of 1 L/min/m².

8. Carbon Adsorption

    Method: A packed-bed throwaway system to remove organic pollutants from oily waste stream.

    System component: Contactor system, and a pump station designed for contact time of 30 min and hydraulic loading of 162 L/min/m² (4 gpm/ft²).

Unit costs shown in Table 10.16 are for the complete treatment options described previously. Unit costs are computed for a model plant where flows are contributed by several waste streams as follows:

- 30% oily waste stream
- 4% cyanide waste stream
- 9% chromium waste stream
- 52.5% common metals stream
- 4.5% complex metal stream.

TABLE 10.16
Total Annual Unit Cost (USD/m³ in 2014 Dollars)

Flow (m³/h)	Option 1 Continuous	Option 1 Batch	Option 2 Continuous	Option 2 Batch	Option 3 Continuous	Option 3 Batch
2.36	—	16.56	—	27.77	—	32.90
11.81	7.06	5.85	11.21	9.74	13.15	11.94
59.07	2.92	—	4.62	—	6.09	—
118.16	2.44	2.44	5.36	4.40	4.87	5.12

Source: USEPA, *Treatability Manual, Volume II. Industrial Descriptions*, report # EPA-600/2-82-001b, U.S. Environmental Protection Agency, Washington, DC, September 1981.

Note: Costs were converted from 1979 USD to 2014 USD using U.S. ACE Yearly Average Cost Index for Utilities [9].

10.7 CFR FOR METAL FINISHING EFFLUENT DISCHARGE MANAGEMENT

This section introduces the CFR Title 40, Part 433 (40 CFR Part 433), for effluent discharge management of metal finishing point source category.

The topics introduced in this section include the following: (1) the applicability and description of the metal finishing point source category; (2) the monitoring requirements of metal finishing effluent discharges; (3) the effluent limitations representing the degree of effluent reduction attainable by applying the best practicable control technology currently available (BPT); (4) the effluent limitations representing the degree of effluent reduction attainable by applying the best available technology economically achievable (BAT); (5) the pretreatment standards for existing sources (PSES); (6) the new source performance standards (NSPS); and (7) the pretreatment standards for new sources (PSNS).

10.7.1 Applicability and Description of the Metal Finishing Point Source Category

Except as noted in the following two paragraphs of this section, the provisions of this subpart apply to plants that perform any of the following six metal finishing operations on any basis material: electroplating, electroless plating, anodizing, coating (chromating, phosphating, and coloring), chemical etching and milling, and printed circuit board manufacture. If any of these six operations are present, then this part applies to discharges from those operations and also to discharges from any of the following 40 process operations: cleaning, machining, grinding, polishing, tumbling, burnishing, impact deformation, pressure deformation, shearing, heat treating, thermal cutting, welding, brazing, soldering, flame spraying, sand blasting, other abrasive jet machining, electric discharge machining, electrochemical machining, electron beam machining, laser beam machining, plasma arc machining, ultrasonic machining, sintering, laminating, hot-dip coating, sputtering, vapor plating, thermal infusion, salt bath descaling, solvent degreasing, paint stripping, painting, electrostatic painting, electropainting, vacuum metalizing, assembly, calibration, testing, and mechanical plating.

In some cases, effluent limitations and standards for the following industrial categories may be effective and applicable to wastewater discharges from the metal finishing operations listed above. In such cases, the 40 CFR, Part 433, limits shall not apply, and the following regulations shall apply:

- Nonferrous metal smelting and refining (40 CFR, Part 421)
- Coil coating (40 CFR, Part 465)
- Porcelain enameling (40 CFR, Part 466)
- Battery manufacturing (40 CFR, Part 461)

- Iron and steel (40 CFR, Part 420)
- Metal casting foundries (40 CFR, Part 464)
- Aluminum forming (40 CFR, Part 467)
- Copper forming (40 CFR, Part 468)
- Plastic molding and forming (40 CFR, Part 463)
- Nonferrous forming (40 CFR, Part 471)
- Electrical and electronic components (40 CFR, Part 469)

The 40 CFR, Part 433, does not apply to (1) metallic platemaking and gravure cylinder preparation conducted within or for printing and publishing facilities, and (2) existing indirect discharging job shops and independent printed circuit board manufacturers that are covered by 40 CFR, Part 413.

10.7.2 Monitoring Requirements of Metal Finishing Effluent Discharges

In lieu of requiring monitoring for total toxic organics (TTO), the permitting authority (or, in the case of indirect dischargers, the control authority) may allow dischargers to make the following certification statement:

> Based on my inquiry of the person or persons directly responsible for managing compliance with the permit limitation [or pretreatment standard] for total toxic organics (TTO), I certify that, to the best of my knowledge and belief, no dumping of concentrated toxic organics into the wastewaters has occurred since filing of the last discharge monitoring report. I further certify that this facility is implementing the toxic organic management plan submitted to the permitting [or control] authority.

For direct dischargers, this statement is to be included as a "comment" on the Discharge Monitoring Report required by 40 CFR 122.44(i), formerly 40 CFR 122.62(i).

For indirect dischargers, the statement is to be included as a comment on the periodic reports required by 40 CFR 403.12(e). If monitoring is necessary to measure compliance with the TTO standard, the industrial discharger need analyze only those pollutants that would reasonably be expected to be present.

In requesting the certification alternative, a discharger shall submit a solvent management plan that specifies to the satisfaction of the permitting authority (or, in the case of indirect dischargers, the control authority) the following: the toxic organic compounds used; the method of disposal used instead of dumping, such as reclamation, contract hauling, or incineration; and procedures for ensuring that toxic organics do not routinely spill or leak into the wastewater. For direct dischargers, the permitting authority shall incorporate the plan as a provision of the permit.

Self-monitoring for cyanide must be conducted after cyanide treatment and before dilution with other streams. Alternatively, samples may be taken of the final effluent, if the plant limitations are adjusted based on the dilution ratio of the cyanide waste stream flow to the effluent flow.

10.7.3 Effluent Limitations Based on the BPT

Except as specifically provided in the CFR, any existing point source subject to the 40 CFR, Part 433, must achieve the effluent limitations shown in Table 10.17, which represents the degree of effluent reduction attainable by applying the BPT. Alternatively, for metal finishing industrial facilities with cyanide treatment, and upon agreement between a source subject to those limits and the pollution control authority, the amenable cyanide limit shown in Table 10.18 may apply in place of the total cyanide limit specified in Table 10.17 [32]. No user subject to the provisions of these regulations shall augment the use of process wastewater or otherwise dilute the wastewater as a partial or total substitute for adequate treatment to achieve compliance with this limitation.

TABLE 10.17
U.S. Best Practicable Control Technology Currently Available (BPT) Effluent Limitations for the Metal Finishing Point Source Category

Pollutant or Pollutant Property	Maximum for Any 1 Day	Monthly Average Shall Not Exceed
	(mg/L Except for pH)	
Cadmium (T)	0.69	0.26
Chromium (T)	2.77	1.71
Copper (T)	3.38	2.07
Lead (T)	0.69	0.43
Nickel (T)	3.98	2.38
Silver (T)	0.43	0.24
Zinc (T)	2.61	1.48
Cyanide (T)	1.20	0.65
TTO	2.13	
Oil and grease	52	26
TSS	60	31
pH	6–9 units	6–9 units

Source: USEPA, Code of Federal Regulations, Metal finishing point source category, Title 40, Volume 27, Part 433, U.S. Environmental Protection Agency, Washington, DC, Revised on July 1, 2003.

TABLE 10.18
Alternative U.S. Best Practicable Control Technology Currently Available (BPT) Effluent Limitations on Cyanide (A) for the Metal Finishing Point Source Category

Pollutant or Pollutant Property	Maximum for Any 1 Day	Monthly Average Shall Not Exceed
	(mg/L)	
Cyanide (A)	0.86	0.32

Source: USEPA, Code of Federal Regulations, Metal finishing point source category, Title 40, Volume 27, Part 433, U.S. Environmental Protection Agency, Washington, DC, Revised on July 1, 2003.

10.7.4 EFFLUENT LIMITATIONS BASED ON THE BAT

Except as specifically provided in the CFR, any existing point source subject to this subpart must achieve the effluent limitations shown in Table 10.19, which represents the degree of effluent reduction attainable by applying the BAT. Alternatively, for the metal finishing industrial facilities with cyanide treatment, and upon agreement between a source subject to those limits and the pollution control authority, the amenable cyanide limit shown in Table 10.20 may apply in place of the total cyanide limit specified in Table 10.19. No user subject to the provisions of these regulations shall augment the use of process wastewater or otherwise dilute the wastewater as a partial or total substitute for adequate treatment to achieve compliance with this limitation.

10.7.5 PRETREATMENT STANDARDS FOR EXISTING SOURCES

Except as specifically provided in the CFR, any existing source subject to this 40 CFR, Part 433, that introduces pollutants into a publicly owned treatment works must also comply with 40 CFR,

TABLE 10.19
U.S. Best Available Technology Economically Achievable (BAT) Effluent Limitations for the Metal Finishing Point Source Category

Pollutant or Pollutant Property	Maximum for Any 1 Day	Monthly Average Shall Not Exceed
	(mg/L)	
Cadmium (T)	0.69	0.26
Chromium (T)	2.77	1.71
Copper (T)	3.38	2.07
Lead (T)	0.69	0.43
Nickel (T)	3.98	2.38
Silver (T)	0.43	0.24
Zinc (T)	2.61	1.48
Cyanide (T)	1.20	0.65
TTO	2.13	

Source: USEPA, Code of Federal Regulations, Metal finishing point source category, Title 40, Volume 27, Part 433, U.S. Environmental Protection Agency, Washington, DC, Revised on July 1, 2003.

TABLE 10.20
Alternative U.S. Best Available Technology Economically Achievable (BAT) Effluent Limitations on Cyanide (A) for the Metal Finishing Point Source Category

Pollutant or Pollutant Property	Maximum for Any 1 Day	Monthly Average Shall Not Exceed
	(mg/L)	
Cyanide (A)	0.86	0.32

Source: USEPA, Code of Federal Regulations, Metal finishing point source category, Title 40, Volume 27, Part 433, U.S. Environmental Protection Agency, Washington, DC, Revised on July 1, 2003.

Part 403, and achieve the PSES. Table 10.21 indicates the PSES for all metal finishing plants except job shops and independent printed circuit board manufacturers. Alternatively, for industrial facilities with cyanide treatment, upon agreement between a source subject to those limits and the pollution control authority, the amenable cyanide limit shown in Table 10.22 may apply in place of the total cyanide limit specified in Table 10.21. No user introducing wastewater pollutants into a publicly owned treatment works under the provisions of this subpart shall augment the use of process wastewater as a partial or total substitute for adequate treatment to achieve compliance with this standard. An existing source submitting a certification in lieu of monitoring pursuant to this regulation must implement the toxic organic management plan approved by the control authority. An existing source subject to this subpart shall comply with a daily maximum pretreatment standard for TTO of 4.57 mg/L.

10.7.6 NEW SOURCE PERFORMANCE STANDARDS

Any new metal finishing point source subject to the 40 CFR, Part 433, regulations must achieve the NSPS shown in Table 10.23. Alternatively, for the metal finishing industrial facilities with cyanide treatment, and upon agreement between a source subject to those limits and the pollution control authority, the amenable cyanide limit shown in Table 10.24 may apply in place of the total cyanide

TABLE 10.21
U.S. Pretreatment Standards for Existing Sources (PSES) for All Metal Finishing Plants except Job Shops and Independent Printed Circuit Board Manufacturers

Pollutant or Pollutant Property	Maximum for Any 1 Day	Monthly Average Shall Not Exceed
	(mg/L)	
Cadmium (T)	0.69	0.26
Chromium (T)	2.77	1.71
Copper (T)	3.38	2.07
Lead (T)	0.69	0.43
Nickel (T)	3.98	2.38
Silver (T)	0.43	0.24
Zinc (T)	2.61	1.48
Cyanide (T)	1.20	0.65
TTO	2.13	

Source: USEPA, Code of Federal Regulations, Metal finishing point source category, Title 40, Volume 27, Part 433, U.S. Environmental Protection Agency, Washington, DC, Revised on July 1, 2003.

TABLE 10.22
Alternative U.S. Pretreatment Standards for Existing Sources (PSES) on Cyanide (A) for All Metal Finishing Plants except Job Shops and Independent Printed Circuit Board Manufacturers

Pollutant or Pollutant Property	Maximum for Any 1 Day	Monthly Average Shall Not exceed
	(mg/L)	
Cyanide (A)	0.86	0.32

Source: USEPA, Code of Federal Regulations, Metal finishing point source category, Title 40, Volume 27, Part 433, U.S. Environmental Protection Agency, Washington, DC, Revised on July 1, 2003.

limit specified in Table 10.23. No user subject to the provisions of this subpart shall augment the use of process wastewater or otherwise dilute the wastewater as a partial or total substitute for adequate treatment to achieve compliance with this limitation.

10.7.7 Pretreatment Standards for New Sources

Except as provided in the CFR, any new source subject to this subpart that introduces pollutants into a publicly owned treatment works must comply with 40 CFR, Part 403, and achieve the PSNS, shown in Table 10.25. Alternatively, for industrial facilities with cyanide treatment, and upon agreement between a source subject to these limits and the pollution control authority, the amenable cyanide limit shown in Table 10.26 may apply in place of the total cyanide limit specified in Table 10.25.

No user subject to the provisions of this subpart shall augment the use of process wastewater or otherwise dilute the wastewater as a partial or total substitute for adequate treatment to achieve compliance with this limitation. An existing source submitting a certification in lieu of monitoring pursuant to Section 433.12(a) and (b) of this regulation must implement the toxic organic management plan approved by the control authority.

TABLE 10.23
U.S. New Source Performance Standards (NSPS) for the Metal Finishing Point Source Category

Pollutant or Pollutant Property	Maximum for Any 1 Day	Monthly Average Shall Not Exceed
	(mg/L Except for pH)	
Cadmium (T)	0.11	0.07
Chromium (T)	2.77	1.71
Copper (T)	3.38	2.07
Lead (T)	0.69	0.43
Nickel (T)	3.98	2.38
Silver (T)	0.43	0.24
Zinc (T)	2.61	1.48
Cyanide (T)	1.20	0.65
TTO	2.13	
Oil and Grease	52	26
TSS	60	31
pH	6–9 units	6–9 units

Source: USEPA, Code of Federal Regulations, Metal finishing point source category, Title 40, Volume 27, Part 433, U.S. Environmental Protection Agency, Washington, DC, Revised on July 1, 2003.

TABLE 10.24
Alternative U.S. New Source Performance Standards (NSPS) on Cyanide (A) for the Metal Finishing Point Source Category

Pollutant or Pollutant Property	Maximum for Any 1 Day	Monthly Average Shall Not Exceed
	(mg/L)	
Cyanide (A)	0.86	0.32

Source: USEPA, Code of Federal Regulations, Metal finishing point source category, Title 40, Volume 27, Part 433, U.S. Environmental Protection Agency, Washington, DC, Revised on July 1, 2003.

10.8 SPECIALIZED DEFINITIONS

The definitions set forth in the CFR for the metal finishing point source category are incorporated in this section for reference.

1. The term "T," as in "Cyanide T," shall mean total.
2. The term "A," as in "Cyanide A," shall mean amenable to alkaline chlorination.
3. The term "job shop" shall mean a facility that owns not >50% (annual area basis) of the materials undergoing metal finishing.
4. The term "independent" printed circuit board manufacturer shall mean a facility that manufactures printed circuit boards principally for sale to other companies.

TABLE 10.25
U.S. Pretreatment Standards for New Sources (PSNS) for the Metal Finishing Point Source Category

Pollutant or Pollutant Property	Maximum for Any 1 Day	Monthly Average Shall Not Exceed
	(mg/L)	
Cadmium (T)	0.11	0.07
Chromium (T)	2.77	1.71
Copper (T)	3.38	2.07
Lead (T)	0.69	0.43
Nickel (T)	3.98	2.38
Silver (T)	0.43	0.24
Zinc (T)	2.61	1.48
Cyanide (T)	1.20	0.65
TTO	2.13	

Source: USEPA, Code of Federal Regulations, Metal finishing point source category, Title 40, Volume 27, Part 433, U.S. Environmental Protection Agency, Washington, DC, Revised on July 1, 2003.

TABLE 10.26
Alternative U.S. Pretreatment Standards for New Sources (PSNS) on Cyanide (A) for the Metal Finishing Point Source Category

Pollutant or Pollutant Property	Maximum for any 1 day	Monthly Average Shall Not Exceed
	(mg/L)	
Cyanide (A)	0.86	0.32

Source: USEPA, Code of Federal Regulations, Metal finishing point source category, Title 40, Volume 27, Part 433, U.S. Environmental Protection Agency, Washington, DC, Revised on July 1, 2003.

5. The term "TTO" shall mean total toxic organics, which is the summation of all quantifiable values >0.01 mg/L for the following toxic organics:
Acenaphthene
Acrolein
Acrylonitrile
Benzene
Benzidine
Carbon tetrachloride (tetrachloromethane)
Chlorobenzene
1,2,4-Trichlorobenzene
Hexachlorobenzene
1,2,-Dichloroethane
1,1,1-Trichloroethane

Hexachloroethane
1,1-Dichloroethane
1,1,2-Trichloroethane
1,1,2,2-Tetrachloroethane
Chloroethane
Bis (2-chloroethyl) ether
2-Chloroethyl vinyl ether (mixed)
2-Chloronaphthalene
2,4,6-Trichlorophenol
Parachlorometa-cresol
Chloroform (trichloromethane)
2-Chlorophenol
1,2-Dichlorobenzene
1,3-Dichlorobenzene
1,4-Dichlorobenzene
3,3-Dichlorobenzidine
1,1-Dichloroethylene
1,2-Trans-dichloroethylene
2,4-Dichlorophenol
1,2-Dichloropropane
1,3-Dichloropropylene (1,3-dichloropropene)
2,4-Dimethylphenol
2,4-Dinitrotoluene
2,6-Dinitrotoluene
1,2-Diphenylhydrazine
Ethylbenzene
Fluoranthene
4-Chlorophenyl phenyl ether
4-Bromophenyl phenyl ether
Bis (2-chloroisopropyl) ether
Bis (2-chloroethoxy) methane
Methylene chloride (dichloromethane)
Methyl chloride (chloromethane)
Methyl bromide (bromomethane)
Bromoform (tribromomethane)
Dichlorobromomethane
Chlorodibromomethane
Hexachlorobutadiene
Hexachlorocyclopentadiene
Isophorone
Naphthalene
Nitrobenzene
2-Nitrophenol
4-Nitrophenol
2,4-Dinitrophenol
4,6-Dinitro-o-cresol
N-nitrosodimethylamine
N-nitrosodiphenylamine
N-nitrosodi-n-propylamine
Pentachlorophenol
Phenol

Bis (2-ethylhexyl) phthalate
Butyl benzyl phthalate
Di-*n*-butyl phthalate
Di-*n*-octyl phthalate
Diethyl phthalate
Dimethyl phthalate
1,2-Benzanthracene (benzo(a)anthracene)
Benzo(a)pyrene (3,4-benzopyrene)
3,4-Benzofluoranthene (benzo(b)fluoranthene)
11,12-Benzofluoranthene (benzo(k)fluoranthene)
Chrysene
Acenaphthylene
Anthracene
1,12-Benzoperylene (benzo(ghi)perylene)
Fluorene
Phenanthrene
1,2,5,6-Dibenzanthracene (dibenzo(a,h)anthracene)
Indeno(1,2,3-cd) pyrene (2,3-o-phenylene pyrene)
Pyrene
Tetrachloroethylene
Toluene
Trichloroethylene
Vinyl chloride (chloroethylene)
Aldrin
Dieldrin
Chlordane (technical mixture and metabolites)
4,4-DDT
4,4-DDE (*p,p*-DDX)
4,4-DDD (*p,p*-TDE)
Alpha-endosulfan
Beta-endosulfan
Endosulfan sulfate
Endrin
Endrin aldehyde
Heptachlor
Heptachlor epoxide
BHC (hexachlorocyclohexane)
Alpha-BHC
Beta-BHC
Gamma-BHC
Delta-BHC
PCB (polychlorinated biphenyl)
PCB-1242 (Arochlor 1242)
PCB-1254 (Arochlor 1254)
PCB-1221 (Arochlor 1221)
PCB-1232 (Arochlor 1232)
PCB-1248 (Arochlor 1248)
PCB-1260 (Arochlor 1260)
PCB-1016 (Arochlor 1016)
Toxaphene
2,3,7,8-Tetrachlorodibenzo-*p*-dioxin

REFERENCES

1. Federal Register. Resource Conservation and Recovery Act (RCRA), 42 U.S. Code s/s 6901 et seq. 1976, U.S. Government, Public Laws, https://www.epa.gov/history/epa-history-resource-conservation-and-recovery-act, Accessed, 2017.
2. USEPA. Resource Conservation and Recovery Act (RCRA)—Orientation manual, U.S. Environmental Protection Agency, report # EPA530-R-02-016, Washington, DC, January (2003).
3. USEPA. Federal Hazardous and Solid Wastes Amendments (HSWA), U.S. Environmental Protection Agency, Washington, DC, November 1984, https://www.congress.gov/bill/98th-congress/house-bill/2867, Accessed April 2017.
4. Federal Register. Clean Water Act (CWA), 33 U.S.C. ss/1251 et seq. 1977, U.S. Government, Public Laws, https://www.epa.gov/laws-regulations/summary-clean-water-act, Accessed April, 2017.
5. USEPA. Meeting hazardous waste requirements for metal finishers, report # EPA/625/4-87/018, U.S. Environmental Protection Agency, Cincinnati, OH (1987).
6. USEPA. Development document for effluent limitations guidelines and standards for the metal finishing point source category, report # EPA-440/1-80/091, U.S. Environmental Protection Agency, Washington, DC (1980).
7. USEPA. *Treatability Manual, Volume II. Industrial Descriptions*, report # EPA-600/2-82-001b, U.S. Environmental Protection Agency, Washington, DC, September (1981).
8. PRC Environmental Management, Inc. *Hazardous Waste Reduction in the Metal Finishing Industry*. Noyes Data Corporation, Park Ridge, NJ (1989).
9. US ACE. Yearly average cost index for utilities. In: *Civil Works Construction Cost Index System Manual*, 1110-2-1304, U.S. Army Corps of Engineers, Washington, DC, www.nww.usace.army.mil/Missions/CostEngineering.aspx (2016).
10. Patterson, J. W. *Industrial Wastewater Treatment Technology*, 2nd Edition. Butterworths (1985).
11. Wang, L. K., Vaccari, D. A., Li, Y., and Shammas, N. K. Chemical precipitation. In: *Physicochemical Treatment Processes*, Wang, L. K., Hung, Y. T., and Shammas, N. K. (Eds). Humana Press, Totowa, NJ, pp. 141–198 (2005).
12. Shammas, N. K., Kumar, I. J., and Chang, S. Y. Sedimentation. In: *Physicochemical Treatment Processes*, Wang, L. K., Hung, Y. T., and Shammas, N. K. (Eds). Humana Press, Totowa, NJ, pp. 379–430 (2005).
13. Chen, J. P., Chang, S. Y., Huang, J. Y. C., Baumann, E. R., and Hung, Y. T. Gravity filtration. In: *Physicochemical Treatment Processes*, Wang, L. K., Hung, Y. T., and Shammas, N. K. (Eds). Humana Press, Totowa, NJ, pp. 501–544 (2005).
14. Wang, L. K., Hung, Y. T., and Shammas, N. K. (Eds). *Physicochemical Treatment Processes*. Humana Press, Totowa, NJ, 723 p. (2005).
15. Wang, L. K., Hung, Y. T., and Shammas, N. K. (Eds). *Advanced Physicochemical Treatment Processes*. Humana Press, Totowa, NJ, 690 p. (2006).
16. Chen, J. P., Mou, H., Wang, L. K., and Matsuura, T. Membrane filtration. In: *Advanced Physicochemical Treatment Technologies*, Wang, L. K., Hung, Y. T., and Shammas, N. K. (Eds). Humana Press, Totowa, NJ, pp. 203–260 (2007).
17. Wang, L. K., Hung, Y. T., and Shammas, N. K. (Eds). *Advanced Physicochemical Treatment Technologies*. Humana Press, Totowa, NJ, 710 p. (2007).
18. Wang, L. K., Pereira, N., and Hung, Y. T. (Eds) and Shammas, N. K. (Consulting Editor). *Biological Treatment Processes*. Humana Press, Totowa, NJ (2008).
19. Wang, L. K., Shammas, N. K., and Hung, Y. T. (Eds). *Advanced Biological Treatment Processes*. Humana Press, Totowa, NJ (2008).
20. Wang, L. K., Fahey, E. M., and Wu, Z. Dissolved air flotation. In: *Physicochemical Treatment Processes*, Wang, L. K., Hung, Y. T., and Shammas, N. K. (Eds). Humana Press, Totowa, NJ, pp. 431–500 (2005).
21. USEPA. Managing cyanide in metal finishing. Capsule report # EPA 625/R-99/009, U.S. Environmental Protection Agency, Cincinnati, OH, December (2000).
22. Hartinger, L. *Handbook of Effluent Treatment and Recycling for the Metal Finishing Industry*, 2nd Edition. Finishing Publications Ltd., Warrington, UK (1994).
23. Eilbeck, W. J. and Mattock, G. *Chemical Processes in Wastewater Treatment*. Ellis Horwood Limited, Sussex, UK (1987).
24. Wang, L. K., Shammas, N. K., and Hung, Y. T. (Eds). *Biosolids Treatment Processes*. Humana Press, Totowa, NJ, 820 p. (2007).
25. Singh, I. B., Chaturvedi, K., Singh, D. R., and Yegneswaran, A. H. Thermal stabilization of metal finishing waste with clay. *Environmental Technology*, 26(8), 877–884 (2005).

26. Roy, C. H. *Operation and Maintenance of Surface Finishing Wastewater Treatment Systems.* American Electroplaters and Surface Finishers Society, Orlando, FL (1988).
27. Altmayer, F. *Plating and Surface Finishing, Advice & Council.* AESF, Orlando, FL (1997).
28. Wang, L. K., Li, Y., Shammas, N. K., and Sakellaropoulos, G. P. Drying beds. In: *Biosolids Treatment Processes*, Wang, L. K., Shammas, N. K., and Hung, Y. T. (Eds). Humana Press, Totowa, NJ, pp. 403–430 (2007).
29. Wang, L. K., Kurylko, L., and Wang, M. H. S. Sequencing batch liquid treatment, U.S. Patent No. 5354458, U.S. Patent and Trademark Office, Washington, DC, October (1994).
30. Wang, L. K. and Wang, M. H. S. (Eds). *Handbook of Industrial Waste Treatment.* Marcel Dekker, New York, pp. 127–172 (1992).
31. Wang, L. K., Hung, Y. T., Lo, H. H., and Yapijakis, C. (Eds). *Hazardous Industrial Waste Treatment.* CRC Press, New York, pp. 289–360 (2007).
32. USEPA, Code of Federal Regulations, Metal finishing point source category, Title 40, Volume 27, Part 433, U.S. Environmental Protection Agency, Washington, DC, Revised on July 1, 2003.
33. Forbes, S. Environmental issues in metal finishing and industrial coatings. In: *Metal Processing and Metal Working, Encyclopedia of Occupational Health and Safety*, International Labor Organization, Geneva, Switzerland (2011).
34. Durkee, J. Cleaning times: Examining the past, present, and future of metal finishing. *Metal Finishing*, 108(1), 44–46 (2010).
35. NMFRC. Technical Data Base, The National Metal Finishing Resource Center, http://nmfrc.org (2016).
36. ICT. Industrial Cleaning Technologies Media Guide, *International Paint & Coating Magazine* (2017). https://issue.com/ipcm/docs/lg9_icpm_ict_media_guide_2017_web
37. Wang, L. K. and Li, Y. Sequencing batch reactors. In: *Biological Treatment Processes*. Wang, L. K., Pereira, N. C., Hung, Y. T., and Shammas, N. K. (Eds). Humana Press, Totawa, NJ, pp. 459–512 (2009).
38. Shammas, N. K. and Wang, L. K. Safety and control for toxic chemicals and hazardous wastes. In: *Waste Treatment of the Service and Utility Industries*. Hung, Y. T., Wang, L. K., Wang, M. H. S., Shammas, N. K., and Chen, J. P. (Eds). CRC Press, Boca Raton, FL, pp. 1–26 (2017).
39. Chen, J. P., Wang, L. K., Wang, M. H. S., Hung, Y. T., and Shammas, N. K. (Eds). *Remediation of Heavy Metals in the Environment.* CRC Press, Boca Raton, FL, 528 pp. (2017).

11 Environment-Friendly Activated Carbon Processes for Water and Wastewater Treatment

Wei-chi Ying, Wei Zhang, Juan Hu, Liuya Huang, and Wenxin Jiang
East China University of Science and Technology

Bingjing Li
Shanghai LIRI Technologies Co. Ltd

CONTENTS

11.1 Introduction	354
11.2 Development of Cost-Effective AC Water Treatment Processes	355
11.2.1 Understand the Fundamentals of AC Adsorption Technology	355
11.2.2 Employ Efficient Methods for the Adsorption Isotherm and Breakthrough Experiments	356
11.2.3 Perform Cost-Effective Treatability Studies	359
11.2.3.1 Case 1—TCE Removal from Groundwater	359
11.2.3.2 Case 2—MTBE Removal from Groundwater	361
11.2.4 Establish Biological Activated Carbon (BAC) Capability of the Adsorber	363
11.2.4.1 Examples of BAC Capability for Water and Wastewater Treatment	363
11.2.5 Advantage of GAC's Catalytic Capability: Reduction of Free Chlorine in Water	369
11.2.6 Reactivate the Spent Carbon	370
11.3 Conclusions	370
Acknowledgments	370
Acronyms	371
References	371

Abstract

Activated carbon (AC) adsorption treatment for removing water pollutants has many environment-friendly advantages relative to alternative methods. Extensive treatability studies should be conducted for developing a cost-effective AC treatment process to realize such advantages. This chapter presents the simple capacity indicator method of AC selection, the fast microcolumn rapid breakthrough method of performance simulation, the efficient two-bed-in-series series mode of treatment, the lasting biological activated carbon (BAC) and/or catalytic capabilities of the adsorber to enable its long-term service without AC replacement, and the steps to be taken in the studies. The recommended steps are: (1) understand the fundamentals of carbon adsorption technology, (2) employ efficient methods to conduct the adsorption isotherm and breakthrough experiments, (3) perform cost-effective treatability studies to choose the best AC and to define the optimal process scheme, (4) establish BAC capability to enable long-term organic removal without AC replacement, (5) take advantage of the AC's catalytic capability, and (6) reactivate the spent AC.

11.1 INTRODUCTION

Activated carbon (AC) is a powerful adsorbent; it is very effective for removing a wide variety of organic pollutants from aqueous influents. In the U.S. and many other developed countries, AC adsorption technology is often selected for removing persistent organic pollutants (POPs) from the influent to meet stringent effluent limits for direct discharge or reuse [1–3]. However, its application potential is still undeveloped in China and many developing countries for four main reasons: (1) lack of stringent effluent limits on individual POPs, (2) the concept of using either granular AC (GAC) or powder AC (PAC) is not generally understood, (3) simple and effective methods for AC selection and adsorption studies are not commonly available [4–6], and (4) commercial GAC reactivation service is unknown to most users.

PAC can be readily introduced to the existing water/wastewater treatment plant system to achieve immediate pollutant removal [6]; it can be employed quickly to improve the quality of influent to the water treatment plant and to solve the environmental problems due to accidental release of chemicals [7]. Relative to the biodegradation and chemical destruction methods, GAC water and wastewater treatment processes have five environment-friendly advantages: (1) capable of removing the pollutants totally to meet the stringent discharge limits [8], (2) taking up less space because additional unit of the high-loading GAC adsorber can be added in the future when there is a need [9], (3) produce no residual treatment chemicals or toxic intermediates, (4) less costly to construct and to operate, and (5) extended service likely due to the possible biological activated carbon (BAC) capability [10], catalytic reactions (such as free chlorine removal), and spent GAC regeneration. Table 11.1 summarizes the environment-friendly advantages of AC treatment processes.

Extensive treatability studies should be conducted for developing a cost-effective AC-based treatment process to realize such advantages. This chapter presents the simple capacity indicator method of AC selection, the fast micro column rapid breakthrough (MCRB) method of performance simulation, the efficient two-bed-in-series mode of treatment, the lasting BAC and/or catalytic capabilities of the adsorber to enable its long-term service without AC replacement, and the recommended steps for the treatability studies.

TABLE 11.1
Environment-Friendly Advantages of Activated Carbon Water Treatment Processes

- PAC treatment can be practiced in the existing water treatment system.
- GAC treatment can meet low effluent limits such as <1 ppb for some toxic pollutants.
- Less space requirement because of modular design and higher organic loading.
- Less costly to construct and to operate due to less consumption of nutrients and DO.
- No residual chemicals or toxic intermediates.
- Long-term service resulting from BAC's catalytic capabilities and regeneration.

Source: ANSI/AWWA, AWWA standard for powdered activated carbon, B600-96, Denver, CO, 1997; D3860-98, Standard practice for determination of adsorptive capacity of activated carbon by aqueous phase isotherm technique, ASTM Standards, 839–842, 2000; Ying, W., et al., *Environ. Prog.*, 9(1), 1–9, 1990; U.S. EPA, *Process Design Manual for Carbon Adsorbers*, 1973; Ying, W. et al., *Proceedings of 49th Purdue Industrial Waste Water Conference*, Lewis Publisher, 1994.

11.2 DEVELOPMENT OF COST-EFFECTIVE AC WATER TREATMENT PROCESSES

11.2.1 Understand the Fundamentals of AC Adsorption Technology

AC is employed for aqueous phase adsorption applications, either as PAC in a mixed reactor or as GAC in an adsorber, to remove a wide variety of organic water pollutants. The fundamentals and applicability of the two adsorption treatment processes are quite different; a certain dose (g/L) of PAC is introduced to a mixed reactor to achieve the desired degree of pollutant removal, while a well-designed GAC adsorber can produce an effluent meeting very stringent limits for specific pollutants.

For all potential applications of AC-based treatment processes, a series of adsorption isotherm experiments should be conducted for several candidate ACs in the preliminary study (also known as a feasibility study); the resulting adsorptive capacities and the final residual (equilibrium) concentrations are correlated by the Freundlich adsorption isotherm model to allow selection of a few high-potential ACs to be evaluated in the treatability study. Table 11.2 presents calculation methods for the important adsorption study parameters; Table 11.3 presents two cases of adsorption treatment to illustrate methods for estimating the minimum AC requirements to accomplish the treatment

TABLE 11.2
Equations of Adsorption Study Parameters

Parameter	Equations
Equilibrium adsorptive capacity	$X/M \, (\text{mg/g}) = (C_0 - C_f) \times V / \text{carbon dose (g)}$ C_0 and C_f: initial and final concentration (mg/L), V: sample volume (L)
Organic loading of GAC in the adsorber	$Q \, (\text{mg/g}) = C_{in} \times V \times f / \text{total carbon in the adsorber (g)}$ C_{in}: influent concentration (mg/L), V: total volume treated (L), f: average removal rate of the breakthrough experiment
PAC dose requirement of the mixed reactor	$P_C \, (\text{g/L}) = (C_{in} - C_{eff}) / (X/M \text{ at } C_{eff})$ C_{in} and C_{eff}: influent and effluent concentration (mg/L), X/M at C_{eff}: adsorptive capacity at C_{eff} (mg/g)
GAC exhaustion rate of the adsorber	$G_C \, (\text{g/L}) = C_{in} / X/M \text{ at } C_{in}$ X/M at C_{in}: adsorptive capacity at C_{in} (mg/g)

Source: Ying, W., et al., *Environ. Pollut. Control (Chinese)*, 27(6), 430–439, 2005.

TABLE 11.3
Two Cases of PAC and GAC Requirement Calculations

Case 1: Reduce organic content of the source water from COD_{Mn} of 5.2–4.0 mg/L; the Freundlich isotherm of the adsorptive capacity study is: $X/M \, (\text{mg/g}) = 0.894 \, C_e^{1.356}$

  A. PAC dose requirement of the mixed reactor:
$$P_C \, (\text{g/L}) = (C_{in} - C_{eff}) / (X/M \text{ at } C_{eff}) = (5.2 - 4.0) / (0.894 \times 4.0^{1.356}) = 0.200$$

  B. GAC exhaustion rate of the adsorber:
$$G_C \, (\text{g/L}) = C_{in} / X/M \text{ at } C_{in} = 5.2 / (0.894 \times 5.2^{1.356}) = 0.622$$

Case 2: Remove a water pollutant from 100 to 1 mg/L, given $X/M \, (\text{mg/g}) = 13.0 \, C_f^{0.355}$

$$P_C \, (\text{g/L}) = (C_{in} - C_{eff}) / (X/M \text{ at } C_{eff}) = (100 - 1) / 13.0 \times 1^{0.355} = 7.62$$

$$G_C \, (\text{g/L}) = C_{in} / X/M \text{ at } C_{in} = 100 / 13.0 \times 100^{0.355} = 1.50$$

Source: Ying, W., et al., *Environ. Pollut. Control (Chinese)*, 27(6), 430–439, 2005.

objectives [7,8]. It is clear that PAC treatment is more efficient when only a fraction of the influent's organic constituents are to be removed while GAC treatment has the potential of removing them totally in a cost-effective manner. Indeed, GAC treatment process has been successfully employed for meeting such effluent limits of <1 ppb for some toxic pollutants [9,10].

Figure 11.1 presents the effects of breakthrough curve pattern and effluent limit on capacity utilization rate of GAC and Figure 11.2 shows GAC columns of three sizes (micro, mini, and small) that can be employed in the lab to obtain the desired breakthrough curves [10,11]. For early breakthrough (Curves C and D) and/or situations requiring a high degree of target compound removal, two-adsorbers-in-series mode of treatment is often the process of choice since GAC in the leading adsorber can be heavily loaded to achieve a high utilization rate (C_1 is higher than the maximum allowable effluent concentration for a large single adsorber), while the polishing treatment of the second adsorber ensures a high-quality (C_2) effluent [11,12].

11.2.2 Employ Efficient Methods for the Adsorption Isotherm and Breakthrough Experiments

The AC's adsorptive capacity for a specific water pollutant is dependent on both the surface chemistry and its pore structure (i.e., surface area and volume vs. pore size distribution), which are governed by the AC's raw material and activation method. Because of the high cost and long time required to obtain the desired pore size distribution data (such as Figure 11.3) several adsorptive capacity indicators have been proposed to identify a few high-potential ACs for a specific application [4,12,13].

We have proposed the four-indicator method of AC selection consisting of phenol, iodine, methylene blue, and tannic acid numbers to cover water pollutants of different sizes; a high phenol number indicates that the AC has a large internal surface area of very small micropores (diameter <1 nm) and a low surface acidity which enhances the adsorption of polar organic compounds, while high values of iodine number, methylene blue number, and tannic acid number indicate abundant

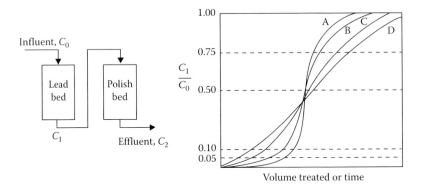

FIGURE 11.1 Effects of breakthrough pattern and effluent limit on capacity utilization rate. (From Ying, W., et al., *Environ. Pollut. Control (Chinese)*, 27(6), 430–439, 2005.)

Environment-Friendly Activated Carbon Processes for Water and Wastewater Treatment 357

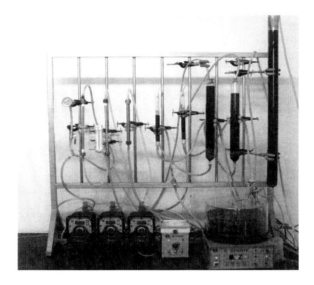

FIGURE 11.2 Apparatus for adsorption breakthrough experiments; *l to r*: micro (5, 5, 8 mm ID), mini (15, 15 mm), small (36, 36 mm) columns; lower: piston and tubing pump, auto sampler. (From Ying, W., et al., *Environ. Pollut. Control (Chinese)*, 27(6), 430–439, 2005.)

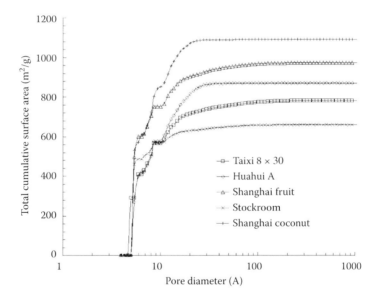

FIGURE 11.3 Cumulative surface area vs. pore diameter for five GACs. (From Ying, W., et al., *Environ. Pollut. Control (Chinese)*, 27(6), 430–439, 2005.)

small micropores (diameter <1.5 nm), large micropores (diameter = 1.5–2.8 nm), and larger pores (diameter >2.8 nm), respectively [12–14]; definitions and applications of the four carbon adsorptive capacity indicators are summarized in Table 11.4. A collection of the capacity indicators of popular commercial ACs is a useful database for selecting a few candidates to be evaluated in the treatability study. This method has been successfully employed to select high-potential GACs for removing a wide variety of water pollutants, such as phenol, dichlorophenols, nitrobenzene, gasoline ingredients [benzene, toluene, ethylbenzene, xylene (BTEX) and methyl-tert-butyl ether (MTBE)], red dye X3B, chloroform, trichloroethylene (TCE), and biphenyl [12,13,15–17].

TABLE 11.4
Definition and Application of the Four Carbon Adsorptive Capacity Indicators

Indicator	Definition	Application	Note
Phenol number	The amount (mg) of phenol adsorbed by 1 g of carbon in a solution having an equilibrium phenol concentration of 20 mg/L	Capacity for small compounds	180/phenol value[a]
Iodine number	The amount (mg) of iodine adsorbed by 1 g of carbon in a solution having an equilibrium iodine concentration of 0.02 N.	Total surface area; capacity for small–medium compounds	Initial concentration = 0.1 N
Methylene blue number	The amount (mg) of methylene blue adsorbed by 1 g of carbon in a solution having an equilibrium methylene blue concentration of 1 mg/L.	Capacity for medium–large compounds	Initial concentration = 20,000 mg/L
Tannic acid number	The amount (mg) of tannic acid adsorbed by 1 g of carbon in a solution having an equilibrium tannic acid concentration of 2 mg/L.	Capacity for large compounds	18,000/tannin value[b]

Source: Ying, W., et al., *Environ. Pollut. Control (Chinese)*, 27(6), 430–439, 2005.
[a] Phenol value: The carbon dose (g/L) required to reduce phenol concentration from 200 to 20 mg/L [6].
[b] Tannin value: The carbon dose (mg/L) required to reduce tannic acid concentration from 20 to 2 mg/L [6].

To conclude the essential breakthrough experiment in a small fraction (1%–5%) of time by using small or mini conventional carbon columns in an ordinary laboratory, we have modified the Calgon's HPMC [18] and U.S. Environmental Protection Agency's RSSCT [19,20] methods with the use of less costly test equipment and simplified procedure to develop a more efficient MCRB method. The validity and the benefits of this improved method have been well demonstrated by the results of many isotherm and breakthrough experiments conducted for the four indicator compounds (Figure 11.4 for phenol) and many typical water pollutants [12,21,22].

Depending on the nature of the pollutants, sample availability, and type and scale of the study, the most effective breakthrough experiments may employ small (>100 g) and mini (>5 g) conventional or MCRB (<2 g) columns (Table 11.5) to select the best GAC, to verify the treatment effectiveness, and to estimate the adsorption treatment cost based on the observed capacity utilization rate of the adsorber.

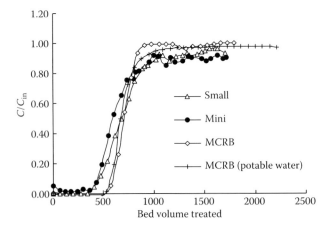

FIGURE 11.4 Phenol breakthrough curves for four GAC columns of three sizes (C_{in} = 100 mg/L). (From Chang, Q., et al., *Environ. Prog.*, 26(3), 280–288, 2007.)

TABLE 11.5
GAC Columns for Adsorption Breakthrough Experiments

Column Size	EBCT	GAC Amount	GAC Size	Pressure
MCRB	3–30 s	<2 g	100–200 mesh	2–4 atm
Mini conventional	5–100 min	>5 g	As shipped	Normal
Small conventional	5–100 min	>100 g	As shipped	Normal

Source: Chang, Q., et al., *Environ. Prog.*, 26(3), 280–288, 2007.

11.2.3 Perform Cost-Effective Treatability Studies

11.2.3.1 Case 1—TCE Removal from Groundwater

An efficient treatability study to define the best treatment process scheme for removing the target pollutant(s) from an influent stream involves two steps: (1) conduct adsorption isotherm experiments or employ a table of the four capacity indicators to identify few high-potential GACs and (2) conduct the necessary breakthrough experiments to obtain a valid breakthrough curve for each of the GACs under conditions simulating the full-scale treatment of the influent stream.

A case of employing four Chinese GACs of different raw materials (coal, coconut shell, fruit nut/shell, and bamboo of Shanghai Activated Carbon Company) and three popular U.S. GACs (coconut-based Norit GCN830 and coal-based Calgon F300 and Calgon F400) to conduct the efficient GAC adsorption treatability study for removing TCE is presented to illustrate the procedures of selecting the best GAC and defining the optimal treatment scheme [22]. Table 11.6 presents the four capacity indicators of the seven and two new domestic bamboo GACs (JHBG1 and JHBG2), and Figure 11.5 presents their capacities for TCE in well water. Figure 11.6 presents the MCRB curves of six GACs in removing TCE from tap water (solid lines) and well water (broken lines); Table 11.7 summarizes the TCE removal data of the MCRB runs. The following can be deduced from the experimental data:

1. The poor mass transfer condition of the filled bottles employed for the isotherm experiments required a long batch contact time of 24 h to measure the adsorptive capacity of GAC for TCE.

TABLE 11.6
Values of the Four Adsorptive Capacity Indicators of Nine GAC Samples

GAC Sample	Phenol Number	Iodine Number	Methylene Blue Number	Tannic Acid Number
Norit coconut GCN830	116	1204	279	12.7
Coconut	112	999	164	13.2
Fruit	102	1033	307	61.4
Calgon F400	100	1175	279	26.5
Calgon F300	98.0	1125	281	29.6
Bamboo	95.0	1163	237	20.0
Bamboo JHBG1	112	984	277	20.0
Bamboo JHBG2	107	1138	290	24.0
Coal	62.5	998	262	31.7

Source: Huang, L., et al., *AIChE J.* 57(2), 542–550, 2011; Hu, J., Biological activated carbon process for removing MTBE from groundwater, PhD dissertation, East China University of Science and Technology, 2010.

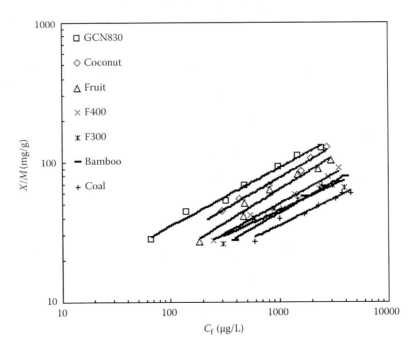

FIGURE 11.5 Adsorption isotherms of TCE in well water. (From Huang, L., et al., *AIChE J.*, 57(2), 542–550, 2011.)

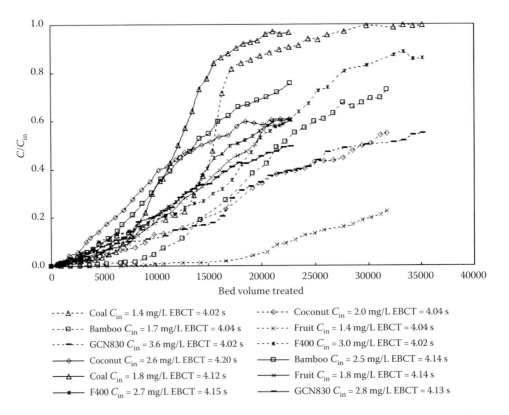

FIGURE 11.6 Adsorption breakthrough curves of TCE in tap water and well water. (From Huang, L., et al., *AIChE J.*, 57(2), 542–550, 2011.)

TABLE 11.7
Summary of MCRB Runs for Removing TCE from Tap Water and Well Water Influents

MCRB Columns	C_{in} (mg/L)	EBCT (s)	TCE Removed[a] (mg) $C/C_{in} = 0.05$	TCE Removed[a] (mg) $C/C_{in} = 0.50$	Adsorptive Capacity[b] (mg)	Capacity Utilization[c] (%) $C/C_{in} = 0.05$	Capacity Utilization[c] (%) $C/C_{in} = 0.50$
Coconut[d]	2.6	4.20	4.6	23.8	38.4	12.0	62.0
Bamboo[d]	2.5	4.14	7.0	23.1	28.0	25.1	82.6
Coal[d]	1.8	4.12	8.5	16.1	19.6	43.3	82.2
Fruit[d]	1.8	4.14	8.3	25.7	29.9	27.6	86.0
F400[d]	2.7	4.15	9.5	29.0	33.4	28.4	86.9
GCN830[d]	2.8	4.13	7.7	37.8	57.1	13.5	68.4
Coconut[e]	2.0	4.04	6.9	32.0	53.6	12.9	59.7
Bamboo[e]	1.7	4.04	12.8	23.9	28.3	45.5	84.6
Coal[e]	1.4	4.02	4.1	18.4	23.5	16.6	78.4
Fruit[e]	1.4	4.04	19.2	>36.7[f]	40.8	47.1	>90.0[f]
F400[e]	3.0	4.02	5.3	32.0	38.1	13.8	83.9
GCN830[e]	3.6	4.02	8.6	49.8	78.6	10.9	63.4

Source: Huang, L., et al., *AIChE J.* 57(2), 542–550, 2011.
[a] Amount removed = Total amount supplied × (area above the breakthrough curve/total area).
[b] Estimated from the 24-h adsorption isotherms.
[c] Capacity utilization (%) = (TCE removed (mg)/adsorptive capacity (mg)) × 100%.
[d] Tap water runs.
[e] Well water runs.
[f] Projected based on the available breakthrough curve.

2. The GACs' adsorptive capacities for TCE were in the same order as their phenol numbers, and their capacity utilization rates (availability) were indicated by their tannic acid numbers.
3. GACs' capacities for TCE in pure water were reduced by competitive adsorption of other organic constituents of the sample; small organic compounds present in tap water were more competitive than the NOM in the well water sample containing higher total organic carbon (TOC).
4. The MCRB data confirmed the GACs' available TCE capacities and that the serial bed mode of treatment is essential for the cost-effective GAC adsorption process because most of the capacity can be utilized for removal of the pollutant with the high allowable effluent concentration of the first bed.
5. The GAC made from toxicant-free, low-cost, and renewable bamboo is of great importance because of its relatively high TCE capacity in actual adsorption treatment.

11.2.3.2 Case 2—MTBE Removal from Groundwater

A similar experimental program was conducted to determine the feasibility of MTBE removal by GAC adsorption To evaluate the relative effectiveness of the environment-friendly bamboo-based GACs, which are produced from a renewable natural resource in a less pollution-intensive process, JHBG1 and JHBG2 of Jinhu Carbon Co. were included in the test program replacing Fruit and F400 of the TCE study, respectively. The MTBE isotherms of Figure 11.7 show that the two new bamboo GACs had higher adsorptive capacities than the others. The MCRB curves of Figure 11.8 provide the simulated treatment data confirming the outstanding performance of the new bamboo GACs, especially JHBG1. It is reasonable to assume that with further R&D to improve the physical properties of such bamboo GACs, they will likely be the most desirable GACs for many water and wastewater treatment applications.

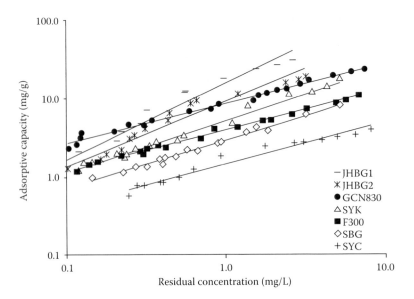

FIGURE 11.7 Adsorption isotherms of MTBE in pure water. (From Hu, J., Biological activated carbon process for removing MTBE from groundwater, PhD dissertation, East China University of Science and Technology, 2010.)

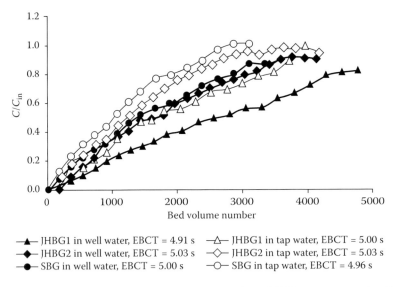

FIGURE 11.8 MCRB curves of three bamboo GACs for removing MTBE in tap water and well water. (From Hu, J., Biological activated carbon process for removing MTBE from groundwater, PhD dissertation, East China University of Science and Technology, 2010.)

Since MTBE is not well adsorbed, using GAC to remove MTBE by adsorption alone will be too costly except for cleaning up a small body of newly contaminated groundwater having a relatively high MTBE concentration. In actual adsorption remediation of an MTBE-contaminated groundwater, the MTBE-degrading soil bacteria retained in the adsorber may become more acclimated and grow actively there turning the adsorber to a long-lasting BAC system after several months of operation [24]. The adsorber's BAC capability will make the GAC treatment much more cost effective when MTBE is removed by adsorption only [25,26].

11.2.4 Establish Biological Activated Carbon (BAC) Capability of the Adsorber

The BAC process was promoted in the early 1970s as a lower cost alternative to such biological treatment process as the activated sludge process for treating domestic wastewater [25]; however, it was not well accepted because of many operational problems associated with the large amount of biomass produced in the adsorber. Although it did not achieve the original objective as a viable secondary wastewater treatment alternative, it has found many beneficial applications for removing POPs of biotreated effluents or organic contaminants of groundwater because the amount of biomass produced is small and that most of the adsorbed POPs are eventually consumed by the acclimated bacteria in the adsorber, which can thus produce a high-quality effluent for a very long time without the periodic replacement of the spent GAC of the conventional adsorption process [10,24].

11.2.4.1 Examples of BAC Capability for Water and Wastewater Treatment

11.2.4.1.1 Case 1—Removing Sulfolane from Groundwater at a Pesticide Plant in California

In the mid-1990s, the groundwater remediation system at the former Occidental Chemical plant (Lathrop, California) employing two pulse-bed 20-ton GAC adsorbers for removing ethylene dibromide and dibromochloropropane was mandated to also remove 1–2 ppm of sulfolane ($C_4H_8O_2S$). Using the sulfolane-degrading bacteria originated from a Shell Chemical wastewater treatment plant, a lab study was conducted to determine the feasibility of adding the BAC capability to the existing GAC adsorbers to accomplish the desired sulfolane removal. Figure 11.9 presents the experimental results demonstrating the long-term sulfolane removal capability of the three small BAC columns (30 g of Calgon F400 in each) following the inoculation of the lab-acclimated bacteria. The successful laboratory-scale BAC treatment performance was verified next year in three inoculated pilot-scale BAC columns (4.5 kg) at the Lathrop site. The two large adsorbers were inoculated with the field-grown acclimated bacteria in early October 1992; Figure 11.10 shows that sulfolane in the feed was mostly removed in the two adsorbers shortly after the inoculation [10]. This case illustrates

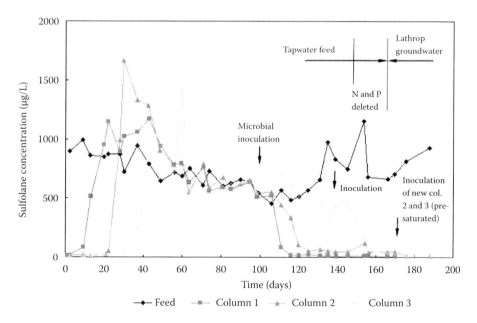

FIGURE 11.9 GAC/BAC performance of three small columns (C_{in} = 1 mg/L, EBCT till Day 100 = 10, 22, and 36 min in Col. 1, 2, and 3, respectively, and then all 22 min). (From Ying, W., et al., *Proceedings of 49th Purdue Industrial Waste Water Conference*, Lewis Publisher, 1994.)

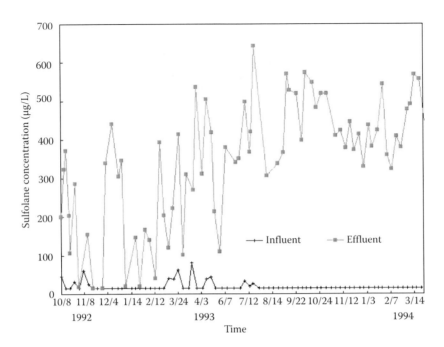

FIGURE 11.10 BAC performance data of sulfolane removal from Lathrop groundwater. (From Ying, W., et al., *Proceedings of 49th Purdue Industrial Waste Water Conference*, Lewis Publisher, 1994.)

the successful full-scale inoculation of mass-produced effective bacteria to establish the BAC capability in the existing adsorbers to accomplish the desired sulfolane removal; therefore, a difficult environmental problem was solved at virtually no extra cost.

11.2.4.1.2 Case 2—Removing MTBE

MTBE is a common gasoline additive; it has become a groundwater pollutant in many countries. GAC adsorption treatment is not cost effective for removing MTBE because it is not well adsorbed. Acclimated MTBE degraders actively growing in the GAC adsorber will make it a BAC system capable of long-term MTBE removal without periodic GAC replacement [24,26]. The MTBE degraders originally present in the spent GAC sample from a BAC system of a southern California MTBE remediation site were successfully inoculated to the new GAC filled on top of the spent GAC in three small columns making them effective BAC systems for long-term (>6 months) removal of MTBE (Figure 11.11). As an evidence of the BAC's lasting treatment capability, each of the five BAC systems removed far more MTBE than the respective adsorptive capacity of their GAC (Table 11.8) [27]. In the subsequent study, several small BAC systems started up employing more effective MTBE degraders (obtained by augmenting the spent GAC bacteria with lab-acclimated soil-based MTBE degraders) that produced stable effluents containing <5 µg/L of MTBE under a variety of treatment conditions (Figure 11.12) [28,29]. Such results have demonstrated that a well-designed and properly operated BAC system is likely to be the least costly option for removing MTBE from groundwater.

11.2.4.1.3 Case 3—Removing BTEX

BTEX are common groundwater pollutants. BAC treatment is attractive since BTEX are both well adsorbed on GAC and easily biodegraded in the adsorber. Results of our recent BAC research have demonstrated the very high BTEX removal capability of the small BAC columns. Figure 11.13 presents results showing that benzene broke through the small column (5 g of coal GAC) on Day 11 as

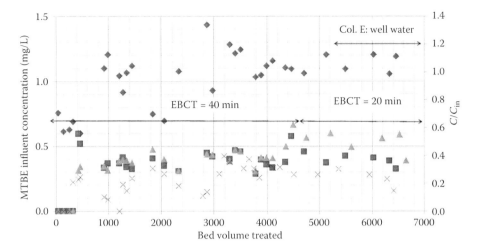

FIGURE 11.11 GAC/BAC treatment performance for removing MTBE (Table 11.8). (From Li, B., Removal of MBTE by adsorption and biodegradation in granular activated carbon columns, PhD dissertation, East China University of Science and Technology, 2010.)

TABLE 11.8
Summary of BAC Treatment Runs for Removing Low Concentration of MTBE in Influent

	Spent (%)	ID (cm)	Weight (g)	Length (cm)	Volume (mL)	Flow Rate (mL/min)	EBCT (min)	MTBE Capacity[a] (mg)	MTBE Removed (mg)
						0.50	249.2		85.7
A*	100	2.52	62.9	25	124.6	0.91	136.3	49.1	36.6
						1.74	71.6		16.5
			62.9	25	124.6	0.50	249.2	49.1	15.3
				10	50.6	0.50	101.2		57.4
B*	100	2.52						19.9	
			25.27	10	50.6	1.00	50.7		36.1
				10	50.6	1.94	26.0		16.8
						0.51	40.0		57.4
C*	90	1.8	10.15	8	20.3			9.33	
						0.85	23.9		4.70
						0.53	38.1		56.8
D*	70	1.8	10.15	8	20.3			11.9	
						0.92	22.1		3.88
						0.51	40.0		68.8
E*	50	1.8	10.15	8	20.3			14.5	
						0.92	22.0		6.04

Source: Li, B., Removal of MBTE by adsorption and biodegradation in granular activated carbon columns, PhD dissertation, East China University of Science and Technology, 2010.

[a] Theoretical adsorptive capacities of GAC in the column were calculated from the Freundlich isotherms of the GACs.

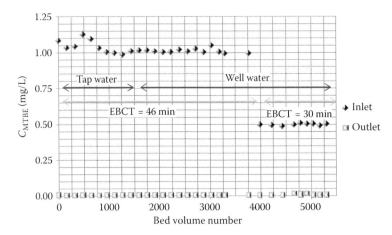

FIGURE 11.12 Outstanding performance of BAC system in removing MTBE. (From Hu, J., Biological activated carbon process for removing MTBE from groundwater, PhD dissertation, East China University of Science and Technology, 2010.)

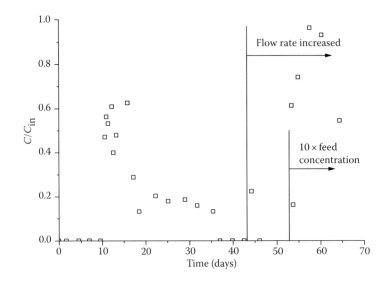

FIGURE 11.13 GAC/BAC treatment performance for removing benzene (EBCT = 3.2–1.6 min, C_{in} = 6–60 mg/L). (From Zhang, W., et al., Biological activated carbon treatment for removing BTEX from groundwater. *Journal of Environmental Engineering*, Vol. 139, No. 10, 1246–1254, 2012.)

expected (according to the benzene isotherm data) and that its concentration in the effluent began to decline on Day 17 as more benzene was biodegraded in the column which became a BAC by Day 20; the 100% increase in the flow rate on Day 43 resulted in a slight increase in the effluent benzene concentration for a few days, while the 10-fold increase in the benzene loading on Day 53 resulted in a longer duration before the subsequent decline. The accumulation of active benzene degraders in the column caused >200% volume expansion of the carbon section from 9.7 mL initially to 24.8 mL. The outstanding performance of this column and three others employed for removing toluene, ethylbenzene, and xylene has demonstrated the lasting treatment capability of the BAC systems in removing BTEX from the contaminated groundwater. Relative to the coconut GAC, the coal-based GAC was more desirable for the BAC treatment because of the lower cost and the higher degree of bioregeneration due to its more abundant large pores [30].

Table 11.9 summarizes the performance of another series of rapid treatment runs (EBCT = 1.2 min based on initial bed volume) for removing BTEX; the results show that the four small columns (2.5 g of coal GAC) became stable BAC systems under very high organic loadings that were much higher than the conventional biological wastewater treatment processes (37.5–1200 mg TOC/L-days [31]), the high sludge age SBR process (up to 2000 mg TOC/L-days [32]), and the pilot-scale aerobic BAC treatment of a phenolic wastewater (up to 4500 mg TOC/L-days) operating at a much longer EBCT of 12 h [33]. Assuming a very conservative TOC reduction of 30% accomplished in those BAC columns, the calculated dissolved oxygen (DO) consumption rates (Table 11.10) were far less than the theoretical oxygen demands noted earlier [30,34]. The BAC treatment is therefore very cost effective for removing readily biodegradable water pollutants [25,35].

11.2.4.1.4 Case 4—Advanced Treatment of Biotreated Coking Plant Effluent for Recycling

A treatability study was performed for advanced treatment of biotreated coking plant effluent to produce a final effluent of <50 mg/L in COD_{Cr} and <0.5 mg/L in total cyanide (TCN) for recycling at Shanghai Coking plant. Small GAC columns were employed for treating the Fenton-oxidized/precipitated feed samples; after the start-up period (<3 weeks) to establish the effective BAC capability in each column, the effluents became stable and met the treatment objectives [36,37]. Figure 11.14 presents the long-term chemical oxygen demand (COD) and TCN removal capability of two BAC

TABLE 11.9
Summary of Rapid BAC Treatment Runs for Removing BTEX (GAC = 2.5 g, Initial Bed Volume = 4.8 mL, Average Feed Concentration = 22 mg/L, flow rate = 4.0 mL/min)

BTEX in the Feed	BAC Bed Volume[a] (mL)	Amount Fed (mg)	Amount Removed (mg)	Total Capacity[b] (mg)	Organic Loading[c] (mg TOC/L-days)
Benzene	12.4	2,056	1,486	246	10,900
Toluene	11.5	1,531	1,463	1,199	8,680
Ethylbenzene	10.6	1,640	1,537	2,200	9,990
o-Xylene	8.8	1,776	1,719	1,263	12,900

Source: Zhang, W., et al., *J. Environ. Eng.*, 139(10), 1246–1254, 2012.

[a] The expanded bed volume after the BAC capability was established.
[b] Isotherm adsorptive capacity (mg/g) × GAC amount (g).
[c] Amount fed (mg)/[run time (days) × expanded bed volume (L)].

TABLE 11.10
Calculations of DO Consumption in BAC Columns for Removing BTEX

	Benzene	Toluene	Ethylbenzene	o-Xylene
DO reduction[a] (mg/L)	2.3	2.7	2.1	3.3
TOC reduction[a] (mg/L)	7.4	7.5	6.2	6.3
Theoretical DO consumption[b] (mg/L)	12	12	9.9	10
Actual/theoretical DO consumption[c] (%)	19	22	21	33

Source: Zhang, W., et al., *J. Environ. Eng.*, 139(10), 1246–1254, 2012.

[a] Based on the feed and effluent DO/BTEX concentrations and 30% TOC removal (Day 14 of Table 9 runs).
[b] $(Y \times 0.53 + (1 - Y) \times 2.67) \times$ TOC reduction (mg/L). Y is the microbial yield coefficient (0.5 assumed); 0.53 and 2.67 are the mass ratios of oxygen to organic carbon for biomass and complete mineralization, respectively.
[c] DO reduction (mg/L)/theoretical DO consumption (mg/L).

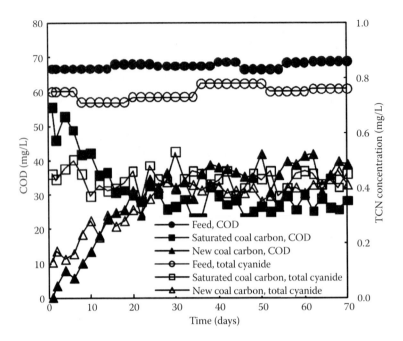

FIGURE 11.14 Performance of two small BAC columns in treating the Fenton-pretreated coking plant effluent. (From Jiang, W., et al., *J. Hazard. Mater.*, 189, 308–314, 2011.)

columns (one started with new coal GAC and the other started with saturated GAC) in treating the Fenton-pretreated biotreated coking plant effluent. After the BAC capability was fully established, the treatment became more effective and the need for the Fenton or other oxidative pretreatment step was reduced or even eliminated [38].

GAC's adsorptive capacity for TCN may be substantially enhanced with preloading of copper and KI treatment [39] and using the Cu/KI-GAC to fill a portion of the column for treating a high-TCN feed sample achieved the effluent objectives even during the start-up period as illustrated in Figure 11.15. Similar to the COD removal, the BAC capability was responsible for much greater amount of TCN removal in each small column than its adsorptive capacity for TCN (Table 11.11).

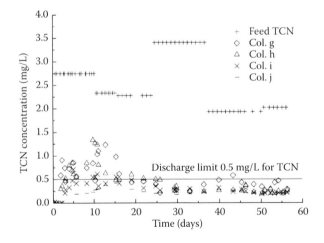

FIGURE 11.15 GAC/BAC treatment performance for removing TCN from Fenton-pretreated coking plant effluent. (From Zhang, W., et al., *J. Hazard. Mater.*, 184, 135–140, 2010.)

TABLE 11.11
Performance of BAC Columns in Treating Fenton-Pretreated Coking Plant Effluent (Feed pH = 6.5, DO = 7 mg/L)

GAC (Figure 15.15 Columns)	Flow Rate (mL/min)	GAC Filled (g)	EBCT (min)	Isotherm Capacity (mg/g)[a]	TCN Removed (mg/g)[b]	Removal Ratio (%)[c]
100% coal (g)	0.94	10.0	26	3.9	14.0	359
10% Cu/KI-GAC +90% coal (h)	1.00	10.2	25	4.1	15.7	383
30% Cu/KI-GAC +70% coal (i)	0.92	10.2	26	4.6	15.3	333
100% Cu/KI-GAC (j)	0.98	10.1	22	6.1	17.1	280

Source: Jiang, W., Enhanced activated carbon adsorption process for advanced treatment of coking plant effluent, PhD dissertation, East China University of Science and Technology, 2008.
[a] 1-h removal capacity at the feed concentration from the batch data.
[b] Cumulative TCN removal at the end.
[c] b/a.

The case study results have demonstrated that the outstanding BAC treatment effectiveness may be further enhanced with the use of higher-capacity commercial or custom GAC.

11.2.5 ADVANTAGE OF GAC'S CATALYTIC CAPABILITY: REDUCTION OF FREE CHLORINE IN WATER

GAC has been commonly employed for removing free chlorine due to its catalytic reduction capability. Most spent GACs from commercial carbon dechlorination systems are still useful for the intended purpose; Figure 11.16 shows that the spent coconut and fruit GACs after 1 year of field service continue to have high capacities for removing free chlorine by adsorption (2 h) and catalytic reduction (difference between 5 and 2 h) [40].

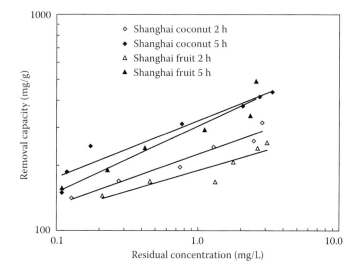

FIGURE 11.16 Free chlorine removal capacities of two GACs after 1 year of service. (From Li, B., et al., *Asian Pacific J. Chem. Eng.*, 5(5), 714–720, 2010.)

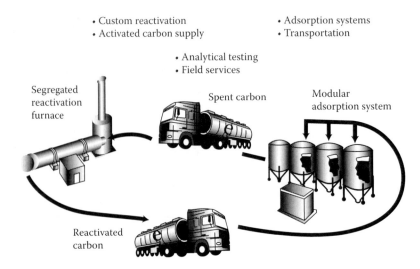

FIGURE 11.17 Typical commercial GAC reactivation service. (From Envirotrol Corp., *Adsorption Service Brochure,* Pittsburgh, PA, 2002.)

11.2.6 REACTIVATE THE SPENT CARBON

Without the benefits of BAC capability and the GAC's catalytic properties, the GAC treatment cost can be significantly reduced if the spent GAC loaded with organic pollutants can be thermally regenerated (reactivated) at a reasonable cost. Commercial spent GAC reactivation service, which includes picking up the spent GAC, reactivating it in a large efficient furnace, and supplying the regenerated GAC for adsorption service, such as illustrated in Figure 11.17 [41], is likely to be the most cost-effective way of carbon regeneration to achieve multiple reuses of GAC.

11.3 CONCLUSIONS

AC adsorption treatment for removing water pollutants has many environment-friendly advantages relative to alternative methods. Extensive treatability studies should be conducted for developing a cost-effective AC treatment process to realize such advantages. This chapter presents the simple capacity indicator method of AC selection, the fast MCRB method of performance simulation, the efficient two-bed-in-series mode of treatment, the lasting BAC and/or catalytic capabilities of the adsorber to enable its long-term service without AC replacement, and the steps to be taken in the studies. The recommended steps are: (1) understand the fundamentals of carbon adsorption technology, (2) employ efficient methods to conduct the adsorption isotherm and breakthrough experiments, (3) perform cost-effective treatability studies to choose the best AC and to define the optimal process scheme, (4) establish BAC capability to enable long-term organic removal capability without AC replacement, (5) take advantage of the AC's catalytic capability, and (6) reactivate the spent AC.

ACKNOWLEDGMENTS

Several research projects covered in this chapter were conducted under the joint support of the Fundamental Research Funds for the Central Universities (WB1014038), Shanghai Natural Science Funds (11ZR409400), and National Natural Science Foundation of China (41201302).

ACRONYMS

AC	Activated carbon
BAC	Biological activated carbon
BTEX	Benzene, toluene, ethylbenzene and xylene
COD	Chemical oxygen demand
DO	Dissolved oxygen
GAC	Granular activated carbon
MCRB	Micro column rapid breakthrough
MTBE	Methyl-tert-butyl ether
PAC	Powder activated carbon
POPs	Persistent organic pollutants
ppb	Parts per billion
R&D	Research and development
TCE	Trichloroethylene
TCN	Total cyanide
TOC	Total organic carbon

REFERENCES

1. Martin, R. (1980) Activated carbon product selection for water and wastewater treatment. *Industrial and Engineering Chemistry Product Research and Development*, Vol. 19, 435–441.
2. Köseoğlu, E.; Akmil-Başar, C. (2016) Preparation, structural evaluation and adsorptive properties of activated carbon from agricultural waste biomass. *Advanced Powder Technology*, Vol. 26, 811–818.
3. Sayğılı, H.; Güzel, F.; Önal, Y. (2015) Conversion of grape industrial processing waste to activated carbon sorbent and its performance in cationic and anionic dyes adsorption. *Journal of Cleaner Production*, Vol. 93, 84–93.
4. Ying, W.; Tucker, M. (1990) Selecting activated carbon for adsorption treatment. In: *Proceedings of 44th Purdue Industrial Waste Conference*, Lewis Publishers Inc., Chelsea, MI, pp. 313–324.
5. Shih, T., Wangpaichitr, M., Suffet, M. (2003) Evaluation of granular activated carbon technology for the removal of MTBE from drinking water. *Water Research*, Vol. 37, 375–385.
6. ANSI/AWWA. (1997) AWWA standard for powdered activated carbon, B600-96. Denver, CO.
7. D3860-98. (2000) Standard practice for determination of adsorptive capacity of activated carbon by aqueous phase isotherm technique, ASTM Standards, pp. 839–842.
8. Ying, W.; Dietz, E.; Woehr, C. (1990) Adsorption capacities of activated carbon for organic constituents of wastewaters. *Environmental Progress*, Vol. 9, No. 1, 1–9.
9. U.S. EPA. (1973) *Process Design Manual for Carbon Adsorbers*.
10. Ying, W.; Hanan, S.; Tucker, M. (1994) Successful application of biological activated carbon process for removing sulfolane from groundwater. In: *Proceedings of 49th Purdue Industrial Waste Water Conference*, Lewis Publisher.
11. Ying, W.; Chang, Q.; Zhang, W.; Jiang, W. (2005) Improved methods for selecting activated carbon for water and wastewater treatment. *Environmental Pollution and Control (Chinese)*, Vol. 27, No. 6, 430–439.
12. Ying, W.; Zhang, W.; Chang, Q. (2006) Improved methods for carbon adsorption studies for water and wastewater treatment. *Environmental Progress*, Vol. 25, No. 2, 110–120.
13. Zhang, W.; Chang, Q.; Liu, W. (2007) Selecting activated carbon for water and wastewater treatment studies. *Environmental Progress*, Vol. 26, No. 3, 289–299.
14. Han, X.; Qiu, Z.; Hu, J.; Lin, L.; Zhang, W.; Ying, W. (2013) Improved methods for measuring activated carbon adsorptive capacity indicators. *Environmental Pollution and Control (Chinese)*, Vol. 35, No. 1, 54–59.
15. Li, B.; Zhang, W.; Liu, W.; Ying, W.; Liu, W. (2008) Selecting of activated carbon for removing THMs and humic acid from water. *Environmental Pollution and Control (Chinese)*, Vol. 30, No. 4, 48–51, 55.

16. Li, Y.; Zhang, W.; Lu, Y.; Lin, K.; Ying, W. (2009) Activated carbon adsorption for removing MTBE and BTEX from Water. *Journal of East China University of Science and Technology (Natural Science Edition) (Chinese)*, Vol. 35, No. 6, 866–871.
17. Zhang, W.; Ying, W.; Chang, Q.; Jiang, W. (2007) Adsorptive capacity indicator based method of carbon selection for treatability. *China Environmental Science (Chinese)*, Vol. 27, No. 3, 289–294.
18. Rosen, M.; Diethorn, R.; Lutchko, J. (1980) High pressure technique for rapid screening of activated carbons for use in potable water. In: *Activated Carbon Adsorption of Organics from the Aqueous Phase, Volume I* (Editors I. Suffet and M. McGuire), Ann Arbor Science, pp. 309–316.
19. Crittenden, J.; Berrigan, J. (1986) Design of rapid small-scale adsorption tests for a constant diffusivity. *Journal (Water Pollution Control Federation)*, Vol. 58, 312–319.
20. D6586-00 (2000) Standard practice for the prediction of contaminant adsorption on GAC in aqueous systems using rapid small-scale column tests, ASTM Standards, pp. 882–886.
21. Chang, Q.; Zhang, W.; Jiang, W. (2007) Efficient micro carbon column rapid breakthrough technique for water and wastewater. *Environmental Progress*, Vol. 26, No. 3, 280–288.
22. Zhang, W.; Chang, Q.; Ying, W.; Xu, B.; Zhang, R.; Ling, L.; Dai, W.; Jiang, J. (2006) New carbon selection method for water treatment applications. *Environmental Pollution and Control (Chinese)*, Vol. 28, No. 8, 499–503.
23. Huang, L.; Hu, J.; Zhang, W.; Yang, Z.; Lin, K.; Ying, W. (2011) Granular activated carbon adsorption process for removing trichloroethylene from groundwater. *AIChE Journal*, Vol. 57, No. 2, 542–550.
24. Sun, P. (2002) The treatment of MTBE and TBA contaminated groundwater in bio-augmented GAC beds. In: *Proceedings of Petroleum Hydrocarbons and Organic Chemical in Ground Water Conference*, Atlanta, GA.
25. Ying, W.; Weber, W. (1979) Bio-physicochemical adsorption systems model. *Journal (Water Pollution Control Federation)*, Vol. 51, 2661–2677.
26. Li, B.; Hu, J.; Huang, L.; Lv, Y.; Zuo, J.; Zhang, W.; Ying, W.; Matsumoto, M. (2012) Removal of MTBE in biological activated carbon adsorbers. *Environmental Progress & Sustainable Energy*, Vol. 32, No. 2, 239–248.
27. Li, B. (2010) Removal of MBTE by adsorption and biodegradation in granular activated carbon columns. PhD dissertation. East China University of Science and Technology.
28. Hu, J. (2010) Biological activated carbon process for removing MTBE from groundwater. PhD dissertation. East China University of Science and Technology.
29. Hu, J.; Huang, L.; Li, B.; Zhang, W.; Ying, W. (2013) Granular activated carbon adsorption process for removing methyl tert-butyl ether from groundwater. *Asian Journal of Chemistry*, Vol. 25, No. 5, 2647–2650.
30. Zhang, W. (2008) Biological activated carbon process for removing BTEX and MTBE from groundwater. PhD dissertation. East China University of Science and Technology.
31. Metcalf & Eddy; Tchobanoglous, G.; Burton, F.; Stensel, H. (2003) Chapter 8. In: *Wastewater Engineering: Treatment and Reuse* (4th Edition), McGraw-Hill, New York.
32. Ying, W.; Bonk, R.; Lloyd, V.; Sojka, S. (1986) Biological treatment of a landfill leachate in sequencing batch reactors. *Environmental Progress*, Vol. 5, No. 1, 41–50.
33. Weber, A.; Lai, M.; Lin, W.; Goeddertz, J.; Ying, W.; Duffy, J. (1992) Anaerobic/Aerobic biological activated carbon treatment of a high strength phenolic wastewater. *Environmental Progress*, Vol. 11, No. 4, 310–331.
34. Ying, W. (1978) Investigation and modeling of bio-physicochemical processes in activated carbon columns. PhD dissertation. University of Michigan.
35. Zhang, W.; Ding, W.; Ying, W. (2012) Biological activated carbon treatment for removing BTEX from groundwater. *Journal of Environmental Engineering*, Vol. 139, No. 10, 1246–1254.
36. Ying, W.; Jiang, W.; Zhang, W.; Chang, Q.; Fu, L.; Zhang, H.; Li, B.; Li. G. (2010) Enhanced carbon adsorption process for recycling coking plant effluent. *International Journal of Environmental Engineering Science*, Vol. 1, 107–123.
37. Jiang, W.; Zhang, W.; Li, B.; Duan, J.; Lv, Y.; Liu, W.; Ying. W. (2011) Combined Fenton oxidation and biological activated carbon process for recycling of coking plant effluent. *Journal of Hazardous Materials*, Vol. 189, 308–314.
38. Jiang, W. (2008) Enhanced activated carbon adsorption process for advanced treatment of coking plant effluent. PhD dissertation. East China University of Science and Technology.
39. Zhang, W.; Liu, W.; Lv, Y.; Li. B.; Ying, W. (2010) Enhanced carbon adsorption treatment for removing cyanide from coking plant effluent. *Journal of Hazardous Material*, Vol. 184, 135–140.
40. Li, B.; Zhang, H.; Zhang, W.; Huang, L.; Duan, J.; Hu, J.; Ying, W. (2010) Cost effective activated carbon treatment process for removing free chlorine from water. *Asian Pacific Journal of Chemical Engineering*, Vol. 5, No. 5, 714–720.
41. Envirotrol Corp. (2002) *Adsorption Service Brochure*, Pittsburgh, PA.

12 Treatment of Wastes from the Organic Chemicals Manufacturing Industry

Debolina Basu
Motilal Nehru National Institute of Technology Allahabad

Sudhir Kumar Gupta
Indian Institute of Technology Bombay

Yung-Tse Hung
Cleveland State University

CONTENTS

12.1 Introduction 374
12.2 Classification of the Organic Chemical Manufacturing Industry 375
12.3 Organic Chemicals Manufacturing Industry and Wastes Generation 375
    12.3.1 Chemical Manufacturing Processes 375
        12.3.1.1 Chemical Reaction Processes 375
        12.3.1.2 Purification of Products 376
        12.3.1.3 Specific Industrial Organic Chemicals 376
12.4 Pollution Prevention in the Organic Chemical Industry 378
    12.4.1 Process Modification 379
12.5 End-of-Pipe Waste Treatment Technologies for Organic Chemicals Manufacturing Industry 381
    12.5.1 Case Example: Waste Treatment in the Organic Dye and Pigment Manufacturing Industry 383
12.6 Federal Regulations for the Organic Manufacturing Industry 385
12.7 Technical Terminologies of Organic Chemicals Manufacturing Industry 385
References 387

Abstract

Organic chemical industries produce a variety of products derived from crude oil, natural gas, or wood. The organic compounds obtained are further used in the manufacture of plastics, synthetic fibers, drugs, surface coatings, solvents, detergents, insecticides, herbicides, explosives, and countless specialty chemicals. Process contaminated water comprises of 10%–20% of the total water intake in the plant, whereas about 80% are being used for cooling purpose and can be reused after providing primary treatment in most cases. The organic chemical manufacturing industries emit chemicals to air (fugitive and direct), water (direct and runoff), and land, many of which may be categorized as hazardous manufacturing wastes. The complexity of the waste generated depends on the feedstock used, process or equipment, storage/handling, and housekeeping practices employed within the plant. This chapter covers the classification of organic chemicals manufacturing industry, manufacturing process description for the various products, sources of emission in the manufacturing plant, suitable pollution prevention measures adopted by the organic chemical manufacturing industry, and available end-of-pipe treatment technology for treatment of the manufacturing effluents generated from such industries.

12.1 INTRODUCTION

The demand to attain improved economy and modern life has created a growing pressure on the chemical industry with respect to increased production of many essential materials like plastics, pharmaceuticals, and agricultural chemicals. The organic chemicals industry plays a pivotal role by converting the basic raw materials into intermediate necessary to create such desired end products. The chemicals industries comprising of the organic and inorganic sector produces about 15,000 chemicals in the U.S. alone, in quantities >10,000 pounds [11]. The industries obtain their raw materials from petroleum and mined products which are converted to intermediate materials or finished chemicals. The chemical industry sector is involved in the manufacturing of various synthetic and plastic materials; drugs, soaps and cleaners; paints and allied products; agricultural chemicals and other miscellaneous chemical products. However, the organic chemical industry has been separately classified into: gum and wood chemicals and cyclic organic crudes, intermediates organic dyes and pigments, and industrial organic chemicals. Plastics, drugs, soaps and detergents, agricultural chemicals or paints, and allied products are typical endproducts manufactured from industrial organic chemicals [9]. Gum and wood chemicals are wood-based products consisting of charcoal, tall oil, rosin, turpentine, pine tar, acetic acid, and methanol. Whereas, the cyclic organic crudes and intermediates processed from petroleum, natural gas, and coal produce benzene, toluene, xylene, and naphthalene [11]. Table 12.1 summarizes examples of the various organic products and their industrial users.

TABLE 12.1
Examples of Major Organic Chemical Products and Their End Uses

Category	Chemicals	End Uses
Aliphatic and other acyclic organic chemicals	Ethylene, butylene, and formaldehyde	Polyethylene plastic, plywood
Solvents	Butyl alcohol, ethyl acetate, ethylene glycol ether, perchloroethylene	Degreasers, dry cleaning fluid
Polyhydric alcohols	Ethylene glycol, sorbitol, synthetic glycerin	Antifreeze, soaps
Synthetic perfume and flavoring materials	Saccharin, citronellal, synthetic vanillin	Food flavoring, cleaning product scents
Rubber processing chemicals	Thiuram, hexamethylene tetramine	Tires, adhesives
Plasticizers	Phosphoric acid, phthalic anhydride, and stearic acid	Rain coats, inflatable toys
Synthetic tanning agents	Naphthalene sulfonic acid condensates	Leather coats and shoes
Chemical warfare gases	Tear gas, phosgene	Military and law enforcement
Esters and/or amines of polyhydric alcohols and fatty and other acids	Allyl alcohol, diallyl maleate	Paints, electrical coatings
Cyclic crudes and intermediates	Benzene, toluene, mixed xylenes, naphthalene	Eyeglasses, foams
Cyclic dyes and organic pigments	Nitro dyes, organic paint pigments	Fabric and plastic coloring
Natural gum and wood chemicals	Methanol, acetic acid, rosin	Latex, adhesives

Source: USEPA, Profile of the organic chemical industry, U.S. Environmental Protection Agency, EPA/310-R-02-001, Cincinnati, OH, 2002.

12.2 CLASSIFICATION OF THE ORGANIC CHEMICAL MANUFACTURING INDUSTRY

The organic chemical industry has been divided into the following three categories [11]:

- Gum and wood chemicals are distilled from wood. The most common products are charcoal, turpentine, pine tar, acetic acid, methanol, etc.
- Cyclic organic crudes and intermediates are processed from petroleum, natural gas, and coal. Important products are benzene, toluene, xylene, and naphthalene. Synthetic dyes and organic pigments manufacturing are also included in this class.
- Industrial organic chemicals not elsewhere classified, e.g., industrial gas manufacturing and ethyl alcohol manufacturing.

12.3 ORGANIC CHEMICALS MANUFACTURING INDUSTRY AND WASTES GENERATION

12.3.1 Chemical Manufacturing Processes

The organic chemicals industry receives raw materials from industries like petroleum refineries and processes them either as finished products or as intermediates requiring further processing by other manufacturers as demonstrated in Figure 12.1. However, the raw materials should be received in sufficient purity. The two major steps in chemical manufacturing are: chemical reaction and purification of reaction products.

12.3.1.1 Chemical Reaction Processes

The two basic types of chemical reactions are batch and continuous. Most reactions take place at high temperatures and involve metal catalysts. The yield of the reaction determines the kind and quantity of by-products and releases. Batch reactions are carried out by adding reactant chemicals to the reaction vessel at the same time and withdrawing the products only after the reaction is complete. The reaction vessel called reactor is made up of stainless steel or glass-lined carbon steel with a size in the range of 50 to several thousand gallons. Batch reactors, also called stirred tank reactors or autoclaves, have appropriate pipes and valves to control the reaction conditions [11]. Batch processes are generally used for small-scale or experimental processes and are easier to operate, maintain, and repair. The advantage of using batch equipment is that it can be adapted to multiple

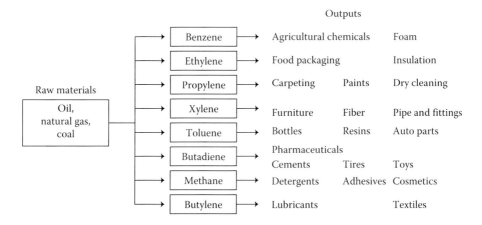

FIGURE 12.1 Common organic chemicals production chains. (From USEPA, Profile of the organic chemical industry, U.S. Environmental Protection Agency, EPA/310-R-02-001, Cincinnati, OH, 2002.)

uses, producing many specialty chemicals. In general, facilities producing less than four million pounds of a particular product per year use a batch process [11]. Toll manufacturing is an important subcategory of the batch process. Many organic chemicals require multi-step manufacturing processes involving precise operating conditions, with specialized equipment and trained employees. Hence, a company outsources one or more steps in the manufacturing process to a contractor, who may send the product to yet another contractor for completing the production process.

Continuous processes occur either in a continuous stirred tank reactor or in a pipe reactor. In this process, the reactants are added and products are removed at a constant rate from the reactor to maintain constant volume of reacting material within the reactor. A continuous stirred tank reactor is quite similar to the batch reactor described above whereas a pipe reactor is a piece of tubing arranged in a coil or helix shape that is jacketed in a heat transfer fluid. Reactants enter one end of the pipe, and the materials mix under the turbulent flow and react as they pass through the system. Pipe reactors are well suited for reactants that do not mix well because the turbulence in the pipes causes all materials to mix thoroughly. Continuous processes require a substantial amount of automation and capital expenditures, and the equipment is dedicated to manufacture of a single product. As a result, this type of process is used primarily for large-scale operations producing >20 million pounds per year of a particular chemical [11]. For facilities producing between 4 and 20 million pounds of a chemical per year, the choice of a batch or continuous process depends on the particular chemical and other site-specific considerations. In some cases, a hybrid reaction process, called a semi-batch reactor, is needed if the reaction is very fast and potentially dangerous. One reactant is placed in the vessel at the beginning of the reaction (like in a batch process) and the other reactant(s) is added gradually [11].

12.3.1.2 Purification of Products

It is quite rare to obtain reaction products in a pure form from a reaction. Often they are mixed with by-products and unreacted inputs. Therefore, the desired product must be isolated and purified in order to be supplied to customers or downstream manufacturers. Common separation methods consist of filtration, distillation, and extraction. Depending on the desired purity and particular mixture, multiple separation methods may also be adopted. Filtration is used to separate solids from liquids. Slurry is passed through a filter which traps the solids and allows liquid to pass through. Another form of filtration is called centrifugation in which the slurry is placed in a porous basket spun rapidly. The centrifugal force pushes the liquid through the filter on the sides of the basket where the fluid is reclaimed. Diffusional processes like distillation separates liquids that have differing boiling points. A mixture of liquids is heated to the boiling point of the most volatile compound that becomes gaseous and condenses back to a liquid form in an attached vessel. The remaining compounds can be similarly isolated from the mixture by increasing the temperature incrementally to the appropriate boiling point. In extraction process, a mixture is added to a fluid such as organic solvents, in which the desired product is insoluble and the undesired materials are soluble. Other separation techniques include absorption of oily-type floating organic materials by foam or by adsorption on solid (i.e., silica, alumina, activated carbon, polymers) and liquid surfaces (bubble adsorption). The final product may be further processed by spray drying or pelletizing to produce the saleable item. Frequently, by-products are also sold and hence their value also alters the process economics.

12.3.1.3 Specific Industrial Organic Chemicals

This section presents the manufacturing steps using three high-volume chemicals namely ethylene, propylene, and benzene, which are the primary building blocks and their reaction products are used to produce other chemicals.

The major uses for ethylene are in the synthesis of polymers (polyethylene) and in ethylene dichloride, a precursor to vinyl chloride. Other important products are ethylene oxide (a precursor to ethylene glycol) and ethylbenzene (a precursor to styrene). Though ethylene itself is not generally considered a health threat, several of its derivatives, such as ethylene oxide and vinyl chloride, have been shown to cause cancer. Figure 12.2 illustrates the manufacturing processes that use ethylene as

Treatment of Wastes

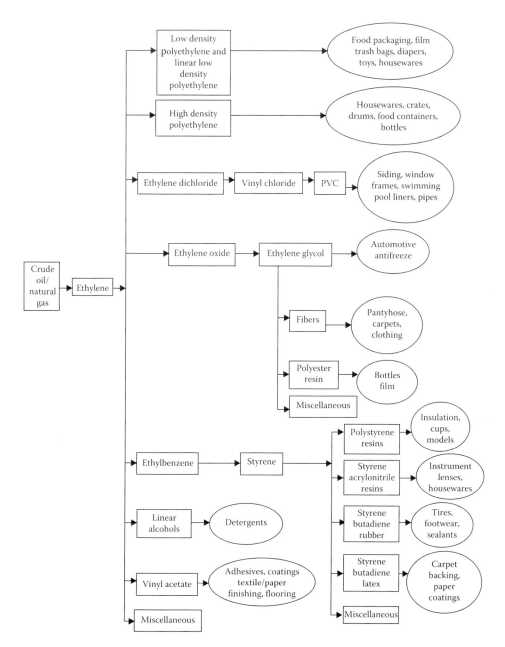

FIGURE 12.2 Flowchart of intermediates from ethylene and major finished products. (From USEPA, Profile of the organic chemical industry, U.S. Environmental Protection Agency, EPA/310-R-02-001, Cincinnati, OH, 2002.)

a feedstock. The primary products of propylene are polypropylene, acrylonitrile, propylene oxide, and isopropyl alcohol. Acrylonitrile and propylene oxide both cause cancer, while propylene alone is not considered a health threat. Benzene is an important intermediate in the manufacture of industrial chemicals like ethyl e itself is not generally considered a health threat. Figure 12.3 shows the major intermediates and finished products associated with benzene, cumene, cyclohexane, nitrobenzene, and various chlorobenzenes. United States Environmental Protection Agency (USEPA) has identified benzene as a human carcinogen. Figure 12.4 summarizes the primary benzene intermediates and products.

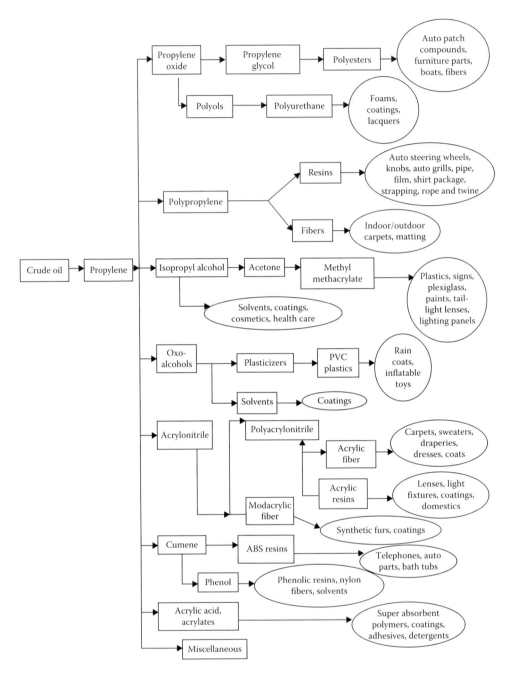

FIGURE 12.3 Intermediates and products obtained from propylene. (From USEPA, Profile of the organic chemical industry, U.S. Environmental Protection Agency, EPA/310-R-02-001, Cincinnati, OH, 2002.)

12.4 POLLUTION PREVENTION IN THE ORGANIC CHEMICAL INDUSTRY

It has been recommended by EPA that the waste loads discharged from the organic chemical industry to receiving streams should be reduced to desirable levels, including no discharge of pollutants, with proper water management by recycle and reuse, in-plant waste control, process modification, and waste treatment systems. Given below are the measures that can be adopted in the organic chemical industry to minimize pollution [11].

Treatment of Wastes

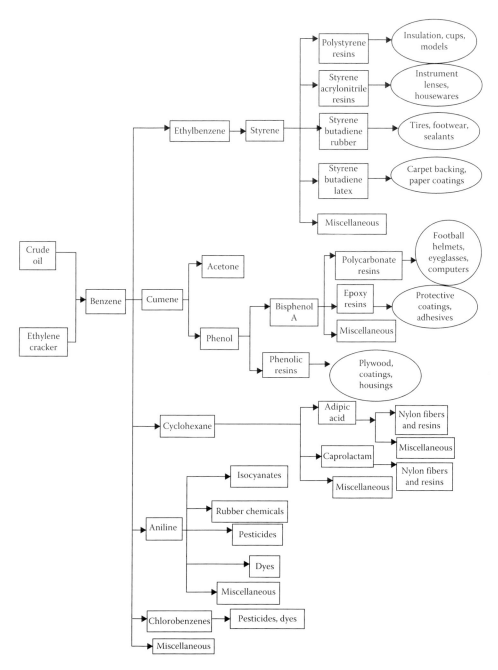

FIGURE 12.4 Intermediates and products obtained from benzene. (From USEPA, Profile of the organic chemical industry, U.S. Environmental Protection Agency, EPA/310-R-02-001, Cincinnati, OH, 2002.)

12.4.1 Process Modification

Catalyst preparation and handling

- Catalyst activation or replacement of spent catalyst leads to loading and unloading emissions; release of effluents during offsite regeneration; or hazardous waste generation due to disposal of spent catalyst.

- De-ashing facilities have been found necessary to take care of the catalyst attrition and carryover into product, which are a likely source of wastewater and solid waste.
- Special treatment and disposal measures may be required for contaminated process wastewaters discharged during catalyst handling and separation. This is due to the presence of heavy metals in catalysts which lead to contamination of process wastewater from catalyst handling and separation. Treatment of the process wastewater in conventional biological treatment units may fail to yield desired efficiency due to the inhibitory or toxic effect of heavy metals to the microbial consortium thriving within the treatment system.
- The hazardous heavy metal content in sludge may pose problems for handling and treatment.
- Pyrophoric catalysts are to be kept wet, resulting in wastewater contaminated with metals.
- Shortening of catalyst life due to thermal or chemical deactivation.
- Catalytic reaction may lead to incomplete conversion and result in by-product formation.

Modification

- Provide in-situ activation/regeneration. Alternatively, catalyst may be obtained in active form.
- Due to the high cost, catalysts are recycled.
- Use a nonpyrophoric catalyst.
- Identify measures to delay catalyst deactivation mechanism. Extending catalyst life helps to reduce emissions and effluents from catalyst handling and regeneration.
- Enhanced mixing/contacting or increased surface area.

Process condition and configuration

- High temperatures in heat exchange tube cause decomposition of many chemicals. These lower molecular weight by-products are a source of "light ends" and fugitive emissions.
- Higher operating temperatures in furnaces and boilers are a source of combustion emissions. Similarly, higher temperature increases vapor pressure leading to more emissions.
- Preventing corrosion by neutralization or addition of inhibitors increase waste generation.
- In batch processes, cleaning of process equipment between production batches to ensure reduced cross-contamination, tend to increase waste generation.
- Chemicals like solvents, absorbants, etc., are added to the waste stream from unit operations or technologies creating effluent of complex composition.
- Nonregenerative treatment systems, i.e., activated clay, result in increased waste versus regenerative systems, i.e., activated carbon. However, emissions and release of by-products due to side reactions during bed activation and regeneration can be significant.
- Process wastewater associated with washing or phase separation (hydrocarbon/water) will affect wastewater treatment.

Modification

- Use hot process streams to reheat feeds and reduce the amount of combustion.
- Use staged heating to minimize product degradation and unwanted side reactions.
- Replace furnace with superheated high-pressure steam.
- Provide anti-corrosion coating or metallurgy in equipment.
- Reuse wash water and identify alternative separation technologies, i.e., membrane, distillation, etc.
- Reduce spills, leaks, and breakdown of equipment during operation.
- Provide measures for flow and concentration equalization.
- Segregation of waste streams bearing toxic/nonbiodegradable pollutants like halogenated hydrocarbons, polynuclear aromatics, and heavy metals for specialized treatment.

Needless to say, the preventive measures adopted to reduce emission and effluent discharge is both process and site specific. Incorporating operational changes has been found to be more economically feasible as compared to process modification. Since contaminated water makes up a major share of the waste generated by a chemical processing industry, Borden Chemical Company segregates wastewater coming from the phenol rail car unloading area and reused the water in resin batches, thereby saving the treatment cost of the entire waste stream. Du Pont, New Jersey uses a high-pressure water-jet system to clean polymer reaction vessels instead of the conventional organic solvent cleaning, eliminating the cost of solvent waste stream treatment [11].

12.5 END-OF-PIPE WASTE TREATMENT TECHNOLOGIES FOR ORGANIC CHEMICALS MANUFACTURING INDUSTRY

A major volumetric contribution to the waste stream is due to the waste cooling water stream, which can be recirculated in the loop with suitable treatment. Whereas, the comparatively lower volume process waste is highly contaminated with many hazardous and nonhazardous chemicals needing special attention in terms of treatment before disposal. Table 12.2 outlines the different categories of wastes from potential sources in the organic chemical industry.

Though the wastewater treatment technology in the organic chemicals manufacturing industries depends on the use of biological treatment methods, they should be supplemented with initial treatment like pH control and equalization. Removal of solids is not found to be a routine requirement for these industrial wastewaters. Generally, wastewaters containing oil are passed through either API (American Petroleum Institute) separators or parallel plate separators [4]. Clarifiers may also be used to remove oil and solids. The dissolved air floatation (DAF) systems are used to remove suspended solids with a specific gravity close to that of water.

The high concentration of organic matter in wastewater from such industries makes it suitable for their treatment by anaerobic processes. Though anaerobic lagoons are not widely used in this industry, they may be a first step in the treatment sequence. Biological oxygen demand (BOD) removal efficiency over 50% may be achieved in lagoons. This system is low cost, easy to operate, and capable to handle shock loads. Anaerobic filters are found to be useful in treating wastewaters with dilute soluble wastes and in denitrifying oxidized effluents for nitrogen control. Chemical oxygen demand (COD) removal efficiency over 90% has been achieved in this system. Aerobic lagoons may also be used in treatment as they allow suspended and colloidal solids to settle, equalize, and control the flow and stabilize organic matter by aerobic and facultative microorganisms. However, they are usually the last stage in secondary treatment following anaerobic or anaerobic-aerated lagoons. Aerated lagoons are fixed with mechanical aerators thus increasing power consumption and in itself this lagoon does not make contributions to reduce the BOD or suspended solids. The

TABLE 12.2
Potential Sources of Emission from the Organic Chemical Industry

Waste	Source of Emission
Process liquid wastewater	Cooling water, surplus chemicals, product purification/wash, equipment wash solvent/water, scrubber blowdown, steam jet, leaks, spills, spent solvent, floor/pad washdown, waste lubricants/oil from maintenance
Solid waste	Spent catalyst, spent filter aid, sludges, biological sludge from effluent treatment, reaction by-products, spent carbon/resins, drying aids
Emission to air	Stack/vent from storage tank or distillation unit, material loading/unloading, fugitive emissions from pumps or relief devices, secondary emissions from treatment units/cooling tower, spill/leak areas

Source: Guyer, H. H., *Industrial Processes and Waste Stream Management*, John Wiley & Sons, New York, 247–258, 1998.

conventional activated sludge process can reduce BOD and suspended solids up to 95% but fails to handle shock loads requiring upstream flow equalization. Many variations of the activated sludge process like tapered aeration process, step aeration, contact stabilization process, and the Pasveer or oxidation ditch may also be applied to wastewater treatment. Activated carbon adsorption may also be applied to remove dissolved organics but wastewater with high suspended solid concentration (>60 mg/L) should be first treated in dual-media filter or chemical coagulation before application to carbon bed.

The end-of-pipe treatment system such as biological processes may be severely affected by intermittent highly concentrated waste loads due to the nature of certain pollutant generating operations or unavoidable spills and leaks. This initiates compulsory installation of equalization tanks of proper volume and retention time. A combination of treatment methods may be used depending on the nature of process operation and availability of land area for construction work. The pollution prevention in the chemical industry is process specific as well as site specific. The age, size, and purpose of the plant generally influence the choice of the most effective pollution prevention strategy [11]. Changes in operational practices have found to yield immediate financial gains, for example, a building and construction chemicals factory producing special building chemicals; concrete add mixture, painting and coating materials, and bitumen emulsion produces 11–15 m^3/day of wastewater. Analysis of the end-of-pipe showed that the wastewater was highly contaminated with nonbiodegradable and toxic organic matter with a BOD/COD ratio of 6% (average). The analysis detected the presence of phenol, oil, and grease. The average value of total suspended solids (TSS) concentration was 200 mg/L. Biological treatment of the end-of-pipe wastewater using activated sludge was carried out. Analysis of the wastewater indicated deficiency in the nitrogen and phosphorous concentration. Nitrogen and phosphorous salts were added to adjust their concentration to meet the requirements of the biomass in the biological treatment unit. However, characteristics of the treated effluent were found not to comply with the permissible limits. This result attributed to the low biodegradability as indicated by the BOD/COD ratio. The chemical coagulation–sedimentation process for the end-of-pipe wastewater was carried out using lime aided with ferric chloride and lime aided with aluminum sulfate. Significant removal of COD, TSS, and oil and grease were achieved [6].

The wash and rinse water from many organic chemical manufacturing processes can be successfully purified by ultrafiltration (UF) membrane systems. The wash water from an organic chemical manufacturing process can contain chemicals, such as amines, glycols, plasticizers, fatty acids, etc. The wastewater generated by the plant may have high concentrations of fats, oils and grease, TSS, BOD, COD, metals, and volatile organics. The pH may also require adjustment. The UF is observed to remove fats, oils and grease, TSS, metals, and reduce the other contaminants such as BOD and COD. The UF system will volumetrically reduce the waste in the range of 96%–97%. The concentrate from UF system can be treated; followed by disposal of the dewatered sludge (in case of only non-hazardous solid waste). The UF permeate can be further purified by nanofiltration, reverse osmosis, or membrane bioreactor (MBR) post treatment. These systems will remove soluble organics and soluble inorganics present in the water phase, metals, and oils. In most cases, the post treatment of ultrafilter permeate makes it suitable for reuse within the plant.

An advanced treatment solution for wastewater treatment is the Aqua-EMBR (enhanced MBR) which was applied for the treatment of effluent generated from a polymer-based chemical manufacturing plant. After proper sludge acclimatization, it was possible to reduce the COD by more than 80% [8]. The Aqua-EMBR system consists of an activated sludge extended aeration biological treatment process and a UF-based membrane system for the separation of activated sludge from treated effluent. The membrane components are not submerged in the biological mixed liquor thereby helping in the independent operation and optimization of both the biological and membrane systems.

Other waste treatment technologies developed include the Wet Air Oxidation process where nonbiodegradable organic compounds in aqueous solution are converted to hydroperoxides, ketones,

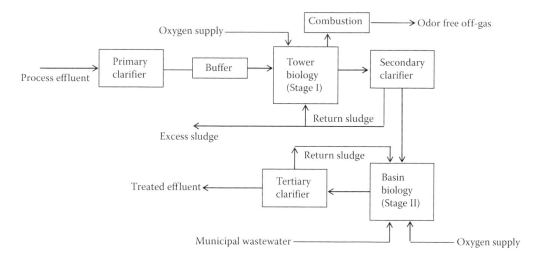

FIGURE 12.5 Treatment flow diagram for the chemical production facility at Bayer factory, Leverkusen. (Adapted from Zlokarnik, M., *Biotechnology, Volume 2: Fundamentals of Biochemical Engineering*, VCH Publishers, Weinheim, Germany, 537–569, 1985.)

aldehydes, and alcohol in the presence of oxygen with carbon dioxide, water, and various carboxylic acids, i.e., acetic acid, as end products [2]. The high temperature enhances oxidation of organic compounds, while high pressure increases concentration of dissolved oxygen. LOPROX (low pressure oxidation technology for treating high COD in wastewater) has been found to successfully treat heavily contaminated production effluent ahead of the biological treatment unit. The LOPROX technology has been successfully applied to agro-chemical (Makhteshim Chemical Works, Israel), pharmaceutical (La Felguera, Spain), and dyestuff manufacturing (Cilegon, Indonesia) waste streams. In this process, the oxidation reaction is carried out in either acidic or alkaline conditions using Fe^{2+} ions and a quinone generating substance as catalyst. The operating temperature is below 200°C and pressure varies from 75 to 300 psi [1]. The residence time is not more than 2–3 h. The best example is the successful application of LOPROX in the production of ethylene from caustic scrubbers, important for the removal of sulfur compounds and carbon dioxide. The spent caustic wastewater containing high concentration of sulfides and mercaptants is not suitable for biological treatment and must be pretreated [5].

Tower Biology (biological treatment with low energy consumption) is a technology successfully applied for the treatment of chemical wastewater in the U.S., Germany, Israel, and other developing countries. The unit consists of an activation tank which receives a supply of oxygen and includes an integrated or separate cyclone secondary clarifier with recirculation facility. Alternatively, the secondary clarification may include BayFlotech, a floatation technique enhanced by adding polyelectrolytes. This unit has also been applied for the treatment of ammonia-rich wastewater from the petrochemical industry by providing Tower Biology in the nitrification stage followed by denitrification and organic content removal by activated sludge process. Figure 12.5 shows a combined treatment scheme for the process and domestic wastewater from a chemical facility with first stage Tower Biology followed by secondary biological treatment in the Basin Biology with multiple basins in series [12].

12.5.1 Case Example: Waste Treatment in the Organic Dye and Pigment Manufacturing Industry

The dye and pigment industries are comprised of three separate types of manufacturers, dye manufacturers, pigment manufacturers, and Food, Drug, and Cosmetic colorant manufacturers. The manufacture of dyestuffs requires synthesis of a chain of intermediates mainly derived from aromatic hydrocarbons. The three classes of organic dyes and pigments are azos, anthraquinones, and triarylmethanes. Another classification of dyes is based on the application process. Basic, acid, direct, and reactive dyes are ionic and water-soluble compounds. Basic dyes are cationic dyes used to dye modified acrylics, nylons, and polyesters. Acid dyes are anionic and used on nylon, wool, silk, etc. Direct dyes are anionic azo compounds, applied to cellulose fibers in presence of electrolytes. Reactive dyes are applied to cellulosic fibers like cotton and contain a reactive component that forms a covalent bond with the fiber, resulting in high color fastness.

Another common method of classifying dyes is based on the application process. Common application process classes include: acid, basic, direct, reactive, disperse, vat, and solvent. Acid dyes are water-soluble, anionic dyes are used primarily on nylon, wool, silk, and modified acrylics. Basic, direct, and reactive dyes are also ionic, water-soluble compounds. Basic dyes are cationic dyes that are generally used to dye modified acrylics, nylons, and polyesters. Direct dyes, which are generally anionic azo compounds, are applied to cellulose fibers in the presence of electrolytes. Reactive dyes, developed for application to cellulosic fibers such as cotton, contain a reactive component that forms a covalent bond with the fiber, resulting in high color fastness. Disperse, vat, and solvent dyes, as well as pigments, are insoluble in water. Disperse dyes are applied to hydrophobic fibers such as cellulose acetate, nylon, polyesters, etc. Solvent dyes are soluble in an organic solvent and are used in varnishes, lacquers, and inks. Disperse, vat, and solvent dyes, as well as pigments, are insoluble in water. Disperse dyes are applied to hydrophobic fibers such as cellulose acetate, nylon, polyesters, etc. Solvent dyes are soluble in an organic solvent and used for varnishes, lacquers, inks, etc.

An important pollution prevention measure in the industry is the use of high-quality input raw material. As an example, some U.S. azo dye and pigment manufacturers detected toluene in the waste stream above prescribed limits. Though toluene was not added in the process, arylide couplers used in the process were found to be manufactured using toluene as a reaction solvent during their syntheses. The residual solvent adhered to the coupler made way into the wastewater from azo manufacturing process [7].

The dye and pigment industry operates batch processes. The solid wastes generated can be divided into two general types: commingled wastes and process-specific wastes. Commingled wastes are wastes combined from various processes. Process-specific wastes are wastes from a specific process and may be managed independently, e.g., spent filter aids, diatomaceous earth, or adsorbents. These wastes adsorb unreacted raw materials, by-products, and impurities. The spent filter aids then are collected in a filter press and the press cake is called clarification sludge.

The wastes from dyes and intermediates manufacture may be listed as follows [10]:

- Clarification sludges
- Still bottoms or heavy ends from the production of triarylmethane dyes or pigments
- Process effluent from the production of azo pigments/dyes
- Mercury amalgams
- Iron sludge generated in iron-acid reduction
- Naphthalene bearing sludges
- Gypsum, sodium sulfate/sulfite sludge
- Cyanide detoxification sludge
- Distillation residue/bottom draining
- Aluminum hydroxide sludge
- Spent catalysts
- Reactor still overheads
- Vacuum system condensate
- Spent adsorbent
- Equipment cleaning sludge

- Product mother liquor
- Dust collector filter fines
- Wastewater treatment sludge

Clarification sludge is mixed with the effluent stream and sent for treatment or the clarification sludge may be incinerated. Manufacturers should opt for recovery of mercury, naphthalene, aluminum hydroxide, and sodium sulfate/sulfite from their respective sludges. Iron and gypsum sludge containing considerable quantity of organics should also be subjected to recovery of value-added products rather than adopting other disposal measures, i.e., landfilling [10]. In plant conventional treatment of liquid effluent consists of both primary and secondary biological treatment units. The wastewater treatment plant receives the following liquid effluent from various sources as given in Figure 12.6.

For understanding the typical waste treatment scheme, an example of the triarylmethane manufacturing process can be considered. The process waste streams may be classified as equipment washdown, plant runoff, spent scrubber liquid, and mother liquor. The wastewater treatment sludge is generated from neutralization to adjust pH, clarification, and biological treatment, which is further processed through filtration and dewatering, prior to disposal.

Technologies based on bisulfite catalyzed borohydride reduction are applied to water-soluble dyes, i.e., azos. This treatment cleaves the dyes into smaller virtually colorless nonchromophoric amine molecules, which can be metabolized by active sludge, chemical oxidizers, or adsorbed on activated carbon or precipitated by cationic coagulant [7]. However, in the ozonation of dye wastewater, it is observed that aromatic ring structured dyes are reactive toward ozone as compared to azo dyes.

12.6 FEDERAL REGULATIONS FOR THE ORGANIC MANUFACTURING INDUSTRY

The organic chemical industry is affected by nearly all federal environmental statutes. The industry is subjected to numerous laws and regulations from state and local government. The major federal regulations applicable to the organic chemical industry include [3]:

- Resource Conservation and Recovery Act
- Clean Water Act
- Superfund Amendments and Reauthorization Act
- Toxic Substances Control Act
- Comprehensive Environmental Response, Compensation, and Liability Act
- Emergency Response and Community Right-to-Know Act
- Stormwater discharge rule

The laws and regulations authorize USEPA to test contaminated samples from the industry; prohibit or ban manufacture of chemicals or conditioning of products, and emission of wastes specially listed in the hazardous category. Consequently, these provisions will inevitably affect the activities of the organic chemical manufacturing industries.

12.7 TECHNICAL TERMINOLOGIES OF ORGANIC CHEMICALS MANUFACTURING INDUSTRY

> Halogenation: A process of adding a halogen atom on an organic compound. (Halogen is the collective name for fluorine, chlorine, bromine, and iodine.) This is an important step in producing chlorinated solvents such as ethylene dichloride.

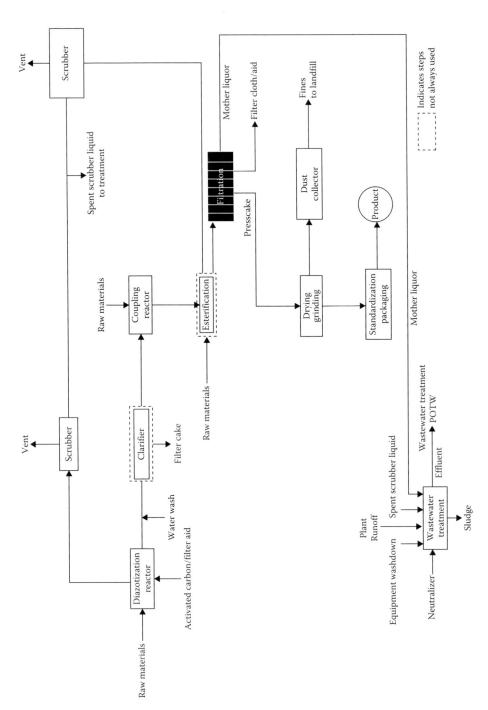

FIGURE 12.6 Sources of wastes from dyes/pigment manufacturing plant. (From USEPA, Background document for identification and listing of the deferred dye and pigment wastes, U.S. Environmental Protection Agency, Cincinnati, OH, 1999.)

Catalyst: A material that facilitates a reaction but is not actually consumed in the process, e.g., in the halogenation of ethylene to form ethylene dichloride, the reaction is conducted with an iron chloride catalyst.

Pyrolysis: A process of breaking down a large compound into smaller components by heating it (in the absence of oxygen) and exposing it to a catalyst. This process is also referred to as cracking. Vinyl chloride is produced in this way by pyrolyzing ethylene dichloride. Because pyrolysis can result in a variety of products, the catalyst and temperature must be carefully selected and controlled in order to maximize the yield of the desired product. The following equation shows the formation of vinyl chloride in the presence of heat and a catalyst.

$$2\ ClCH_2CH_2Cl \rightarrow 2\ CH_2CHCl + 2\ HCl \qquad (12.1)$$
Ethylene Dichloride   Vinyl chloride

Oxidation: The addition of an electron-donating atom (such as oxygen) and/or the removal of hydrogen from a compound. For example, formaldehyde is formed by removing two hydrogen atoms from methanol, as shown in the following equation.

$$CH_3OH \rightarrow H \cdot CHO + H_2 \qquad (12.2)$$
Methanol   Formaldehyde

Oxygen and a metal catalyst, such as silver, typically are used in the reaction.

Hydrolysis: The addition or substitution of water into a compound. This process is used in the manufacturing of ethylene glycol, the main component of antifreeze. The following equation shows how ethylene oxide is hydrolyzed to form ethylene glycol.

$$C_2H_4O + H_2O \rightarrow HO - CH_2CH_2 - OH \qquad (12.3)$$
Ethylene oxide   Ethylene glycol

REFERENCES

1. Cheremisinoff, N. P., and Davletshin, A., *Hydraulic Fracturing Operations: Handbook of Environmental Management Practices*, Dayal, M. (ed.), John Wiley & Sons, Hoboken, NJ and Scrivener Publishing LLC, Salem, MA, p. 265 (2015).
2. Debellefontaine, H., and Foussard, J. N., Wet air oxidation for the treatment of industrial wastes. Chemical aspects, reactor design and industrial applications in Europe, *Waste Management*, 20(1), pp. 15–25 (2000).
3. Guyer, H. H., *Industrial Processes and Waste Stream Management*, John Wiley & Sons, New York, pp. 247–258 (1998).
4. Hackman, E. E. III, *Toxic Organic Chemicals: Destruction and Waste Treatment*, Noyes Data Corporation, Park Ridge, NJ (1978).
5. Leonhauser, J., Pawar, J., and Birkenbeul, U., Novel technologies for the elimination of pollutants and hazardous substances in the chemical and pharmaceutical industries. In: *Industrial Wastewater Treatment, Recycling and Reuse*, 1st Edition, Ranade, V. V., and Bhandari, V. M. (eds), Butterworth-Heinemann, Oxford, UK, pp. 215–234 (2014).
6. Nasr, F. A., Doma, H. S., Abdel-Halim, H. S., and El-Shafai, S. A., Chemical industry wastewater treatment. *Environmentalist*, 27(2), pp. 275–286 (2007).
7. Reife, A., and Freeman, H. S. (eds), *Environmental Chemistry of Dyes and Pigments*, John Wiley & Sons, New York, pp. 33–59 (1995).
8. Tantak, N., Chandan, N., and Raina, P., An introduction to biological treatment and successful application of the Aqua EMBR system in treating effluent generated from a chemical manufacturing unit: A case study. In: *Industrial Wastewater Treatment, Recycling and Reuse*, 1st Edition, Ranade, V. V., and Bhandari, V. M. (eds), Butterworth-Heinemann, Oxford, UK, pp. 369–397, (2014).

9. USEPA, Profile of the organic chemical industry, U.S. Environmental Protection Agency, EPA/310-R-95-012, Cincinnati, OH (1995).
10. USEPA, Background document for identification and listing of the deferred dye and pigment wastes, U.S. Environmental Protection Agency, Cincinnati, OH (1999).
11. USEPA, Profile of the organic chemical industry, U.S. Environmental Protection Agency, EPA/310-R-02-001, Cincinnati, OH (2002).
12. Zlokarnik, M., Tower shaped reactors for aerobic biological wastewater treatment. In: *Biotechnology, Volume 2: Fundamentals of Biochemical Engineering*, Rehm, H. J., and Reed, G. (eds), VCH Publishers, Weinheim, Germany, pp. 537–569 (1985).

13 Management, Recycling, and Disposal of Electrical and Electronic Wastes (E-Wastes)

Lawrence K. Wang
Lenox Institute of Water Technology, Krofta Engineering Corporation, and Zorex Corporation

Mu-Hao Sung Wang
Lenox Institute of Water Technology Krofta Engineering Corporation

CONTENTS

13.1 Introduction 390
    13.1.1 Problems of Electrical and Electronic Wastes (E-Wastes) 391
13.2 Handling, Management, and Disposal of Electrical and Electronic Wastes—The Switzerland Experience 391
13.3 Handling, Management, and Disposal of Electrical and Electronic Wastes—The U.S. Experience 391
13.4 General Requirements for Collection, Separation, and Disposal of Electrical and Electronic Wastes Containing Particularly Hazardous Substances 393
13.5 Practical Examples 394
    13.5.1 General Management and Disposal of Electronic Waste Appliances 394
    13.5.2 General Management and Disposal of Large Electrical Waste Appliances 395
    13.5.3 General Management and Disposal of Small Electrical Waste Appliances 396
    13.5.4 General Management and Disposal of Refrigeration and Air-Conditioning Waste Appliances 396
    13.5.5 General Management and Disposal of Universal Wastes 397
    13.5.6 Management and Disposal of a Specific Electronic Waste—CRTs 400
    13.5.7 Management and Disposal of Mercury-Containing Equipment Including Lamps 401
    13.5.8 Management, Reuse, Recycle, and Disposal of Vehicle Batteries 401
    13.5.9 Management, Reuse, Recycle, and Disposal of Household Batteries 404
    13.5.10 Management of Electronic Wastes—Waste Computers 407
    13.5.11 Nanotechnology for Mercury Removal 408
    13.5.12 Solidification (Cementation) Technology for Hazardous E-Waste Disposal 408
13.6 Recent Environmental Awareness, Actions, and Recommendations 409
    13.6.1 Recent Environmental Awareness 409
        13.6.1.1 Legal Terms of Electrical Wastes and Electronic Wastes 409
        13.6.1.2 Breakdown and Generation of Electrical and Electronic Wastes 409
        13.6.1.3 Contents, Values, and Recycle of E-Waste 410
    13.6.2 Environmental Problems and Actions 410
        13.6.2.1 Global Environmental Problems and International Actions 410
        13.6.2.2 Actions in the United States and the United Nations 412
        13.6.2.3 Summary and Recommendations 413
References 413

Abstract

Electrical waste and electronic waste are defined and their potential environmental problems are identified. This chapter introduces mainly the Switzerland experience and the US experience in handling, management, and disposal of e-wastes. After general requirements for collection, separation, and disposal of e-wastes containing hazardous substances are introduced by the authors, they give many practical examples for disposal of the following e-wastes: (1) electronic waste appliances, (2) large electrical waste appliances, (3) small electrical waste appliances, (4) refrigeration and air-conditioning waste appliances, (5) universal wastes, (6) specific electronic waste—CRTs, (7) mercury-containing equipment including lamps, (8) vehicle batteries, (9) household batteries, (10) waste computers, (11) mercury removal by nanotechnology, and (12) general e-wastes disposal by solidification (cementation) technology. The authors further present their views on environmental awareness and suggest some actions to be taken by the federal and local government agencies.

13.1 INTRODUCTION

The broken, outdated, and unwanted electrical and electronic appliances are hazardous e-wastes or universal wastes, which must be recycled or disposed of in an environmentally sound way. This technical chapter introduces and discusses (1) the guidelines of selected industrial nations for the ordinance on the return, taking back, and disposal of hazardous universal wastes with special emphasis on electrical and electronic wastes containing heavy metals; (2) general requirements for e-waste disposal or reuse; (3) typical examples of recycling and disposing of electronic appliances, electrical appliances, refrigeration and air-conditioning appliances, household batteries, vehicle batteries, lamps, mercury-containing equipment, and cathode ray tubes (CRTs); (4) a new environmental nanotechnology developed by the U.S. Department of Energy, Pacific Northwest National Laboratory (PNNL), for mercury removal from wastes or contaminated sites, and (5) a new cementation technology developed by the Lenox Institute of Water Technology, Lenox, MA, for solidification of mercury-containing batteries and equipment, making mercury environmentally harmless. All universal wastes may be found to contain components of particular hazardous substances. It is essential that the following components are removed (i.e., stripping of hazardous materials): (1) batteries and accumulators; (2) condensers and ballasts; (3) mercury switches, mercury relays, and mercury vapor lamps; (4) parts containing chlorofluorocarbons (CFCs); (5) parts containing polychlorinated biphenyls (PCBs); (6) selenium drums in photocopying machines; and (7) components that release asbestos fibers. Besides the environmentally sound disposal of hazardous components, the recovery of ferrous, nonferrous, and precious metals should be the main priority in the disposal of universal wastes. It is important to ensure that the requirements relating to scrap quality are met. During the processing of appliances in shredders, components highly contaminated with hazardous substances should not end up in fractions that are originally intended for recycling. The disposal of treatment residues should not be hindered by the presence of any hazardous substances. Components containing hazardous substances usually should be removed manually. Future disposal processes, such as pyrolysis, may allow recycling of appliances without prior manual removal of hazardous substances, or may even be possible without the manual disassembling of hazardous components at all. Fractions containing halogenated flame retardants should be incinerated if recycling is not possible. Examples of electrical waste recycling and disposal, electronic waste recycling and disposal, nanotechnology treatment, and cementation treatment are presented. E-waste's global environmental problem and international actions are introduced. Recommendations for proper collection, treatment, recycle, and disposal of e-waste are made.

13.1.1 PROBLEMS OF ELECTRICAL AND ELECTRONIC WASTES (E-WASTES)

The disposal of electrical and electronic wastes around the world has not been very satisfactory. The rapid evolution of electrical, electronic, information, and communication technologies leads to an increased production of such wastes in the future. It is our ideal objective that we do not dispose of electrical and electronic wastes together with municipal solid wastes. We should try our best to create separate disposal paths for electrical and electronic wastes. Some electrical and electronic wastes contain hazardous, but recyclable, components, particularly in metals. These can be recovered at a justifiable expense only if the appliances are collected separately and treated by suitable processes. In addition, there are often problematic legal and managerial issues on waste labeling, handling, packaging, transportation, and disposition. Different countries have established their national policies to solving the problems of hazardous wastes and universal wastes [1–55].

13.2 HANDLING, MANAGEMENT, AND DISPOSAL OF ELECTRICAL AND ELECTRONIC WASTES—THE SWITZERLAND EXPERIENCE

Each country establishes its own ordinance on the handling, disposal, and general management of electrical and electronic wastes. Switzerland government has established the Ordinance on Return, Taking Back, and Disposal of Electrical and Electronic Appliances (ORDEA), which forms the legal framework allowing the industrial and commercial sectors to establish tailored and efficient return and recycling schemes [4]. Switzerland's Ordinance takes into account the regulations on cooperation between the country's federal Council and private sectors that the Parliament has included in the revised Law Relating to the Protection of the Environment. Their ORDEA came into force on July 1, 1998. Its provisions are short and primarily regulate the following:

1. Users of electrical and electronic appliances must bring worn-out appliances back to the manufacturers, importers, dealers, or to specialized disposal firms.
2. Manufacturers, importers, and dealers of electrical and electronic appliances are obliged to take back worn-out appliances.
3. Worn-out appliances must be recycled or finally disposed of in an environmentally sound way, by the most technically up-to-date means. The ORDEA also contains criteria for the environmentally sound disposal of worn-out appliances.
4. Anyone who accepts appliances for disposal in Switzerland requires a permit. Export of appliances for disposal must be authorized by the government.

Switzerland's authorities and economic sector are working closely together to implement the ORDEA. A uniform enforcement practice and substantial input from the companies are important prerequisites for success, to which their present guidelines will contribute.

13.3 HANDLING, MANAGEMENT, AND DISPOSAL OF ELECTRICAL AND ELECTRONIC WASTES—THE U.S. EXPERIENCE

In the United States, electrical and electronic appliances, when old are considered to be wastes, are sent to the sanitary landfill sites for dismantling, separation, resource recovery, and disposal. Commercial companies are formed for the waste handling, packaging, transportation, resource recovery, and disposition operations, aiming at profit making [1–3,5–11].

Fluorescent lamps, fluorescent lamp ballasts, batteries, pesticides, mercury-containing thermostats, and other mercury-containing equipment are singled out for special consideration. Specifically, these electrical and electronic wastes outfall into a regulated category called "Universal Wastes" in the United States.

By a strict definition, these electrical and electronic wastes are hazardous. Fluorescent lamps contain mercury, and almost all fluorescents fail the U.S. Environmental Protection Agency (U.S. EPA) toxicity test for hazardous wastes. Fluorescent lamp ballasts manufactured in the mid-1980s contain PCBs, a carcinogen; most of these ballasts are still in service. Batteries can contain any of a number of hazardous materials, including cadmium (nickel–cadmium batteries), the explosive lithium (lithium-ion batteries), and lead (lead-acid batteries). Some household nonrechargeable batteries still in use also contain mercury, although mercury has been phased out of batteries that are in wide circulation.

In the United States, the Universal Waste Regulations so far have streamlined hazardous waste management standards for the abovementioned U.S. federal universal wastes (batteries, pesticides, thermostats, and lamps). The regulations govern the collection and management of these widely generated wastes. This facilitates the environmentally sound collection and increases the proper recycling or treatment of the abovementioned universal wastes.

These U.S. regulations have eased the regulatory burden on American retail stores and others that wish to collect or generate these wastes. In addition, they also facilitate programs developed to reduce the quantity of these wastes going to municipal solid waste landfills or combustors. It also ensures that the wastes subject to this system will go to appropriate treatment or recycling facilities pursuant to the full hazardous waste regulatory controls.

According to a strict reading of the characteristics established by the U.S. EPA and the state environmental agencies, all of these items are hazardous wastes when disposed of, and should therefore be subject to the whole onerous spectrum of handling, transportation, and disposition requirements that have been established for toxins, carcinogens, mutagens, explosives, and other wastes that are threatening to health and the environment.

But batteries and fluorescents and generated by almost every company and every household in the country (hence the name "universal"). If they were defined as a hazardous waste, that would make practically every company and every household in the United States a hazardous waste generator, with the accompanying burden of reporting, record-keeping, handling, and management requirements (not to mention outrageous waste management costs). The state and federal agencies would be flooded with mountains of paperwork and information to track, sort, store (and ultimately throw away).

Recognizing that the full hazardous waste approach would be overkill for batteries and fluorescents, the U.S. EPA created the "universal waste" regulatory category in the mid-1990s, and it has been adopted since then by almost all states. The universal waste requirements are straightforward. First, batteries and fluorescents are banned from disposal in landfills and incinerators. But, as long as they are handled, packed, and transported in a way that prevents their breakage and possible release to the environment, and are recycled through a licensed facility, they are exempt from definition and regulation as a hazardous waste. Instead, they are subject to a much less onerous (and much less costly) set of requirements specifically crafted to ensure their convenient, but safe, management, transportation, and ultimate disposition.

Fluorescents and batteries need to be handled and packaged in a way that prevents breakage and potential release of hazardous materials, on a site and throughout the chain of custody to the ultimate disposition facility. A commercial company can provide packaging for all types of fluorescents (4′ and 8′ straight tubes, U-tubes, and others) to be delivered to a receiver. Straightforward handling and packaging procedures will prevent spills and breakage and their associated cleanup costs.

Handling and packaging needs for batteries are different. Batteries need to be handled and packed to prevent short circuits and minimize transportation costs. Again, a commercial company can provide appropriate packaging materials and instructions designed to minimize handling requirements and costs and eliminate possible liabilities associated with mispackaged materials.

The universal waste transportation requirements are not onerous. Because they are not defined as hazardous wastes, universal wastes in the United States do not need to be accompanied by a hazardous waste manifest, or shipped by a hazardous waste transporter. Even so, transportation is where many generators lose money, and where many recyclers make their margins.

The problem with transporting universals is the volume. Fluorescents are too light to make a cost-effective load. A generator rarely generates a truckload which leaves the generator at the mercy of less-than-load freight rates, or even higher on-call or "convenience" rates charged by some shippers and recyclers. Batteries are the opposite—too heavy and too bulky to cube out an efficiently loaded box trailer.

There are several possible solutions. If both electronic wastes and universal wastes are handled at the same time, they may be on the same truck and be cross-docked to the correct end markets. The generator may save money on both sets of materials. A commercial company can routinely set up "milk run" pickups from multiple generators, building to that critical truckload volume and dividing transportation charges among multiple generators, with savings for all individual small generators.

The universal waste regulatory requirement is that all universals must be handled by a licensed recycler. There are, however, only a few licensed recyclers in the United States available for services.

13.4 GENERAL REQUIREMENTS FOR COLLECTION, SEPARATION, AND DISPOSAL OF ELECTRICAL AND ELECTRONIC WASTES CONTAINING PARTICULARLY HAZARDOUS SUBSTANCES

All electrical and electronic wastes may be found to harbor components containing particularly hazardous substances. It is essential that these be removed (stripping of hazardous materials). Below are some examples of such components. Batteries and accumulators, notably, include the following:

1. Nickel–cadmium batteries and accumulators
2. Batteries and accumulators containing mercury
3. Lithium batteries and accumulators
4. Condensers and ballasts (preswitches)
5. Mercury switches/mercury relays/mercury vapor lamps
6. Parts containing CFCs (refrigeration cycle in refrigerators/insulation materials)
7. Selenium drums in photocopying machines
8. Components that release asbestos fibers

Stripping of electrical and electronic waste appliances must be done properly. During the processing of waste appliances (e.g., in shredders), the components highly contaminated with hazardous substances should not end up in fractions that are intended for recycling. It is furthermore necessary to ensure that the disposal of treatment residues (e.g., shredder residues) is not impeded by the presence of hazardous substances. As a rule, components containing particularly hazardous substances are to be removed manually. Future disposal processes, such as pyrolysis, may allow recycling of appliances without prior removal of hazardous substances, in which case it will be possible to do without the disassembly of hazardous components [4,11].

It is the responsibility of the concerned disposal company to identify and separate novel components containing hazardous substances. However, the disposal company can only do this provided the manufacturers or importers assume their responsibility as producers by making a corresponding declaration.

Fractions containing halogenated flame retardants (e.g., from printed circuit boards, cable insulation, and plastic housings) must be incinerated in suitable plants if recycling is not possible.

Besides the environmentally sound disposal of hazardous components, the recovery of ferrous, nonferrous, and noble metals is the main priority in the disposal of electrical and electronic appliances. Here, it is important to ensure that the requirements relating to scrap quality are met.

13.5 PRACTICAL EXAMPLES

13.5.1 General Management and Disposal of Electronic Waste Appliances

All appliances and modules consisting mainly of electronic components fall under the category of electronic waste appliances. This group comprises the following categories: entertainment electronics, office, information and communication appliances, and electronic components of appliances.

Owing to the rapid pace of technical developments, the composition of appliances is subject to continual change. Particular attention must be paid to the following:

1. Batteries and accumulators
2. Mercury switches/mercury relays
3. Condensers containing PCBs
4. Photoconductive drums of copying machines coated with selenium arsenate or cadmium sulfide
5. Cathode ray tubes
6. Printed circuit boards
7. Wood treated with paints, varnishes, and preservatives
8. Plastics containing halogenated flame retardants

Furthermore, appliances also contain valuable constituents such as gold (from connectors), nickel, copper, iron, aluminum, and permanent magnets, which are worth recovering.

The objectives for disposal of electronic waste appliances are (1) stripping of hazardous substances; (2) reduction of pollutant and metal content in the plastic fraction, thus permitting recycling or incineration in waste incineration plants or cement works; (3) recovery of nonferrous metals; and (4) attainment of commercially recyclable scrap quality.

The requirements for disposal of electronic waste appliances are (1) appliances may only be broken up (shredded) if the components containing particularly hazardous substances have previously been removed; and (2) since in disposing of electronic appliances the main emphasis is on the recovery of nonferrous metals, nonstripped appliances must not be shredded together with scrap cars. As a rule, electronic appliances are dismantled manually to achieve effective separation of the components containing hazardous substances.

Typical examples for disposal of electronic waste appliances include the following steps:

1. Stripping of hazardous components: In an initial step, components containing particularly hazardous substances are for the most part removed manually.
2. Shredding of appliances and separation of fractions: The stripped appliances are, as a rule, ground in a fine shredder (e.g., rotary cutter). The material resulting from this can be further processed by several methods. Possible processes are air classification, riddle screening, cyclone, turbo-rotor, sink-float, eddy current, or magnetic separation. The separated fractions are handed on in workable lots for further processing or recycling, or to resellers.
3. Recycling and disposal of waste fractions.
4. Handling and processing of stripped components containing particularly hazardous substances: Batteries and accumulators are classified as hazardous waste even if they are recycled. Mercury is classified as hazardous waste and can be recovered in special plants. Condensers containing PCBs must be incinerated in a hazardous waste incineration plant.
5. Separation of ferrous and nonferrous metals, copper, aluminum for separate recovery: The scrap material and scrap metal dealers sort these metals (in part very finely) and send them to steelworks at home and abroad.
6. Handling and processing of CRTs: CRTs are handed on for special processing.

7. Processing of printed circuit boards: Printed circuit boards are subjected to special treatment in order to recover their entire metal content.
8. Recycling of plastic-sheathed cables: Electrical cables are sent to cable recycling plants that separate the plastic and copper components.
9. Disposal of residual fraction: Depending on their quality and on the specific requirements applicable, residual fractions are disposed of in municipal solid waste incinerators, hazardous waste incinerators, and cement works, or they are recycled.

13.5.2 General Management and Disposal of Large Electrical Waste Appliances

Large electrically powered domestic waste appliances, such as cookers, ovens, washing machines and other cleaning appliances, mobile electrical heaters, and ventilators (see List of appliances) come under the category of large electrical waste appliances. The electricity for the large waste electrical appliances is supplied by the electrical mains.

These large electrical waste appliances consist mainly of iron, copper, aluminum, and insulation materials. The insulation materials are mostly inorganic. The electronic controllers contained in the appliances are classified as electronic scrap (see separate fact sheet). They may contain particularly hazardous components (accumulators, batteries, condensers, mercury switches, etc.).

The objectives for management and disposal of large electrical waste appliances are (1) stripping of hazardous substances; (2) reduction of pollutant and metal content in the shredder residue; (3) recycling and recovery of ferrous metals; and (4) attainment of commercially recyclable scrap quality (e.g., low copper content in the scrap iron).

The requirements for management and disposal of large electrical waste appliances are that appliances may only be shredded if the particularly hazardous components have previously been removed.

Older appliances (such as ovens) still sometimes contain asbestos. Waste from which asbestos fibers may be released is classified as hazardous waste and must be disposed of as specified in the appropriate environmental laws. The heat-transfer oils of older types of mobile convector heaters still sometimes contain PCBs. These fluids must be disposed of as hazardous waste.

Typical examples for management and disposal of large electrical waste appliances include the following steps:

1. Stripping of hazardous substances: In an initial step, components containing particularly hazardous substances are removed.
2. Breaking up of appliances and separation of fractions: After stripping, the large electrical appliances are, as a rule, ground in a shredder (hammer mill for scrap cars). The resulting fragments are separated by means of special equipment, such as air classifiers, magnetic separators, electrostatic separators, eddy current separators, and sink-float separators. The main fractions are fractions of ferrous or nonferrous metals, printed circuit boards (if applicable), and residual fraction (shredder residue).
3. Recycling and disposal of waste fractions.
4. Handling and processing of stripped components containing particularly hazardous substances: Batteries and accumulators are classified as hazardous waste even if they are to be recycled. Mercury is classified as hazardous waste and can be recovered in special plants. Condensers containing PCBs must be incinerated in a hazardous waste incineration plant.
5. Separation of ferrous and nonferrous metals, copper, aluminum for separate recovery: The scrap material and scrap metal dealers sort these metals (in part very finely) and send them to steelworks at home and abroad.
6. Processing of printed circuit boards: Printed circuit boards are subjected to special treatment in order to recover their entire metal content.
7. Recycling of plastic-sheathed cables: Cables are sent to cable recycling plants that separate the plastics and copper components.

8. Disposal of residual fraction: Depending on their quality and on the specific requirements applicable, residual fractions are disposed of in municipal solid waste incinerators, hazardous waste incinerators, and cement works, or they are recycled.

13.5.3 General Management and Disposal of Small Electrical Waste Appliances

The category of small electrical waste appliances comprises electrical appliances such as electric razors, music players, hair-removing appliances, hair dryers, egg boilers, immersion water heaters, and coffee grinders. They are generally composed of plastics, ferrous, and nonferrous metals. A large proportion of these small appliances is powered by batteries or accumulators.

The objectives of disposal of small electrical waste appliances are simple: (1) stripping of hazardous substances; (2) recycling and recovery of ferrous and nonferrous metals; (3) reduction of pollutant and metal content in the plastic fraction; and (4) recovery of the copper fraction.

The only requirement for disposal of small electrical waste appliances is that appliances may be shredded if the components containing particularly hazardous substances have previously been removed. In the case of small cordless electrical appliances, the greater part of the hazardous substances can be eliminated by prior removal of batteries and accumulators.

The following are typical operational steps for disposal of small waste appliances:

1. Stripping of hazardous substances: In an initial step, components containing particularly hazardous substances are removed from most of the parts manually.
2. Breaking up of appliances and separation of fractions: The stripped appliances are, for example, finely shredded (in a rotary cutter). Using an air classifier, plastics, nonmetallic components, etc. are removed. The ferrous metals are separated from nonferrous ones in a magnetic separator. An eddy current separator is used for fine separation of nonferrous metals. Copper and aluminum are separated in sink–float separators. The material resulting from the fine shredding can be processed by various means. Possible processes are air classification, riddle screening, cyclone, turbo-rotor, sink–float, eddy current, or magnetic separation. The separated fractions are handed on in workable lots for further processing or recycling, or to resellers.
3. Recycling and disposal of waste fractions.
4. Handling and processing of stripped components containing particularly hazardous substances: Batteries and accumulators are classified as hazardous waste even if they are to be recycled.
5. Separation of ferrous and nonferrous metals, copper, and aluminum for separate recovery: The scrap material and scrap metal dealers sort these metals (in part very finely) and send them to steelworks at home and abroad.
6. Recycling of plastic sheathed cables: Cables are handed on to cable recycling plants that separate the plastic and copper components.
7. Disposal of residual fraction: Depending on their quality and on the specific requirements applicable, residual fractions are disposed of in municipal solid waste incinerators, hazardous waste incinerators, and cement works, or they are recycled.

13.5.4 General Management and Disposal of Refrigeration and Air-Conditioning Waste Appliances

Refrigerators, deep-freezers, ice machines equipped with a circulation system, mobile air conditioners, dehumidifiers, etc. are discussed in this section.

The cooling circuit of these appliances contains refrigerants. The most common are CFCs, ammonia, or pentane. In many types of refrigerators, the circulation system also contains oil. Other components are metals (steel, aluminum, and copper), plastics (housings, drawers, and shelves), polyurethane (PU) insulation, polystyrene (PS) insulation, glass, etc. In older appliances, the insulation material also generally contains CFCs. The following components are removed prior to shredding: compressors, cooling coils, glass, cables, and switches.

The objectives of disposal of refrigeration and air-conditioning waste appliances are (1) separate disposal of the CFCs from the circulation system and the insulating material; (2) further stripping of hazardous substances (e.g., mercury switches); and (3) recovery of ferrous metals to be the priority in metal recycling.

The requirements for disposal of refrigeration and air-conditioning waste appliances are very stringent: (1) mercury switches and condensers containing PCBs must be removed in advance and disposed of; (2) 90% of the CFCs, both from the circulation system and the insulation, must be recovered and disposed of in an environmentally sound manner, as specified in the regulations; (3) the amount of residual CFCs in the pressed-out foam must not exceed 0.5% if it is to be reused; (4) the government emission standard for CFCs (20 mg/m^3 at a flow rate >100 g/h, for instance) must be complied with, and therefore, the emission flow rate must be measured and recorded continuously; (5) recovered CFCs or recovered components containing CFCs (e.g., foam containing >0.5% CFCs) must be disposed of in suitable plants; (6) chrome-plated ferrous scrap (chromium VI) must not be mixed with unsorted scrap but must be delivered direct to the steelworks, in compliance with the relevant workplace protection and safety regulations.

Since pentane is a flammable gas that can form explosive mixtures in combination with air or oxygen, suitable safety precautions must be taken.

Typical operational steps for disposal of refrigerators and similar appliances are listed below:

1. Stripping of hazardous substances: Mercury switches and other components containing particularly hazardous substances must be removed. CFCs are recovered from the cooling circuit PU foam is removed with special equipment and appliances are taken apart with varying degrees of automation; ammonia is dissolved in water; and waste oil (from compressors) is disposed of separately.
2. Breaking up of appliances and separation of fractions.
3. Removal of special components: Loose fittings are mostly removed. They include plastic accessories and trays, steel racks, glass shelves, and doors made of plastic, metal, and insulation material.
4. Handling of the main unit: The first step is to extract the refrigerant. It must be recovered as completely as possible by means of suitable plants and equipment. The refrigerants and foaming agents are condensed by refrigeration and sent to be destroyed.
5. Shredding and fractionation of the main unit take place under partial vacuum in a special shredder. PU foamed with CFCs is pressed out as completely as possible. The vitiated air from the shredder and the press is cleaned through activated carbon and passed through a condensation cooling system. By this means, the foaming agent may be almost entirely recovered.
6. Separation of the residual fractions as follows: Separation of CFCs by condensation; separation of expanded PS and PU foam by air classification; separation of iron with a magnetic separator; and separation of nonferrous metals with an eddy current separator.

Following stripping of hazardous substances, air conditioners and dehumidifiers can be further dismantled either manually or in a shredder. Figure 13.1 shows the flow diagram for management, separation, recycle, and disposal of used refrigeration appliances [4].

13.5.5 General Management and Disposal of Universal Wastes

Universal waste is a legal, environmental term used in the United States. The universal waste regulations in the U.S. streamline collection requirements for certain hazardous wastes in the specific categories decided by the federal and the state governments. The Universal Waste Regulations ease

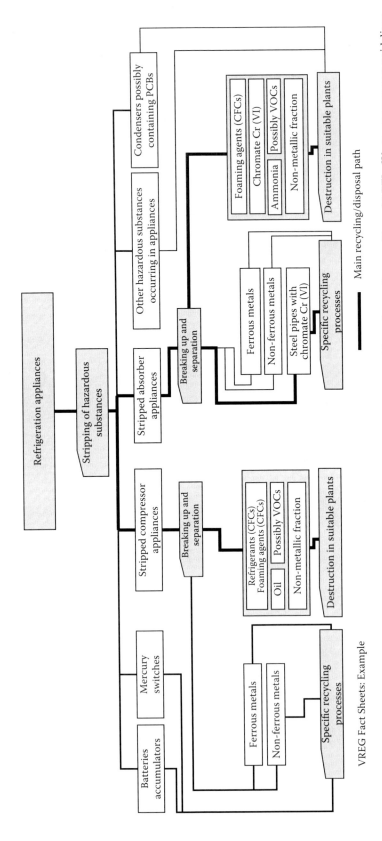

FIGURE 13.1 Flow diagram for management, separation, recycling, and disposal of waste refrigeration appliances. (From SAEFL, Waste management guidelines for the ordinance on the return, the taking back and the disposal of electrical and electronic appliances (ORDEA), Swiss Agency for the Environment, Forests and Landscape, 76 p., Bern, Switzerland, 2000.)

regulatory burdens on businesses; promote proper recycling, treatment, or disposal; and provide for efficient, proper, and cost-effective collection opportunities.

The U.S. EPA federal universal wastes are (1) batteries such as nickel–cadmium (Ni–Cd) and small sealed lead-acid batteries, which are found in many common items in the business and home setting, including electronic equipment, mobile telephones, portable computers, and emergency backup lighting; (2) agricultural pesticides that are recalled under certain conditions and unused; pesticides that are collected and managed as part of a waste pesticide collection program; and the pesticides that are unwanted for a number of reasons, such as being banned, obsolete, damaged, or no longer needed due to changes in cropping patterns or other factors; (3) thermostats which can contain as much as 3 g of liquid mercury and are located in almost any building, including commercial, industrial, agricultural, community, and household buildings; (4) lamps which are the bulb or tube portion of electric lighting devices that have a hazardous component (Note: Examples of common universal waste electric lamps include, but are not limited to, fluorescent lights, high-intensity discharge, neon, mercury vapor, high pressure sodium, and metal halide lamps. Many used lamps are considered hazardous wastes under the Resource Conservation and Recovery Act [RCRA] because of the presence of mercury, or occasionally lead); and (5) mercury-containing equipment is proposed as a new universal waste category because mercury is used in several types of instruments that are common to electric utilities, municipalities, and households. Some of these devices include switches, barometers, meters, temperature gauges, pressure gauges, and sprinkler systems.

It is important to note that each state in the United States can add different wastes and does not have to include all the U.S. federal universal wastes. In other words, the states can modify the federal universal waste rule and add additional universal waste in individual state regulations. A waste generator should check with the state for the exact regulations that apply to the generator.

For proper management and disposal of the aforementioned universal wastes, a waste generator, a waste handler, a transporter, or a destination facility must understand the legal definitions of wastes and their legal status. The following is an overview of legal definitions and related requirements:

1. Universal Waste: A waste must be a hazardous waste before it can be considered a universal waste. A waste must also meet certain criteria to qualify as a universal waste. For instance, it must be widespread, commonly found in medium to large volumes, and exhibit only low-level hazards or be easily managed.
2. Federal Universal Wastes: In the United States, the universal wastes (such as batteries, pesticides, thermostats, lamps, and mercury-containing wastes) are decided and legally defined by the U.S. EPA.
3. State Universal Wastes: In the United States, the states do not include all of the federal universal wastes when they use (adopt) the program and the states can make them more stringent and add their own universal wastes (like antifreeze).
4. Universal Waste Battery: Battery means a device consisting of one or more electrically connected electrochemical cells, which are designed to receive, store, and deliver electric energy. An electrochemical cell is a system consisting of an anode, cathode, and an electrolyte, plus such connections (electrical and mechanical) as may be needed to allow the cell to deliver or receive electrical energy. The term "battery" also includes an intact, unbroken battery from which the electrolyte has been removed.
5. Universal Waste Pesticide: Pesticide means any substance or mixture of substances intended for preventing, destroying, repelling, or mitigating any pest, or intended for use as plant regulator, defoliant, or desiccant.
6. Universal Waste Thermostat: Thermostat means a temperature control device that contains metallic mercury in an ampule attached to a bimetal sensing element.
7. Universal Waste Lamp: Lamp, also referred to as "universal waste lamp," is defined as the bulb or tube portion of an electric lighting device. A lamp is specifically designed to produce radiant energy, most often in the ultraviolet, visible, and infrared regions of the

electromagnetic spectrum. Examples of common universal waste electric lamps include, but are not limited to, fluorescent, high-intensity discharge, neon, mercury vapor, high-pressure sodium, and metal halide lamps.

8. Universal Waste Handlers: These could be (1) a business that generated (needs to dispose of) a universal waste (fluorescent lights for instance), (2) a take-back program, and (3) a collection program.
9. Small-Quantity Handlers of Universal Waste (SQHUW): A handler that accumulates less than 5000 kg (11,000 lbs) of universal waste at any one time.
10. Large-Quantity Handlers of Universal Waste (LQHUW): A handler that accumulates 5000 kg (11,000 lbs) or more of universal waste at any one time.
11. Universal Waste Transporter: A transporter that transports universal waste from handlers to other handlers, destination facilities, or foreign destinations.
12. Universal Waste Destination Facilities: The facilities that recycle, treat, or dispose of universal wastes as hazardous waste (no longer universal waste). Note: This does not include facilities that only store universal waste since those facilities qualify as a universal waste handler.

13.5.6 Management and Disposal of a Specific Electronic Waste—CRTs

CRTs, shown in Figure 13.2, are the video display components of televisions and computer monitors. CRT glass typically contains enough lead to be classified as hazardous waste when it is being recycled or disposed of. Currently, businesses and other organizations that recycle or dispose of their CRTs are confused about the applicability of hazardous waste management requirements to their computer or television monitors. The federal government is proposing to revise regulations to encourage opportunities to safely collect, reuse, and recycle CRTs [4].

To encourage more reuse and recycling, intact CRTs being sent for possible reuse are considered to be products rather than wastes and therefore not regulated unless they are being disposed of. If CRT handlers disassemble the CRTs and send the glass for recycling, the U.S. EPA has proposed to exclude them from being a waste, provided they comply with simplified storage, labeling, and

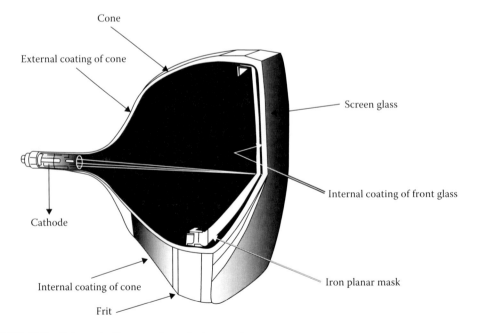

FIGURE 13.2 Schematic diagram of a cathode ray tube.

transportation requirements. Furthermore, the U.S. EPA believes that if broken CRTs are properly containerized and labeled when stored or shipped before recycling, they resemble commodities more than waste.

Finally, processed glass being sent to a CRT glass manufacturer or a lead smelter is excluded from hazardous waste management under most conditions. If the glass is being sent to any other kind of recycler, it must be packaged and labeled the same as broken CRTs. The U.S. EPA believes that these proposed changes will encourage the recycling of these materials, while minimizing the possibility of releasing lead into the environment. Figure 13.3 shows a flow diagram for management, separation, recycle, and disposal of CRTs [4].

13.5.7 Management and Disposal of Mercury-Containing Equipment Including Lamps

Mercury is contained in several types of instruments that are commonly used by electric utilities, municipalities, and households. Among others, these devices include barometers, meters, temperature gauges, pressure gauges, sprinkler system contacts, and parts of coal conveyor systems. U.S. EPA has received data on mercury-containing equipment since 1995, when it issued the first federal universal waste rule. The Agency believes that adding mercury-containing devices to the universal waste stream will facilitate better management of this waste [10].

The universal waste rule tailors management requirements to the nature of the waste in order to encourage collection (including household collections) and proper management. Universal waste generators, collectors, and transporters must follow specific record-keeping, storage, and transportation requirements. The U.S. EPA is proposing the same tailored requirements for all mercury-containing equipment.

U.S. EPA initiated a mercury-containing lamp recycling outreach program in 2002 to promote mercury lamp recycling by commercial and industrial users. The outreach program aims to increase awareness of the proper disposal methods of these lamps in compliance with federal and state universal waste rules. This outreach effort will be effective in increasing the amount of lamps recycled in the short term, as well as have lasting impact in the long term. The U.S. EPA's goal is to raise the national recycling rate for mercury lamps from the current to 20%–40% by 2005, and to 80% by 2009.

U.S. EPA awarded funds in the form of 10 cooperative agreements for the development and implementation of a coordinated nationwide mercury-containing lamp recycling outreach program. This program is currently being implemented in two phases. Recipients of phase one cooperative agreements are developing outreach materials such as fact sheets, recycling database, websites, public service announcements, and educational materials.

While phase one cooperative agreement recipients focused on developing outreach materials, the recently selected phase two recipients will focus on outreach program implementation. They will conduct outreach to segments of the lamp-disposing population by adapting outreach materials developed in phase one to target specific audience (i.e., industry-specific lamp users or lamp users within a certain geographic location).

13.5.8 Management, Reuse, Recycle, and Disposal of Vehicle Batteries

Every year in the United States, billions of batteries are bought, used, and thrown out. In 1998 alone, over 3 billion industrial and household batteries were sold. The demand for batteries can be traced largely to the rapid increase in automobiles, and cordless and portable products such as cellular phones, video cameras, laptop computers, and battery-powered tools and toys.

Because many batteries contain toxic constituents such as mercury and cadmium, they pose a potential threat to human health and the environment when improperly disposed. Though batteries generally make up only a tiny portion of municipal solid waste (MSW), <1%, they account for a

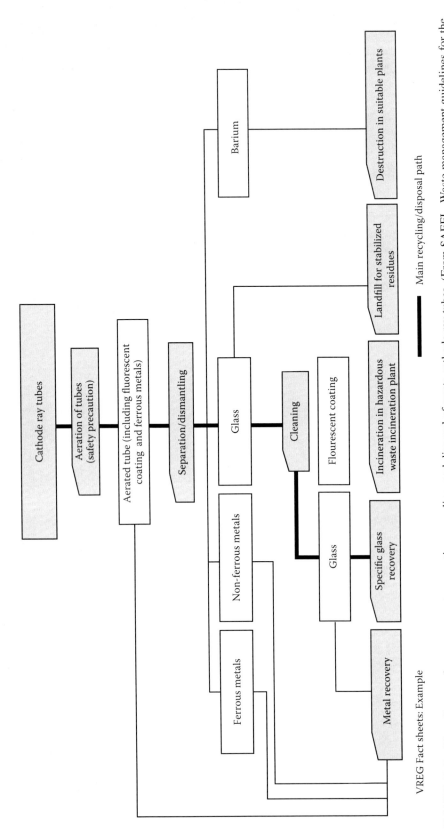

FIGURE 13.3 Flow diagram for management, separation, recycling, and disposal of waste cathode ray tubes. (From SAEFL, Waste management guidelines for the ordinance on the return, the taking back and the disposal of electrical and electronic appliances (ORDEA), Swiss Agency for the Environment, Forests and Landscape, 76 p., Bern, Switzerland, 2000.)

disproportionate amount of the toxic heavy metals in MSW. For example, the U.S. EPA has reported that, as of 1995, nickel–cadmium batteries accounted for 75% of the cadmium found in MSW. When MSW is incinerated or disposed of in landfills, under certain improper management scenarios, these toxics can be released into the environment.

Over the past decade, the battery industry, partly in response to public concerns and legislation, has played an active role in finding solutions to these problems. Industry efforts have touched on every stage of the product life cycle.

Seventy million vehicle batteries are produced each year in the Unites States. About 80% of discarded lead-acid batteries are being collected and recycled. Lead-acid batteries contain about 15–20 lb of lead per battery and about 1–2 gal of sulfuric acid. Vehicle batteries have been banned from disposal in Nebraska landfills since September 1, 1994.

Environmental hazards of batteries can be briefly summarized as below. A battery is an electrochemical device with the ability to convert chemical energy to electrical energy to provide power to electronic devices. Batteries may contain lead, cadmium, mercury, copper, zinc, manganese, nickel, and lithium, which can be hazardous when incorrectly disposed. Batteries may produce the following potential problems or hazards: (1) they pollute the lakes and streams as the metals vaporize into the air when burned, (2) they contribute to heavy metals that leach from solid waste landfills, (3) they expose the environment and water to lead and sulfuric acid, (4) they contain strong acids that are corrosive, and (5) they may cause burns or danger to eyes and skin.

Heavy metals have the potential to enter the water supply from the leachate or runoff from landfills. It is estimated that nonrecycled lead-acid batteries produce about 65% of the lead in the municipal waste stream. When burned, some heavy metals such as mercury may vaporize and escape into the air, and cadmium and lead may end up in the ash, making the ash a hazardous material for disposal.

Vehicle batteries may be recycled by trading in an old battery when replacing a battery. Most battery distribution centers, automotive garages, and repair centers have collection points. Batteries are also accepted at some scrap yards, automobile dismantlers, and some retail chain stores. Batteries should be stored in a secure area, locked, or kept away from children and sources of sparks. All old batteries should be recycled.

Prolonging battery life is another method for environmental protection. To reduce waste, a consumer should buy longer-life batteries that may result in fewer batteries to recycle and follow recommended maintenance procedures to lengthen battery life.

Good maintenance of a vehicle battery can prolong a battery's life, if the following procedures can be followed: (1) check the battery for adequate water level if the battery is not a sealed battery and check the battery and the vehicle charge system, if the battery is low on water; (2) do not over fill a battery; (3) make sure all connections are clean; (4) if the vehicle is seldom used, charge the battery at least every 2 months to maintain the battery charge, because in a discharge state, the battery might freeze; (5) if the battery must be stored out of the vehicle, store in cool dry place; (6) do not jump-start a battery when the battery is extremely cold; (7) when jump-starting, connect the jumper cables first to the power source, then connect the positive cable to positive cable on the battery to be jumped, and the negative to a solid ground on the vehicle (e.g., bracket on alternator). This avoids going directly to the battery to be charged to prevent sparking.

Redesign, reuse, and recycling will be the best management practice for waste vehicle battery management. Some battery manufacturers are redesigning their products to reduce or eliminate the use of toxic constituents. For example, since the early 1980s, manufacturers have reduced their use of mercury by over 98%. Many manufacturers are also designing batteries for a longer life.

Most states have passed legislation prohibiting the disposal of lead-acid batteries (which are primarily vehicle batteries) in landfills and incinerators and requiring retailers to accept used batteries for recycling when consumers purchase new batteries. For example, Maine, USA, has adopted legislation that requires retailers to either (1) accept a used battery upon sale of a new battery, or (2) collect a US$10 deposit upon sale of a new battery, with the provision that the deposit shall be returned to the customer if the buyer delivers a used lead-acid battery within 30 days of the date of sale. This legislation is based on a model developed by the lead-acid battery industry. Lead-acid batteries are collected for recycling through a reverse distribution system. Spent lead batteries are returned by consumers to retailers, picked up by wholesalers or battery manufacturers, and finally taken to secondary smelters for reclamation. These recycling programs have been highly successful: the nationwide recycling rate for lead-acid batteries stands at roughly 95+%, making them one of the most widely recycled consumer products. Automotive and other industrial batteries are, more and more, being recycled and better designed now.

13.5.9 Management, Reuse, Recycle, and Disposal of Household Batteries

More and more household batteries are being used today. The average person owns about two button batteries, ten normal (A, AA, AAA, C, D, 9V, etc.) batteries, and throws out about eight household batteries per year. About three billion batteries are sold annually in the United States averaging about 32 per family or 10 per person [5–9].

Table 13.1 shows the typical types of household batteries.

Battery manufacturers are producing more rechargeable batteries each year. The National Electrical Manufacturers Association has estimated that U.S. demand for rechargeable batteries is growing twice as fast as demand for nonrechargeable batteries.

The Rechargeable Battery Recycling Corporation (RBRC) started a nationwide take-back program in 1994 for collection and recycling of used nickel–cadmium batteries. The RBRC expanded in 2001 to include all portable rechargeable batteries in its take-back program. This is the first nationwide take-back program that involves an entire U.S. industry. Much of this progress has come in response to far-reaching legislation at the state and federal levels in the United States. Starting in 1989, 13 states took the lead by adopting laws (including battery labeling requirements) to facilitate the collection and recycling of used rechargeable batteries. In 1996,

TABLE 13.1
Typical Types of Household Batteries

Primary Cells (Nonrechargeable)	
Alkaline	Cassette players, radios, appliances
Carbon-zinc	Flashlights, toys, etc
Lithium	Cameras, calculators, watches, computers, etc.
Mercury	Hearing aids, pacemakers, cameras, calculators, watches, etc.
Silver	Hearing aids, watches, cameras, calculators
Zinc	Hearing aids, pagers

Secondary Cells (Rechargeable)	
Nickel–cadmium	Cameras, rechargeable appliances such as portable power tools, handheld vacuums, etc.
Small sealed lead-acid	Camcorders, computers, portable radios and tape players, cellular phones, lawn mower starters, etc.

the U.S. Congress passed the Mercury-Containing and Rechargeable Battery Management Act, which removed barriers to and helped facilitate the RBRC's nationwide take-back program. In addition, many states have passed legislation prohibiting incineration and landfilling of mercury-containing and lead-acid batteries.

The following are important legal terminologies for this section. The term "mercuric-oxide battery" means a battery that uses a mercuric-oxide electrode.

The term "rechargeable battery" (1) means one or more voltaic or galvanic cells, electrically connected to produce electric energy, that are designed to be recharged for repeated uses; and (2) includes any type of enclosed device or sealed container consisting of one or more such cells, including what is commonly called a battery pack (and in the case of a battery pack, for the purposes of the requirements of easy removability and labeling under law, means the battery pack as a whole rather than each component individually), but it does not include a lead-acid battery used to start an internal combustion engine, a lead-acid battery used for load leveling or for storage of electricity, a battery used as a backup power source for memory or program, or a rechargeable alkaline battery.

The term "rechargeable consumer product" (1) means a product that, when sold at retail, includes a regulated battery as a primary energy supply, and that is primarily intended for 1 kW personal or household use; but (2) does not include a product that only uses a battery solely as a source of backup power for memory or program instruction storage, time-keeping, or any similar purpose that requires uninterrupted electrical power in order to function if the primary energy supply fails or fluctuates momentarily.

The term "regulated battery" means a rechargeable battery that (1) contains a cadmium electrode, a lead electrode, or any combination of cadmium and lead electrodes; or (2) contains other electrode chemistries and is the subject of a determination by the Administrator of the U.S. EPA under environmental laws.

The term "remanufactured product" means a rechargeable consumer product that has been altered by the replacement of parts, repackaged, or repaired after initial sale by the original manufacturer.

As stated previously, a battery is an electrochemical device with the ability to convert chemical energy to electrical energy to provide power to electronic devices. Household batteries may also contain cadmium, mercury, copper, zinc, lead, manganese, nickel, and lithium, which may create a hazard when disposed incorrectly. The potential problems or hazards of household batteries are similar to that of vehicle batteries.

In landfills, heavy metals have the potential to leach slowly into soil, groundwater, or surface water. Dry cell batteries contribute about 88% of the total mercury and 50% of the cadmium in the municipal solid waste stream. In the past, household batteries accounted for nearly half of the mercury used in the United States and over half of the mercury and cadmium in the municipal solid waste stream. When burned, some heavy metals such as mercury may vaporize and escape into the air, and cadmium and lead may end up in the ash.

Controversy exists about reclaiming household batteries. Currently, most batteries collected through household battery collection programs are disposed of in hazardous waste landfills. There are no known recycling facilities in the United States that can practically and cost-effectively reclaim all types of household batteries, although facilities exist that reclaim some button batteries. Currently battery collection programs typically target button and nickel–cadmium batteries, but may collect all household batteries because of the consumers' difficulty in identifying battery types.

There are two major types of household batteries: (1) primary batteries are those that cannot be reused; they include alkaline/manganese, carbon-zinc, mercuric-oxide, zinc-air, silver-oxide, and other types of button batteries; and (2) secondary batteries are those that can be reused; they (rechargeable) include lead-acid, nickel–cadmium, and potentially nickel–hydrogen.

Mercury reduction in household batteries began in 1984 and continues to date. During the past 5 years, the industry has reduced the total amount of mercury usage by about 86%. Some batteries such as the alkaline battery have had about a 97% mercury reduction in the product. Newer alkaline batteries may contain about one-tenth the amount of mercury previously contained in the typical alkaline battery. Some alkaline batteries have zero-added mercury, and several mercury-free, heavy-duty, carbon-zinc batteries are on the market.

Mercuric-oxide batteries are being gradually replaced by new technology such as silver-oxide and zinc-air button batteries that contain less mercury.

Nickel–cadmium rechargeable batteries are being researched. Alternatives such as cadmium-free nickel and nickel-hydride system are being researched, but nickel–cadmium is unlikely to be totally replaced. Nickel–cadmium batteries can be reprocessed to reclaim the nickel. However, currently, approximately 80% of all nickel–cadmium batteries are permanently sealed in appliances. Changing regulations may result in easier access to the nickel–cadmium batteries for recycling.

To reduce waste, start with pollution prevention. Starting with prevention creates less or no leftover waste to become potentially hazardous wastes. Rechargeable batteries result in a longer lifespan and use fewer batteries. However, rechargeable batteries still contain heavy metals such as nickel–cadmium. When disposing of rechargeable batteries, recycle if possible.

The use of rechargeable nickel–cadmium batteries can reduce the number of batteries entering the waste stream, but may increase the amount of heavy metals entering the waste stream unless they are more effectively recycled. As of 1992, the percentage of cadmium in nickel–cadmium batteries was higher than the percentage of mercury in alkaline batteries, so substitution might only replace one heavy metal for another, and rechargeable batteries do use energy resources in recharging. Rechargeable alkaline batteries are available along with rechargers.

Recycle waste batteries if possible. Batteries with high levels of mercury or silver can be recovered for the refining process. The mercuric oxide batteries can be targeted for recollection and mercury recovery. There are a few mercury-refining locations in the United States that accept mercury batteries, and they could be contacted about battery recycling.

Mercury-oxide and silver-oxide button batteries are sometimes collected by jewelers, pharmacies, hospitals, and electronic or hearing aid stores for shipping to companies that reclaim mercury or silver. Some batteries cannot be recycled. If recycling is not possible, batteries should be saved for a hazardous waste collection. Battery recycling and button battery collection may be good options at present, but may change as the mercury concentration in the majority of button batteries continues to decrease.

Batteries may be taken to a household hazardous waste collection, local battery collection program. One can also contact the battery manufacturer for other disposal options or for information on collection programs. If disposal is the only option, and the household batteries are not banned from the area permitted landfill, one should protect the batteries for disposal by placing them in a sturdy plastic bag in a sturdy container to help guard against leakage. Waste batteries should not be burned because of the metals that could explode. When burned, some heavy metals such as mercury may vaporize and escape into the air, and cadmium and lead may end up in the ash.

In the United States, federal initiatives and state initiatives are assisting businesses and consumers in managing, reusing, recycling, and disposal of household batteries. These include the Universal Waste Rule and the Mercury-Containing and Rechargeable Battery Management Act.

The Universal Waste Rule, promulgated in 1995, was designed to encourage recovery and recycling of certain hazardous wastes (including batteries, thermostats, and some pesticides) by removing some of the regulatory barriers. Under the rule, batteries recovered and properly managed are exempt from some RCRA provisions, no matter who generates the waste. Promulgation of the Universal Waste Rule facilitated the battery industry's take-back system for Ni–Cd batteries in states that adopted the rule through state rulemaking.

The Mercury-Containing and Rechargeable Battery Management Act (the "Battery Act"), which was signed into law on May 13, 1996, removed previous barriers to Ni–Cd battery recycling programs resulting from varying individual state laws and regulatory restrictions governing the labeling, collection, recycling, and transportation of these batteries. The Act facilitated and encouraged voluntary industry programs for recycling Ni–Cd batteries, such as the national "Charge Up to

Recycle" program. The Act also established national labeling requirements for rechargeable batteries, ordered that rechargeable batteries be easy to remove from consumer products, and restricted the sale of certain batteries that contain mercury.

The 1996 Battery Act eased the burden on battery recycling programs by mandating national, uniform labeling requirements for Ni–Cd and certain small sealed lead-acid batteries and by making the Universal Waste Rule effective in all 50 states. The Battery Act indicates (1) the state labeling requirements for these battery types and (2) the state legislative and regulatory authority for the collection, storage, and transportation of Ni–Cd and other covered batteries. States can, however, adopt standards for battery recycling and disposal that are more stringent than existing federal standards. States can also adopt more stringent requirements concerning the allowable mercury content in batteries.

Several states have passed legislation mandating additional reductions in mercury beyond those in the Battery Act and prohibiting or restricting the disposal in MSW of batteries with the highest heavy metal content (i.e., Ni–Cd, small sealed lead-acid, and mercuric-oxide batteries). A andful of states have gone further, placing collection and management requirements on battery manufacturers and retailers to ensure that certain types of batteries are recycled or disposed of properly.

Many states and regional organizations have developed far-reaching legislation for battery management, which is beyond the scope of the federal law. Only the following two organizations are introduced here: (1) Northeast Waste Management Officials' Association (NEWMOA); and (2) New England Governors' Conference.

The NEWMOA, a coalition of state waste program directors from New England and New York, has developed a model legislation meant to reduce mercury in waste. The model legislation proposes a variety of approaches that states can use to manage mercury-containing products (such as batteries, thermometers, and certain electronic products) and wastes, with a goal of instituting consistent controls throughout the region. The proposed approaches focus on notification; product phase-outs and exemptions; product labeling; disposal bans; collection and recycling programs; and a mechanism for interstate cooperation. Bills based on the model legislation have been under consideration by legislators in New Hampshire and Maine. In April 2000, NEWMOA released a revised version of the model legislation following a series of public meetings and the collection of comments from stakeholders.

The New England Governors' Conference passed a resolution in September 2000 recommending, among other things, that each New England state work with its legislature to adopt mercury legislation based on the NEWMOA model (see above). The NEWMOA model legislation is meant to reduce the amount of mercury in waste through strategies such as product phase-outs, product labeling, disposal bans, and collection and recycling programs. Certain types of mercury-containing batteries are among the products targeted by the model legislation.

13.5.10 Management of Electronic Wastes—Waste Computers

In the early 1980s, the world witnessed the sale of the first personal computers. Its transition from the relatively bulky and slow first units to the sleek, speed demons has made the computer truly revolutionary. With each improvement in computers, however, comes the increasing problem of what to do with the ever-increasing number of computer e-wastes. The U.S. EPA estimates that nearly 250 million computers will become obsolete in the next 5 years in the United States alone. Unfortunately, only approximately 10% of these old computers that are retired each year are being recycled. This presents a substantial concern because of (1) the presence of toxic elements such as lead, cadmium, mercury, barium, chromium, beryllium as well as flame retardant, and phosphor in a typical computer and (2) the potential harm if there was a release of these elements into the environment [1].

The Town of Colonie, County of Albany, New York, USA, has a good management policy. The Town residents can bring their old computers to the Town Solid Waste Management Facility's

"Residential Recyclables Drop Off Area" for recycling. The Town collects old computers from residents and packages them to be shipped out to a private recycling firm, SR Recycling, who separates the salvageable components for reuse, removes the special metals/materials that have recyclable value, and only disposes of the remaining waste materials. The Town charges the residents a fee, US$10 per computer system (monitor, CPU, printer, keyboard, mouse, as a set or parts of set) to pay for the recycling of these units. When the Town collects sufficient units to make up a shipment, the vendor is called to collect the computers [1].

Through the Town's recycling system, the residents are provided an environmentally and economically sound means of managing the e-wastes. This ensures that the materials of concern within these e-wastes are effectively and appropriately managed.

13.5.11 Nanotechnology for Mercury Removal

When the mercury-containing equipment is improperly disposed of on land, the mercury will eventually leach out from the waste equipment. Once released into the environment, mercury remains there indefinitely, contaminating the soil, sediment, and groundwater. This contamination eventually enters the food chain, exposing local populations to mercury's harmful effects [2].

Until now there has been no effective technology for reducing groundwater mercury to two parts per billion, as required by the maximum contamination limit for drinking water established by the U.S. Food and Drug Administration and the U.S. EPA.

According to the U.S. Department of Energy's PNNL, a new nanotechnology has been developed by PNNL for mercury removal without producing harmful byproducts or secondary waste. The technology is an advanced adsorption technology involving the use of a powder adsorbent, called SAMMS. SAMMS stands for "Self-Assembled Monolayers on Mesoporous Supports," which is critically important in constantly changing industries and environment. It has broad applications in environmental cleanup where mercury contamination is prevalent, or for mercury removal in radiological hazardous waste.

Technically speaking, SAMMS is a hybrid of two frontiers in materials science: molecular self-assembly techniques and nanoporous materials. SAMMS is created by attaching a monolayer of contaminant-specific molecules to nanoporous ceramic supports. The nanoporous materials arranging from 2 to 20 nm, with high surface areas (about 600–1000 m^2/g), are functionalized with a self-assembled monolayer, resulting in an extremely high density of binding sites. The functionalized material exhibits fast kinetics, high loading, and excellent selectivity for contaminants.

Both the monolayer and the nanoporous support can be tailored for a specific application. For example, the functional group at the free end of the monolayer can be designed to selectively bind targeted molecules, while the pore size, monolayer length, and density can be adjusted to give the material specific adsorptive properties. This monolayer will seek out and adsorb specific contaminants. When tested on 160 L of waste solution containing about 11 ppm of mercury, or a total of 1.76 g, mercury concentration in the solution was reduced by about 99.5%. Estimates indicate that it will cost about US$200 (October 2004 cost), including material, analysis, and labor, to treat similar volumes of this waste solution, resulting in a saving of US$3200 over more traditional polymeric adsorbent (resin) or activated carbon disposal methods.

13.5.12 Solidification (Cementation) Technology for Hazardous E-Waste Disposal

Cementation technology is one of the solidification technologies, involving the use of a solidifying agent (i.e., cement, in this case) for solidifying the hazardous solid e-wastes (such as mercury-containing batteries or equipment). Conventional cementation technology has problems: (1) the solidified cement or concrete is still porous, and eventually the hazardous substances may leak out; and (2) the solidified cement or concrete blocks are not strong enough, and may break upon impact or earthquake.

An improved solidification (cementation) technology has been used by Dr. Lawrence K. Wang of the Lenox Institute of Water Technology, Massachusetts, USA, for successful solidification of mercury-containing batteries in concrete blocks. The concrete blocks, which are friendly to the environment, can then be properly buried in the government-approved hazardous waste landfill sites [3].

Specifically, the improved solidification (cementation) technology involves the use of (1) a special dry powder admixture for generation of a nonsoluble crystalline formation deep within the pores and capillary tracts of the concrete—a crystalline structure that permanently seals the concrete against the penetration or movement of water and other hazardous liquids from any direction; (2) a special nonmetal reinforced bar for enhancing the concrete block's tensile and compressive strengths; and (3) a unique chemical crystallization treatment for the waterproofing and protection of the concrete block's surface.

To create its crystalline waterproofing effect, the special solidifying agent must become an integral part of the concrete mass. It does so by taking advantage of the natural and inherent characteristics of concrete; concrete is both porous (capillary tract system) and chemical in nature. By means of diffusion, the reactive chemicals in the agent use water as a migrating medium to enter and travel through the capillary tracts in the concrete. This process precipitates a chemical reaction between the agent, moisture, and the natural chemical by-products of cement hydration (calcium hydroxide, mineral salts, mineral oxides, and unhydrated and partially hydrated cement particles). The end result is crystallization and, ultimately, a nonsoluble crystalline structure that plugs the pores and capillary tracts of the concrete is thereby rendered impenetrable by water and other liquids from any direction.

The chemical treatment is permanent. Its unique, crystalline growth will not deteriorate under wide conditions. The treated concrete block is structurally strong and is not affected by a wide range of aggressive chemicals including acids, solvents, chlorides, and caustic materials in the pH range of 3–11.

13.6 RECENT ENVIRONMENTAL AWARENESS, ACTIONS, AND RECOMMENDATIONS

13.6.1 RECENT ENVIRONMENTAL AWARENESS

13.6.1.1 Legal Terms of Electrical Wastes and Electronic Wastes

The definition of e-waste is unclear. Some government agencies define e-waste being the combination of (1) electrical waste (electrical washing machines, dryers, air conditioners, humidifiers, dehumidifiers, vacuum cleaners, toasters, ovens, irons, coffee machines, lawn mowers, stoves, electrical exercise machines, toys, lamps, batteries, power tools, refrigerators, etc.); and (2) electronic waste (VCR players, DVD/CD players, Blu-ray disc players, game consoles, radios, Hi-Fi equipment, camcorders, computers, tablets, terminals, CPUs, keyboards, scanners, calculators, telephones, fax machines, printers, photocopiers, external storage drives, modems, speakers, cell phones, pagers, cable/satellite receivers, televisions, monitors, etc.) [1]. Some government agencies, however, define e-waste as being electronic waste only. Still some government agencies define e-waste being the selected certain electrical and electronic waste. For instance, the City of Jacksonville considers only the following are e-wastes: televisions, computer monitors, computer terminals, CPUs, keyboards, printers, scanners, stereo equipment, radios, VCRs, DVDs, camcorders, desk phones, mobile phones, pagers, power tools, small kitchen appliances (microwaves, toaster ovens), and health and beauty appliances [57]. Legally, large appliances are not included in the e-waste definition in many states.

Still further, some government agencies consider e-waste to be a part of universal wastes.

13.6.1.2 Breakdown and Generation of Electrical and Electronic Wastes

It has been known that the breakdown of electrical waste and e-waste is about (1) 30% electrical washing machines, dryers, air conditioners, humidifiers, dehumidifiers, vacuum cleaners, small appliances (toasters, microwave ovens, irons, coffee machines, etc.), lawn mowers, stoves, electrical exercise machines, toys, lamps, batteries, power tools, etc.; (2) 20% refrigerators; (3) 15% VCR players, DVD/CD players, Blu-ray disc players, game consoles, radios, hi-fi equipment, camcorders, etc.; (4) 15% computers, terminals, CPUs, keyboards, scanners, calculators, telephones, fax machines, printers, photocopiers, external storage drives, modems, speakers, cell phones, pagers, cable/satellite receivers, etc.; (5) 10% televisions; and (6) 10% monitors [1].

In 2009, the United States generated a total of 3.19 million tons of e-wastes, among which 82.3% was trashed and only 17.7% was recycled. Also in 2009, the entire world generated about 53 million metric tons of e-wastes, of which 77% was trashed, and only 13% was recycled.

13.6.1.3 Contents, Values, and Recycle of E-Waste

E-waste is hazardous because it contains toxic heavy metals and toxic organics. The toxic heavy metals (such as lead, antimony, arsenic, beryllium, mercury, etc.) and toxic organics (such as PCB, CFC, etc.) originally in the e-waste and the toxic substances produced during the e-waste recycling process (such as cyanide, acid, dioxins) require high cost for proper treatment and disposal in order to protect public health and the environment.

E-waste is also valuable because its precious metals (such as platinum, palladium, silver, gold, copper, cobalt, etc.), plastic pellets, processed glass, base metals (such as iron, aluminum, etc.) can be recycled for reuse or profit. The scrap value of e-waste is much higher than regular solid wastes. Circuit boards, for instance, contain small, but significant, amounts of gold, silver, copper, and other valuable elements. The recycling value of base metals, plastic pellets, and processed glass is also high [1].

Environmental technologies and managerial skills are readily available in industrial countries (such as the United States, the United Kingdom, Germany, Japan, France, etc.) as well as some developing countries (such as China, India, etc.) for proper e-waste collection, treatment, separation, recycle, and disposal. Figure 13.4 shows how old computers can be easily processed for reuse in new computer production. Unfortunately, the e-waste problem is similar to the global warming problem. Most politicians and businessmen understand the solution, but refuse to face the problem or to pay for any problem-solving costs.

13.6.2 Environmental Problems and Actions

13.6.2.1 Global Environmental Problems and International Actions

European industrial nations ban exporting e-wastes to foreign countries, and require proper e-waste collection, treatment, and disposal within the nations. Inspections of 18 European seaports in 2005 found that as much as 47% of waste (including e-waste) destined for export was illegal according to the International law. In the UK alone, government computers have been illegally dumped in African countries, Pakistan, Malaysia, India, and China [17,51–54].

Although the United States is a global leader in designing and developing new and improved electrical and electronic technologies, there is no U.S. federal mandate to recycle e-waste. There have been numerous attempts to develop a federal law. However, to date, there is no consensus on a U.S. federal approach due to objections of some politicians and businessmen who place economics ahead of environment [24,49–50]. Since export of e-waste is legal in the United States, most of the e-waste collected in this country for recycling is actually exported to developing countries, such as China and Malaysia, where labor costs are lower and environmental regulations are lax or not enforced, resulting in major environmental pollution and serious health problems. China is now the electronic wastebasket of the world because about 70% of e-waste globally generated ends up in China [12–13,17]. Figure 13.5 shows the routes of (1) illegal e-waste export from Western Europe to Africa, Pakistan, and India, and (2) legal but immoral e-waste export from North America to China and Vietnam.

Management, Recycling, and Disposal of Electrical and Electronic Wastes (E-Wastes)

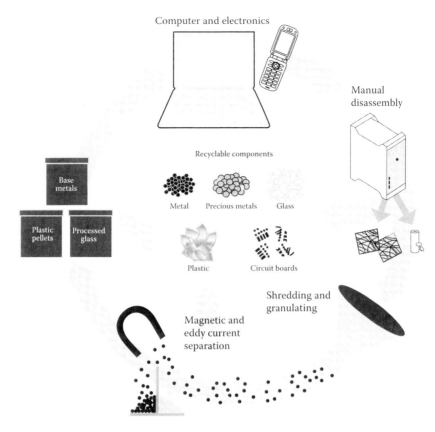

FIGURE 13.4 Flow diagram for management, separation, recycling, and disposal of e-waste in general. (From IAGS, Electronics Recycling FAQ, IAGS Newsletter, April 2013, Interior Alaska Green Star, Fairbanks, AK, www.iagreenstar.org/resources/e-recycling-faq.)

FIGURE 13.5 Exports of e-waste from North America and Western Europe to Africa and Asia. (From Greenpeace, Where does e-waste end up? Greenpeace, Amsterdam, the Netherlands, www.greenpeace.org/international/en/campaigns/toxics/electronics/the-e-waste-problem/where-does-e-waste-end-up/, February 24, 2009.)

E-wastes that are shipped overseas are not properly treated with the Best Available Technology for its valuable substances recovery and for its safe residue disposal. Instead, the e-waste is improperly burned, soaked in acid baths, dumped into receiving waters, or piled into mountains for scrap recovery using primitive methods. These primitive practices risk the release of toxic heavy metals, toxic volatile organic compounds, and acids into the surrounding air, water, and land, thereby risking significant negative impact to the public health and the environment. The United Nations [17,45–46], largest e-waste basket China [13], largest e-waste producer United States [24,49–50], European Union [53], Greenpeace [53], Interior Alaska Green Star [55], etc., are all taking actions in order to ensure that all collected e-waste can be processed and recycled according to safe and responsible procedures.

13.6.2.2 Actions in the United States and the United Nations

The U.S. EPA encourages recycling of CRTs under the Cathode Ray Tubes Final Rule. Circuit boards are subject to a special exemption from federal hazardous waste rules. The following are the U.S. federal regulatory requirements for disposal of CRTs and other e-wastes that test "hazardous":

1. Large Quantities Sent for Disposal: Wastes from facilities that generate over 100 kg (about 220 lb) per month of hazardous waste are regulated under federal law when disposed. CRTs from such facilities sent for disposal (as opposed to reuse, refurbishment, or recycling) must be manifested and sent as "hazardous waste" to a permitted hazardous waste landfill;
2. Small Quantities Exempt: Businesses and other organizations that send for disposal (as opposed to reuse, refurbishment, or recycling) less than 100 kg (about 220 lb) per month of hazardous waste are not required to handle this material as hazardous waste. If a "small quantity generator" wishes to dispose of a small quantity of CRTs or other used electronics that test hazardous under federal law, these materials can go to any disposal facility authorized to receive solid waste (e.g., a municipal landfill), unless state law requires more stringent management (e.g., CA); and
3. Household Exemption for Electronics Sent to Disposal: Used computer monitors or televisions generated by households are not considered hazardous waste and are not regulated under federal regulations. State laws may be more stringent.

The state and local regulatory requirements for e-waste in the United States can be more stringent than the U.S. federal requirements, and vary from state to state, and local to local [25–39,47–48,56–58]. For instance, the State of California considers CRTs to be spent materials and regulates all CRTs as hazardous waste, and CRTs are banned from landfills [56]. Other states, such as Massachusetts and Florida, have taken steps to streamline hazardous waste regulations for CRTs, reducing special handling requirements if these products are directed to recycling [57–60]. Many states are developing universal waste exemptions for CRTs, which also streamline management of CRTs bound for recycling. If you are planning on disposing used CRTs (or other electronics that test "hazardous" under state or federal law), check relevant state requirements, which might be different from the U.S. federal regulatory requirements.

At the international level, the United Nations University (UNU), Bonn, Germany, has had a Solving the e-waste Problem (STEP) initiative, which comprised manufacturers, academics, governments, recycling companies, and other organizations committed to solving the world's problems of waste electrical and electronic equipment (WEEE), or e-waste [61–68]. By providing a forum for discussion and management among stakeholders, STEP may be effectively sharing technical information, seeking feasible answers and implementing solutions.

The UNU STEP initiative's prime objectives are (1) research and piloting: by conducting and sharing scientific research, STEP may help to shape effective policy-making; (2) strategy and goal setting: a key strategic goal is to empower pro-activity in the marketplace through expanded membership and to secure a robust funding base to support activity; (3) training and development: STEP's global overview of e-waste issues makes it the obvious provider of training on e-waste issues; and (4) communication and branding: one of STEP's priorities is to ensure that UN members, prospective members, and legislators are all made aware of the nature and scale of the problem, its development opportunities, and how STEP is contributing to solving the e-waste problem at global level [61–68].

According to UNU, there are four core principles for its STEP initiative: (1) STEP views the e-waste issue holistically, focusing on its social, environmental, and economic impact, locally, regionally, globally; (2) STEP follows the lifecycle of electrical and electronic equipment and its component materials from sourcing natural resources, through distribution and usage, to final disposal; (3) STEP's research and pilot projects are "steps to e-waste solutions"; (4) STEP vigorously condemns the illegal activities that exacerbate e-waste issues, such as the illegal shipments, recycling practices, and disposal methods that are hazardous to people and the environment; and (5) STEP encourages and supports best-practice reuse and recycling worldwide. The readers are encouraged to contact UNU directly for more technical information, financial support, and/or managerial assistance [61–68].

13.6.2.3 Summary and Recommendations

WEEE is also commonly called "e-waste," which includes all old, end-of-life, or discarded appliances, instruments, apparatus, or equipment, using electricity. Common e-wastes include computers, audio/video equipment, refrigerators, air-conditioning equipment, televisions, cell phones, lamps, batteries, analytical instruments, copiers, fax machines, etc. [60,69,70,71]. Many of these e-wastes can be reused, refurbished, or partially recycled. The following are some recommendations for solving the e-waste problem:

1. A federal bill of each nation shall be passed requiring electronics manufacturers to recycle old goods with a surcharge (similar to the waste tire recycle program and waste car battery recycle program). It is important to note that passing a similar state bill would be less sufficient because the prices of goods would vary from state to state creating unfair competition. A state bill is better than no bill at all.
2. A federal bill of each nation shall be passed to ban any export of e-waste to any foreign countries. It is noted that e-waste is a valuable commodity. Keeping the e-waste within the country for proper treatment, recycle, and disposal would create jobs using the surcharge collected from Recommendation No. 1, and would eliminate environmental pollution (caused by improper e-waste treatment) in a foreign developing or underdeveloped country. A federal e-waste export ban is effective. A similar state e-waste export ban is useless.

REFERENCES

1. Wang, L.K. Recycling and disposal of electrical and electronic wastes. *Proceedings of the 2004 Modern Engineering and Technology Seminar*, Taipei, Taiwan (2004).
2. Mattigod, S. Mercury remediation: A tiny solution to a big problem—Using nanotechnology for adsorbing mercury. *Water & Wastewater Products*, 9, 20–24 (2004).
3. Wang, L.K. *Advanced Solidification Technology for Disposal of Hazardous Wastes*. Zorex Corporation, Pittsfield, MA, Technical Report 29008-3-90-52, 79 p. Co-sponsored by the Lenox Institute of Water Technology, Lenox, MA (1990).
4. SAEFL. Waste management guidelines for the ordinance on the return, the taking back and the disposal of electrical and electronic appliances (ORDEA). Swiss Agency for the Environment, Forests and Landscape, 76 p. Bern, Switzerland (2000).
5. Bleicher, S.A. The mercury-containing and rechargeable battery management act of 1996: A new direction for recycling. *Environment Reporter*, Vol. 27, The Bureau of National Affairs, Inc. (1996).
6. Fishbein, B. Extended product responsibility: A new principle for product-oriented pollution prevention, industry program to collect and recycle nickel-cadmium (Ni-Cd) batteries. US Environmental Protection Agency, Washington, DC, EPA530-R-97-009, pp. 6-1–6-32 (1997).

7. U.S. EPA. Mercury-Containing and Rechargeable Battery Management Act. Public Law No. 104-142. US Environmental Protection Agency, Washington, DC (1996).
8. U.S. EPA. Characterization of municipal solid waste in the United States: 1996 update. US Environmental Protection Agency, Washington, DC, EPA530-R-97-015. www.epa.gov/epaoswer/non-hw/muncpl/msw96.htm (2016).
9. U.S. EPA. *Decision-Maker's Guide to Solid Waste Management*, 2nd Edition. US Environmental Protection Agency, Washington, DC, EPA530-R-95-023. www epa.gov/epaoswer/non-hw/muncpl/dmg2.htm (2015).
10. U.S. EPA. Characterization of products containing lead and cadmium in municipal solid waste in the United States, 1970 to 2000. US Environmental Protection Agency, Washington, DC, EPA530-SW-89-015B (1989).
11. U.S. EPA. Contaminated scrap metal. US Environmental Protection Agency, Washington, DC. http://www.epa.gov/rpdweb00/source-reduction-management/scrapmetal.html (2008).
12. Watson, I., and Young, C. China: The electronic wastebasket of the world. May 30, 2013. www.cnn.com/2013/05/30/world/asia/china-electronic-waste-e-waste.
13. Conference Manager. e-Waste world, Shanghai, China. March 20–21, 2013. www.demand-led.com/ewaste/.
14. Ongondo, F.O., Williams, I.D., and Cherrett, T.J. How are WEEE doing? A global review of the management of electrical and electronic wastes. *Waste Management*, 31 (4), 714–730 (2011).
15. Lau, W.K.Y., Chung, S.S., and Zhang, C. A material flow analysis on current electrical and electronic waste disposal from Hong Kong households. *Waste Management*, 33 (3), 714–721 (2013).
16. Kiddee, P., Naidu, R., and Wong, M.H. Electronic waste management approaches: An overview. *Waste Management*, 33 (5), 1237–1250 (2013).
17. Wang, F., Kuehr, R., Ahlquist, D., and Li, J. e-Waste in China: A country report. United Nations University, Japan and Tsinghua University, China, 60 p. (2013).
18. STEP. Solving the e-waste problem: e-Waste country study Ethiopia, Bonn, Germany, April (2013).
19. Kuehr, R., and Williams, E. (eds). *Computers and the Environment: Understanding and Managing Their Impacts*. Dordrecht, Boston, and London, 285 p. (2004).
20. Hilty, L.M. *Information Technology and Sustainability*. BOD, Norderstedt, Germany, 180 p. (2008).
21. Huisman, J. 2007 Review of Directive 2002/96 on Waste Electrical and Electronic Equipment (WEEE), European Commission, Bonn, Germany, 377 p. (2007).
22. Laissaoui, S.E., and Rochat, D. *Assessment of e-Waste Management in Morocco*. Hewlett Packard, DSF, Empa, St. Gallen, Switzerland, 67 p. (2008).
23. Waema, T., and Mureithi, M. *e-Waste Management in Kenya*. Hewlett Packard, DSF, Empa, Nairobi, Kenya, 67 p. (2008).
24. U.S. EPA. Cleaning up electronic waste (e-waste). US Environmental Protection Agency, Washington, DC. May 2013. www.epa.gov/international/toxics/ewaste/index.html.
25. TMC. e-Waste collection. Texas Medical Center (TMC), Sustainability Advisory Council, Houston, TX. April 2013. http://tmcsustainability.org/home.html.
26. UU. e-Waste recycling day. University of Utah, Salt Lake City, UT, and Samsung. April 2013. http://facilities.utah.edu/news/2013/e-waste-day.php.
27. NYSDEC. Management of used electronic equipment in New York State. NYS Department of Environmental Conservation, Albany, NY (2013).
28. NYSDEC. Guidance for municipal collection activities. NYS Department of Environmental Conservation, Albany, NY (2013).
29. NYSDEC. Guidance for business, institutions and government. NYS Department of Environmental Conservation, Albany, NY (2013).
30. NYSDEC. Guidance for collectors, dismantlers and recyclers. NYS Department of Environmental Conservation, Albany, NY (2013).
31. NYSDEC. C7 notifications for electronic waste. NYS Department of Environmental Conservation, Albany, NY (2013).
32. NYSDEC. Used electronic equipment FAQs. NYS Department of Environmental Conservation, Albany, NY (2013).
33. Capital District Junk King. e-Waste recycling in Albany, New York. New York Recycling. Capital District Junk King, Ravena, NY. September 19, 2012.
34. Capital District Junk King. Lawn mower disposal in Albany, New York. Albany Appliance Disposal. Capital District Junk King, Ravena, NY. April 30, 2013.
35. Capital District Junk King. Albany NY bulk pickup. Albany, Appliance Disposal. Capital District Junk King, Ravena, NY. February 15, 2013.

36. Capital District Junk King. Albany, NY old washing machine and dryer disposal. Albany Appliance Disposal. Capital District Junk King, Ravena, NY. December 18, 2012.
37. Capital District Junk King. Albany treadmill and exercise machine disposal. Albany Appliance Disposal. Capital District Junk King, Ravena, NY. October 29, 2012.
38. Capital District Junk King. Albany, NY appliance disposal. Albany Appliance Disposal. Capital District Junk King, Ravena, NY. March 30, 2012.
39. City of Albany. Hazardous waste and electronics. City of Albany Landfill, NY. June 2013. www.albanyny.org.
40. Fizz, R. *Multiple Choice: How Do You Handle e-Waste at MIT.* Massachusetts Institute of Technology, Cambridge, MA, 2013.
41. USD. San Diego's only non-profit, one-stop e-waste collection center. University of San Diego, CA. May 28, 2013.
42. Green Deals. Recycling electronics. June 2013. http://blog.greendeals.org.
43. Kopytoff, V. The complex business of recycling e-waste. *Businessweek*, New York. January 8, 2013.
44. Satariano, A. e-waste recycling becomes techie's mission. Bloomberg, New York. December 28, 2012. http://www.sfgate.com.
45. Kuehr, R. First STEP open meeting in Africa in Addis Ababa, Ethiopia. United Nations University, Japan. October 2011. http://isp.unu.edu/news/2011/first-step-open-meeting-in-africa-in-addis-ababa-ethiopia.html.
46. UNU and UNIDO. Implementing Ethiopian e-waste management project. United Nations University, Japan, and United Nations Industrial Development Organization, Austria. May 24, 2013.
47. NYC. *eCycling: The Future Is Now.* New York City EPA, New York (2012).
48. Miss DEQ. Solid waste: Electronics waste (e-waste) program. Mississippi Department of Environmental Quality, Jackson, MS. June 2013. www.deq.state.ms.us.
49. U.S. EPA. e-Cycling regulations and standards. US Environmental Protection Agency, Washington, DC. June 2013. http://www.epa.gov.
50. U.S. EPA. EPA's international priorities. US Environmental Protection Agency, Washington, DC. June 2013. www.epa.gov.
51. BBC. Europe breaking electronic waste export ban. BBC, London. August 4, 2010. www.bbc.co.uk/news/world-10851645.
52. Gray, R. Government computers illegally exported as waste. The Telegraph. September 11, 2010. www.telegraph.co.uk/earth/earthnews/7996458.
53. Greenpeace. Where does e-waste end up? Greenpeace, Amsterdam, the Netherlands. February 24, 2009. www.greenpeace.org/international/en/campaigns/toxics/electronics/the-e-waste-problem/where-does-e-waste-end-up/.
54. Land, G. UK Government and European e-waste illegally dumped in Africa. Politics, pollution, recycling. September 13, 2010. www.greenfjdge.org.
55. IAGS. Electronics Recycling FAQ. IAGS Newsletter, April 2013. Interior Alaska Green Star, Fairbanks, AK. www.iagreenstar.org/resources/e-recycling-faq.
56. CAS-DTSC. Electronic hazardous waste (e-waste). California State Department of Toxic Substances Control, Sacramento, CA. www.dtsc.ca.gov (2010).
57. City of Jacksonville. Electronic wastes. City of Jacksonville, FL. June 2013. www.coj.net/departments/public-works/solid-waste/electronic-wastes.aspx.
58. Editor. Bill would require electronics manufacturers to recycle old goods. WWLP-22 News, Chicopee, MA. March 26, 2013. www.wwlp.com/dpp/news/politics/state_politics/.
59. Wang, L.K., and Wang, M.H.S. The story of mercury-containing e-wastes and batteries. New York State Cultural and Environmental Education Programs, Albany, NY. July 17–August 25, 2006.
60. California State. What is e-waste? California State Government. January 8, 2016. www.calrecycle.ca.gov/Electronics/.
61. UNU. e-Waste indicators, United Nations University, Institute for the Advanced Study of Sustainability, Bonn, Germany. September 15, 2011.
62. UNU. Recommendations on standards for collection, storage, transport and treatment of e-waste. United Nations University, Institute for the Advanced Study of Sustainability, Bonn, Germany. Green Paper issued on June 22, 2012; White Paper issued on June 2, 2014.
63. UNU. e-Waste take-back system design and policy approaches. United Nations University, Institute for the Advanced Study of Sustainability, Bonn, Germany. White Paper issued on January 28, 2009.
64. UNU. One global understanding of reuse—Common definitions. United Nations University, Institute for the Advanced Study of Sustainability, Bonn, Germany. White Paper issued on March 5, 2009.

65. UNU. Transboundary movements of discarded electrical and electronic equipment. United Nations University, Institute for the Advanced Study of Sustainability, Bonn, Germany. Green Paper issued on March 25, 2013.
66. UNU. e-Waste in China. United Nations University, Institute for the Advanced Study of Sustainability, Bonn, Germany. Green Paper issued on April 5, 2013.
67. UNU. e-Waste country study—Ethiopia. United Nations University, Institute for the Advanced Study of Sustainability, Bonn, Germany. Green Paper issued on April 10, 2013.
68. UNU. Differentiating EEE products and wastes. United Nations University, Institute for the Advanced Study of Sustainability, Bonn, Germany. Green Paper issued on January 14, 2014.
69. Wang, L. K., and Wang, M. H. S. *Environmental Engineering Glossary*, 2nd Edition. Lenox Institute of Water Technology, Newtonville, NY (2015).
70. Wang, M. H. S and Wang, L. K. Glossary of Land and Energy Resources Engineering. In Natural Resources and Control Processes. Wang, L. K., Wang, M. H. S., Hung, Y. T. and Shammas, N. K. (eds.). 493–623. Springer, Switzerland (2016).
71. Chen, J. P., Wang, L. K., Wang, M. H. S., Hung, Y. T., and Shammas, N. K. (eds). *Remediation of Heavy Metals in the Environment*. CRC Press, Boca Raton, FL, 528 p (2017).

Index

Page numbers followed by *f* indicate figures; those followed by *t* indicate tables.

A

Abrasive jet machining, 310
Accident prevention and emergency response, 16–19
Acid dyes, 384
Acoustic Doppler velocimeter, 170
Acrylonitrile, 377
Activated carbon (AC) water treatment processes, development of
    adsorption treatability studies
        for methyl-tert-butyl ether (MTBE) removal, 361–362
        for trichloroethylene (TCE) removal, 359–361
    advantages of, 354*t*
    apparatus for adsorption breakthrough experiments, 357*f*
    biological activated carbon (BAC) capability of adsorber, 363–369
        biotreated coking plant effluent, treatment, 367–369
        for methyl-tert-butyl ether (MTBE) removal, 364
        for sulfolane removal, 363–364, 363–364*f*
        for xylene (BTEX) removal, 364, 366–367, 367*t*
    breakthrough curve pattern and effluent limit on capacity utilization rate, 358*f*
    equations of adsorption study parameters, 355*t*
    fundamentals of AC adsorption technology, 355–356
    GAC's catalytic capability, advantage, 369, 369*f*
    methods for adsorption isotherm and breakthrough experiments, 356–359, 360*f*
    PAC and GAC requirement calculations, 355*t*
    reactivating spent carbon, 370, 370*f*
Activated sludge processes, in poultry waste treatment, 38–40
    contact stabilization process, 40, 40*f*
    conventional process, 38–39, 39*f*
    extended aeration process, 40, 40*f*
    high-rate process, 39, 39*f*
Adsorption techniques, 228–231
    activated carbon and bentonite for various dyes, 244*t*
    color adsorption from reactive and basic dyeing wastewaters, 245*f*
Aerated lagoon, 42
Aerobic/anaerobic lagoons, 42
Aerobic degradation, of wastewater constituents, 29
Agricultural waste
    other solid wastes, 249, 253*t*
    rice husk, 249, 251–252*t*
    sawdust, 248, 250*t*
Air-conditioning waste appliances, management and disposal of, 396–397
    flow diagram for, 398*f*
Air dissolving tube (ADT), 70

Air emissions
    produced by chlor-alkali plants, 126–127
        treatment technology, 133
    standards, for mercury, 140, 141*t*
Air knives, 325–326
Air, solubility of, 157–159
Alar Auto-Vac filter drum, 87
Algae, 29
Alkaline batteries, 405–406
Alkaline chlorination, 333
American Petroleum Institute (API), 79, 381
Ammonia stripping, 16
Anaerobic digestion process, food scraps in, 189–194
Animals, food scraps for, 188
Anionic dyes, 384
Anodizing, 309
Anthraquinone dyes, 214
"Apparently fit grocery product," 186
"Apparently wholesome food," 186
Aqua-EMBR system, 382
Arsenic, 6*t*
Arsenic- and selenium-containing wastes, 336–337
Asbestos, 104, 119, 124, 128
    occupational health and safety issues, 130–131
Assembly, 312
Autoclaves, see Batch reactors
Auto-Vac vacuum filter, 86, 86*f*
    analysis of sludge cake and filtrate from, 87*t*
Auxochromes, 214
Azo disperse dye, 214
Azos, 385

B

Barrel finishing, 309
Barrel plating, 324
Basic dyes, 384
Batch reactors, 374
Bath evaporation rate, 327
Batteries, 399
    environmental hazards of, 403
    management/reuse/recycle/disposal of household and vehicle, 401, 403–407
Battery Act, 406–407
Battery pack, 405
BayFlotech technique, 383
Benzene, 6*t*, 377
Best available technology (BAT)
    for chlor-alkali production, 134–135
    effluent limitations for metal finishing point source category, 343, 344*t*
    effluent limitations on cyanide, 343, 344*t*

417

Best management practices (BMPs)
   stormwater quality controls
      BMP control performance assessment, 93, 94t
      detention basins, 90–91
      grassed swales, 92–93
      infiltration basins, 91–92
      infiltration trenches, 91
      retention basins, 91
      sand filters, 92
      water quality inlet, 92
Best practical control technology (BPT)
   effluent limitations for metal finishing point source category, 342, 343t
   effluent limitations on cyanide, 342, 343t
Bhopal calamity, in India, 16–17
Bioaccumulation, of methyl mercury, 128, 129f
Biochemical oxygen demand (BOD) determination, 29, 40, 286
   BOD bottles and glassware cleaning, 292–293
   BOD testing wastewater, 293
      small quantity generator status, 293
      treatment containing chromium, 293–294
      treatment containing no heavy metals, 294
   interferences, 286–287
   laboratory waste management and disposal, 293–294
   on-site waste treatment, 293–294
   pretreatment of wastewater samples for BOD testing, 288–289
   procedures with seeding
      BOD calculations, 291–292
      reliability observation, 292
      sample preparation and DO determinations, 291
   procedure without seeding
      BOD calculations, 290–291
      reliability observation, 291
      sample preparation and DO determinations, 289–290
   testing apparatus, 287
   testing reagents, 287
Biogas, 193
Biohydrogen, 188
   production, 197–198
Biological activated carbon (BAC) capability of adsorber, 363–369
Biological technologies
   combined sewer overflow (CSO) treatment by, 68
   wastewater treatment by, 69
Biological treatment
   of poultry processing wastewater, see Poultry processing wastewater
Biological treatment, on dyeing wastewater, 225–226
   from acid chrome dye on wool, 230t
   from acid dye on polyamide, 232t
   from after-copperable direct dye on cotton, 229t
   from basic dyes on polyacrylic, 231t
   from basic dyes on polyester, 236t
   from direct developed dye on rayon, 233t
   from direct dye on rayon, 232t
   from disperse, acid, and basic dyes on polyamide carpet, 233t, 236t
   from disperse dyes on polyamide carpet, 230t
   from disperse dyes on polyester, 228t, 234t
   from disperse dyes on polyester carpet, 231t
   effect on nitrification, 238t
   experimental design, 226f
   list of samples from industrial and municipal wastewaters, 227t
   from napthol on cotton on polyamide carpet, 237t
   from 1:2 metal complex dye on polyamide, 228t
   and physical–chemical treatment, 237, 249t
   from reactive dyes on cotton, 229t, 235t
   from sulfur dye on cotton, 234t
   from vat and disperse dyes on 50/50 cotton-polyester blend, 235t
   from vat dyes on cotton, 227t
Biological waste treatment processes, 16
Bio-magnification process, 128
Biomass
   adsorption capacities of, 261–262t
   for dye removal, 256
   gasification process, 194–197
Bleach, 113, 137t
BMP control performance, assessment of, 93
BOD_5 (5 days biochemical oxygen demand), 37, 40
Borden Chemical Company, 381
Brazing, 310
Breakdown and generation of E-wastes, 409–410
Brilliance of the Seas, 201, 202f
Brine manufacturing process, 112–113
Brownfields program, 13
Bubble formation, processes in, 151
Bubble generation, 159–160
Bubble–particle agglomerates, flotation of, 151
Bubble–particle attachment mechanisms, 151
Bubble–particle system, geometry of, 162f
Burnishing, 309

C

Cadmium, 6t
Calcium chloride ($CaCl_2$) solution, 124–125, 287
Calcium sulfate ($CaSO_4$), 125
Calibration, 312
Capture/concentration techniques, 326–327
Carbon adsorption, 340
Carbon cycle, 29
Carbon sorption, 16
Carbon tetrachloride, 131
Carboxymethyl cellulose (CMC), 213
Carpet finishing subcategory, 218
   wastewater treatment cost for, 269t
Catalyst, 387
Cathode ray tubes (CRTs)
   management and disposal of, 400–401
      flow diagram for, 402f
   schematic diagram of, 400f
Caustic scrubber system, and gas emissions, 126
Caustic soda
   effluents/solid/hazardous waste streams from, 122t
   energy consumption/carbon emissions for, 114, 116t
   manufacturing process, 112
   occupational health and safety issues, 131
   production of, 101t
   as products of Olin Corporation, 137t
Cementation technology, for hazardous E-waste disposal, 408–409
Centrifugation, 376
Char, 194

Index

"Charge Up to Recycle" program, 406–407
Chemical accidents
    approaches minimizing, 17, 18f, 19t
    categorical causes of, 17, 17t
Chemical coagulation method
    characteristics of sludge resulting from, 243t
    for dye removal, 226
    physical–chemical treatability studies, 239–242t
Chemical conversion coatings, 309
Chemical finishes
    in textile industry, 214–215
    wastewater from, 223
Chemical oxygen demand (COD) determination
    analytical procedure, 303–304
    apparatus, 301
    COD testing wastewater treatment
        combined removal of toxic chromium and Hg by NaOH neutralization, 304–305
        combined removal of toxic silver and Hg by Na_2S precipitation, 305
        effluent discharge and sludge disposal, 305
        neutralization process and decision-making process, 304
        on-site/off-site treatment, 304
    interference and chemical equation, 301
    principles
        chemical oxidation equation, 299–300
        $Cr_2O_7^{2-}$ determination and FAS standardization, 300–301
        example of chemical oxidation, 300
    regents, 301–303
Chemical precipitation, 337
Chemical risk assessment, 7–8, 17
Chemical substitution
    in metal finishing industry, 321–322, 322–323t
    in textile manufacturing, 271
Chemical waste treatment processes, 16
Chemisorption, 229
Chitin and chitosan
    adsorption capacities of, 257–260t
    for dye removal, 256
Chlor-alkali industries
    contaminants of concern
        asbestos, 128
        lead, 128
        mercury, 127–128
    end use of products, 101–102
    energy requirements for, 114, 115–116t
    environmental impacts
        mercury, 127–128
        sodium chloride, 130
    global overview of discharge requirements
        air emissions, 140, 141t
        liquid effluents, 138, 140
    industry description, 100–101
    manufacturing process
        brine processing, 112–113
        caustic soda processing, 112
        chlorine processing, 108, 110–111t, 112f
        diaphragm cell process, 104
        hydrogen processing, 108, 112
        membrane cell process, 104, 107–108
        mercury cell process, 105–106, 108–109
        sodium chlorate processing, 113–114
        sodium hypochlorite processing, 113
    monitoring and reporting, 138
    occupational health and safety issues
        asbestos, 130–131
        caustic soda, 131
        chlorine, 130
        mercury, 130
    operation and maintenance costs, 136–138
    pollution prevention and abatement
        material substitution, 131–132
        mercury recovery, 132–133
        process related measures, 132
        waste segregation, 132
    treatment technology
        air emission, 133
        BAT, 134–135
        disposal option, 135
        liquid effluents, 133–134
        solid waste, 133
    waste characterization
        air emissions, 126–127
        liquid effluents, 123–126
        process flow for diaphragm cell, 118f
        process flow for membrane cell, 119f
        process flow for mercury cell, 117f
        solid wastes, 116, 118–119, 121–123, 122–123t
        waste streams from diaphragm cell, 121t
        waste streams from mercury cell, 120t
Chlorination water treatment system, 237, 248f
Chlorine
    gaseous emissions of, 126–127
    occupational health and safety issues, 130
Chlorine cell technology, 103f
Chlorine gas, 104
Chlorine production
    energy consumption/carbon emissions for, 114, 115t
    by Olin Corporation, 137t
    by plant and cell type, in U.S., 108, 110–111t, 112f
Chloroform, 6t
Chromatic acid solution, see Dichromate acid cleaning solution
Chromium-containing wastes, 336
Chromium reduction, 340
Chromophores, 214
Chrysolite fibers, 130
Circular flotation tank, 149
Clarification sludge, 384
Clays, for dye removal, 263
    adsorption capacities of, 262t
    advantages/disadvantages of methods, 263t
Clean Air Act (CAA) Amendments, 13–14
Cleaning, 309
Clean plating baths, 325
Clean Water Act, 67
Coarse screens, 89
Code of Federal Regulations (CFR), 341–346
Collection or gleaning of donations, 187
Collision, 160–163
Combined sewer overflow (CSO) treatment
    by biological technologies, 68
    Control Policy of U.S. EPA, 67
    by disinfection technologies, 68
    by physicochemical technologies, 67–68

420 Index

Combined sewer overflow (CSO) treatment (*cont.*)
    summary, 83–84
    waste sludge management system, 84, 84*f*
Commingled wastes, 384
Complexation, 337–338
Complexed metal wastes, 331
Composting process, 198–200
    carbon to nitrogen ratios for, 199*t*
Comprehensive Environmental Response, Compensation, and Liability Act (CERCLA), 11–13
Computer E-wastes, management of, 407
Contact stabilization process, in poultry waste treatment, 40, 40*f*
Continuous stirred tank reactor, 376
Conventional activated sludge process, in poultry waste treatment, 38–39, 39*f*
Costs
    for metal finishing industry wastes
        carbon adsorption, 340
        chromium reduction, 340
        cyanide oxidation, 339–340
        emulsion breaking and oil separation, 339
        lime precipitation and sedimentation, 340
        multimedia filter, 340
        sludge drying beds, 340
        typical treatment options, 339
        ultrafiltration, 340
    for textile wastewater treatment, 263, 266
        carpet finishing, 269*t*
        knit fabric finishing, 268*t*
        low water use processing, 266*t*
        stock and yarn finishing subcategories, 270*t*
        wool finishing, 265*t*
        wool scouring, 264*t*
        woven fabric finishing, 267*t*
Cotton Producers Association, 47
Cracking, 387
Cruise Lines International Association (CLIA), 200–201
Cruise ship's waste treatment system, 202
Crustaceans, 29
Cyanide-containing wastes, 333–336
Cyanide oxidation, 333, 339–340
Cyanide plating processes, 323*t*
Cyanide reduction, 333, 334*f*
Cyanides, 332
Cyanide subcategory, 315, 317–318
    pollutants concentrations, 318*t*

D

Dechlorination, procedure for, 289
Deep tunnels, 89
Demister, 132
Detention basins, 90–91
Dewatered dry sludges, 85
    heavy metal limitations of, 86*t*
Dewatering, 338
Diaphragm cell
    effluent pretreatment standards, 126*t*
    energy requirements, 114
    flow diagram, 105*f*
    process, 101, 104, 134
        investment, 136
        standards for, 141*t*

process flowcharts for wastes generated in, 118*f*
waste streams from, 121*t*
wastewater quantities of plant sources in, 123*t*
Dichromate acid cleaning solution, 292, 293
Dichromate ($Cr_2O_7^{2-}$) determination, 300–301
Dichromate reflux method, procedure for COD determination using, 299–305
Diffusional processes, 376
Diffusion, interception and, 163–166
Dilution water, 287
Direct dyes, 214, 384
Disinfection, 90
Dispersed air flotation, 147
    processes, 147
Disperse dyes, 214, 384
Disposal technologies, 16
Dissolved air flotation (DAF), 332
    chemical dosage after optimization, 73*t*
    clarifier, 82
    DAF application, advantages and process
        in landfill leachate treatment, 173–176
        in wastewater treatment, 170–173, 174–175*t*
    with filtration (DAFF), 71–72
    floated sludge
        analysis, 85
        extraction procedure toxicity, 88
    flow diagram of Lee WWTF with pilot plant location, 72*f*
    investigation, 72–74
    operational parameters of treatment units, 73*t*
    by physicochemical technologies, 70
    raw influent wastewater parameters, 74*t*
    and sand filtration
        pilot plant, 76–77
        treatment efficiency of, 77, 79*t*
    theory
        bubble formation, 151
        bubble generation, 159–160
        bubble–particle attachment, 151
        collision, 160–163
        flotation of bubble–particle agglomerate, 151
        interception and diffusion, 163–166
        kinetics of flotation, 151–157
        solubility of air, 157–159
        tank design, 166–170
    unit, 70
    for wastewater treatment
        evaluation, 74*t*
        model parameters for DAF facilities, 166*t*
        models developed for, 155*t*
        process description, 148–150
        types, 147–148
Distilled water, 287
Donation of food, 186–188
Double drag-out for full reuse, 326*f*, 327*f*
Drag-out
    concentration, 327
    drainage rates, 324*f*
    loss, 323–326
    return rate, 327
Drag-out reduction techniques
    clean plating baths, 325
    fog sprays, 325
    high-temperature baths, 325

low-concentration plating solutions, 324–325
low-viscosity conducting salts, 325
racking, proper, 326
rinse elimination, 324
Drain times, longer, 324
Dye and pigment industries, waste treatment in, 383–385, 386f
Dyebath reuse, 273
costs and savings of, 271t
Dyeing operation
in textile industry, 213–214
wastewater from, 221, 222–223t
Dyes removal by low-cost adsorbents, 237, 243
agricultural waste
other solid wastes, 249, 253t
rice husk, 249, 251–252t
sawdust, 248, 250t
industrial by-products, 249, 250, 252
fly ash, 250, 252, 254–255t
metal hydroxide sludge, 250, 253t
red mud, 252, 255t
natural materials
biomass, 256, 261–262t
chitin and chitosan, 256, 257–260t
clays, 262t, 263, 263t
peat, 255–256, 256t

E

East Bay Municipal Utilities District (EBMUD)
of California, 193
food scrap digester system, 194f
Education program, of chemical safety and hazardous waste management, 20
Egg-shaped anaerobic digester, 191f
Electrical and electronic wastes (e-wastes), management/recycling/disposal of
actions in US and UN, 412
environmental problems and actions, 410–413
flow diagram for, 411f
general requirements for, 393
global environmental problems and international actions, 410–412
practical examples, 394–409
problems, 391
recommendations for solving, 413
recent environmental awareness
breakdown and generation E-wastes, 409–410
contents/values/recycle of E-wastes, 410
legal terms, 409
Switzerland experience, 391
U.S. experience, 391–393
Electrical discharge machining, 310
Electrochemical machining, 310
Electroflotation, see Electrolytic flotation
Electroless plating, 309
Electrolytic decomposition, 333, 334f
Electrolytic flotation, 146, 147f
Electron beam machining, 310–311
Electropainting, 312
Electroplating, 309
Electrostatic painting, 312
Emersion Good Samaritan Food Donation Act, 186, 187
Emulsion breaking, 339

End-of-pipe waste treatment technologies, for organic chemicals manufacturing industry, 381–385
Energy consumption costs, for chlor-alkali plant, 136–138
Energy requirements, for chlor-alkali industry, 114, 115–116t
Environmental managers, 14
Environmental Protection Agency (EPA), see U.S. Environmental Protection Agency (U.S. EPA)
Environmental protection programs, 20
Etching and chemical milling, 309
Ethylene, 376
Ethylene diamine tetraacetic acid, 338
Ethylene dibromide (EDB), 8
Evaporation process, of wastes, 16
E-wastes, see Electrical and electronic wastes (e-wastes)
Extended aeration process, in poultry waste treatment, 40, 40f
Extraction procedure toxicity, of DAF floated sludges, 88
Extraction process, 376

F

Fabrication operations, in textile industry, 216
Fabric formation, 212–213
wastewater from, 219
Fabric preparation, 213
wastewater from, 221, 222t
Fair market value, 188
Federal Insecticide, Fungicide, and Rodenticide Act (FIFRA), 8
Federal Universal Wastes, 399
Felted fabric processing subcategory, 219
Fenton's reagent (H_2O_2–Fe(II) salts), 233–235
Ferric chloride ($FeCl_3$) solution, 287
Ferroin indicator solution, 301
Ferrous ammonium sulfate (FAS) standardization, 300–301, 302
Ferrous sulfate precipitation, 336
Filtration, 376
devices, 331
Fixed-bed gasifier, 196
Flame spraying, 310
Floated sludges
DAF analysis, 85
DAF extraction procedure toxicity, 88
thickening and dewatering of, 86–87
Fluidized bed gasifier, 196
Fly ash
adsorption capacities of, 254–255t
for dye removal, 250, 252
Foam flotation process, 146, 147f
FOG sprays, 325–326
Food waste, from restaurants, 184
Food waste recovery
donation of food, 186–188
food scraps for animal feed, 188
industrial uses
for food scraps, 188–199
for waste FOG, 199–200
restaurant waste treatment on board cruise ships, 200–202
source reduction, 184–186
Forced aeration composting process, 198
Formaldehyde, 387

Free fatty acid (FFA) content, 200
Freundlich adsorption isotherm model, 355
Froth flotation process, 146, 148f
Fugitive emissions, 126
Full-flow pressure flotation, 148
Fuming H_2SO_4 solution, 293
Fungi, 29
Furmano Foods' production plant, in Northumberland, 194

G

Gasification, of biomass, 194–197
Glassware and BOD bottles cleaning, 292–293
Gleaner, 187
Gold Kist project (poultry processing wastewater, case history)
    component adequacy review
        aeration tanks, 60
        aerobic digester, 61
        by-products collector, 59
        final clarifier, 61
        lift station, 59
        outfall sewer and Cl_2 facilities, 62
        sludge drying beds, 61
        stabilization pond, 61
    current operating difficulties
        condenser cooling water, 55, 57
        hydraulic overflows, 57
        odor problems, 58
        solids control, 57–58
    current wastewater loads, 55
    design criteria, 50–53
    expansion at, 54–63
    flow diagram, 50
    future expansion provisions, 53–54
    operating arrangements, 54
    proposed modifications
        air supply, 62
        by-products reclamation system, 62
        condenser cooling water, 63
        grease removal, 63
        plant hydraulics, 62
        sludge drying facilities, 62
    proposed wastewater treatment system loads
        biological loads, 59
        hydraulic loads, 58–59
    selection of treatment process, 49–50
    site selection, 47–48
    waste treatment system costs, 54
    wastewater survey and criteria, 48–49
Granular-activated carbon (GAC) treatment, 354t
    for adsorption breakthrough experiments, 82f, 359t
    adsorptive capacity indicators, values of, 359t
    advantages, 354
    beds, 77, 82
    reactivation service, 370
    for removing arsenic, 77–79
    for removing benzene, 366
    for removing color, O&G and SCOD, 80t
    for removing free chlorine, 369
    for removing MTBE, 361–362
    for removing sulfolane, 363–364
    for removing TCE, 359–361
    for removing TCN, 367–368
    requirement calculations, 355t
Graphite, 106
Grassed swales, 92–93
Grinding, 309
Gross negligence, 187

H

Halogenation, 385
Hazardous and Solid Waste Amendments (HSWA), 308, 337
    hierarchy, 337
Hazardous E-waste disposal, solidification technology for, 408–409
Hazardous waste, from chlor-alkali manufacture, 123t
Hazardous waste minimization plan (HWMP), 22
Hazardous Waste Reduction, conferences on, 22
Heat treating, 310
Henry's law, 157
Hexavalent chromium, 331–332
    reduction, 336
    subcategory, 318–319
        pollutants concentrations, 318t
High-rate activated sludge process, in poultry waste treatment, 39, 39f
High rate trickling filters, 41, 41f
High-temperature baths, 325
Hoover, T., 146
Hot dip coating, 311
Household batteries
    management/reuse/recycle/disposal of, 404–407
    types of, 404t
Hybrid reaction process, 376
Hydrochloric acid (HCl), 137t
    cleaning solution, 292
Hydrogen, 137t
Hydrogen gas, 104, 106
Hydrogen manufacturing process, 108, 112
Hydrogen peroxide (H_2O_2), 232–233, 335
Hydrogen production, 197–198
Hydrolysis, 387
    mechanisms, 335
Hydroxide precipitation, 331
Hypochlorite, 108, 113

I

Impact deformation, 309
Incineration, 338
Industrial by-products, for dye removal, 249, 250, 252
    fly ash, 250, 252, 254–255t
    metal hydroxide sludge, 250, 253t
    red mud, 252, 255t
Industrial technology, and environmental abuse, 6
Industrial uses
    for food scraps
        anaerobic digestion process, 189–194
        biohydrogen production, 197–198
        biomass gasification process, 194–197
        composting process, 198–200
    for waste FOG, 199–200

Index

Industrial wastewater survey, 36
Infiltration basins, 91–92
Infiltration trenches, 91
Intentional misconduct, 187
Interception and diffusion, 163–166
Iodine number, 358t

K

Kinetics of flotation, 151–157
Knit fabric finishing subcategory, 218
    wastewater treatment cost for, 268t
Knitting process, 212
Kushner method
    double drag-out for full reuse, 326f

L

Lagoons
    aerobic/anaerobic, 42
    upgrading existing systems, 42–43, 44–45f
Laminating, 311
Lamps, 399
    management and disposal of, 401
Land disposal, 16
Landfill leachate treatment
    DAF application process in, 173–176
Large-Quantity Handlers of Universal Waste (LQHUW), 400
Laser beam machining, 311
Lead, 6t, 128
Lead-acid batteries, 392, 399, 403–404
Levich, V. G., 155–156
Liability for damages from donated food, 187
Life-cycle cost analysis (LCA), 93–94
Lime precipitation and sedimentation, 340
Liquid effluents, 123–126
    diaphragm cell plant, 134
    mercury cell process, 133–134
    standards, for chlor-alkali industry, 140t
Liquid solubility of gases, 158t
Liquid waste, from restaurants, 184
Longer drain times, 324
LOPROX (low pressure oxidation technology for treating high COD in waste water), 383
Low-concentration plating solutions, 324–325
Low-cost adsorbents, dyes removal by, 237, 243, 248–263
Low-viscosity conducting salts, 325
Low water use processing
    in textile industry, 217
    wastewater treatment cost for, 266t

M

Machining, 309
Magnesium sulfate (MgSO$_4$) solution, 287
Man-made fibers, 212
Maquiladora textile and apparel industry, 210
Mean cell residence time (MCRT), 192, 194
Mechanical finishes, in textile industry, 215
Membrane cell
    energy requirements, 114
    flow diagram, 107f
    process, 101, 104, 107–108
        investment, 136
    process flowcharts for wastes generated in, 119f
    wastewater quantities of plant sources in, 123t
Mercuric-oxide batteries, 405, 406, 407
Mercuric sulfate (HgSO$_4$), 302
Mercury, 6t, 125
    in aquatic environments, 129t
    contaminants of concern, 127–128
    environmental impacts, 128–129
    MITI banned use of, 140
    monitoring techniques, 139t
    occupational health and safety issues, 130
    removal, from waste gases, 132–133
        BDAT process, 133
    standards, for industrial discharge, 138, 140, 141t
Mercury cell
    effluent pretreatment standards, 125t
    energy requirements, 114
    flow diagram, 109f
    process, 101, 105–106, 108–109, 133–134
        investment, 136
    process flowcharts for wastes generated in, 117f
    waste streams from, 120t
    wastewater quantities of plant sources in, 123t
Mercury Code of Practice, 130
Mercury-Containing and Rechargeable Battery Management Act, 405, 406
Mercury-containing wastes, 399
    management and disposal of, 401
Mercury emissions
    from chlorine/caustic soda manufacture, 127t
Mercury removal, nanotechnology for, 408
Metal finishing effluent discharges, monitoring requirements of, 342
Metal finishing industry wastes
    Code of Federal Regulations, 341–346
        effluent limitations based on BAT, 343, 344t
        effluent limitations based on BPT, 342, 343t
        metal finishing point source category, 341–342
        monitoring requirements, 342
        new source performance standards, 344–345, 346t
        pretreatment standards for existing sources, 343–344, 345t
        pretreatment standards for new sources, 345, 347t
    costs, 338–341
        carbon adsorption, 340
        chromium reduction, 340
        cyanide oxidation, 339–340
        emulsion breaking and oil separation, 339
        lime precipitation and sedimentation, 340
        multimedia filter, 340
        sludge drying beds, 340
        typical treatment options, 339
        ultrafiltration, 340
    industry description
        general description, 309–312
        subcategory descriptions, 312–313
        subcharacterization of unit operations, 314t
        summary for, 312t

Metal finishing industry wastes (*cont.*)
   pollutant removability, 328–333
      common metals, 329, 331
      complexed metal wastes, 331
      cyanide, 332
      hexavalent chromium, 331–332
      oils, 332
      precious metals, 331
      solvents, 332–333
   source reduction, 319, 321–328
      capture/concentration techniques, 326–327
      chemical substitution, 321–322, 322–323t
      process modifications to reducing drag-out loss, 323–326
      waste reduction costs and benefits, 327–328
      waste segregation, 323
   specialized definitions, 346–349
   treatment flow diagram, 339f
   treatment methods, 330t
   treatment technologies
      arsenic- and selenium-containing wastes, 336–337
      chromium-containing wastes, 336
      cyanide-containing wastes, 333–336
      neutralization, 333
   wastewater characterization, 313–319
      common metals subcategory, 314
      complexed metals subcategory, 315, 318t
      hexavalent chromium subcategory, 318–319
      oils subcategory, 319
      pollutants concentrations, 316–321t
      precious metals subcategory, 315, 317t
      solvent subcategory, 319
Metal finishing plants
   U.S. PSES, 345t
      on cyanide, 345t
Metal finishing point source category
   applicability and description, 341–342
   U.S. BAT, 344t
      on cyanide, 344t
   U.S. BPT, 343t
      on cyanide, 343t
   U.S. NSPS, 346t
      on cyanide, 346t
   U.S. PSNS, 347t
      on cyanide, 347t
Metal hydroxide
   adsorption capacities of, 253t
   for dye removal, 250
Metals
   in dyeing wastewaters, 221, 222–223t
   in printing processes, 222
Metals subcategory
   aromatics, 316t
   classical, 317t
   common, 314
   complexed, 315, 318t
   halogenated aliphatics, 316t
   metals and inorganics, 316t
   nitrogen compounds, 316t
   pesticides and metabolites, 316t
   phenols, 316t
   phthalates, 316t
   pollutants concentrations, 316–321t
   polycyclic aromatic hydrocarbons, 316t
   precious, 315, 317t
Methane formers, 29
Methane gas, conversion of food scraps to, by anaerobic process, 188f
Methanogens, 189
Methyl chloride, 15
Methylene blue number, 358t
Methyl esters, 200
Methyl mercury, 128–129
Methyl-tert-butyl ether (MTBE), 361–362, 364
Microorganisms, 29
Millipore Filtration technique, 85
Minamata disease, 129
Multimedia filter, 340
Multimedia pollution prevention programs, 15–16, 20, 21–23
Municipal solid waste facilities, 184

N

N-acetyl-N,N,N-trimethylammoniumbromide (CTAB), 171
Nanotechnology for mercury removal, 408
National Environmental Policy Act (NEPA), 15, 18, 20
Natural fibers, 211–212
Neutralization, 16, 333
New England Governors' Conference, 407
New source performance standards (NSPS)
   for metal finishing point source category, 344–345, 346t
   on cyanide, 347t
New York State Department of Environmental Conservation (NYSDEC), 22–23
Niagara Falls Wastewater Treatment Plant, 69
Nickel–cadmium batteries, 392, 399, 403, 406
Nickel solution concentration limits, standard, 325t
Nitric acid cleaning solution, 292
Nitrification inhibitor (CTCMP), 287
N-methyl-2-pyrrolidone, 15
Nochromix cleaning solution, 293
Noncyanide plating processes, 323t
Nonpolluting baths, 323
Nonpolluting streams, 323
Nonprofit organization, 187
Nonstructural best management practices (BMPs), 90
Nonwoven manufacturing subcategory, 219
Northeast Waste Management Officials' Association (NEWMOA), 407
N-Serve, 287

O

Occidental Petroleum, 2
Occupational health and safety issues
   asbestos, 130–131
   caustic soda, 131
   chlorine, 130
   mercury, 130
Office of Management and Budget (OMB), 209
Offline storage-release systems, 88
Oil company stormwater runoff treatment (case history)
   DAF and sand filtration
      pilot plant, 76–77
      pollutant removal efficiencies of, 76

Index

design and operation of full-scale wastewater treatment system
    DAF clarifier, 82
    oil–water separators, 79, 81
    optional GAC carbon beds, 82
    sludge management technologies, 84–88
    summary of CSO liquid stream treatment, 83–84
posttreatment by GAC adsorption, 77–79
Oils, separation, 339
Oils subcategory, 319
    pollutants concentrations, 320–321t
        aromatics, 320t
        classical, 321t
        ethers, 320t
        metaganated hydrocarbons, 320t
        nitrogen compounds, 320t
        pesticides and metabolites, 320–321t
        phenols, 320t
        phthalates, 320t
        polychlorinated hydroxis, 321t
        polynuclear aromatic hydrocarbons, 320t
Olin Corporation, 136–137
Ordinance on Return, Taking Back, and Disposal of Electrical and Electronic Appliances (ORDEA), Switzerland, 391
Organic chemicals manufacturing wastes treatment, 374
    categories, 374
    chemical manufacturing processes
        chemical reaction processes, 375–376
        purification of products, 376
        specific industrial organic chemicals, 376–377, 378f
    end-of-pipe waste treatment technologies, 381–385
        case example, 383–385, 386f
    federal regulations, 385
    organic products and industrial users, 374t
    pollution prevention, 378–381
    production chains, 375f
    sources of emission, 381t
    technical terminologies, 385, 387
Oxidation, 387
Oxidation methods
    chlorine, 237, 248f
    Fenton's reagent, 233–235
    hydrogen peroxide, 232–233
    ozonation, 235–237, 246f, 246t, 247f, 248t
Oxidation–reduction processes, 16
Ozonation, 333
Ozonation treatment, on dyeing wastewaters, 235–237
    apparatus for, 246f
    concentration of textile pollutants, 246t
    decolorization by, 247f, 248t

P

PacBio plant, 199
Pad dyeing, 214
Painting, 312
Paint stripping, 311
Partial compliance, 187
Peat
    adsorption capacities of, 256t
    for dye removal, 255–256
Pentane, 397

Perchlorate, 6t
Perkin Elmer Model Atomic Absorption Spectrophotometer, 85
Person, defined, 186
Pesticides, 6, 399
Peterson, Niels, 146
Pharmaceuticals, 6
Phenol number, 358t
Phosphate buffer solution, 287
Physical waste treatment processes, 16
Physicochemical technologies
    combined sewer overflow (CSO) treatment by, 67–68
    wastewater treatment by, 68–69
Physisorption, 228
Pigment printing, 214
Piranha cleaning solution, 293
Plasma arc machining, 311
Plasma process, for gasification, 196–197
Plastics, 6
Polishing, 309
Pollutant removability
    metal finishing industry wastes, 328–333
        common metals, 329, 331
        complexed metal wastes, 331
        cyanide, 332
        hexavalent chromium, 331–332
        oils, 332
        precious metals, 331
        solvents, 332–333
    of textile industry wastewater
        adsorption, 228–231, 244t, 245f
        biological treatment, 225–226, 227–238t
        chemical coagulation, 226, 239–243t
        combined biological/physical–chemical treatment, 237, 249t
        cost for wastewater treatments, 263, 264–266t, 266, 267–270t
        dyes removal by low-cost adsorbents, 237, 243, 248–263
        oxidation methods, 231–237, 246–248f, 246t, 248t
        predicted hazardous nature of dyes, 224f, 225t
        untreated wastewater concentrations of pollutants, 224t
Pollution prevention
    from chlor-alkali plants, 131–133
    in organic chemical industry, 378–381
        catalyst preparation and handling, 379–381
        modification, 380–381
        process condition modification, 380
    of toxic chemicals and hazardous wastes, 14–15, 20, 21–23
Pollution prevention opportunities, in textile industry
    chemical substitution, 271
    equipment modification, 273
    good operating practices, 273–274
    process modification, 271–272, 271t
    process water reuse and recycle, 271t, 272–273
    quality control for raw materials, 266
Polychlorinated biphenyls (PCBs), 6t
Polycyclic aromatic hydrocarbons (PAHs), 6t
Polyvinyl alcohol (PVA), 213
Potassium dichromate, 300
Potassium hydroxide, 137t

Poultry processing wastewater
  biological treatment of, 27–63
  in California, 31
  characteristics of poultry plant wastewater, 30*t*
  description of operations, 31
  first processing operations, 32*f*
    chilling, 35
    evisceration, 33–34
    killing and bleeding, 32–33
    packaging and freezing, 35
    receiving areas, 31–32
    scalding and defeathering, 33
  further processing operations, 35–36
  Gold Kist project (case history)
    component adequacy review, 59–62
    current operating difficulties, 55, 57–58
    current wastewater loads, 55
    design criteria, 50–53
    expansion at, 54–63
    flow diagram, 50
    future expansion provisions, 53–54
    operating arrangements, 54
    proposed modifications, 62–63
    proposed wastewater treatment system loads, 58–59
    selection of treatment process, 49–50
    site selection, 47–48
    waste treatment system costs, 54
    wastewater survey and criteria, 48–49
  operating wastewater treatment
    daily checklist, 46
    monthly checklist, 47
    weekly checklist, 46–47
    yearly checklist, 47
  planning for wastewater treatment
    activated sludge processes, 38–40
    aerobic and anaerobic lagoons, 42
    initial planning, 37–38
    selection, 38–42
    trickling filters, 40–41
    upgrading existing lagoons, 42–45
    wastewater surveys and waste minimization, 36–37, 37*f*
  slaughtering, 30
Powder activated carbon (PAC) treatment, 354
  requirement calculations, 355*t*
Precious metals, 331
Precipitation, reagents, 337
Preconsumer kitchen waste, 185
Pressure deformation, 310
Pressure flotation, in wastewater treatments, 147–148
Pressure swing adsorption (PSA) process, 196
Pretreatment standards for existing sources (PSES), 343–344
  for metal finishing plants, 345*t*
  on cyanide, 345*t*
Pretreatment standards for new sources (PSNS)
  for metal finishing point source category, 345, 347*t*
  on cyanide, 347*t*
Printing operations
  in textile industry, 214
  wastewater from, 222
Process-specific wastes, 384
Product fabrication, in textile industry, 216
Propylene, 377
Protozoa, 29
Publicly owned treatment works (POTWs), 38, 274
Pulmonary edema, 130
Pyrolysis, 387
  process, 195

Q

Quality control, for raw materials, 266

R

Racking, proper, 326
Reactive dyes, 214, 384
Reagents, for BOD testing, 287
Rechargeable battery, 405
Rechargeable Battery Recycling Corporation (RBRC), 404
Rechargeable consumer product, 405
Rectangular flotation tank, 149
  with recycle flow system, 150*f*
Recycle flow pressure flotation, 148
Red mud
  adsorption capacities of, 255*t*
  for dye removal, 252
Reflex apparatus, for chemical oxygen demand (COD) determination, 301
Refrigeration and air-conditioning waste appliances, management and disposal of, 396–397
  flow diagram for, 398*f*
Regulated battery, 405
Remanufactured product, 405
Resource Conservation and Recovery Act (RCRA), 8, 9–11, 308, 399
  Corrective Action Program, 11, 12*t*
  essential requirements
    for transporters, 10
    for treatment, storage, and disposal facilities (TSDFs), 10
    for waste generators, 9–10
  inadequacies in regulations, 10–11
  location of RCRA Corrective Action 2020 Universe sites, 11, 13*f*
Restaurant waste treatment and management
  on board cruise ships, 200–202
  food waste recovery
    donation of food, 186–188
    food scraps for animal feed, 188
    industrial uses for food scraps and waste FOG, 188–199
    source reduction, 184–186
Retention basins, 91
Reverse osmosis (RO), 327
Reynolds numbers, 153–156
Rice husk, adsorption capacities of, 249, 251–252*t*
Rinse elimination, 324
Risk assessment and management
  toxic chemicals and hazardous wastes, 7–8
Rotifers, 29

S

Salt bath descaling, 311
SAMMS (Self-Assembled Monolayers on Mesoporous Supports), 408
San Antonio Water System's Biosolids Facilities, 193
Sand blasting, 310
Sand filters, 92
Sawdust, 248
  adsorption capacities of, 250t
Sedimentation, 331, 337, 340
Sedimentation basins, 90
Semi-batch reactor, 376
Shearing, 310
Silver-oxide button batteries, 406
Single-stage high-rate anaerobic digester system, 190f
Sintering, 311
Slaughtering, poultry, 30
Sludge
  drying beds, 340
  management technologies, 84–88
Small-Quantity Handlers of Universal Waste (SQHUW), 400
Soda ash, 102
Sodium carbonate, emissions from the use of, 127
Sodium chlorate, 137t
  manufacturing process, 113–114
Sodium chloride, 112
  effluent pretreatment standards, 126t
  environmental impacts, 130
Sodium hydrosulfite, 137t
Sodium hydroxide (NaOH), 101–103, 287, 303
Sodium hypochlorite
  manufacturing process, 113
  as products of Olin Corporation, 137t
Sodium–mercury amalgam, 106
Sodium metabisulfite, 336
Sodium sulfite solution (Na_2SO_3), 287
Soldering, 310
Solidification technology, for hazardous E-waste disposal, 408–409
Solid retention time (SRT), 192, 194
Solid wastes
  in chlor-alkali plants, 116, 118–119, 121–123, 122–123t, 133
  from restaurants, 184
Solubility of air, 157–159
  in water, 158f
Soluble dyes, 214
Solvent degreasing, 311
Solvent dyes, 384
Solvent subcategory, 319
Solving the E-Waste Problem (STEP) initiative, UNU's, 412–413
Source reduction
  of food waste, 184–186
  from metal finishing operations, 319, 321–328
Spilt-flow pressure flotation, 148
Sputtering, 311
SR Recycling firm, 407
Stabilization, 338
Standard industrial classification (SIC) codes within textile industry, 209–210t
Standard rate trickling filters, 41, 41f
Starch, 213
State universal wastes, 399
Stirred tank reactors, see Batch reactors
Stock and yarn finishing subcategory, 218–219
  wastewater treatment cost for, 270t
Stokes' law, 152
Storage-release systems
  deep tunnels, 89
  surface storage, 88
Storm runoff and process wastewater
  pollutants removal by oil-water separation (case history), 76–88
Stormwater quality controls, cost-effectiveness of
  best management practices (BMPs)
    control performance, assessment of, 93, 94t
    detention basins, 90–91
    grassed swales, 92–93
    infiltration basins, 91–92
    infiltration trenches, 91
    retention basins, 91
    sand filters, 92
    water quality inlet, 92
  disinfection, 90
  offline storage-release systems, 88
  operations and maintenance costs for controls, 93–95
  screens, 89–90
  sedimentation basins, 90
  swirl concentrators, 89
Structural best management practices (BMPs), 90
Stunning, 32
Substantial use of chemicals, 6
Sulfamic acid, 303
Sulfide precipitation, 16
Sulfide reagents, 338
Sulfolane, 363–364
Sulfur dioxide, 336
Sulfuric acid-silver sulfate (H_2SO_4–Ag_2SO_4) solution, 301
Sulfuric acid (H_2SO_4) solution, 106, 287, 292
Superfund Amendments and Reauthorization Act (SARA), 11
Superfund program, 11
Surface storage, 88
Sveen, Carl, 146
Swirl concentrators, 89
Syngas, 194–196

T

Tail gases, 106, 124, 127, 132
Tank design, for flotation unit, 166–170
Tannic acid number, 358t
Tap water characteristics, 81t
Tax Reform Act 1976, 188
Testing, 312
Tetrachloroethylene (PCE), 6t
Textile industry waste treatment
  effluent limitations of nine subcategories, 272–274t
  establishment sizes, 210t

Textile industry waste treatment (cont.)
  new source performance standards (NSPS) of nine subcategories, 275t
  overview, 208–210
  pollutant removability
    adsorption, 228–231, 244t, 245f
    biological treatment, 225–226, 227–238t
    chemical coagulation, 226, 239–243t
    combined biological/physical–chemical treatment, 237, 249t
    cost for wastewater treatments, 263, 264–266t, 266, 267–270t
    dyes removal by low-cost adsorbents, 237, 243, 248–263
    oxidation methods, 231–237, 246–248f, 246t, 248t
    predicted hazardous nature of dyes, 224f, 225t
    untreated wastewater concentrations of pollutants, 224t
  pollution prevention opportunities
    chemical substitution, 271
    equipment modification, 273
    good operating practices, 273–274
    process modification, 271–272, 271t
    process water reuse and recycle, 271t, 272–273
    quality control for raw materials, 266
  process descriptions
    dyeing, 213–214
    fabrication, 216
    fabric formation, 212–213
    fabric preparation, 213
    final finishing, 214–215
    flowchart, 211f
    printing, 214
    yarn formation, 211–212
  standard industrial classification (SIC) codes, 209–210t
  subcategory descriptions
    carpet finishing, 218
    felted fabric processing, 219
    knit fabric finishing, 218
    low water use processing, 217
    nonwoven manufacturing, 219
    stock and yarn finishing, 218–219
    wool finishing, 216–217
    wool scouring, 216
    woven fabric finishing, 217–218
  U.S. environmental regulations, 272–275t, 274
  wastewater characterization
    contents and characteristics, 221t
    from dyeing operation, 221, 222–223t
    from fabric formation, 219
    from fabric preparation, 221, 222t
    from finishing operation, 223
    from printing operation, 222
    release of pollutants, with various processes, 220t
Thermal cutting, 310
Thermal infusion, 311
Thermal oxidation, 335
Thermostats, 399
Thickening and dewatering of floated sludges, 86–87
Thixography, 214
Threshold limit value (TLV) of mercury, 130
Total suspended solids (TSS), 125

Total toxic organics (TTO), 342, 347
Tower Biology technology, 383
Town's recycling system, 407–408
Toxic chemicals and hazardous wastes
  accident prevention and emergency response, 16–19
  Clean Air Act (CAA) Amendments, 13–14
  Comprehensive Environmental Response, Compensation, and Liability Act (CERCLA), 11–13
  education and training, 20
  legislation and regulation, 8–14
  management and implementation strategies, 14–16
  methods for expressing toxicity, 7
  objectives of management of, 2
  problem, 2–7, 3–6t
  Resource Conservation and Recovery Act (RCRA), 8, 9–11, 12t, 13f
  risk assessment and management, 7–8
  safety and control for, 1–23
  Toxic Substance Control Act (TSCA), 8–9
  waste treatment and disposal technologies, 16
Toxics Release Inventory (TRI), U.S. EPA's, 8, 128
Toxic Substance Control Act (TSCA), 8–9, 17
Transesterification process, 200
Transporters, essential requirements for, 10
Treatment, storage, and disposal facilities (TSDFs), essential requirements for, 10
Trichloroethylene (TCE), 6t, 359–361
Trickling filters, 40–41
Triglycerides, 200
Two-stage high-rate anaerobic digester system, 190f

U

Ultrafiltration (UF), 340
Ultrasonic machining, 311
Ultraviolet (UV) radiation, 332, 334–335
Underground Storage Tanks program, 13
United Nation University (UNU)
  Solving the E-Waste Problem (STEP), 412–413
Universal waste battery, 399
Universal waste destination facilities, 400
Universal waste handlers, 400
Universal waste lamp, 399–400
Universal waste pesticide, 399
Universal waste rule, 406
Universal Wastes, 391–393, 399
  management and disposal of, 397, 399–400
Universal waste thermostat, 399
Universal waste transporter, 400
U.S. Army Corps of Engineers (U.S. ACE), 328
U.S. chlorine industry, 108, 110–111t, 112f
U.S. Department of Energy's PNNL, 408
U.S. Environmental Protection Agency (U.S. EPA)
  in chemical regulatory program, 17
  Clean Air Act (CAA) Amendments, 13–14
  CSO Control Policy of, 67
  Federal Universal Wastes, 399
  food recovery program, 184
  identified benzene as human carcinogen, 377
  listed as hazardous waste from chlor-alkali manufacture, 123t

Index

mercury-containing lamp recycling outreach program, 401
poultry processing operations, 31
practices to reducing toxic contaminant flow from chlor-alkali plants, 131
regulation of toxic chemicals and hazardous wastes, 8
solvent subcategory, 319
standard for diaphragm cell process, 134, 141*t*
standard for K071 sludges, 133
and state agencies, 11
Superfund program, 11
Toxics Release Inventory (TRI), 8, 128
Toxic Substance Control Act (TSCA), 8–9
U.S. environmental regulations, for textile nine subcategories, 272–275*t*, 274
U.S. Federal Swine Health Protection Act, 188
U.S. Manufacturing Energy Consumption Survey, 114
U.S. National and Community Service Act, 186

V

Vacuum metalizing, 312
Vapor plating, 311
Vat dyes, 214
Vehicle batteries, management/reuse/recycle/disposal of, 401, 403–404
Vinyl chloride, 387
Visqueen, 51

W

Wang, Lawrence K., 409
Waste computers, management of, 407–408
Waste detoxification, 321
Waste electrical and electronic equipment (WEEE), see Electrical and electronic wastes (E-wastes)
Waste exchange program, 23
Waste generators, 6
  essential requirements for, 9–10
Waste reduction
  annual cost savings, 328*t*
  costs and benefits, 327–328
Waste Reduction Guidance Manual, NYSDEC, 22
Waste reduction impact statement (WRIS), 22
Waste Reduction Information Clearinghouse, maintenance of, 22

Waste segregation, 323
Waste sludge management, 84–88
Wastewater treatment and characterization
  by biological technologies, 69
  in chlor-alkali industries, see Chlor-alkali industries
  and disposal technologies, 16
  dissolved air flotation (DAF) for, see Dissolved air flotation (DAF)
  in metal finishing industry, see Metal finishing industry wastes
  by physicochemical technologies, 68–69
  in poultry processing, see Poultry processing wastewater
  in textile industry, see Textile industry waste treatment
Water and Waste Control for the Plating Shop, 326
Water jet weaving process, 217
Water quality inlet, 92
Weaving mills, 212
Welding, 310
Wet Air Oxidation process, 382–383
Wet ponds, see Retention basins
Wetting agents, 323
Windrows composting process, 198
Wool finishing subcategory, 216–217
  wastewater treatment cost for, 265*t*
Wool scouring subcategory, 216
  wastewater treatment cost for, 264*t*
Woven fabric finishing subcategory, 217–218
  wastewater treatment cost for, 267*t*

X

Xylene, 364, 366–367, 367*t*

Y

Yarn formation
  man-made fibers, 212
  natural fibers, 211–212
  in textile industry, 211–212

Z

Zones, of flotation tank, 150